IFIP Advances in Information and Communication Technology 339

IFIP – The International Federation for Information Processing

IFIP was founded in 1960 under the auspices of UNESCO, following the First World Computer Congress held in Paris the previous year. An umbrella organization for societies working in information processing, IFIP's aim is two-fold: to support information processing within its member countries and to encourage technology transfer to developing nations. As its mission statement clearly states,

> *IFIP's mission is to be the leading, truly international, apolitical organization which encourages and assists in the development, exploitation and application of information technology for the benefit of all people.*

IFIP is a non-profitmaking organization, run almost solely by 2500 volunteers. It operates through a number of technical committees, which organize events and publications. IFIP's events range from an international congress to local seminars, but the most important are:

- The IFIP World Computer Congress, held every second year;
- Open conferences;
- Working conferences.

The flagship event is the IFIP World Computer Congress, at which both invited and contributed papers are presented. Contributed papers are rigorously refereed and the rejection rate is high.

As with the Congress, participation in the open conferences is open to all and papers may be invited or submitted. Again, submitted papers are stringently refereed.

The working conferences are structured differently. They are usually run by a working group and attendance is small and by invitation only. Their purpose is to create an atmosphere conducive to innovation and development. Refereeing is less rigorous and papers are subjected to extensive group discussion.

Publications arising from IFIP events vary. The papers presented at the IFIP World Computer Congress and at open conferences are published as conference proceedings, while the results of the working conferences are often published as collections of selected and edited papers.

Any national society whose primary activity is in information may apply to become a full member of IFIP, although full membership is restricted to one society per country. Full members are entitled to vote at the annual General Assembly, National societies preferring a less committed involvement may apply for associate or corresponding membership. Associate members enjoy the same benefits as full members, but without voting rights. Corresponding members are not represented in IFIP bodies. Affiliated membership is open to non-national societies, and individual and honorary membership schemes are also offered.

Harris Papadopoulos
Andreas S. Andreou
Max Bramer (Eds.)

Artificial Intelligence Applications and Innovations

6th IFIP WG 12.5 International Conference, AIAI 2010
Larnaca, Cyprus, October 6-7, 2010
Proceedings

Volume Editors

Harris Papadopoulos
Frederick University
Computer Science and Engineering Department
1036 Nicosia, Cyprus
E-mail: h.papadopoulos@frederick.ac.cy

Andreas S. Andreou
Cyprus University of Technology
Department of Electrical Engineering and Information Technology
P.O. Box 50329, 3603 Limassol, Cyprus
E-mail: andreas.andreou@cut.ac.cy

Max Bramer
University of Portsmouth
School of Computing
Portsmouth, PO1 2UP, UK
E-mail: max.bramer@port.ac.uk

CR Subject Classification (1998): I.2, H.3, H.4, F.1, I.4, I.5

ISSN 1868-4238
ISBN-10 3-642-42361-2 Springer Berlin Heidelberg New York
ISBN-13 978-3-642-42361-1 Springer Berlin Heidelberg New York

springer.com

Softcover re-print of the Hardcover 1st edition 2010

Typesetting: Camera-ready by author, data conversion by Scientific Publishing Services, Chennai, India
Printed on acid-free paper 219/3180

Preface

The abundance of information and increase in computing power currently enable researchers to tackle highly complicated and challenging computational problems. Solutions to such problems are now feasible using advances and innovations from the area of Artificial Intelligence. The general focus of the AIAI conference is to provide insights on how Artificial Intelligence may be applied in real-world situations and serve the study, analysis and modeling of theoretical and practical issues.

This volume contains papers selected for presentation at the 6th IFIP Conference on Artificial Intelligence Applications and Innovations (AIAI 2010) and held in Larnaca, Cyprus, during October 6–7, 2010. IFIP AIAI 2010 was co-organized by the University of Cyprus and the Cyprus University of Technology and was sponsored by the Cyprus University of Technology, Frederick University and the Cyprus Tourism Organization. AIAI 2010 is the official conference of the WG12.5 "Artificial Intelligence Applications" working group of IFIP TC12, the International Federation for Information Processing Technical Committee on Artificial Intelligence (AI).

AIAI is a conference that grows in significance every year attracting researchers from different countries around the globe. It maintains high quality, standards and welcomes research papers describing technical advances and engineering and industrial applications of intelligent systems. AIAI 2010 was not confined to introducing how AI may be applied in real-life situations, but also included innovative methods, techniques, tools and ideas of AI expressed at the algorithmic or systemic level.

The AIAI 2010 conference encouraged the submission of research papers describing prototypes, innovative systems, tools and techniques of AI, as well as applications of AI in real-world problems. General survey papers indicating future directions and professional work-in-progress reports were also of equal interest. The response to the call for papers was quite satisfactory, with 111 papers initially submitted and 48 finally accepted for presentation after peer reviewing of each paper by at least two referees; where needed, a third referee was consulted to resolve any reviewing conflicts. The acceptance rate of the AIAI 2010 conference was 43%, while the authors of accepted papers came from 16 countries. The collection of papers included in this volume of proceedings offers the latest advances in the AI field and describes innovative applications of AI on a number of challenging issues and real-life situations. Therefore, the proceedings of AIAI 2010 will be useful not only to researchers of the AI field but also to IT professionals and consultants, as well as to the general scientific community, that are interested in new technologies of the artificial world and their practical applications/benefits.

Three keynote speakers were invited to lecture about innovative and state-of-the-art subjects in AI:

1. Zbigniew Michalewicz, University of Adelaide, Adelaide, Australia. Title of the talk: "How Artificial Intelligence May Be Applied in Real World Situations."
2. Alexander Gammerman, Royal Holloway, University of London, UK. Title of the talk: "Modern Machine Learning Techniques and Their Applications to Medical Diagnostics."
3. Masoud Mohammadian, Faculty of Information Sciences and Engineering, University of Canberra ACT 2601.
   Title of the talk: "Innovative Applications of Artificial Intelligence Techniques in Software Engineering."

The AIAI 2010 conference consisted of the following main thematic sessions:

- Machine Learning
- Fuzzy Logic Techniques
- Evolutionary Computation
- Medical Informatics and Biomedical Engineering
- Text Mining and Natural Language Processing
- Knowledge Representation and Reasoning
- Planning and Scheduling
- Feature Selection and Dimensionality Reduction
- Engineering Intelligent Systems
- Intelligent User Environments and HCI
- Environmental Modeling

Additionally, the technical program featured one workshop and two tutorials that covered advances on AI and applications in various areas, including Software Engineering, Multi-Agent Technologies and Information Detection.

The wide range of topics covered, the high quality of contributions and the innovative ideas proposed by the authors guaranteed a successful conference. We would like to express our thanks to the Program Committee Chair, Lazaros Iliadis, the Workshops Chair, Nicos Mateou and the Tutorials Chair, Anastasios Sofokleous. Special thanks are also due to the Proceedings Co-editors, Efthyvoulos Kyriakou and Ilias Maglogiannis.

Andreas S. Andreou
Harris Papadopoulos

In Memoriam

On 9th August 2010 the Department of Computer Science at the University of Cyprus was filled with profound sadness following the news of the untimely death of one of its most beloved undergraduate students, Polyvios Polyviou. Polyvios was part of the AIAI 2010 team and played an active role in the creation and maintenance of the conference's website. Though we are forced to say goodbye to him, his courage and fighting spirit will forever remind us to never submit nor surrender in hard times.

Organization of the AIAI 2010 Conference

General Co-chairs

Andreas S. Andreou	University of Cyprus & Cyprus University of Technology
Max Bramer	University of Portsmouth, UK

Organizing Chair

Harris Papadopoulos	Frederick University, Cyprus

Program Committee Chair

Lazaros Iliadis	Democritus University of Thrace, Greece

Proceedings Co-editors

Efthyvoulos Kyriakou	Frederick University, Cyprus
Ilias Maglogiannis	University of Central Greece

Workshops Chair

Nicos Mateou	Ministry of Defense, Cyprus

Tutorials Chair

Tasos Sofokleous	Cyprus University of Technology

Program Committee

A.S. Andreou	University of Cyprus & Cyprus University of Technology
P. Agelov	Lancaster University, UK
N. Bassiliades	Aristotle University of Thessaloniki, Greece
V. Camps	Université Paul Sabatier Toulouse, France
C. Charalambous	Frederick University, Cyprus
C. Chrysostomou	Frederick University, Cyprus
J. Debenham	University of Technology, Sydney, Australia
Y. Demazeau	CNRS, LIG Laboratory, France

T. Dillon	University of Technology Sydney, Australia
C. Georgiadis	University of Macedonia Thessaloniki, Greece
I. Hatzilygeroudis	University of Patras, Greece
L. Iliadis	Democritus University of Thrace, Greece
V. Karkaletsis	NCSR Demokritos, Greece
I. Katakis	Aristotle University of Thessaloniki, Greece
P. Kefalas	City College, Thessaloniki, Greece
A. Konstantinidis	Frederick University, Cyprus
D. Kosmopoulos	NCSR Demokritos, Greece
M. Koubarakis	University of Athens, Greece
E. Kyriakou	Frederick University, Cyprus
S. Lecoeuche	Ecole des Mines de Douai, France
G. Leonardi	University of Pavia, Italy
S. Likothanasis	University of Patras, Greece
I. Maglogiannis	University of Central Greece
F. Makedon	University of Texas Arlington, USA
M. Maragoudakis	Aegean University, Greece
S. Montani	University del Piemonte Orientale, Italy
M. Oprea	University of Ploiesti, Romania
D. Palmer-Brown	London Metropolitan University, UK
H. Papadopoulos	Frederick University, Cyprus
C. Pattichis	University of Cyprus
W. Pedrycz	University of Alberta, Canada
E. Pimenidis	University of East London, UK
C. Schizas	University of Cyprus
S. Spartalis	Democritus University of Thrace, Greece
A. Sofokleous	University of Cyprus
C. Spyropoulos	NCSR Demokritos, Athens, Greece
A. Stafylopatis	National Technical University of Athens, Greece
P. Y. Schobbens	Institut d'Informatique, Belgium
S. Senatore	University di Salerno, Italy
A. Tsadiras	TEI of Thessaloniki, Greece
D. Tsaptsinos	Kingston University, UK
V. Verykios	University of Thessaly, Greece
Z. Voulgaris	Georgia Institute of Technology, USA
G. Vouros	Aegean University, Greece

Additional Reviewers

G. Beligiannis
R. Bellazzi
A. Bottrighi
G. Demetriou
D. Fierens
D. Koutsomitropoulos
V. Lombardo
S. Martello
J. Natwichai
V. Plagianakos
M. Vozalis
N. Yorke-Smith

Website Administrators

Efi Papatheocharous	University of Cyprus
Polyvios Polyviou	University of Cyprus

Table of Contents

Invited Talks

Machine Learning

Fuzzy Logic Techniques

Evolutionary Computation

Medical Informatics and Biomedical Engineering

Text Mining and Natural Language Processing

Knowledge Representation and Reasoning

Planning and Scheduling

Feature Selection and Dimensionality Reduction

Engineering Intelligent Systems

Intelligent User Environments and HCI

Environmental Modeling

How Artificial Intelligence May Be Applied in Real World Situations

Zbigniew Michalewicz

University of Adelaide, Adelaide, Australia
zbyszek@cs.adelaide.edu.au

Abstract. In the modern information era, managers must recognize the competitive opportunities represented by decision-support tools. New family of such systems, based on recent advances in Artificial Intelligence, combine prediction and optimization techniques to assist decision makers in complex, rapidly changing environments. These systems address the fundamental questions: What is likely to happen in the future? and what is the best course of action? These modern AI systems include elements of data mining, predictive modelling, forecasting, optimization, and adaptability and aim at providing significant cost savings and revenue increases for businesses. The talk introduces the concepts behind construction of such systems and indicates the current challenging research issues. Several real-world examples will be shown and discussed.

H. Papadopoulos, A.S. Andreou, and M. Bramer (Eds.): AIAI 2010, IFIP AICT 339, p. 1, 2010.

Modern Machine Learning Techniques and Their Applications to Medical Diagnostics

Alexander Gammerman

Computer Learning Research Centre,
Royal Holloway, University of London, UK
A.Gammerman@cs.rhul.ac.uk

Abstract. The talk presents several machine learning techniques and their applications to clinical decision-making. In many problems of computer-aided medical diagnosis and treatment a program must be capable of learning from previously accumulated past patients data records, and extrapolating to make diagnosis for new patient by considering their symptoms. Many machine learning and statisitical techniques have been developed to help in clinical decision making. Among them decision trees, the Bayesian techniques, dicriminant analysis, neural networks and many others. These techniques usually deal with conventional, small-scale, low-dimensional problems, and the application of these techniques to modern high-dimensional data sets with many thousand attributes (symptoms) usually leads to serious computational problems. Several new techniques such as Support Vector Machine (SVM) have been developed to tackle the problem of dimensionality by transferring the problem into high-dimensional space, and solving it in that space. They based on so-called kernal methods and can very often solve some high-dimensional problems. These techniques perform very well with good accuracy. However, a typical drawback of techniques such as the SVM is that they usually do not provide any useful measure of confidence of new, unclassified examples (new pattients). Recently a new set of techniques, called Conformal Predictors, have been developed that allows to make predictions with valid measures of confidence. The approach is based on approximations to the universal measures of confidence given by the algorithmic theory of randomness and allows us to compute diagnostic classes and estimate confidence of the diagnostics for high-dimensional data. The talk will present Conformal Predictors and their applications in medicine.

H. Papadopoulos, A.S. Andreou, and M. Bramer (Eds.): AIAI 2010, IFIP AICT 339, p. 2, 2010.

Innovative Applications of Artificial Intelligence Techniques in Software Engineering

Masoud Mohammadian

University of Canberra, ACT, Australia
Masoud.mohammadian@canberra.edu.au

Abstract. Artificial Intelligence (AI) techniques have been successfully applied in many areas of software engineering. The complexity of software systems has limited the application of AI techniques in many real world applications. This talk provides an insight into applications of AI techniques in software engineering and how innovative application of AI can assist in achieving ever competitive and firm schedules for software development projects as well as Information Technology (IT) management. The pros and cons of using AI techniques are investigated and specifically the application of AI in IT management, software application development and software security is considered.

Organisations that build software applications do so in an environment characterised by limited resources, increased pressure to reduce cost and development schedules. Organisations demand to build software applications adequately and quickly. One approach to achieve this is to use automated software development tools from the very initial stage of software design up to the software testing and installation. Considering software testing as an example, automated software systems can assist in most software testing phases.

On the hand data security, availability, privacy and integrity are very important issues in the success of a business operation. Data security and privacy policies in business are governed by business requirements and government regulations. AI can also assist in software security, privacy and reliability. Implementing data security using data encryption solutions remain at the forefront for data security. Many solutions to data encryption at this level are expensive, disruptive and resource intensive. AI can be used for data classification in organizations. It can assist in identifying and encrypting only the relevant data thereby saving time and processing power. Without data classification organizations using encryption process would simply encrypt everything and consequently impact users more than necessary. Data classification is essential and can assist organizations with their data security, privacy and accessibility needs. This talk explores the use of AI techniques (such as fuzzy logic) for data classification and suggests a method that can determine requirements for classification of organizations' data for security and privacy based on organizational needs and government policies. Finally the application of FCM in IT management is discussed.

H. Papadopoulos, A.S. Andreou, and M. Bramer (Eds.): AIAI 2010, IFIP AICT 339, p. 3, 2010.

Linear Probability Forecasting

Fedor Zhdanov and Yuri Kalnishkan

Computer Learning Research Centre and Department of Computer Science,
Royal Holloway, University of London, Egham, Surrey, TW20 0EX, UK
{fedor,yura}@cs.rhul.ac.uk

Abstract. In this paper we consider two online multi-class classification problems: classification with linear models and with kernelized models. The predictions can be thought of as probability distributions. The quality of predictions is measured by the Brier loss function. We suggest two computationally efficient algorithms to work with these problems, the second algorithm is derived by considering a new class of linear prediction models. We prove theoretical guarantees on the cumulative losses of the algorithms. We kernelize one of the algorithms and prove theoretical guarantees on the loss of the kernelized version. We perform experiments and compare our algorithms with logistic regression.

Keywords: Online prediction, classification, linear regression, Aggregating Algorithm.

1 Introduction

Online prediction provides efficient algorithms which adapt to a predicted sequence "on the fly" [1]. In the online regression framework one assumes the existence of some input at each step; the goal is to predict the outcome on this input. This process is repeated step by step. We consider the multidimensional case where outcomes and predictions come from the probability simplex and can be thought of as probability distributions on the vertices of the simplex. If the outcomes are identified with the vertices of the simplex, this problem can be thought of as the multi-class classification problem of the given input. We prove upper bounds on the cumulative square loss of the learner in comparison with the loss of the best linear function (we say the learner *competes* with these functions).

We develop two algorithms to solve the problem of multidimensional prediction. The first algorithm applies a variant of the AAR (see [2]) to predict each component of the outcome separately, and then combines these predictions in a certain way to get probability prediction. The other algorithm is designed to give probability predictions directly. In order to develop this algorithm we consider an original class of experts which are asymmetrical by the components of their predictions. We are not aware of any other computationally efficient online regression algorithms designed to solve linear and non-linear multi-class classification problems. We come to an unexpected conclusion that the upper bound

H. Papadopoulos, A.S. Andreou, and M. Bramer (Eds.): AIAI 2010, IFIP AICT 339, pp. 4–11, 2010.

on the loss of the component-wise algorithm is better than the upper bound on the loss of the second algorithm asymptotically, but worse in the beginning of the prediction process.

One component of the prediction of the second algorithm has the meaning of a remainder. In practice this situation is quite common. For example, it is popular in financial applications to predict the direction of the price: up, down, or close to the current value. We perform classification experiments with linear algorithms and compare them with logistic regression. The full version of our paper with proofs and detailed explanations can be found in [3].

2 Framework

We are interested in the generalisation of the Brier game from [4]. The set of possible outcomes $\Omega = \mathcal{P}(\Sigma)$ is the set of all probability measures on a finite set Σ with d elements, the prediction set $\Gamma := \{(\gamma_1, \ldots, \gamma_d) : \sum_{i=1}^d \gamma_i = 1, \gamma_i \in \mathbb{R}\}$ is a hyperplane in d-dimensional space containing all the outcomes. For any $y \in \Omega$ we define the Brier loss

$$\lambda(y, \gamma) = \sum_{\sigma \in \Sigma} (\gamma\{\sigma\} - y\{\sigma\})^2 .$$

The game of prediction is being played repeatedly by a learner receiving some input vectors $x_t \in \mathbf{X} \subseteq \mathbb{R}^n$, and follows Protocol 1.

Protocol 1. Protocol of forecasting game

for $t = 1, 2, \ldots$ **do**
  Reality announces a signal $x_t \in \mathbf{X} \subseteq \mathbb{R}^n$.
  Learner announces $\gamma_t \in \Gamma \subseteq \mathbb{R}^d$.
  Reality announces $y_t \in \Omega \subseteq \mathbb{R}^d$.
end for

Learner competes with all linear functions (experts) $\xi_t = (\xi_t^1, \ldots, \xi_t^d)'$ of x:

$$\begin{aligned} \xi_t^1 &= 1/d + \alpha_1' x_t \\ &\ldots \\ \xi_t^{d-1} &= 1/d + \alpha_{d-1}' x_t \\ \xi_t^d &= 1 - \xi^1 - \cdots - \xi^{d-1} = 1/d - \left(\sum_{i=1}^{d-1} \alpha_i \right)' x_t, \end{aligned} \tag{1}$$

where $\alpha_i = (\alpha_i^1, \ldots, \alpha_i^n)'$, $i = 1, \ldots, d-1$. In the model (1) the prediction for the last component of the outcome is calculated from the predictions for other components. Denote $\alpha = (\alpha_1', \ldots, \alpha_{d-1}')' \in \Theta = \mathbb{R}^{n(d-1)}$. Then any expert can be represented as $\xi_t = \xi_t(\alpha)$. Let also $L_T(\alpha) = \sum_{t=1}^T \lambda(y_t, \xi_t(\alpha))$ be the cumulative loss of the expert α over T trials and L_T denote the cumulative loss of the learner. The goal of the learner is to ensure that $L_T \le L_T(\alpha) + R_T$ for all T and α, and some small R_T.

3 Derivation of the Algorithms

The component-wise algorithm gives predictions for each component of the outcome separately and then combines them in a special way. To derive this algorithm, we use component-wise experts

$$\xi_t^i = 1/d + \alpha_i' x_t, \quad i = 1, \ldots, d \tag{2}$$

and follow [2]. The algorithm cAAR (component-wise Aggregating Algorithm for Regression) gives its preliminary prediction at the step T

$$\gamma_T^i = \frac{1}{d} + \left(\sum_{t=1}^{T-1} (y_t^i - 1/d) x_t' + \frac{d-2}{2d} x_T' \right) \left(aI + \sum_{t=1}^{T} x_t x_t' \right)^{-1} x_T \tag{3}$$

for $i = 1, \ldots, d$. Then it projects the preliminary prediction onto the probability prediction simplex such that the loss does not increase. This is done by the means of the projection algorithm (Algorithm 1) suggested in [5].

Algorithm 1. Projection of a point from $\mathbb{R}^n$ onto probability simplex

Initialize $I = \emptyset$, $x = \mathbf{1} \in \mathbb{R}^d$.
Let γ_T be the prediction vector and $|I|$ is the dimension of the set I.
while 1 **do**
 $\gamma_T = \gamma_T - \frac{\sum_{i=1}^d \gamma_T^i - 1}{d - |I|}$;
 $\gamma_T^i = 0, \forall i \in I$;
 If $\gamma_T^i \geq 0$ for all $i = 1, \ldots, d$ then break;
 $I = I \bigcup \{i : \gamma_T^i < 0\}$;
 If $\gamma_T^i < 0$ for some i then $\gamma_T^i = 0$;
end while

We apply the Aggregating Algorithm (AA) proposed in [6] to mix experts (1) and make predictions. The resulting algorithm is Algorithm 2. We will further call it the mAAR (the multidimensional Aggregating Algorithm for Regression).

4 Theoretical Guarantees

We derive upper bounds for the losses of Algorithm 2 and of the component-wise algorithm predicting in the same framework.

The loss of the component-wise algorithm by each component is bounded as in the following theorem.

Theorem 1. *Let the outcome set in a prediction game be* $[A, B]$, $A, B \in \mathbb{R}$. *Assume experts' predictions at each step are* $\xi_t = C + \alpha' x_t$, *where* $\alpha \in \mathbb{R}^n$ *and* $C \in \mathbb{R}$ *is the same for all experts. Let* $\|x_t\|_\infty \leq X$ *for all* t *and take* $a > 0$. *There exists a prediction algorithm producing predictions* γ_t *such that for every*

Algorithm 2. mAAR for the Brier game

Fix n, $a > 0$. $C = 0, h = 0$.
for $t = 1, 2, \dots$ **do**
Read new $x_t \in \mathbf{X}$.

$$C = C + x_t x_t',\ A = aI + \begin{pmatrix} 2C & \cdots & C \\ \vdots & \ddots & \vdots \\ C & \cdots & 2C \end{pmatrix}$$

Set $b_i = h + (x_t', \dots, x_t', 0, x_t', \dots, x_t')'$, where 0 is a zero-vector from $\mathbb{R}^n$ is placed at i-th position, $i = 1, \dots, d-1$.
Set $z_i = (-x_t', \dots, -x_t', -2x_t', -x_t', \dots, -x_t')'$, where $-2x_t'$ is placed at i-th position, $i = 1, \dots, d-1$.
Calculate $r_i := -b_i' A^{-1} z_i, r_d := 0,\ i = 1, \dots, d-1$.
Solve $\sum_{i=1}^{d} (s - r_i)^+ = 2$ in $s \in \mathbb{R}$.
Set $\gamma_t^i := (s - r_i)^+/2,\ \omega \in \Omega,\ i = 1, \dots, d$.
Output prediction $\gamma_t \in \mathcal{P}(\Omega)$.
Read outcome y_t.
$h_i = h_i - 2(y_t^i - y_t^d) x_t, h = (h_1', \dots, h_{d-1}')'$.
end for

positive integer T, every sequence of inputs and outcomes of the length T, and any $\alpha \in \mathbb{R}^n$, it achieves

$$\sum_{t=1}^{T} (\gamma_t - y_t)^2 \le \sum_{t=1}^{T} (\xi_t - y_t)^2 + a\|\alpha\|_2^2 + \frac{n(B-A)^2}{4} \ln\left(\frac{TX^2}{a} + 1\right). \quad (4)$$

In the multidimensional case the loss of the component-wise algorithm is bounded as in the following theorem.

Theorem 2. *If $\|x_t\|_\infty \le X$ for all t, then for any $a > 0$, every positive integer T, every sequence of outcomes of the length T, and any $\alpha \in \mathbb{R}^{n(d-1)}$, the loss L_T of the component-wise algorithm satisfies*

$$L_T \le L_T(\alpha) + da\|\alpha\|_2^2 + \frac{nd}{4} \ln\left(\frac{TX^2}{a} + 1\right). \quad (5)$$

The upper bound for the loss of Algorithm 2 is proven in the following theorem. To prove it, we extend the result of [7] for our game and ensure that the Aggregating Algorithm is applicable here.

Theorem 3. *If $\|x_t\|_\infty \le X$ for all t, then for any $a > 0$, every positive integer T, every sequence of outcomes of the length T, and any $\alpha \in \mathbb{R}^{n(d-1)}$, the* mAAR$(2a)$ *satisfies*

$$L_T(\mathrm{mAAR}(2a)) \le L_T(\alpha) + 2a\|\alpha\|_2^2 + \frac{n(d-1)}{2} \ln\left(\frac{TX^2}{a} + 1\right). \quad (6)$$

The upper bound (6) has worse constant $\frac{n(d-1)}{2}$ in front of the logarithm than $\frac{nd}{4}$ of the bound (5) if $d \ge 3$. In this case (6) is worse asymptotically in T than (5), but it is better in the beginning, especially when the norm of the best expert $\|\alpha\|$ is large. This can happen in the important case when the dimension of the input vectors is larger than the size of the prediction set: $n >> T$.

5 Kernelization

We use the popular in computer learning kernel trick (see, e.g., [8]). Our algorithm competes with the following experts:

$$\begin{aligned} \xi_t^1 &= 1/d + f_1(x_t) \\ &\dots \\ \xi_t^{d-1} &= 1/d + f_{d-1}(x_t) \\ \xi_t^d &= 1 - \xi^1 - \dots - \xi^{d-1}. \end{aligned} \tag{7}$$

Here $f_1, \dots, f_{d-1} \in \mathcal{F}$ are any functions from some RKHS $\mathcal{F}$ (Reproducing Kernel Hilbert Space). We start by rewriting the mAAR in the dual form. Denote

$\widetilde{Y}_i = -2(y_1^i - y_1^d, \dots, y_{T-1}^i - y_{T-1}^d, -1/2); \overline{Y}_i = -2(y_1^i - y_1^d, \dots, y_{T-1}^i - y_{T-1}^d, 0)$;
$\widetilde{k}(x_T) = (x_1' x_T, \dots, x_T' x_T)'; \widetilde{K} = (x_s', x_t)_{s,t}$ is the matrix of scalar products

for $i = 1, \dots, d-1$, $s, t = 1, \dots, T$. We show that the predictions of the mAAR can be represented in terms of the variables defined above.

Lemma 1. *On trial T, the values r_i for $i = 1, \dots, d-1$ in the mAAR can be represented as*

$$r_i = \left(\widetilde{Y}_1 \cdots \overline{Y}_i \cdots \widetilde{Y}_{d-1}\right) \cdot \left(aI + \widetilde{K}\right)^{-1} \left(\widetilde{k}(x_T)' \cdots 2\widetilde{k}(x_T)' \cdots \widetilde{k}(x_T)'\right)'. \tag{8}$$

To get predictions one can use the same calculations with r_i as in the mAAR. We call this algorithm the mKAAR (K for Kernelized).

5.1 Theoretical Guarantee for the Kernelized Algorithm

The upper bound on the loss of the mKAAR is proven in the following theorem.

Theorem 4. *Assume $\mathbf{X}$ is an arbitrary set of inputs and $\mathcal{F}$ is an RKHS of functions on $\mathbf{X}$ with the given kernel $K : \mathbf{X}^2 \to \mathbb{R}$. Then for any $a > 0$, any $f_1, \dots, f_{d-1} \in \mathcal{F}$, any positive integer T, and any sequence of input vectors and outcomes $(x_1, y_1), \dots, (x_T, y_T)$, mKAAR achieves*

$$L_T(\text{mKAAR}) \le L_T(f) + a \sum_{i=1}^{d-1} \|f_i\|_{\mathcal{F}}^2 + \frac{1}{2} \ln \det(aI + \widetilde{K}) \tag{9}$$

Here the matrix $\widetilde{K}$ is the matrix of the kernel values $K(x_i, x_j)$, $i, j = 1, \dots, T$.

We can represent the bound (9) in another form which is more familiar from the online prediction literature:

Corollary 1. *Under the assumptions of Theorem 4 and if the number of steps T is known in advance, the mKAAR reaches the performance*

$$L_T(\text{mKAAR}) \le L_T(f) + \left(c_{\mathcal{F}}^2 + \sum_{i=1}^{d-1} \|f_i\|_{\mathcal{F}}^2\right)\sqrt{(d-1)T}, \qquad (10)$$

for any $f_1, \ldots, f_{d-1} \in \mathcal{F}$.

6 Experiments

We run our algorithms on six real world time-series data sets. DEC-PKT,LBL-PKT-4[1] contain network traffic; C4,C9,E5,E8[2] relate to transportation data.

We use ten previous observations as the input vectors for the tested algorithms at each prediction step. We predict whether the next value in a time series will be more than the previous value plus a precision parameter ϵ, less than that value, or will stay in the 2ϵ tube around the previous value. The measure of the quality of predictions is the mean square loss (MSE) over the last two thirds of each time series (test set). We also introduce another quality measure, which is the average of the MSEs up to each step in the test set (AMSE).

We compare the performance of our algorithms with the multinomial logistic regression (mLog), because it is a standard classification algorithm which gives probability predictions:

$$\gamma_{\text{mLog}}^i = \frac{e^{\theta^i x}}{\sum_{i=1}^3 e^{\theta^i x}}$$

for all the components of the outcome $i = 1, 2, 3$. Here the parameters $\theta^1, \theta^2, \theta^d$ are estimated from the training set (the first third for each series). We apply this algorithm in two regimes. The batch regime, where the algorithm is trained only on the training set and is tested on the test set. The online regime (mLog Online), where at each step new parameters θ are found and only one next outcome is predicted. In both regimes logistic regression does not have theoretical guarantees on the square loss. We also compare our algorithms with the simple predictor (Simple) predicting the average of the ten previous outcomes (it thus always gives probability predictions).

In order to make a prediction, the cAAR requires $O(dn^2)$ operations per step (the most time-consuming operation is the inverse of the matrix, and can be done using Sherman-Morrison formula). The mAAR requires $O((d-1)^2n^2 + d^3)$ operations per step (using Woodbury Formula). Batch logistic regression just gives predictions, which requires $O(n)$ operations per step (we ignore the training time here). Online logistic regression has to be retrained at each step, and

[1] Data sets can be found http://ita.ee.lbl.gov/html/traces.html.

[2] Data sets can be found http://www.neural-forecasting-competition.com/index.htm.

the convergence is not guaranteed. Training algorithms usually use Iteratively Reweighted Least Squares, which in general requires at least $O(dn^3)$ operations per each iteration.

We are not aware of other efficient algorithms for online probability prediction, and thus logistic regression and simple predictor are the only baselines. Component-wise algorithms which could be used for comparison (e.g., Gradient Descent, [9], Ridge Regression, [10]), have to use Algorithm 1 to project their predictions. Thus they have to be applied in a different way than they are described in the corresponding papers, and can not be fairly compared with our algorithms.

The ridge for our algorithms is chosen to achieve the best MSE on the training set. The results are shown in Table 1. As we can see from the table, online methods perform better than the batch method. The online logistic regression performs well, but it is very slow. Our algorithms perform similar to each other and comparable to the online logistic regression, but they are much faster.

Table 1. The square losses and prediction time (sec) of different algorithms applied for time series prediction

Algorithm	MSE	AMSE	Time	MSE	AMSE	Time
	DEC-PKT			LBL-PKT		
cAAR	0.45906	0.45822	0.578	0.48147	0.479	0.579
mAAR	0.45906	0.45822	1.25	0.48147	0.479	1.266
mLog	0.46107	0.46265	0.375	0.47749	0.47482	0.391
mLog Online	**0.45751**	0.45762	2040.141	**0.47598**	0.47398	2403.562
Simple	0.58089	0.57883	0	0.57087	0.5657	0.016
	C4			C9		
cAAR	0.64834	0.65447	0.015	**0.63238**	0.64082	0.015
mAAR	**0.64538**	0.65312	0.062	0.63338	0.64055	0.063
mLog	0.76849	0.77797	0.016	0.97718	0.91654	0.031
mLog Online	0.68164	0.7351	4.328	0.71178	0.75558	10.625
Simple	0.69037	0.69813	0.016	0.6509	0.65348	0
	E5			E8		
cAAR	0.34452	0.34252	0.078	0.29395	0.29276	0.078
mAAR	0.34453	0.34252	0.219	0.29374	0.29223	0.25
mLog	0.31038	0.30737	1.109	0.31316	0.30382	0.109
mLog Online	**0.30646**	0.30575	446.578	**0.27982**	0.27068	83.125
Simple	0.58212	0.58225	0	0.69691	0.70527	0.016

7 Conclusions

We present new algorithms which give probability predictions in the online multidimensional Brier game. Both algorithms do not involve any numerical integration and can be easily computed. Both algorithms have theoretical guarantees on their cumulative losses. One of the algorithms is kernelized and a theoretical bound is proven for the kernelized algorithm. We perform experiments with the linear algorithms and show that they perform relatively well. We compare them with the logistic regression: the benchmark algorithm giving probability predictions.

Kivinen and Warmuth's work [11] includes the case when the possible outcomes lie in a more than 2-dimensional simplex and their algorithm competes with all logistic regression functions. They use the relative entropy loss function $\mathcal{L}$ and get a regret term of the order $O(\sqrt{\mathcal{L}_T(\alpha)})$ which is upper unbounded in the worst case. Their prediction algorithm is not computationally efficient and it is not clear how to extend their results for the case when the predictors lie in an RKHS.

We can prove lower bounds for the regret term of the order $O(\frac{d-1}{d}\ln T)$ for the case of the linear model (1) using methods similar to the ones described in [2], and lower bounds for the regret term of the order $O(\sqrt{T})$ for the case of RKHS. Thus we can say that the order of our bounds by time step is optimal. Multiplicative constants may possibly be improved though.

Acknowledgments

Authors are grateful for useful comments and discussions to Alexey Chernov, Vladimir Vovk, and Ilia Nouretdinov. This work has been supported in part by EPSRC grant EP/F002998/1 and ASPIDA grant from the CRPF.

References

1. Cesa-Bianchi, N., Lugosi, G.: Prediction, Learning, and Games. Cambridge University Press, Cambridge (2006)
2. Vovk, V.: Competitive on-line statistics. International Statistical Review 69, 213–248 (2001)
3. Zhdanov, F., Kalnishkan, Y.: Linear probability forecasting. Technical report, arXiv:1001.0879 [cs.LG], arXiv.org e-Print archive (2009)
4. Brier, G.W.: Verification of forecasts expressed in terms of probability. Monthly Weather Review 78, 1–3 (1950)
5. Michelot, C.: A finite algorithm for finding the projection of a point onto the canonical simplex of $\mathbb{R}^n$. Journal of Optimization Theory and Applications 50, 195–200 (1986)
6. Vovk, V.: Aggregating strategies. In: Proceedings of COLT 1990, pp. 371–383 (1990)
7. Vovk, V., Zhdanov, F.: Prediction with expert advice for the Brier game. In: Proceedings of ICML 2008, pp. 1104–1111 (2008)
8. Gammerman, A., Kalnishkan, Y., Vovk, V.: On-line prediction with kernels and the complexity approximation principle. In: Proceedings of UAI 2004, pp. 170–176 (2004)
9. Kivinen, J., Warmuth, M.K.: Exponentiated gradient versus gradient descent for linear predictors. Information and Computation 132, 1–63 (1997)
10. Hoerl, A.E., Kennard, R.W.: Ridge Regression: biased estimation for nonorthogonal problems. Technometrics 42, 80–86 (2000)
11. Kivinen, J., Warmuth, M.K.: Relative loss bounds for multidimensional regression problems. Machine Learning 45, 301–329 (2001)

The Importance of Similarity Metrics for Representative Users Identification in Recommender Systems

Olga Georgiou and Nicolas Tsapatsoulis

Cyprus University of Technology,
P.O. Box 50329, 3036, Lemesos, Cyprus
{olga.georgiou,nicolas.tsapatsoulis}@cut.ac.cy
http://www.cut.ac.cy

Abstract. In this paper we explore the efficiency of recommendation provided by representative users on behalf of cluster members. Clustering is used to moderate the scalability and diversity issues faced by most recommendation algorithms face. We show through extended evaluation experiments that cluster representative make successful recommendations outperforming the K-nearest neighbor approach which is common in recommender systems that are based on collaborative filtering. However, selection of representative users depends heavily on the similarity metric that is used to identify users with similar preferences. It is shown that the use of different similarity metrics leads, in general, to different representative users while the commonly used Pearson coefficient is the poorest similarity metric in terms of representative user identification.

Keywords: representative users, collaborative filtering, recommender systems, similarity metrics, user clustering.

1 Introduction

People rely on recommendations related to products, services and so forth in order to find the product or service that suites them mostly [9]. The enormous increase of web products, web services, e-commerce environments and so forth, increases also the demand for new approaches that will help users by providing intelligent recommendations [10]. Market trends shows that online buyers will double, over the coming years [18]. The need for trusted [4] and accurate [2] recommendation now is stronger than ever. A good example is the challenge declared by the Netflix Inc. which was searching for the most improved performance system, for movie recommendations and awarded the winners with the impressive prize of one million dollars [14].

Recommender systems face several challenges such as the scalability when dealing with enormously large amount of data and the requirements demand immediate reaction to online requests. The data sparsity which is another challenge, appears when a new user enters the system and makes it hard to find

H. Papadopoulos, A.S. Andreou, and M. Bramer (Eds.): AIAI 2010, IFIP AICT 339, pp. 12–21, 2010.

similarities with existing users. The diversity challenge [23], describes the problem were users who have similar preference in one category of items they may have totally different judgment on items of another kind, while the synonymy problem, refers mostly to the tendency to name an item or service with many different names causing the similarity estimation algorithms to treat the same item named differently as multiple items [16].

In this paper we attempt to improve recommendation efficiency and eliminate most of the above mentioned problems. Starting with the use of clustering we aim to overcome the scalability problem by seeking users for recommendation within smaller and highly similar clusters instead of an overwhelming number of users [1]. Each cluster is shaped based on current user's preferences (ratings) towards similar data objects (like movies, books, jokes, etc), therefore the diversity problem is also handled since user judgment is based on a specific item each time. The next step, following clusters' formation, is the identification of a representative user per cluster. This user is the one achieving the highest similarity score among all the members of his/her cluster. Representative users provide recommendations towards the other cluster members, succeeding the significant improvement of scalability problem.

Identification of representative users depends heavily on the metric used to compute similarity scores among the cluster members [3], [13]. Our main concern is that both the Pearson correlation coefficient and Cosine based metrics [18] are biased toward the most experience users of a specific item [17]. We propose two modified versions of the Cosine-based correlation coefficient that alleviate this bias and lead to representative users, whose recommendation accuracy is higher.

The remaining of the paper is organized as follows. In Section 2 we present a short literature review on similar topics. The proposed recommendation algorithm and similarity metrics are described in Section 3. Section 4 presents our experimental framework, the details regarding the datasets used and discussion on testing findings. Finally, conclusions are drawn and further work hints are given in Section 5.

2 Literature Review

The market interest for recommendation systems improvement arise from the need of companies to provide more accurate recommendations to their customers and have an additional advantage toward their competitors. An example of market's constant seeking for recommendation improvement is the Netflix challenge as mentioned earlier. It is significant to emphasize the large number of participants and the interest presented from many scientist toward the specific competition, which was one of the reasons that lead the company to declare a second round, with more impressive prizes [14].

Several researchers investigated and evaluated existing collaborative filtering algorithms and techniques. Some of them stroke the area of trusted user identification which presents similarities with the representative user identification problem. However, the target there is to separate malicious users from trusted ones [15],[22]. Another similarity with the specific research, was the use of clusters. Clustering techniques were initially proposed to address the problem of scalability [5],[21], although their performance were inferior to the straight collaborative filtering approach. This case is not valid anymore, since new clustering based recommender systems show excellent performance [12],[19].

The importance of similarity metrics in collaborative filtering recommender systems is, somehow, underestimated. Emphasis is given either to alternative recommendation techniques like the Top-N recommendation [11] or to the overall performance evaluation [10] and less to the similarity metrics [3]. The most commonly used similarity metrics in collaborative filtering approaches are the Pearson correlation coefficient and variations of the Cosine correlation coefficient [18] and the influence of similarity metrics in recommendation efficiency was examined [17]. The authors proved, through a series of testing experiment, that the performance of these commonly used correlation approaches is rather poor and new similarity metrics must be investigated. This is the area that the current work emphasizes and additionally, it is important to state that the influence of similarity metrics in recommendation performance is much higher in clustering based approaches were clusters are formed and representative users are selected on the basis of these metrics.

3 The Proposed Method

The recommendation problem can be formulated as follows: Let C be the set of users (customers) and let I be the set of all possible items that the users can recommends, such as books, movies, or restaurants. Let also u be a utility function that measures the usefulness (as may expressed by user ratings) of item i to user c_j, i.e., $u : C \times I \rightarrow \Re$. The usefulness of all items to all users can be expressed as a matrix U with rows corresponding to users and columns corresponding to items. An entry $u(c_j, i)$ of this matrix may have either positive value indicating the usefulness (rating) of item i to user c_j or a zero value indicating that the usefulness $u(c_j, i)$ has not been evaluated. The recommendation problem can be seen as the estimation of zero values of matrix U from the non-zero ones.

Recommendation in the collaborative filtering approach requires some similarity $r(c_a, c_b)$ between users c_a and c_b to be computed based on the items that both of them evaluated with respect to their usefulness. The most popular approaches for user similarity computation are Pearson correlation and Cosine-based metrics. Both of these methods produce values $r(c_a, c_b) \in [-1\ 1]$. In this paper we propose two alternatives of the Cosine correlation coefficient for similarity computation. However, unlike the classic collaborative filtering approach, recommendations are given only by the representative user toward the other cluster members. The proposed method involves the following steps: (1) the users of an existing recommendation system are partitioned into clusters based on their profile of choices made in the past, (2) for each cluster a representative user is identified (using a variety of similarity metrics), (3) once an active user seeks for recommendation, its cluster is identified and the corresponding representative user is selected to provide his/her recommendations.

3.1 Clustering Using K-Means Method

A cluster is a collection of data objects that are similar to one another. The distance measure of the similarity between two objects (users $\underline{c}_i$ and $\underline{c}_j$), influences the shape of clusters, as some elements may be close to one another according to one distance and farther away according to another. A typical similarity metric used in clustering is Euclidean distance; however, when clustering is applied in the context of recommender systems, Pearson and Cosine coefficient based distances are the obvious choices.

The K-means clustering algorithm is widely used in clustering based recommendation systems mainly due to its simplicity [12]. It aims to partition N_c users defined by the user profile vectors $\{\underline{c}_1, \underline{c}_2, ..., \underline{c}_{N_c}\}$ into K clusters ($K << N_c$) $\mathbf{S} = \{S_1, S_2, ..., S_K\}$ in which each user profile vector belongs to the cluster with the nearest mean. The user profile vectors correspond to the rows of the matrix U mentioned earlier.

The K-means algorithm is similar to the expectation-maximization algorithm for mixtures of Gaussians in that they both attempt to find the centers of natural clusters in the data. The optimization criterion is to find the partition $\mathbf{S}$ that minimizes the within-cluster sum of squares (WCSS):

$$\mathbf{S}^o = \underbrace{argmin}_{\mathbf{S}} = \sum_{i=1}^{K} \sum_{\underline{c}_j \in S_i} \|\underline{c}_j - \underline{\mu}_i\|^2 \tag{1}$$

where $\underline{\mu}_i$ is the mean vector of datapoints (user profile vectors) in cluster S_i.

3.2 Similarity Metrics

The aim of this paper is to show that the similarity metric that is used for the computation of a representative user r_i of cluster S_i affects the overall recommendation efficiency. Let us denote $\underline{c}_i^{I_c}$ and $\underline{c}_j^{I_c}$ the user profile vectors for users c_i and c_j computed on the set of items I_c that both of them evaluated. A typical metric used for the similarity computation of users c_i and c_j is the Cosine correlation coefficient:

$$r(c_i, c_j) = \frac{\underline{c}_i^{I_c} \cdot \underline{c}_j^{I_c}}{\|\underline{c}_i^{I_c}\| \|\underline{c}_j^{I_c}\|} \tag{2}$$

where "·" denotes the dot product of two vectors and $\|\underline{x}\|$ denotes the Euclidean norm of vector $\underline{x}$.

In an actual situation, different users may use different rating scales, which the vector Cosine similarity cannot take into account. To address this drawback, the adjusted Cosine similarity coefficient known as Pearson correlation coefficient is used:

$$r(c_i, c_j) = \frac{(\underline{c}_i^{I_c} - \overline{c}_i) \cdot (\underline{c}_j^{I_c} - \overline{c}_j)}{\|\underline{c}_i^{I_c} - \overline{c}_i\| \|\underline{c}_j^{I_c} - \overline{c}_j\|} \tag{3}$$

where $\overline{c}_i$ and $\overline{c}_j$ are the average ratings over all rated items of users c_i and c_j respectively.

The use of Pearson or Cosine correlation coefficient for representative user identification creates bias toward very active users, the ones that rated lot of items. For further explanation, consider a recommender system with N_c users, N_I items, and a cluster S_i containing N_{S_i} users. It is common that $N_{S_i} << N_c$ and $N_{S_i} << N_I$; are the users that are classified into many clusters and the number of cluster members is much smaller than the total number of items. As a result, an average user rates a limited number of items, assume that there is a very active user who rated much more items that the average user. Since both the Cosine and Pearson coefficients are computed based on the commonly rated items, there is an increased probability for a very active user to present similarity (even with a low score) with all cluster members. As a result the specific user will be selected as cluster's representative, even though his/her actual similarity with many of the cluster members is low. In order to overcome this limitation we propose the *min_norm* Pearson coefficient as follows:

$$r(c_i, c_j) = \frac{(\underline{c}_i^{I_c} - \overline{c}_i) \cdot (\underline{c}_j^{I_c} - \overline{c}_j)}{\|\underline{c}_i^{I_c} - \overline{c}_i\| \|\underline{c}_j^{I_c} - \overline{c}_j\|} \frac{N_I^{ij}}{min(N_I^i, N_I^j)} \tag{4}$$

where N_I^{ij} is the number of items commonly rated by users i and j, N_I^i is the total number of items rated by user i and N_I^j is the total number of items rated by user j.

In cases where $N_I^i << N_I^j$ or $N_I^j << N_I^i$ the *min_norm* Pearson coefficient creates bias against very active users making it difficult to create dense clusters. The *max_norm* Pearson coefficient is proposed to moderate this problem by decreasing the similarity value in cases where there is a large difference among the total number of ratings per user:

$$r(c_i, c_j) = \frac{(\underline{c}_i^{I_c} - \overline{c}_i) \cdot (\underline{c}_j^{I_c} - \overline{c}_j)}{\|\underline{c}_i^{I_c} - \overline{c}_i\| \|\underline{c}_j^{I_c} - \overline{c}_j\|} \frac{N_I^{ij}}{max(N_I^i, N_I^j)} \tag{5}$$

It is shown in the experimental results session that both representative users selected either by using the *max_norm* or *min_norm* Pearson coefficient perform better that the classic Pearson coefficient.

4 Experimental Evaluation

The aim of experimental evaluation is to prove the efficiency of the proposed similarity metrics method, for representative user identification. The recommendation efficiency of representative users identified, with various similarity metrics, is measured based on the experiments conducted. The quality of the overall recommendation method is also benchmarked against no recommendation.

4.1 Evaluation Framework

Experimental evaluation of the proposed method was done with the aid of two widely used datasets. The first dataset was collected from Jester online

Jokes Recommender System [6] contains over than 4.1 million continuous ratings [−10.00 10.00] of 100 jokes from 73421 users. It is one of the most widely used dataset for evaluating algorithms for recommender systems. The second dataset was collected from the GroupLens research project [8] and consists of 943 users (demographical information's are also provided for each user), 1682 movies (more details related to the type of movie are also provided) and a record file of 100000 ratings with a rating scale 1 to 5, with 5 being the highest score. In both cases we divided the datasets ratings into 60% for training and 40% for testing producing two matrics $T_r \in \Re^{N_c x J_1}$ and $T_s \in \Re^{N_c x J_2}$ respectively.

Our experiments where conducted on a typical PC, running Windows XP, and with the aid of Matlab platform (`http://www.mathworks.com/`). First, the K-means algorithm (Matlab implementation) was used to partition the N_c users into a varying number of clusters. For each cluster the user with the highest similarity with the other cluster members was selected as the representative user of the cluster. Similarity among users was measured by comparing their ratings on the training set with the aid of Pearson coefficient, *max_norm* and *min_norm* similarity metrics (see equations (3), (5), (4) respectively). In general three different representative users per cluster were identified; however, in several cases, especially in the Joke dataset, the identified representative users coincide.

Recommendations of the representative users were considered the rated items (jokes / movies) with value higher than a threshold T. We selected $T = 3$ for the Joke dataset and $T = 2$ for the GroupLens dataset; the main reason for T restriction in both cases, was the ranging ranges of the two datasets. however, the threshold value does not actually affect the experiments, recommendation efficiency will remain. Removing the thresholds will lead to negative recommendations which is out of scope for this paper. Let c_i^j be the i-th member of the j-th cluster and c_r^j be the representative of that cluster. Let us, also, denote the set of items rated by c_i^j as $\mathbf{I}_i^j$ and the set of high rated items of the same user as $\mathbf{H}_i^j$ (obviously $\mathbf{H}_i^j \subseteq \mathbf{I}_i^j$). For each cluster member two values are computed: the satisfaction without recommendation s_i^j and the satisfaction after receiving recommendation by the cluster representative $\widetilde{s}_i^j$. The last value was computed using the three different similarity metrics mentioned earlier. The s_i^j and $\widetilde{s}_i^j$ values are computed as follows:

$$s_i^j = \frac{Car(\mathbf{H}_i^j)}{Car(\mathbf{I}_i^j)} \qquad \widetilde{s}_i^j = \frac{Car(\mathbf{H}_i^j \bigcap \mathbf{H}_r^j)}{Car(\mathbf{I}_i^j) \bigcap \mathbf{H}_r^j)} \tag{6}$$

where $Car(\mathbf{X})$ denotes the cardinality of set $\mathbf{X}$.

4.2 Results and Discussion

Figure 1 shows the recommendation quality, in the GroupLens dataset, as a function of the number of clusters for the three similarity metrics used. For comparison we plot also the user satisfaction without receiving any recommendation. The representative user identified with the *max_norm* metric clearly provides more accurate recommendations than the representatives, of the other two methods. The *min_norm* metric also outperforms the Pearson coefficient in all cases. Nevertheless, Figure 1 shows that recommendation provided by representative users is effective despite the similarity metric used; user satisfaction is always higher, in cases were recommendations was provided, compared to user satisfaction were no recommendations were provided. An exception occurs in cases where very few clusters are created and recommendations provided only from representative users identified by the Pearson correlation coefficient. Partition into small number of clusters, leads to very sparse and non-homogeneous clusters. As a consequence the identification of representative users, is risky and prone to error.

The conclusions drawn above also hold for the Joke dataset, however, it is shown in Figure 2 that the difference in recommendation efficiency for the three methods is smaller. In addition the overall recommendation efficiency of Jokes dataset is much higher than of the GroupLens dataset. Both observations can be explained by carefully examining the nature of the two datasets. In the Joke dataset the number of items N_I is small (100) and the variation in the number of items evaluated by each user is also small (the minimum number of items evaluated by a single user is 60 while the maximum is 100). In contrary the GroupLens dataset contains much more items (1682) and much less users (943), resulting a very large (minimum 20, maximum 737) variation in the number of items evaluated by each user. For this reason the representative users identified by the different similarity metrics rarely coincide in the GroupLens dataset while coincidence is very common in the Joke dataset.

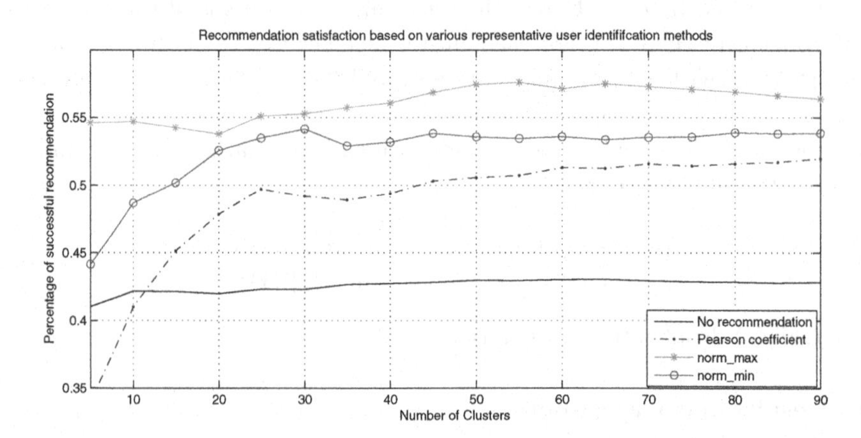

Fig. 1. GroupLens dataset: Recommendation quality using one representative per cluster identified using the Pearson coefficient, the *max_norm* and the *min_norm* similarity metrics. Benchmarking against no recommendation is also shown.

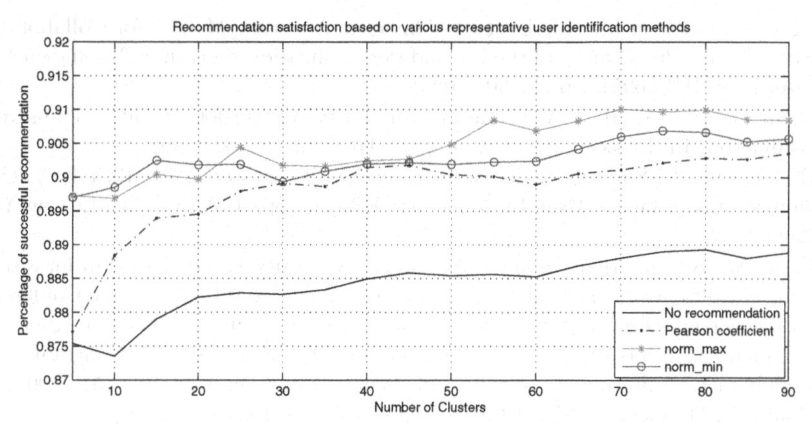

Fig. 2. Jokes Dataset 1: Recommendation quality using one representative per cluster identified using the Pearson coefficient, the *max_norm* and the *min_norm* similarity metrics. Benchmarking against no recommendation is also shown.

5 Conclusion and Further Work

Recommender systems are becoming a valuable tool for E-commerce on the Web as well as for information retrieval in general. They are being stressed by the huge volume of customer data in existing corporate databases, and will be stressed even more by the increasing volume of customer data available on the Web. In order to keep up with this information explosion the scalability of recommender systems must be improved while the time complexity of recommendation algorithms must be kept low. These are the main reasons the clustering based recommendation systems gain attention. In this paper, we have proposed two different similarity metrics for representative user identification. Representative users provide recommendations on behalf and for the other cluster members. It was shown in this paper that the selection of an appropriate similarity metric for representative user identification affects the recommendation efficiency of the overall system. It was experimentally justified, with the aid of two widely used datasets, that the proposed modifications to the Pearson coefficient method, lead to much better recommendation efficiency.

Future work includes testing the proposed method in sparse datasets like the one of the book recommendation system [23]. The creation of dense clusters in such datasets is challenging. In addition Top-N recommendation methods will be investigated using all cluster members recommendations instead of the ones of clusters' representatives. Finally, methods for choosing the best cluster representatives will be also explored.

References

1. Braak, P.t., Abdullah, N., Xu, Y.: Improving the Performance of Collaborative Filtering Recommender Systems through User Profile Clustering. In: The 2009 IEEE/ACM Int'l. Joint Conference on Web Intelligence and Intelligent Agent Technology, vol. 3, pp. 147–150. IEEE Computer Society, Washington (2009)

2. Carenini, G.: User-Specific Decision-Theoretic Accuracy Metrics for Collaborative Filtering. In: Proceedings of the Beyond Personalization Workshop, Intelligent User Interfaces (IUI 2005), pp. 26–30 (2005)
3. Chan, Y.S.: MAXSIM: A Maximum Similarity Metric for Machine Translation Evaluation. In: Proceedings of ACL 2008: HLT, pp. 55–62 (2008)
4. O'Donovan, J., Smyth, B.: Trust in recommender systems. In: 10th Int'l. Conference on Intelligent User Interfaces (IUI 2005), pp. 167–174. ACM, New York (2005)
5. Ester, M., Kriegel, H.-P., Sander, J., Xu, X.: A density-based algorithm for discovering clusters in large spatial databases with noise. In: 2nd International Conference on Knowledge Discovery and Data Mining, Portland, OR, pp. 226–231 (1996)
6. Goldberg, K., Roeder, T., Gupta, D., Perkins, C.: Eigentaste: A Constant Time Collaborative Filtering Algorithm. Information Retrieval 4(2), 133–151 (2001)
7. Goldberg, D.: Genetic Algorithms. Addison-Wesley, Reading (1989)
8. The GroupLens Research Project, `http://www.grouplens.org/`
9. Herlocker, J. L., Konstan, J. A., Riedl, J. T.: Explaining Collaborative Filtering Recommendations. In: Proceedings of the 2000 ACM Conference on Computer Supported Cooperative Work, pp. 241–250 (2000).
10. Herlocker, J.L., Konstan, J.A., Terveen, L.G., Riedl, J.T.: Evaluating collaborative filtering recommender systems. ACM Transactions on Information Systems 22(1), 5–53 (2004)
11. Jamali, M., Ester, M.: Using a trust network to improve top-N recommendation. In: Proceedings of the 3rd ACM Conference on Recommender Systems (RecSys 2009), pp. 181–188.
12. Kim, K., Ahn, H.: A recommender system using GA K-means clustering in an online shopping market. Expert Systems with Applications: An International Journal 34(2), 1200–1209 (2008)
13. McLaughlin, M.R., Herlocker, J.L.: A collaborative filtering algorithm and evaluation metric that accurately model the user experience. In: 27th Annual International ACM Conference on Research and Development in Information Retrieval (SIGIR 2004), pp. 329–336. ACM, New York (2004)
14. Netflix Inc., `http://sacramento.bizjournals.com/sacramento/stories/2009/09/21/daily3.html` (Last visited on May 24, 2010)
15. Sarda, K., Gupta, P., Mukherjee, D., Padhy, S., Saran, H.: A Distributed Trust-based Recommendation System on Social Networks. In: Proceedings of the 14th ACM SIGKDD International Conference on Knowledge Discovery and Data Mining, pp. 160–168 (2008)
16. Sarwar, B., Karypis, G., Konstan, J., Riedl, J.: Recommender systems for large-scale e-commerce: Scalable neighborhood formation using clustering. In: 5th International Conference on Computer and Information Technology. ICCIT (2002)
17. Spertus, E.: Evaluating Similarity Measures: A Large-Scale Study in the Orkut Social Network. In: Proceedings of the 16th International Conference on World Wide Web, pp. 131–140 (2005).
18. Su, X., Khoshgoftaar, T.M.: A Survey of Collaborative Filtering Techniques. In: Advances in Artificial Intelligence. Article ID 421425, vol. 2009, Hindawi Publishing Corporation (2009)
19. Truong, K., Ishikawa, F., Honiden, S.: Improving Accuracy of Recommender System by Item Clustering. IEICE - Transactions on Information and Systems E90-D(9), 1363–1373 (2007)

20. Vogiatzis, D., Tsapatsoulis, N.: Modeling User Networks in Recommender Systems. In: 3rd International Workshop on Semantic Media Adaptation and Personalization (SMAP 2008), pp. 106–111. IEEE Computer Society, Washington (2008)
21. Zhang, T., Ramakrishnan, R., Livny, M.: BIRCH: an efficient data clustering method for very large databases. In: The ACM SIGMOD International Conference on Management of Data, Montreal, Canada, vol. 25, pp. 103–114 (1996)
22. Zhang, F., Bai, L., Gao, F.: A User Trust-Based Collaborative Filtering Recommendation Algorithm. In: Information and Communications Security. LNCS, vol. 5927, pp. 411–424. Springer, Heidelberg (2009)
23. Ziegler, C.-N., McNee, S.M., Konstan, J.A., Lausen, G.: Improving Recommendation Lists Through Topic Diversification. In: 14th International World Wide Web Conference (WWW 2005), Chiba, Japan, pp. 22–32 (2005)

An Optimal Scaling Approach to Collaborative Filtering Using Categorical Principal Component Analysis and Neighborhood Formation

Angelos I. Markos[1], Manolis G. Vozalis[2], and Konstantinos G. Margaritis[2]

[1] Department of Primary Education, Democritus University of Thrace
amarkos@eled.duth.gr
[2] Department of Applied Informatics, University of Macedonia
{mans,kmarg}@uom.gr

Abstract. Collaborative Filtering (CF) is a popular technique employed by Recommender Systems, a term used to describe intelligent methods that generate personalized recommendations. The most common and accurate approaches to CF are based on latent factor models. Latent factor models can tackle two fundamental problems of CF, data sparsity and scalability and have received considerable attention in recent literature. In this work, we present an optimal scaling approach to address both of these problems using Categorical Principal Component Analysis for the low-rank approximation of the user-item ratings matrix, followed by a neighborhood formation step. The optimal scaling approach has the advantage that it can be easily extended to the case when there are missing data and restrictions for ordinal and numerical variables can be easily imposed. We considered different measurement levels for the user ratings on items, starting with a multiple nominal and consecutively applying nominal, ordinal and numeric levels. Experiments were executed on the MovieLens dataset, aiming to evaluate the aforementioned options in terms of accuracy. Results indicated that a combined approach (multiple nominal measurement level, "passive" missing data strategy) clearly outperformed the other tested options.

Keywords: Collaborative filtering, Recommender systems, Low-rank approximation, Categorical Principal Component Analysis, Optimal scaling.

1 Introduction

Recommender Systems (RSs) are intelligent applications that generate personalized recommendations. RSs often rely on Collaborating Filtering (CF), which is based on the premise that users who have agreed in the past, tend to agree in the future. Among the most common and successful approaches to CF are latent factor models, which directly profile both users and products. The goal of latent factor models is to uncover latent features that explain user preferences by transforming both items and users to the same latent factor space, thus

H. Papadopoulos, A.S. Andreou, and M. Bramer (Eds.): AIAI 2010, IFIP AICT 339, pp. 22–29, 2010.

making them directly comparable. Latent factors represent either obvious characteristics of the items or completely uninterpretable dimensions. Each factor measures the degree of preference for each user towards items that score high on the corresponding factor.

Several latent factor models have been successfully applied to CF. Sarwar et al. [1] apply Singular Value Decomposition to reduce the dimensionality of sparse rating matrices and improve the scalability of CF systems. Goldberg et al. [2] use Principal Component Analysis (PCA) as a preprocessing step for low-rank approximation and then utilize clustering to predict user preferences. Hoffman [3] suggests a collaborative prediction method based on a probabilistic latent variable model. Salakhutdinov and Mnih [4] propose a momentum based, probabilistic matrix factorization algorithm with batch learning which scales linearly with the number of observations. Bell and Koren [5] present a neighborhood based approach, which uses alternating least squares and removes the global effect from the data. Tacacs et al. [6] discuss and compare modifications of already published matrix factorization methods. In addition to a straightforward optimization approach for the general low-rank approximation problem, many authors propose extensions of well established algorithms to cope with the missing data problem [7,8].

In this work, we present an optimal scaling approach to address two fundamentals problems of CF, data sparsity and scalability. Optimal scaling is a general approach to treat multivariate data through the optimal transformation of qualitative scales to quantitative values [9,10]. In other words, both nominal and ordinal variables (or features) can be optimally transformed to variables with numeric properties. These optimal transformations of the original variables may be used to overcome the linear assumption underlying many classic dimensionality reduction methods. In this light, PCA has been reviewed and extended to Categorical or Nonlinear Principal Component Analysis (CatPCA) with many potential applications [11].

Our approach uses CatPCA for the low-rank approximation of the user-item ratings matrix, followed by a neighborhood formation step. The combination of matrix factorization and neighborhood formation can lead to accurate predictions [5,6]. CatPCA is based on the Alternating Least Squares (ALS) algorithm utilizing an optimal least squares scaling process where original data are transformed so that their overall variance is maximized. The problem is formulated by means of a loss function and it is solved by the ALS. Eventually, this leads to optimally scaled scores on each factor [12]. Unlike other matrix factorization methods, CatPCA does not assume multivariate normality and linear relationships between variables and provides a flexible framework for parametrization. In particular, we consider different measurement levels for the user ratings on items, starting with a multiple nominal and consecutively applying nominal, ordinal and numeric levels. A series of experiments was executed on the MovieLens dataset in order to evaluate the aforementioned options in terms of accuracy. A combined approach (multiple nominal level, "passive" missing data strategy) clearly outperformed the other tested options.

The remainder of this paper is organized as follows. The next section, Section 2, gives a brief presentation of CatPCA with optimal scaling. In Section 3, the proposed approach is thoroughly described, starting with the low-rank approximation of the ratings matrix, which is achieved through CatPCA, and continuing with a user neighborhood formation step. The efficiency of four distinct variations of this approach is demonstrated in Section 4 through a set of experiments on a publicly available dataset. Finally, we give conclusions and provide a discussion over the results in Section 5.

2 Categorical Principal Component Analysis with Optimal Scaling

CatPCA with optimal scaling or optimal scoring is a general approach to treat multivariate data through the optimal transformation of qualitative scales to quantitative values [9,10]. In the optimal transformation process, an appropriate quantification level has to be chosen for each of the variables. The most restricted transformation level is called *numeric*; it applies a linear transformation to the original integer scale values, so that the resulting variables are standardized. The order of variable categories and the equal distances between category numbers of the original variable are preserved in the optimally scaled variable. When all variables are at a numeric analysis level, the analysis is analogous to standard PCA. The optimal least squares transformations of *ordinal* data can be handled by means of monotonic transformations, which maintain the order in the original data. In the case of *nominal* data, categories are not ordered and the only information that is preserved is the grouping of objects in categories. Finally, when all variables are at a *multiple nominal* level, the only information that is preserved is the grouping of objects in categories. The quantification is called *multiple* because there is a separate quantification for each dimension. With all variables at a multiple nominal level, the analysis is equivalent to multiple correspondence analysis or homogeneity analysis [9,13].

3 Algorithm Description

In this Section we describe the proposed optimal least squares approach for CF, which starts with the low-rank approximation of the user-item ratings matrix and is followed by a user neighborhood formation step.

3.1 Low-Rank Approximation through CatPCA

We start with the following basic definitions. For $i = 1, \ldots, n$ users, ratings on $j = 1, \ldots, m$ items are collected in the $n \times m$ data matrix $\mathbf{R}$. Each of the corresponding items takes on k different rating values (levels or categories) from a given range, i.e. $(1, 2, 3, 4, 5)$. They are coded using m binary indicator or dummy matrices $\mathbf{G}_j$ of size $n \times k$, with entries $\mathbf{G}_j(i,t) = 1$ if user i has given item j a rating of t, and $\mathbf{G}_j(i,t) = 0$ for any of the remaining rating values. The whole set of indicator matrices can be collected in a block matrix $\mathbf{G} = [\mathbf{G}_1 \ldots \mathbf{G}_m]$.

Missing item ratings are coded as complete zero row sums: if user i has not rated item j, then the row sum of $\mathbf{G}_j$ is 0. Otherwise the row sum becomes 1. Let $\mathbf{M}_j$ denote the $n \times n$ binary diagonal matrix with entries $\mathbf{M}_j(i,i) = 1$ if user i has rated item j and 0 otherwise. Based on $\mathbf{M}_j$ we define $\mathbf{M}_\star$ as the sum of the $\mathbf{M}_j$'s and $\mathbf{M}_\bullet$ as their average. This missing data strategy is known in the literature as *missing data passive*, because it leaves the indicator matrix $\mathbf{G}_j$ incomplete [12]. The main advantage of this option is that it ignores the missing ratings without making strong assumptions regarding the pattern of missing data.

Let $\mathbf{X}$ be the unknown $n \times p$ matrix containing the user scores in the latent factor space R^p and $\mathbf{Y}_j$ be the unknown $k \times p$ matrix containing the rating scores (quantifications) of item j in the same p-dimensional space. Both matrices have to be determined during optimization. Based on these definitions, the following loss function can be established:

$$\sigma(\mathbf{X}; \mathbf{Y}_1 \dots \mathbf{Y}_m) = \frac{1}{m} \sum_{j=1}^{m} tr(\mathbf{X} - \mathbf{G}_j\mathbf{Y}_j)' \mathbf{M}_j (\mathbf{X} - \mathbf{G}_j\mathbf{Y}_j) \tag{1}$$

which is optimized under the normalization conditions $u'\mathbf{M}_\bullet\mathbf{X} = 0$, with u denoting an n vector with ones and $\mathbf{X}'\mathbf{M}_\bullet\mathbf{X} = I$ such that the trivial solution of complete 0-scores is avoided. From an analytical point of view the loss function represents the sum of squares of $(\mathbf{X} - \mathbf{G}_j\mathbf{Y}_j)$ and a simultaneous minimization over $\mathbf{X}$ and $\mathbf{Y}_j$. Minimizing equation 1 leads to a low dimensional space where users are positioned as close as possible to the ratings they have given to specific items and where an item's rating is the centroid of the users who have given this rating to the item.

The minimization problem can be solved by means of an iterative ALS procedure as follows:
At iteration 0 we begin with a random starting solution $\mathbf{X}^0$ for the user scores. Each iteration s consists of three steps:

1. Update item rating quantifications: $\mathbf{Y}_j^{(s)} = \mathbf{D}_j^{-1}\mathbf{G}_j'\mathbf{X}^{(s)}$ for $j = 1, \dots m$, where $\mathbf{D}_j = \mathbf{G}_j'\mathbf{G}_j$ the $k \times k$ diagonal matrix with the (marginal) frequencies of item j in its main diagonal.
2. Update user scores: $\widetilde{\mathbf{X}}^{(s)} = \mathbf{M}_*^{-1} \sum_{j=1}^{m} \mathbf{G}_j\mathbf{Y}_j^{(s)}$
3. Normalization: $\mathbf{X}^{(s+1)} = \mathbf{M}_*^{-1/2}\mathbf{orth}()(\mathbf{M}_*^{-1/2}\widetilde{\mathbf{X}}^{(s)})$
   Here $\mathbf{orth}()$ denotes the modified Gram-Schmidt process which is used to compute an orthonormal basis for the column space of a matrix.

This iterative process is continued until the improvement in subsequent loss values is below a convergence criterion. Note that matrix multiplications using indicator matrices can be implemented efficiently as cumulating the sums of rows over $\mathbf{X}$ and $\mathbf{Y}$. The ALS algorithm only computes the first p dimensions of the solution, which leads to an increase in computational efficiency. Moreover, by capitalizing on sparseness of $\mathbf{G}$, the algorithm is able to handle large datasets.

The approach described above is known as homogeneity analysis (homals) with all item ratings at a *multiple nominal* level of measurement and it has the advantage that it can be easily extended to the case when restrictions can be easily imposed. Nominal, numerical and ordinal item ratings can be incorporated with a rank-1 restriction of the form

$$\mathbf{Y}_j = \mathbf{z}_j \mathbf{a}'_j \tag{2}$$

where $\mathbf{z}_j$ is a column vector of length k with item rating quantifications and $\mathbf{a}_j$ a vector of length p with weights. Using this restriction the ratings of a specific item are forced to be on a line in the p dimensional space. However, this restriction is not sufficient to preserve the order for ordinal ratings or even the relative distance for numerical ratings. Therefore, $\mathbf{z}_j$ should be transformed in every ALS iteration to satisfy these restrictions. In the case of ordinal ratings this transformation is achieved by means of a weighted monotone regression of $\mathbf{z}_j$ on $\mathbf{r}_j$ and in the case of numerical ratings $\mathbf{z}_j$ is replaced by the original item rating $\mathbf{r}_j$. For a more detailed description we refer to [12,9].

3.2 User Neighborhood Formation

At this point, we use the $n \times p$ matrix $\mathbf{X}$, which is obtained from the previous step, in order to form each user's neighborhood. Note that $\mathbf{X}$ contains the user scores in the p-dimensionality space. To find the proximity between two users, a and i, we utilize the Pearson product-moment correlation coefficient, as a similarity measure between each user and his closest neighbors. This similarity measure has been evaluated in a number of studies and has been found to be reliable and accurate. The coefficient is calculated as follows:

$$cor_{ai} = \frac{\sum_{j=1}^{l} r_{aj} r_{ij}}{\sqrt{\sum_{j=1}^{l} r_{aj} \sum_{j=1}^{h} r_{ij}}} \tag{3}$$

where r_{ij} denotes the rating of user i on item j. The summations over j are calculated over the l items for which both users a and i have expressed their opinions.

After a user neighborhood of size h has been formed for user a, we proceed with prediction generation. A prediction rating p_{aj} for user a on item j is computed using the following equation:

$$p_{aj} = \frac{\sum_{i=1}^{h} r_{ij} * cor_{ai}}{\sum_{i=1}^{h} |cor_{ai}|} \tag{4}$$

4 Experiments

In this section we provide a brief description of the experiments we executed in order to evaluate and compare the proposed methods, and then we present and comment on their results.

For this purpose, we utilized the MovieLens dataset. It consists of 100,000 ratings which were assigned by 943 users on 1682 movies and is available at `http://www.grouplens.org/`. Ratings follow the 1(bad)-5(excellent) numerical

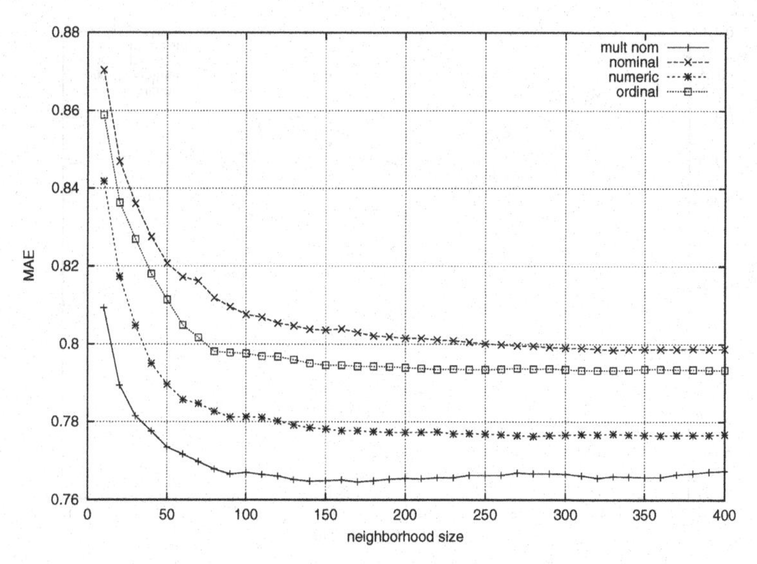

Fig. 1. Comparison of the three CatPCA algorithms for varying neighborhood sizes

scale. The sparsity of the dataset is high, at a value of 93.7%. Starting from the original dataset, five distinct splits of training (80%) and test (20%) data were utilized. The proposed algorithms were applied on each split and the average result across all 5 trials was computed.

Mean Absolute Error (MAE) was the metric we employed to evaluate the accuracy of the methods. MAE measures the deviation of predictions generated by the RSs from the true rating values, as they were specified by the user.

For our first experiment, we kept a fixed number of retained dimensions and tried to evaluate the impact of a varying neighborhood size on prediction accuracy. Figure 1 depicts the MAE values for neighborhood sizes ranging between 1 and 400 users. Based on that figure, it is clear that the multiple nominal and numeric approaches performed better than the nominal and ordinal ones. In particular, *multiple nominal* has generated the most accurate prediction among those tested (MAE=0.7646). This value was achieved for a neighborhood of 170 users, the smallest in size when compared to the number of neighbors required by the rest of the approaches to reach their optimal predictions. Specifically, the lowest MAE values for the numeric, nominal and ordinal approaches were 0.7763, 0.7985, and 0.7993, observed for neighborhoods including 280, 330 and 330 users, respectively.

Once the effect of varying neighborhood sizes was assessed, a second experiment was executed, aiming to evaluate the impact of dimensionality reduction on prediction accuracy. Figure 2 depicts the MAE results for values of retained dimensions d, between 30 and 100. Based on that Figure, the advantage of *multiple nominal* approach over the rest was verified. Furthermore, both multiple nominal and ordinal levels reached their best MAE values for d=70, whereas d was 50 and 60 for numeric and nominal levels, respectively.

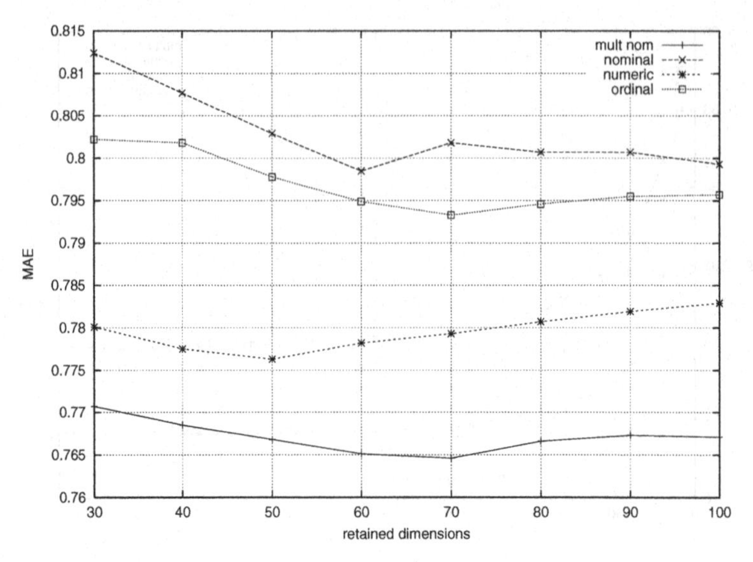

Fig. 2. Comparison of the three CatPCA algorithms for different values of retained dimensions

5 Conclusions

In this paper, we describe an optimal scaling approach for CF using CatPCA, which combines the low-rank approximation of the user-item ratings matrix with user neighborhood formation. We considered four different measurement levels for the user ratings in the Movielens dataset. A combined approach, where the user ratings on each item were handled as multiple nominal, gave the most accurate predictions according to MAE.

CatPCA with optimal scaling can offer a versatile set of options suited to CF problems. The main advantage of the proposed approach stems from the fact that CatPCA is able to account for more of the variance in the data compared to linear PCA, when the variables are (or may be) nonlinearly related to each other. Another advantage of this approach is that the "passive" strategy for handling missing data, ignores missing ratings without making strong assumptions for their pattern as is the case for various imputation methods. Additionally, predictions for new users of the system can be obtained by the model without requiring a complete retraining of the whole dataset. This is achieved after optimization by means of a cone restricted SVD [13], without influencing the existing scores computed by ALS.

In order to further explore the effectiveness of the proposed approach, more applications on datasets of different size and sparsity are needed. Except for the optimal least squares transformations described in Section 3, optimal spline transformations can also be utilized. Splines are usually smoother and more robust, albeit at the cost of less goodness of fit with respect to the overall loss function that is minimized [12,13]. Another direction for future work is to introduce regularization terms into the loss function of CatPCA in order to improve the stability of the algorithm.

References

1. Sarwar, B.M., Karypis, G., Konstan, J.A., Riedl, J.T.: Application of dimensionality reduction in recommender systems - a case study. In: ACM WebKDD 2000 Web Mining for E-Commerce Workshop, pp. 82–90 (2000)
2. Goldberg, K., Roeder, T., Gupta, D., Perkins, C.: Eigentaste: A constant time collaborative filtering algorithm. Information Retrieval Journal 4, 133–151 (2001)
3. Hofmann, T.: Latent semantic models for collaborative filtering. ACM Transactions on Information Systems 22(1), 89–115 (2004)
4. Salakhutdinov, S., Mnih, A.: Probabilistic matrix factorization. In: Platt, J., Koller, D., Singer, Y., Roweis, S. (eds.) Advances in Neural Information Processing Systems, vol. 20, pp. 1257–1264. MIT Press, Cambridge (2008)
5. Bell, M., Koren, Y.: Scalable collaborative filtering with jointly derived neighborhood interpolation weights. In: Proceedings of 2007 Seventh IEEE International Conference on Data Mining (ICDM), pp. 43–52 (2007)
6. Tacacs, G., Pilaszy, I., Nemeth, B., Tikk, D.: Scalable collaborative filtering approaches for large recommender systems. The Journal of Machine Learning Research 10, 623–656 (2009)
7. Kim, D., Yum, B.J.: Collaborative filtering based on iterative principal component analysis. Expert Systems with Applications 28(4), 823–830 (2005)
8. Paterek, A.: Improving regularized singular value decomposition for collaborative filtering. In: Proceedings of 13th ACM International Conference on Knowledge Discovery and Data Mining (KDD 2007), San Jose, CA, USA, pp. 39–42 (2007)
9. de Leeuw, J.: Nonlinear principal component analysis and related techniques. In: Greenacre, M., Blasius, J. (eds.) Multiple Correspondence Analysis and Related Techniques, pp. 107–133. Chapman & Hall, Boca Raton (2006)
10. Costantini, P., Linting, M., Porzio, G.: Mining performance data through nonlinear pca with optimal scaling. Applied Stochastical Models in Business and Industry 26, 85–101 (2010)
11. Meulman, J., van der Kooij, A., Heiser, W.: Principal components analysis with nonlinear optimal scaling transformations for ordinal and nominal data. In: Kaplan, D. (ed.) Handbook of Quantitative Methods in the Social Sciences, pp. 49–70. Sage Publications, Newbury Park (2004)
12. Michailidis, G., de Leeuw, J.: The gifi system of descriptive multivariate analysis. Statistical Science 13(4), 307–336 (1998)
13. de Leeuw, J., Patrick, M.: Gifi methods for optimal scaling in r: The package homals. Journal of Statistical Software 31(4), 1–21 (2009)

A Classroom Observation Model Fitted to Stochastic and Probabilistic Decision Systems

Marios Poulos[1], Vassilios S. Belesiotis[2], and Nikolaos Alexandris[3,*]

[1] Ionian University, Department of Archives and Library Science,
Ioannou Theotoki 72, 49100, Corfu Greece
[2,3] University of Piraeus – Department of Informatics
mpoulos@ionio.gr, vbel@unipi.gr, alexandr@unipi.gr

Abstract. This paper focuses on solving the problems of preparing and normalizing data that are captured from a classroom observation, and are linked with significant relevant properties. We adapt these data using a Bayesian model that creates normalization conditions to a well fitted artificial neural network. We separate the method in two stages: first implementing the data variable in a functional multi-factorial normalization analysis using a normalizing constant and then using constructed vectors containing normalization values in the learning and testing stages of the selected learning vector quantifier neural network.

Keywords: Probabilistic Decision Systems, Normalization Data, Education, Formative Assessment in the Classroom, Classroom Observation.

1 Introduction

1.1 Background and Motivation

The preparation stage of data processing is the most significant procedure in stochastic and probabilistic decision systems [1], an important subclass of which is neural networks. Modeling has an enormous impact on the success of a wide range of such complex data analyses as data mining and feature-extraction procedures [1, 2], mainly because the quality of the input data in neural network models strongly influences their data analysis's results [1, 3] and the efficiency of their performance, as wrongly prepared data is likely to make the data analysis problematic [1]. The appropriate preprocessing of input data is therefore a significant procedure [1, 4]. A study of the procedure for grouping collected data [1] that addresses the selection of the data's variables with genetic algorithms and those algorithms' linkage to significant relevant properties would therefore be valuable.

1.2 Problems with the Preparation Data Stage

Grouping these categories with the specific multi-scale and normalization practices in the non-structural data's category may, however, be overly simplistic [1], as doing so

* N. Alexandris. Visiting Professor at the University of Canberra, Faculty of Information Sciences and Engineering, is Professor of University of Piraeus in the Department of Informatics.

H. Papadopoulos, A.S. Andreou, and M. Bramer (Eds.): AIAI 2010, IFIP AICT 339, pp. 30–36, 2010.

cannot provide solutions for significant multi-factorial unconstructed data that are linked with significant relevant properties. A characteristic application of this method is the domain of classroom observation [5], [6], because data collected from this process fall into the aforementioned category.

Existing solutions have focused on the consolidation phase, during which data management systems consist of a continuous process. The capture procedure stores unconstructed data in a well-constructed database [7], but their basic problem is in the pre-processing training setting, as the number of possible interaction factors grows exponentially with the inclusion of each additional main effect in the logistic regression model. Logistic regression's ability to deal with interactions involving many factors is therefore limited, as having too many independent variables in relation to the number of observed outcome events is problematic [8], [9].

1.3 Aims and Scope

This paper attempts to solve the problems of preparing and normalizing data linked with significant relevant properties and captured from many sources by adapting these factors to a first phase using a Bayesian model that creates normalization conditions with an equitable distribution between the significant relevant properties of such types of data. It also aims to adapt this normalization to a well-fitted neural network so it can accept pre-processing data for training and testing procedures. It does this using two separate procedures, first implementing the data variable in a functional multi-factorial normalization analysis using a normalizing constant and then using constructed vectors containing normalization values in the learning and testing stages of the selected learning vector quantifier neural network.

For application purposes of this model we focus on the educational domain and more specifically on classroom observations in order to construct a decision system able to learn to evaluate the quality of a teaching process given as input a set of observations describing the events that took place during this process. The required responses of the system are boolean (yes or no) and form a statistical model which follows the t-test. The proposed method will focus on the normalization of the input data, enabling the effective training of a neural network in order to provide evaluation results for any given set of data. The evaluation of a teaching process can be any value from the fuzzy set: excellent, good, adequate, poor.

This paper is divided in four sections. Section 2 describes the integrated data preparation method as well as the architecture of the neural network with the original classroom observation model. In Section 3, the fitted model is presented, adapting the model to the classroom observation data. Finally, Section 4 concludes and explains our plans for future work.

2 Proposed Method

Figure 1 outlines the schedule of this approach. Following our method, captured classroom observation data are normalized using a Bayesian model and adapted to a well-fitted neural network.

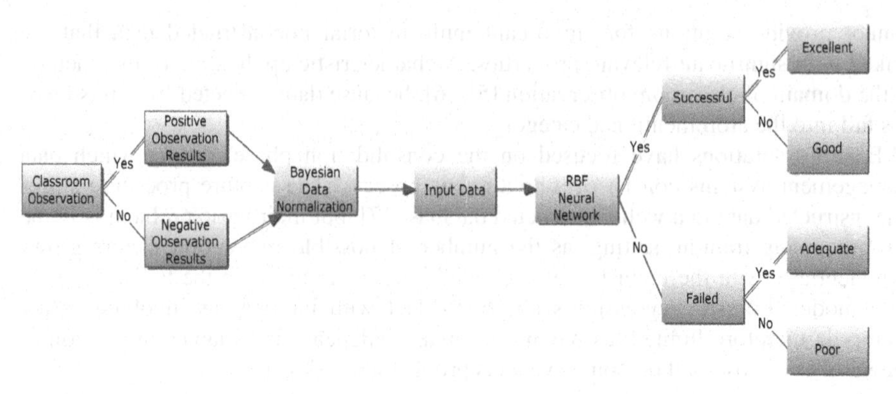

Fig. 1. The schedule of the proposed method is analyzed

2.1 Education Model

The original model is based on Systematic Observation [8], which presents a process for categorizing classroom observations regarding events describing the teacher's and pupils' behavior in a well constructed form. The pioneer of this method Flanders introduced a system [9] which has formed the basis for a number of studies in this field [8]. He defines teaching behavior as 'acts by the teacher which occur in the context of classroom interaction' [9]. Flanders devised a system which is divided into 3 broad areas: teacher talk, pupil talk and silence. This classification is overviewed in Table 1. However, in order to apply our method we consider that the expected value for the number of occurrences of each characteristic is equal to S/n, where S is the total number of observations and n is the number of characteristics. Furthermore, we consider a confidence probability of a=0.025, ranging the area between $x^2_{0,975,v}$ and $x^2_{0,025,v}$ with degree of freedom v=10-1=9.

2.2 Integrated Data Preparation Method

According to previous study [1], the proposed technique to formulate related features of non-structural data is based on extracting related features from text data or documents by semantic analysis and formulates an event-specific summary.

But this technique doesn't support our cases which are based on % related results. In this case, the solution of this conversion could be found in the Bayes's theorem according to the posterior probability measure is proportional to the product of the prior probability measure and the likelihood function. Proportional to implies that one must multiply or divide by a normalizing constant to assign measure 1 to the whole space, i.e., to get a probability measure. In a simple discrete case we have

$$P(H_0|D) = \frac{P(D|H_0)P(H_0)}{P(D)} \quad (1)$$

where $P(H_0)$ is the prior probability that the hypothesis is true; $P(D|H_0)$ is the conditional probability of the data given that the hypothesis is true, but given that the data are known it is the likelihood of the hypothesis (or its parameters) given the data;

Table 1. Flanders' Interaction Analysis categories [9: 34], where T: Teacher, R: Response, I: Initiation, S is the total number of observations and Si (i=1..9) is the number of observations for each characteristic.

Parameters	Characteristics	Groups	Expected Result Total=S	Observation Result Total=S A=0.025
1	**Teacher, accepts feeling**. Accepts and clarifies an attitude or the feeling tone of a pupil in a non-threatening manner. Feelings may be positive or negative. Predicting and recalling feelings are included.	T-R	S/10	S1/S
2	**Teacher, Praises or encourages**. Praises or encourages pupil action or behaviour. Jokes that release tension, but not at the expense of another individual: nodding head, or saying 'Um hm?' or 'Go on' are included.	T-R	S/10	S2/S
3	**Teacher, accepts or uses ideas of pupils**. Clarifying, building or developing ideas suggested by a pupil. Teacher extensions of pupil ideas are included but as the teacher brings more of his own ideas into play, shift to category five.	T-R	S/10	S3/S
4	**Teacher, Asks questions**. Asking a question about content or procedure, based on teacher ideas, with the intent that a pupil will answer.	T-I	S/10	S4/S
5	**Teacher, Lecturing**. Giving facts or opinions about content or procedures: expressing his own ideas, giving his own explanation or citing an authority other than a pupil.	T-I	S/10	S5/S
6	**Teacher, Giving directions**. Directions, commands or orders to which a pupil is expected to comply.	T-I	S/10	S6/S
7	**Teacher, Criticizing or justifying authority**. Statements intended to change pupil behavior from non-acceptable to acceptable pattern; bawling someone out; stating why the teacher is doing what he is doing; extreme self-defense.	T-I	S/10	S7/S
8	**Pupil talk – response**. Talk by pupils in response to teacher. Teacher initiates the contact or solicits pupil statement or structures the situation. Freedom to express own ideas is limited.	P-R	S/10	S8/S
9	**Pupil talk – initiation**. Talk by pupils which they initiate. Expressing own ideas; initiating a new topic; freedom to develop opinions and a line of thought, like asking thoughtful questions: going beyond the existing structure.	P-I	S/10	S9/S
10	**Silence or confusion**. Pauses, short periods of silence and periods of confusion in which communication cannot be understood by the observer.	S		

$P(H_0|D)$ is the posterior probability that the hypothesis is true given the data. P(D) should be the probability of producing the data, but on its own is difficult to calculate, so an alternative way to describe this relationship is as one of proportionality:

$$P(H_0 \mid D)P(D \mid H_0)P(H_0). \tag{2}$$

Since P(H|D) is a probability, the sum over all possible (mutually exclusive) hypotheses should be 1, leading to the conclusion that

$$P(H_0|D) = \frac{P(D|H_0)P(H_0)}{\sum_i P(D|H_i)P(H_i)}. \tag{3}$$

In this case, the reciprocal of the value

$$P(D) = \sum_i P(D|H_i)P(H_i) \tag{4}$$

In our case according to Table 1 the prior probability is given by $P(H_0)$ =S1/S. The first conditional probability is given by $P(D|H_0) = P(H_0)/\alpha*100$.

$$k(j)=1/P(D) \tag{5}$$

is the normalizing constant [10] where j=1..9 is number of the parameters. Vector k(j) is of size 1xj.

2.3 Neural Network Architecture

We adopt as the ideal neural network to accept an appropriate fitted Bayesian vector k(j), since the main applications of RBF have been shown to be in pattern with a Bayesian classifier [11]. For this purpose we select an RBF neural network classifier from a variety of available neural networks architectures, for the following reasons:

The RBF network was preferred to a Multilayer Perceptron (MLP), because it can solve a given problem using fewer neurons and in sorter time, yet with the same success.

The RBF network has the ability to classify incoming vectors into classes that are not linearly separable because it uses static Gaussian function as the nonlinearity for the hidden layer processing elements [12]. The Gaussian function responds only to a small region of the input space where the Gaussian is centered. Successful application of this classifier relies on finding suitable centers for the Gaussian functions. An unsupervised approach usually produces better results than supervised learning.

Using n classes according to the referred problem we design the node characteristics and the RBF network topology.

An RBF neural network can be considered as a special three-layered network. The input nodes pass the input values to the internal nodes that formulate the hidden layer. The nonlinear responses of the hidden nodes are weighted in order to calculate the final outputs of network in the third (output) layer [13].

A typical hidden node in an RBF network is characterized by its center, which is a vector with dimension equal to the number of inputs to the node. The activity vl(x) of the lth node is the Euclidean norm of the difference between the input vector k(j) and the node center and is given by:

$$v_i(x) = \|x - \hat{x}_i\| \tag{1}$$

The output function of the node is a radially symmetric function. A typical choice, which is also used in this work, is the Gaussian function [14]:

$$f(v) = \exp\left(-\frac{v^2}{\sigma^2}\right) \tag{2}$$

where σ is the width of the node.

3 Fitted Model

According to Sections 2.1, 2.2 and 2.3 we constructed the vector of the normalization data k(j). For the training procedure we constructed a satisfied number n vectors k(j), where n-h (h<n h E N) is the number of vectors used in the training procedure. It must be noted that the number of the vectors n is distributed in the four target classes of the RBF neural network. These classes correspond to a decision Liker scale (excellent, good, satisfied, fair). For implementation purposes we used 4 neurons and 2 layers. In the testing procedure we used the sigmoid function and h vectors k(j).

For the statistical evaluation of this model we used t-test criterion in order to evaluate the significant probability by considering as hull hypothesis that the system yields a random results aiming to the significant rejection.

4 Conclusion and Futures

In this paper we focus on a method solving problems related to preparing and normalizing data linked with significant relevant properties, which have been captured from classroom observations by adapting these factors to a first phase using a Bayesian model that creates normalization conditions with an equitable distribution between the significantly relevant properties of such types of data. Moreover, we described how the normalization can aid the training of a well-fitted neural network.

In the future, we would like to perform an extensive statistical evaluation of our model with real classroom observation data obtained through experimental questioners. This process can be supported by the software environment proposed in [14], which provides a system aiding the recording of observed classroom events. Finally, we would like to extend [14] by enhancing it with decision making capabilities in order to help the teacher identify problems and provide formative assessment.

References

1. Yu, L., Wang, S., Lai, K.K.: An integrated data preparation scheme for neural network data analysis. IEEE Transactions on Knowledge and Data Engineering 18(2), 217–230 (2006)
2. Hu, X.: DB-H Reduction: A Data Preprocessing Algorithm for Data Mining Applications. Applied Math. Letters 16, 889–895 (2003)

3. Sattler, K.U., Schallehn, E.: A Data Preparation Framework Based on a Multidatabase Language. In: Proc. Int'l. Symp. Database Eng. & Applications, pp. 219–228 (2001)
4. Pastor-Satorras, R., Vespignani, A.: Epidemics and immunization in scale-free networks. In: Bornholdt, S., Schuster, H.G. (eds.) Handbook (2002)
5. IATEFL, `http://www.ttedsig.iatefl.org/resources/articles/6.doc` (attached 29-3-2010)
6. Flanders, N.A.: Analyzing Teaching Behaviour. Addison Wesley, Reading (1970)
7. Tsumoto, S.: Mining diagnostic rules from clinical databases using rough sets and medical diagnostic model. Information Sciences: An International Journal 162(2), 65–80 (2004)
8. Ritchie, M.D., Motsinger, A.A., Bush, W.S., Coffey, C.S.: Moore J. H.: Genetic Programming neural networks: A powerful bioinformatics tool for human genetics. Applied Soft Computing 7, 471–479 (2007)
9. Moore, J.H., Williams, S.M.: New strategies for identifying gene–gene interactions in hypertension. Ann. Med. 34, 88–95 (2002)
10. Feller, W.: An Introduction to Probability Theory and its Applications, vol. I. John Wiley & Sons, Chichester (1968), ISBN: 0-471-25708-7
11. Traven, H.G.C.: A neural-network approach to statistical pattern classification by semiparametric estimation of a probability density functions. IEEE Trans. Neural Networks 2, 366–377 (1991)
12. Anita, S.Y., et al.: Transvaginal ultrasound measurement ofendometrial thickness as a biomarker for estrogen exposure. Cancer Epidemiology Biomarkers 13(9), 1459–1465 (1968)
13. Sarimveis, H., Doganis, P., Alexandridis, A., et al.: A classification technique based on radial basis function neural networks. Advances in Engineering Soft2ware 37(4), 218–221 (2006)
14. Belesiotis, V.: A software system for the didactic utilization of classroom observation for Informatic courses. In: 5th Panhellenic Conference on Didactic on Informatics, pp. 259–256 (2010) (in Greek)

Prediction with Confidence Based on a Random Forest Classifier

Dmitry Devetyarov and Ilia Nouretdinov

Computer Learning Research Centre, Royal Holloway, University of London, Egham, Surrey, UK
{dmitry,ilia}@cs.rhul.ac.uk

Abstract. Conformal predictors represent a new flexible framework that outputs region predictions with a guaranteed error rate. Efficiency of such predictions depends on the nonconformity measure that underlies the predictor. In this work we designed new nonconformity measures based on a random forest classifier. Experiments demonstrate that proposed conformal predictors are more efficient than current benchmarks on noisy mass spectrometry data (and at least as efficient on other type of data) while maintaining the property of validity: they output fewer multiple predictions, and the ratio of mistakes does not exceed the preset level. When forced to produce singleton predictions, the designed conformal predictors are at least as accurate as the benchmarks and sometimes significantly outperform them.

Keywords: Conformal predictor, confidence machine, region prediction, random forest.

1 Introduction

The new framework of conformal prediction introduced in [1] allows us to output region predictions (a set of predicted labels) with the guaranteed error rate under a simple statistical assumption (this property is called *validity*), as opposed to point predictions when we always produce singleton predictions but the error rate is not guaranteed.

Having a guaranteed error rate, we may still obtain multiple region predictions, that is, predictions that comprise more than one label. Multiple predictions are not mistakes: they indicate that there was no sufficient information provided for predicting one label. The question is how big these region predictions are, and our aim is to decrease a number of multiple predictions. The ability of conformal predictors to produce predictions as certain as possible is called *efficiency*.

Most of known machine learning algorithms can be used as an *underlying algorithm* in a conformal predictor. Efficiency of conformal predictors is usually in line with the accuracy of the underlying algorithm and therefore varies across the range of underlying algorithms and also depends on the type of data analysed. For this reason, we are looking for new nonconformity measures that could result in efficient predictions.

H. Papadopoulos, A.S. Andreou, and M. Bramer (Eds.): AIAI 2010, IFIP AICT 339, pp. 37–44, 2010.

So far various nonconformity measures have been designed. In this paper we propose new conformal predictors based on a random forest. We expect the predictors to inherit random forest advantages and to maintain the property of validity.

Although the main aim of this paper is to elaborate and analyse new methodology of providing region predictions, we can force conformal predictors to output one prediction instead of multiple predictions. Such approach allowed us to compare designed conformal predictors with machine learning methods (random forest, in particular). But one should bare in mind that in this case we do not have the advantages of conformal predictors as region predictors: there is no guaranteed validity.

2 Outline of Conformal Prediction

The framework of conformal prediction is described in detail in [1]. Conformal predictors are based on the only assumption about the data generating mechanism: all the examples have been generated independently by some probability distribution (the i.i.d. assumption).

Let us assume that we are given a training set of examples (x_1, y_1), ..., (x_{n-1}, y_{n-1}), where $x_i \in X$ is a vector of attributes and $y_i \in Y$ is a label out of a finite set of possible labels (classes), and our goal is to predict the classification y_n for remaining example x_n. A combination of an example and a label $z_i = (x_i, y_i) \in Z = X \times Y$ is called an object.

A nonconformity measure is a set of measurable mappings $\{A_n : n \in N\}$ of the type $A_n : Z^{(n-1)} \times Z \to (-\infty, +\infty]$, where $Z^{(n-1)}$ is the set of all bags (multisets) of elements of Z of size $n-1$. For each possible label y, we consider the hypothesis $y_n = y$ and the nonconformity measure assigns some values α_i (*nonconformity scores*) to every example in the sequence $\{z_i, i = 1, \ldots, n\}$ including a new example and evaluates 'nonconformity' $\alpha_i := A_n(\lbag z_1, \ldots, z_{i-1}, z_{i+1}, \ldots, z_n \rbag, z_i)$, $i = 1, \ldots, n$ between a set and its element ($\lbag \ldots \rbag$ denotes a multiset).

For each hypothesis $y_n = y$, we compare α_n to the other α_is and calculate $p(y) = |\{i = 1, \ldots, n : \alpha_i \geq \alpha_n\}|/n$ — the p-value associated with the possible label y for x_n. Thus, we can compliment each label with a p-value that shows how well a new example with this label conforms with the rest of the sequence.

The *conformal predictor* determined by the nonconformity measure $A_n, n \in N$ and a significance level ϵ is then defined as a measurable function $\Gamma : Z^* \times X \times (0,1) \to 2^Y$ (2^Y is a set of all subsets of Y) such that the prediction set $\Gamma^{(\epsilon)}(x_1, y_1, \ldots, x_{n-1}, y_{n-1}, x_n)$ is defined as the set of all labels $y \in Y$ such that $p_n > \epsilon$. Thus, for any finite sequence of examples with labels, $(x_1, y_1, \ldots, x_{n-1}, y_{n-1})$, a new unlabelled object x_n and a significance level ϵ, the conformal predictor outputs a region prediction $\Gamma^{(\epsilon)}$ — a set of possible labels for a new object.

Conformal predictors defined above are *valid*: in the long run the frequency of errors made by a conformal predictor (that is, cases when prediction set Γ^ϵ does not contain a true label) does not exceed ϵ subject to the i.i.d. assumption. Strictly speaking, for any exchangeable probability distribution P on Z^∞ (Z^∞ is the set of all infinite sequences of elements of Z) and any significance level ϵ,

$$\limsup_{n\to\infty} \frac{\sum_{i=1}^{n} err_n^{\epsilon}(\Gamma)}{n} \leq \epsilon \tag{1}$$

with probability one, where $err_n^{\epsilon}(\Gamma)$ is equal to 1 when the prediction set Γ^{ϵ} does not contain a real label y_n, and 0 otherwise. The property of validity is theoretically proven in *the on-line mode* and empirically confirmed in the *off-line mode* [1].

3 Random Forests

In this work we consider the type of random forests described in [4]. Theoretical results [4] demonstrate that random forests do not overfit when more trees are added. They also empirically proved to have the following advantages ([4], [5]): random forests produce high accuracy for many data sets; they can process data with a large number of features where each feature is weak, that is, carries a small amount of information; they are relatively robust to mixed variable types, missing data, outliers and noisy data; constructing random forests is relatively fast (faster than bagging and boosting).

In brief, a random forest is a classifier that consists of decision trees, each of which provides a vote for a certain class. Combining a large number of trees in a random forest can lead to more reliable predictions, while single decision tree may overfit the data.

3.1 Nonconformity Measures Based on Random Forests

In this paper we designed nonconformity measures based on random forests.

We will use the following notation: suppose we are given a bag $\wr(x_1, y_1), (x_2, y_2), \ldots, (x_m, y_m)\wr$, $(x_i, y_i) \in Z$, and we need to define a nonconformity measure $A(x, y) = A(\wr(x_1, y_1), (x_2, y_2), \ldots, (x_m, y_m)\wr; (x, y))$. Alternatively, we can define a conformity measure $B(x, y) = 1 - A(x, y)$ when it is more intuitive.

The nonconformity or conformity measures we propose are the following:

1. A random forest is constructed from a training set $\{(x_1, y_1), (x_2, y_2), \ldots, (x_m, y_m)\}$. The conformity score of a new example (x, y) is then equal to the percentage of correct predictions given for x by decision trees.

2. Conformity measure 1 is the most natural one, however, it is computationally inefficient: when considering example $N + 1$ we have to construct $(N + 1)L$ random forests, where L is the number of labels. We will therefore use another conformity measure, which will require only one random forest when making a prediction for a new object. The random forest is grown for the union of a bag $\wr(x_1, y_1), (x_2, y_2), \ldots, (x_m, y_m)\wr$ and a new example (x, y). Since for each decision tree, the training set is a bootstrap sample, a new example is not included in this training set in about one third of decision trees. For each (x, y) we aggregate the votes for this example only of those decision trees where this example is out-of-bag (not in the training set for the tree). The conformity score is then equal to the proportion of correct votes for (x, y) among these trees.

3. This nonconformity measure was proposed by Huazhen Wang and Fan Yang in our personal communication. It is based on random forest proximities $P(i,j), i,j = 1,\ldots,m+1$, which provide a measure of how close to each other two objects are regardless of their labels and are calculated as a ratio of trees, running through which objects i and j land at the same terminal node. We construct a random forest for the union of a bag $\wr(x_1,y_1),(x_2,y_2),\ldots,(x_m,y_m)\int$ and a new example (x,y) and form the corresponding $(m+1)\times(m+1)$ matrix of proximities for objects $x_1,x_2,\ldots,x_m,x_{m+1}=x$. The nonconformity measure is the ratio of the average proximity of the example with examples of other classes to the average proximity of the example to examples of the same class. In both averages we consider only proximities of those k examples that have the greatest values of proximities among examples of the same class y and among all the other examples. Strictly speaking, $A(x,y) = A(x,y)^-/A(x,y)^+$, where

$$A(x,y)^+ = \sum_{s=1}^{k} P(i_s, m+1), A(x,y)^- = \sum_{s=1}^{k} P(j_s, m+1)\ , \qquad (2)$$

i_s and j_s are the numbers of examples with s-st greatest value of proximity with example (x,y) among examples labelled with the same label y and among all the other examples, respectively.

4 Experiments

The designed nonconformity measures were implemented and applied to different data sets.

4.1 Data

In our experiments we used six proteomic data sets, two medical non-proteomic data sets and two non-medical data sets. Proteomic data sets comprise: ovarian cancer data from the *UKOPS* trial [6]; ovarian cancer (*OC*), breast cancer (*BC*) and heart disease (*HD*) samples collected in the *UKCTOCS* trial[1]; *Competition* data provided by the Leiden Clinical Mass Spectrometry Proteomic Diagnosis Competition and preprocessed as described in [7]; Ciphergen's *7 biomarkers* of UKOPS data [8]. The other medical data sets are shortened *Abdominal pain* data [9], which comprises 33 symptoms of acute abdominal pain, and *Microarray* data of lung cancer, colon cancer and breast cancer patients provided in the ICMLA 2009 Challenge [10]. *Sonar* and *Iris* non-medical data sets were taken from the University of California, Irvine (UCI) Machine Learning Repository.

4.2 Implemented Conformal Predictors

We implemented non-conformity measures 2 (referred to as *CP-RF*) and 3 (referred to as *CP-RF-kNN*, where k is the number of nearest neighbours). These

[1] For more information see www.ukctocs.org.uk

conformal predictors were compared with benchmark predictors based on the kNN algorithm [3] (we will denote them *CP-kNN*, where k is the number of nearest neighbours) and SVM with different kernels [2] (denoted as *CP-SVM (kernel, parameter)*).

The experiments were carried out in two settings: off-line in leave-one-out (LOO) procedure, to show the usage of conformal predictors as conventional classifiers, and on-line, to demonstrate the advantages of region predictions.

We used the following parameters for random forest construction: the number of trees 1000, the number of features selected at each node to split equals a square root of the number of features. These values are recommended in [4], where it is theoretically proven that the results converge when we increase the number of trees in random forests and it is empirically shown that the results are insensitive to the number of features selected at each node.

4.3 Results

For each significance level $\epsilon > 0$, conformal predictors output a set of labels with p-values greater than ϵ. Thus, the predictor may output no label (we call this an *empty prediction*), one label (*certain prediction*) or more than one label (*multiple prediction*).

Firstly, the designed conformal predictors proved to be valid, that is, for a given significance level $\epsilon > 0$ the rate of erroneous predictions, that is, predictions not containing an actual label, is close to ϵ.

The example of the erroneous prediction dynamics is shown in Figure 1a. The figure demonstrates validity of the CP-RF-1NN applied to the Microarray data for significance levels $\epsilon = 5\%$ and 10% : solid lines, which represent the actual number of errors, are close to dotted lines, which demonstrate the expected number of errors for different significance levels.

Figure 1b demonstrates the dynamics of efficiency characteristics at significance level of 10% of the CP-RF-1NN applied to the Microarray data in the on-line mode. The characteristics shown are the number of multiple predictions, the number of certain predictions and the number of empty predictions. The figure demonstrates that while the number of analysed examples is low, they do

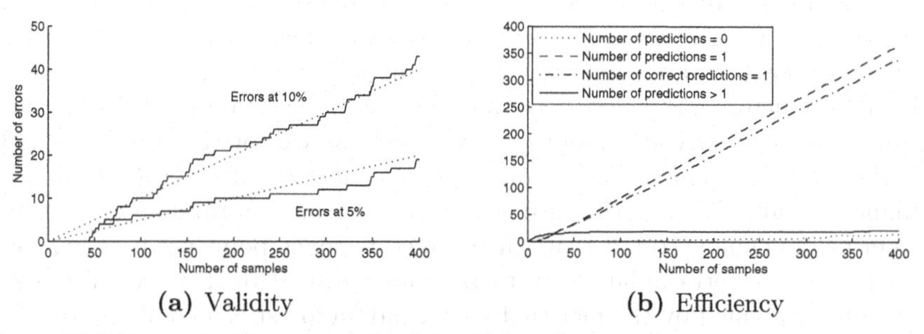

(a) Validity (b) Efficiency

Fig. 1. Validity and efficiency of CP-RF-1NN applied to the Microarray data in the on-line mode

not carry enough information to make certain predictions without losing validity. But starting from example 50, we have accumulated enough information so that multiple predictions cease to occur and most of prediction regions contain exactly one label, which is in most cases correct. The dynamics on the plot also conforms with the empirical fact established in [1] that when multiple predictions disappear, empty predictions start to occur.

As mentioned before, all implemented conformal predictors have a theoretically proven property of validity, and the general aim is to design a nonconformity measure that could improve efficiency, that is, make the algorithm output as few multiple predictions and as many empty predictions as possible. Comparison of efficiency of CP-RF and CP-RF-kNN with the benchmarks demonstrated that CP-RF and CP-RF-kNN produce at least as few multiple predictions and as many correct certain and empty predictions as the benchmarks, and they perform much better in terms of efficiency on all mass spectrometry data sets. This allows us to speculate that conformal predictors based on random forests benefit from the advantages of the underlying algorithm and perform well on noisy data and data with a lot of weak inputs. Table 1 summarizes the multiple prediction rate for different conformal predictors.

Table 1. The rate of multiple predictions for significance level $\epsilon = 10\%$ in the LOO mode

Data	CP-RF	CP-RF -1NN	CP-RF -5NN	CP-1NN	CP-5NN	CP-SVM (rbf, 5)	CP-SVM (poly, 5)
UKOPS	46.1%	47.0%	45.8%	74.8%	72.0%	59.2%	69.8%
UKCTOCS OC	16.0%	16.0%	13.8%	44.6%	30.8%	38.5%	79.8%
UKCTOCS BC	77.8%	78.4%	77.8%	80.9%	80.9%	81.5%	82.7%
UKCTOCS HD	56.0%	58.1%	57.2%	64.7%	59.5%	66.1%	64.0%
Competition	11.1%	18.3%	17.0%	26.8%	19.6%	32.7%	30.7%
7 biomarkers	51.1%	55.4%	53.8%	67.0%	61.2%	97.9%	96.9%
Abdominal pain	0.3%	1.0%	0.0%	3.0%	0.0%	0.0%	1.0%
Microarray	0.0%	1.5%	0.3%	13.5%	3.5%	8.5%	40.4%
Sonar	14.9%	11.1%	13.0%	13.9%	16.4%	32.2%	30.8%
Iris	0.0%	0.0%	0.0%	0.0%	0.0%	86.7%	8.0%

Conformal predictors have been developed to provide region predictions with the preset error rate. However, in order to compare them with bare predictions output by conventional machine learning methods, we can ignore the nature of conformal predictors and force them to always make a certain prediction. After assigning a p-value for each label to every object, we can predict the label with the highest p-value. This is called *forced point prediction*. If several labels have the highest p-value (we call this situation a *tie*), we make a random prediction.

Experiments demonstrated that when forced to make point predictions, conformal predictors perform similarly to random forest algorithm (see Table 2). This can be explained by the fact that each random forest is a combination of a large number of trees constructed randomly and each sample is not included in the training set for about one third of all trees in a random forest.

Table 2. Accuracy of forced point predictions in the LOO mode

Data	RF	CP-RF	CP-RF -1NN	CP-RF -5NN	CP-1NN	CP-5NN	CP-SVM (rbf, 5)	CP-SVM (poly, 5)
UKOPS	72.6%	72.3%	71.5%	72.6%	55.1%	61.7%	66.7%	55.5%
UKCTOCS OC	84.9%	84.8%	83.8%	84.6%	72.3%	80.6%	78.9%	77.9%
UKCTOCS BC	66.0%	66.7%	59.0%	62.4%	50.3%	62.4%	56.2%	54.3%
UKCTOCS HD	71.8%	72.3%	69.2%	71.4%	63.2%	67.9%	62.4%	62.1%
Competition	83.7%	85.3%	83.7%	83.3%	82.0%	84.6%	86.3%	87.6%
7 biomarkers	74.6%	74.8%	72.2%	73.9%	64.5%	73.7%	60.9%	59.0%
Abdominal pain	91.7%	92.7%	91.8%	91.5%	88.0%	92.2%	91.7%	90.7%
Microarray	92.0%	91.3%	92.8%	91.4%	86.1%	89.4%	88.3%	89.5%
Sonar	85.1%	84.6%	88.7%	85.6%	86.3%	82.9%	84.6%	85.3%
Iris	95.3%	94.7%	95.0%	95.3%	93.3%	97.0%	89.3%	89.7%

This implies that we can add the framework of conformal prediction to the random forest algorithm without losing in accuracy, while benefiting from conformal predictions: we can produce valid region predictions and compliment each prediction with confidence.

The results of comparison of forced point prediction accuracy of different conformal predictors (Table 2) were in line with efficiency comparison: CP-RF and CP-RF-kNN significantly outperformed other predictors on certain mass spectrometry datasets and were at least as good as the benchmarks on all data sets.

5 Conclusion

In this paper we worked on further development of conformal predictors, which produce region predictions that make a preset number of errors in the long run. Designed nonconformity measures based on random forests proved to be valid and efficient. First, the ratio of mistakes does not exceed the preset level. Second, CP-RF and CP-RF-kNN produce more efficient predictions on all mass spectrometry data sets and are not beaten on the other data.

When forced to produce singleton predictions, conformal predictors based on random forest result in accuracy similar to random forest accuracy. The accuracy of forced point predictions output by CP-RF and CP-RF-kNN is at least as high as accuracy produced by known conformal predictors and sometimes is significantly higher. This implies that although conformal predictors are designed for producing valid region prediction, they can also be a useful tool when making singleton predictions.

Acknowledgments. We would like to thank our collaborators in MRC project ("Proteomic analysis of the human serum proteome") Ian Jacobs, Usha Menon, Rainer Cramer, John F. Timms, Ali Tiss, Jeremy Ford, Stephane Camuzeaux, Aleksandra Gentry-Maharaj for collecting the original UKCTOCS and UKOPS data and carrying out MS experiments.

This work is supported by EPSRC grant EP/F002998/1 (Practical competitive prediction); MRC grant G0301107 (Proteomic analysis of the human serum proteome); VLA of DEFRA grant on Development and Application of Machine Learning Algorithms for the Analysis of Complex Veterinary Data Sets; EU FP7 grant O-PTM-Biomarkers HEALTH-2007-2.4.1-2; ASPIDA grant "Development of new methods of conformal prediction with application to medical diagnosis" from the Cyprus Research Promotion Foundation; European Union EU FP7 programme "Living with uninvited guests: comparing plant and animal responses to endocytic invasions" (ERASysBio); and MRC Award G0802594.

References

1. Vovk, V., Gammerman, A., Shafer, G.: Algorithmic Learning in a Random World. Springer, New York (2005)
2. Gammerman, A., Vovk, V., Vapnik, V.: Learning by Transduction. In: 14th Conference on Uncertainty in Artificial Intelligence, pp. 148–155 (1998)
3. Proedrou, K., Nouretdinov, I., Vovk, V., Gammerman, A.: Transductive Confidence Machines for Pattern Recognition. Technical report 01-02, Royal Holloway, University of London (2001)
4. Breiman, L.: Random Forests. Mach. Learn. 45, 5–32 (2001)
5. Breiman, L., Cutler, A.: Random Forests, `http://www.stat.berkeley.edu/users/breiman/RandomForests/cc_home.htm#intro`
6. Timms, J.F., Cramer, R., Camuzeaux, S., Tiss, A., Smith, C., Burford, B., Nouretdinov, I., Devetyarov, D., Gentry-Maharaj, A., Ford, J., Luo, Z., Gammerman, A., Menon, U., Jacobs, I.: Peptides Generated Ex Vivo from Abundant Serum Proteins by Tumour-Specific Txopeptidases are Not Useful Biomarkers in Ovarian Cancer. Clin. Chem. 56, 262–271 (2010)
7. Gammerman, A., Nouretdinov, I., Burford, B., Chervonenkis, A., Vovk, V., Luo, Z.: Clinical Mass Spectrometry Proteomic Diagnosis by Conformal Predictors. Stat. Appl. Genet. Mo. B. 7(2), Art. 13 (2008)
8. Nouretdinov, I., Burford, B., Luo, Z., Gammerman, A.: Data Analysis of 7 Biomarkers. Technical report, Royal Holloway, University of London (2008)
9. Gammerman, A., Thatcher, A.R.: Bayesian Diagnostic Probabilities without Assuming Independence of Symptoms. Method Inform Med. 30(1), 15–22 (1991)
10. Nouretdinov, I., Burford, B., Gammerman, A.: Application of Inductive Confidence Machine to ICMLA Competition Data. In: The Eighth International Conference on Machine Learning and Applications, pp. 435–438 (2009)

A Generic Tool for Building Fuzzy Cognitive Map Systems

Maria Papaioannou[1], Costas Neocleous[2], Anastasis Sofokleous[1],
Nicos Mateou[3], Andreas Andreou[1,2], and Christos N. Schizas[1]

[1] 75 Kallipoleos Str, Department of Computer Science,
University of Cyprus, Nicosia, Cyprus
{cs03pm2,asofok,aandreou,schizas}@ucy.ac.cy
[2] 31 Archbishop Kyprianos Str, Cyprus University of Technology, Limassol, Cyprus
{costas.neocleous,andreou}@cut.ac.cy
[3] Ministry of Defence, Cyprus
NMateou@mod.gov.cy

Abstract. A generic system for simulating complex dynamical systems along the paradigm of fuzzy cognitive maps (FCM) has been created and tested. The proposed system enables a user to design appropriate FCM structures, by specifying the desired concepts and the various parameters such as sensitivities, as well as a variety of shaping functions. The user is able to see the results, change the parameters, modify the functions, and rerun the system using an alteration of the final results and make new conclusions. The system is introduced and demonstrated using a simple real case. The results of a usability test of the system suggest that the system is capable of simulating complicated FCM structures in an effective manner, helping the user to reduce the degree of risks during decision making.

Keywords: Fuzzy Cognitive Maps, Soft Computing, Intelligent Systems.

1 Introduction and Background Theory

Fuzzy Cognitive Maps (FCMs) constitute a powerful soft computing modeling method that emerged from the combination of Fuzzy Logic and Neural Networks as an extension of the Cognitive Maps [3]. FCMs were introduced first by Kosko [4] and since then a wide range of applications in modeling complex dynamic systems have been reported such as medical [7], environmental [2], supervisory [9] and political [1, 5, 6] systems. Essentially, a Fuzzy Cognitive Map is developed by integrating the existing experience and knowledge regarding a system. This can be achieved by using a group of human experts to describe the system's structure and behavior in different conditions.

In a graphical form, the FCMs are typically signed fuzzy weighted graphs, usually involving feedbacks, consisting of nodes and directed links connecting them. The nodes represent descriptive behavioral concepts of the system and the links represent cause-effect relations between the concepts [8]. In the context of FCM theory, the fuzzy value of a concept denotes the degree in which the specific concept is active in

H. Papadopoulos, A.S. Andreou, and M. Bramer (Eds.): AIAI 2010, IFIP AICT 339, pp. 45–52, 2010.

the general system, usually bounded in a normalized range of [0, 1]. Furthermore, the weights (or sensitivity) of the system's interrelations reflect the degree of influence between two concepts and they are usually made to vary in the interval [-1,+1][8].

Considering the mathematical formulation of the FCM, an FCM system can be represented by a concept state vector $\boldsymbol{A}$ and an adjacency matrix $\boldsymbol{W}$. The dimension of the concept vector is equal to the total number of distinct concepts n that compose the modeled system. Moreover, the values of its elements correspond to the states of the concepts. The nxn adjacency matrix represents the cause – effect interrelations among the concepts. The value of each cell $[i,j]$ corresponds to the weight of the cause – effect relation between the ith concept and the jth concept.

The basic FCM inference is an iterative process of estimating the new values of the concept vector until they converge. The change of the value of a concept depends on the values of the connected concepts along with the corresponding weights of their interrelations. In practice, at each iteration, the produced concept state vector A^{t+1} emanates from a vector – matrix multiplication, using the previous concept vector A^t and the adjacency matrix W.

After the system interacts for a number of iterations there are 3 possible system behaviors to appear: 1. *Fixed point attractor*, where the activation levels of all concepts become steady and don't alternate through iterations. 2. *Limit Cycle*, where an identical sequence of activation levels appears every fixed number of iterations. 3. *Chaotic attractor* where the activation levels keep changing for ever through iterations.

Additionally, the variation of a concept's impact on another concept may be defined through a manner expressed by a variety of functions (e.g. linear, stepwise, sigmoid, Gaussian and exponential) [11].

A wide range of real life dynamical systems are characterized by causality through their interrelations. Most of the times, experienced and knowledgeable users can identify the parameters and the conditions of their system. However, the complexity of these systems acts as a prohibitive factor on prognosis of their future states. Nevertheless, it is crucial for the users to make a prognosis on the cost of changing a state of a concept, in taking a decision whether they will take such a risk of change or not. FCM modeling constitutes an alternative way of building intelligent systems providing exactly this opportunity to the users; predicting the final states of the system caused by a change on initial concept states. The proposed tool implements the FCM inference mechanism in a simple manner by combining in a simple, easy to use and friendly environment the mechanics of FCM and a user's knowledge. This leads to the modeling of the system in an intelligent framework. An attempt has also been reported in [14].

The main purpose of designing this tool was to create a user-friendly software for building and simulating complex dynamic systems via FCM, with little computer literacy or experience from the part of a user. The only requirement from the user's part is to go through the fuzzification process for the concept values and the weights of their interrelations. The main goal is to help users to design and simulate FCM systems even when they have only the basic knowledge of FCM modeling, provided that they have a good and clear understanding of the parameters and conditions of their system, as well as the corresponding fuzzy values. The long-term objective is to develop a multi-function tool, incorporating certain modeling procedures and algorithms, such as genetic algorithms and neural networks, as well as fuzzification and defuzzification functions. The first edition of the tool implementation, which is limited to simple simulation and testing, is presented in this paper.

2 A Generic Tool for Running FCM Systems

In FCM model, the causality propagation is accomplished through the configuration of the new concept values at each iteration. The proposed system allows the user to choose out of three functions for computing the new concept values.

$$A_i^t = f\left(\sum_{\substack{j=1 \\ j \neq i}}^{n} \left(A_j^{t-1} W_{ji} \right)\right) . \quad (1)[4]$$

$$A_i^t = f\left(\sum_{\substack{j=1 \\ j \neq i}}^{n} \left(A_j^{t-1} W_{ji} + A_i^{t-1} \right)\right) . \quad (2)[9]$$

$$A_i^t = f\left(\sum_{\substack{j=1 \\ j \neq i}}^{n} \left(k_1 A_j^{t-1} W_{ji} + k_2 A_i^{t-1} \right)\right) . \quad (3)[9]$$

In all the above equations A_i^t is the concept value of concept C_i at iteration t, n is the total number of the concepts affecting concept C_i , W_{ji} is the value of the sesnsitivity connecting concept C_j with (to) concept C_i, k_1 is the coefficient which controls the influence of the interconnected concepts in the calculation of the new concept value A_i^t, k_2 is the coefficient that expresses the contribution of the previous concept value in the calculation of the new one, *f(.)* is a transformation function which maps each concept value into the space of either [-1,1] or [0,1]. The user is able to choose between sigmoid, hyperbolic, bivalent or trivalent function *f(.)*.

2.1 Interaction with the System

The graphical environment of the tool is created using the available interface features that MATLAB 7.9 (R2009 b) provides. The software tool incorporates Microsoft Excel for facilitating the storing and retrieving of data from files.

By invoking the tool, a window with a tabbed menu navigation system appears as shown in Figure 1.

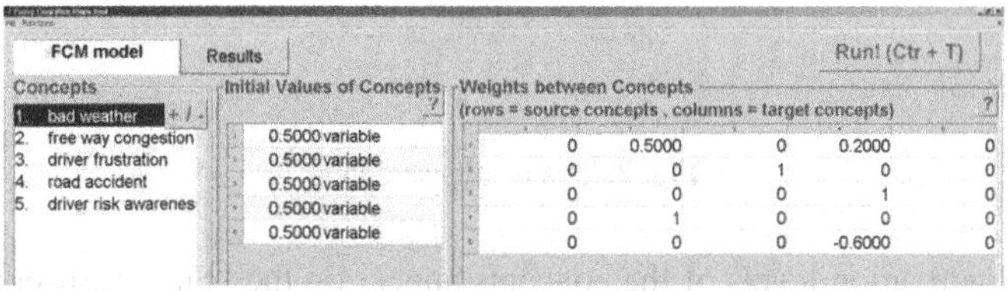

Fig. 1. The initial screen of the main window

The tabs divide the window into two screens. By selecting the firs tab *"Set FCM Model Parameters"* the user is lead to a screen where one can set or modify the FCM model's concepts, their initial values, and the weights of their interrelations. The second tab named *Results* brings out a screen with information about the final values of the concepts. In the case where no results are available for any simulation of any model, the tab *Results* is disabled. The user has the flexibility to switch between *FCM model's parameters settings* and the *Results* screens whenever one desires. A button named *Run* is provided for executing the FCM model.

2.1.1 Model Building and Editing

The user can either build an FCM model from scratch or modify an existing one through the "Set FCM mode's Parameters" screen. The names of the concepts are listed in a list-box, while the initial values of the concepts and their weights are exposed in tables. The first step of building a model is the definition of the names of the concepts. The user can insert, delete, edit the name of a concept or re-order the list of concepts. For each new concept added to the model, the system adds a new row to the "Initial Values of Concepts" and the "Weights between Concepts" table and initializes their numeric values to 0. Each numbered row of the table corresponds to the same numbered concept of the "Concepts" list. The user can edit the numeric values of the tables. There is also a pop-up menu next to each initial concept value that allows the user to lock the corresponding value, by choosing the option "constant" (i.e. execute this concept out of any calculations that alter its initial value).

2.1.2 Running the System and Results

The user can run the FCM model by using the button "Run" located at the right top of the main window. When the algorithm stops executing, the "Results" screen appears (Figure 2).

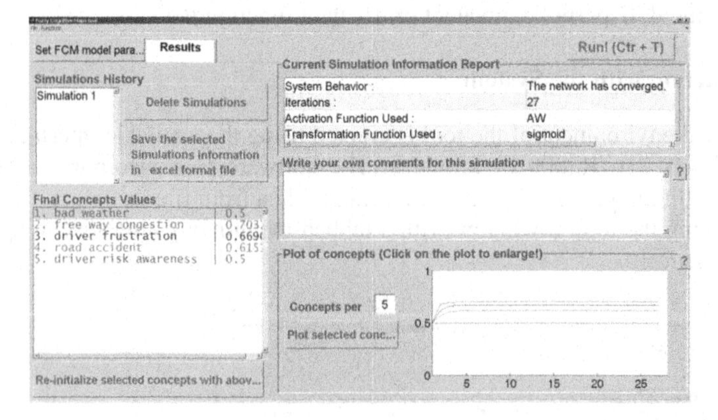

Fig. 2. The results screen

The final activation levels of the concepts appear on the "Final Concepts Values" list. Each concept in that list is colored and so do the values of the corresponding plot on the graph right next to the list. In the "Results" screen, the main graph presents the variation of the states of the model concepts. It is also possible to select the desired concepts to plot, from the "Final Concept Values" list and then click on the "Plot selected concepts" pushbutton right next to the graph. There is also an edit-box where the user can write comments and conclusions about a simulation. There is also a table displaying information about a simulation (functions used and the number of required iterations). There is also another list on this screen, named "Simulations History". Each time the user runs a model, the system saves all the information about the simulation and the "Simulations History" list adds one more "Simulation" at the bottom of the list. When the user selects a saved Simulation, all the corresponding information is retrieved and displayed. One can delete one or more simulations from the list. The user has also the ability to save a group of simulations in excel file format. The user is thus able to visit the results of each simulation of an FCM model easily and quickly, while one can transfer this simulations file to any computer running WINDOWS, without the need of the FCM tool.

3 Evaluation of the Tool

The credibility of the tool was tested by modeling and simulating a real process control problem taken from [9]. The results of the FCM model built with the tool agree with the results reported in [9].

For evaluating the usability of the tool, an evaluation questionnaire was developed. The evaluators had to carry out 5 tasks. The tasks cover all the functions a user is able to execute with the tool. After finishing the tasks the user had to answer 10 usability questions suggested in [12] and one extra question. Twenty senior computer science, mathematics and computer engineering students were invited to evaluate the tool. Seventeen of the selected students had never used tools for simulating graph models. All of the students had no knowledge about FCM modeling so an essential description of the FCM theory took place just before the evaluation.

Analytically, the 10 questions taken from [12] were:

1. I think that I would like to use this system frequently
2. I found the system unnecessarily complex
3. I though the system was easy to use
4. I think I would need the support of a technical person to be able to use this system
5. I found the various functions in this system were well integrated
6. I thought there was too much inconsistency in this system
7. I would imagine that most people would learn to use this system very quickly
8. I needed to learn a lot of things before I could get going with this system
9. I would easily learn how to user this tool
10. I found the system flexible to use

And the extra question included in the questionnaires was:

I think I would save time by using the tool instead of developing the FCM model without the tool.

The user had to express the degree of agreement or disagreement with each statement on a seven point Likert scale. Each questionnaire was first analyzed by calculating the score from each item (statement). For items 1, 3, 5, 7, 9, 10, 11, as listed above, the score was the scale position minus one. On the other hand, for items 2, 4, 6, 8 the score was seven minus the scale position [12]. As a result, there is a mapping

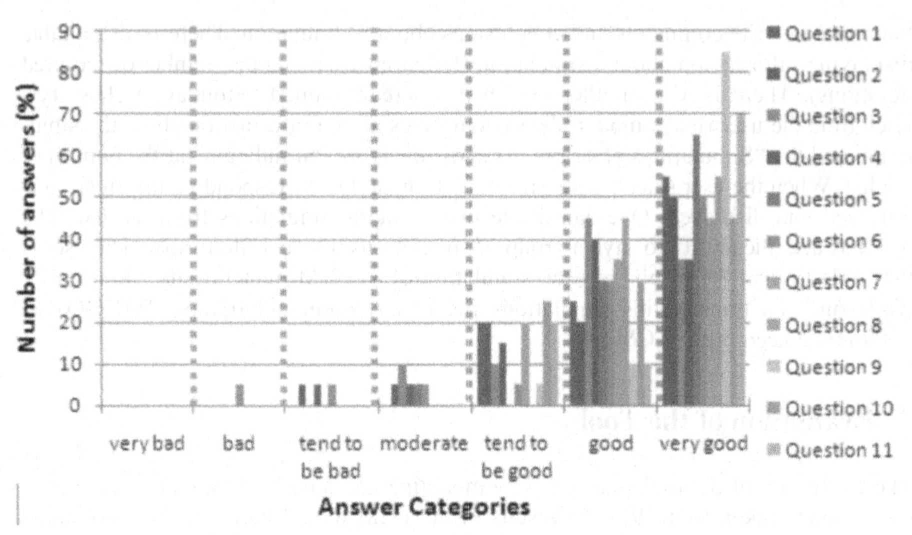

Fig. 3. Histogram presenting the percentage of the answers distribution per each question. The grey dashed lines divide the histogram into the 7 answer categories.had 10% scores.

between the Likert style user's choice, which ranges in [1,7], to a score ranging in [0,6] where, {0,1,2,3,4,5,6} stand for {"very bad", "bad", "tend to be bad", "moderate", "tend to be good", "good", "very good"} system in terms of usability, respectively. As shown in Figure 3, none of the questions was scored as "very bad". Only one question had 5% scores in "bad" category. Only the 5% of four questions were scored below "moderate". Considering the worst scores, *Question 6.* is the only one that had 10% scores falling in both "bad" and "tend to be bad" categories. After the first analysis of the questionnaires the students were asked why they thought there wasn't satisfactory consistency in the system. The users stated that they couldn't understand the word "consistency" and they got confused. *Question 2,* had also answers in the group of "tend to be bad","moderate" and "tend to be good" categories. A possible reason for that could be that the users felt they had to memorize many ways for utilizing data, since the first tab screen is about editing tables and a listbox, while the second one is more about pushing buttons and utilizing plots. 25% of the answers of *Question 4* were scored as "tend to be bad", "moderate" and "tend to be good". A possible reason why users felt that way is that they didn't have any experience with graph models simulation tools or FCM. The same reason applies for the "moderate" and "tend to be good" scores of *Question 3.* Hence, it is absolutely reasonable that some of the users would face some difficulties using for the first time the tool and they would ask for help from a technical person. Besides, 95% of the answers of *Question 9* fall in "good" and "very good" categories, stating that 95% of the users believe they would easily learn how to use the tool. This implies that the users understand that the difficulties they had to resolve during the evaluation came from their inexperience and lack of knowledge. Furthermore, it is noteworthy that all eleven questions had 70% - 100% of their answers falling into categories "good" and "very good" indicating that most of the users thought the system was indeed successful in terms of usability. This fact is also reflected in Figure 4, whereas all question medians

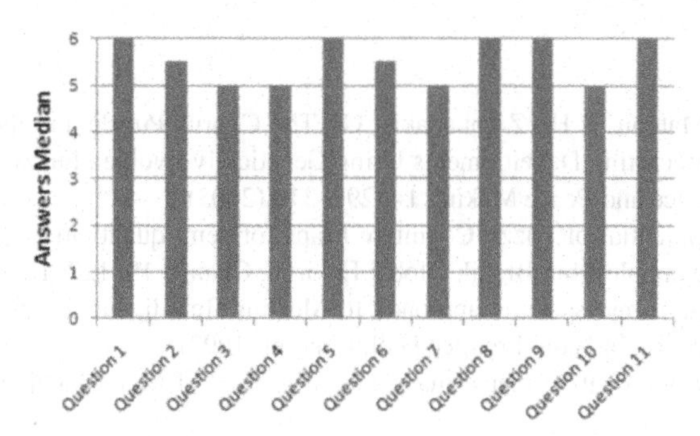

Fig. 4. Histogram presenting the median of the answers of each question

range between five and six (good and very good) and approximately 50% of them equal to very good which is the optimum score.

4 Conclusions and Future Work

Dynamical complex systems are met in most human decision making activities. Decision making problems concerning such systems can be solved by utilizing FCM modeling methodology. The proposed tool enables a user, who is not familiar with computer literacy, to design and simulate an FCM modeled system and benefit from the advantageous features of FCM modeling.

The tool and its capabilities were presented and described in detail in the presented paper. The friendly graphical interface helps the user to easily insert, save, retrieve and display data. The main advantages of the tool is the preservation of the information of multiple simulations of different FCM models at the same time, the display of the plot of each model final states in different windows and the feature of saving information of multiple simulations in an spreadsheet file so that the user is able to go back to the results of any simulation of any model without the need to re-build and re-simulate the system on the tool.

The usability of the system was evaluated by developing an evaluation questionnaire. The conclusions drawn from the results of the evaluation are (1) the tool meets the usability standards as defined in [12] since the majority of the students gave very good scores for the usability of the tool through their answers (2) users having the basic knowledge of FCM theory are capable of using the tool successfully in designing and simulating FCM systems even if they had never interacted with FCM modeling systems in the past (3) the tool is easy to learn and easy to use.

The tool is expected to introduce common people to the challenging world of FCMs, by offering them a user-friendly functional environment which allows them to create and simulate FCM models.

In the future, the tool will be enhanced with pre-processing functions, helping and enabling the user to extract the fuzzy rules which characterize a system and embed this knowledge to the tool. The user will be also able to "run" a demonstration of built examples of the tool. In addition, the notion of time, in terms of time dependencies, as well as an algorithm inspired by the theory of evaluation of species will be introduced in the model.

References

1. Andreou, A., Mateou, N.H., Zombanakis, G.: The Cyprus Puzzle and the Greek-Turkish Arms Race: Forecasting Developments Using Genetically Evolved Fuzzy Cognitive Maps. Journal of Defence and Peace Making 14, 293–310 (2003)
2. Kok, K.: The potential of Fuzzy Cognitive Maps for semi-quantitative scenario development, with an example from Brazil. Global Environ. Change 19, 122–133 (2009)
3. Kosko, B.: A dynamic systems approach to Machine Intelligence. In: Neural Networks and Fuzzy Systems, 2nd edn. Prentice Hall, London (1992)
4. Kosko, B.: Fuzzy cognitive maps. International Journal of Man-Machine Studies 24, 65–75 (1986)
5. Neocleous, C., Schizas, C.: Application of Fuzzy Cognitive Maps to the Political-Economic Problem of Cyprus, pp. 340–349 (2004)
6. Neocleous, C., Schizas, C., Yenethlis, C.: Fuzzy Cognitive Models in Studying Political Dynamics - The case of the Cyprus problem
7. Papageorgiou, E.I., Papandrianos, N.I., Apostolopoulos, D.J., Vassilakos, P.J.: Fuzzy Cognitive Map based decision support system for thyroid diagnosis management, pp. 1204–1211 (2008)
8. Stylios, C.D., Groumpos, P.P.: Modeling Complex Systems Using Fuzzy Cognitive Maps. IEEE Transactions on Systems, Man, and Cybernetics Part A: Systems and Humans 34, 155–162 (2004)
9. Stylios, C.D., Groumpos, P.P.: Fuzzy Cognitive Maps in modeling supervisory control systems. Journal of Intelligent and Fuzzy Systems 8, 83–98 (2000)
10. Papageorgiou, E.I., Groumpos, P.P.: A new hybrid method using evolutionary algorithms to train Fuzzy Cognitive Maps. Applied Soft Computing Journal 5, 409–431 (2005)
11. Koulouriotis, D.E., Diakoulakis, I.E., Emiris, D.M., Antonidakis, E.N., Kaliakatsos, I.A.: Efficiently modeling and controlling complex dynamic systems using evolutionary fuzzy cognitive maps (invited paper). International Journal of Computational Cognition 1, 41–65 (2003)
12. John, B.: SUS – A quick and dirty usability scale. In: Jordan, P.W., Thomas, B., Weerdmeester, B.A., McClelland, A.L. (eds.) Usability Evaluation in Industry. Taylor and Francis, London (1996)
13. Mateou, N.H., Stylianou, C., Andreou, A.S.: Hybrid Fuzzy Cognitive Map Modeller. In: Soft computing as Transdisciplinary Science and Technology, vol. 29, pp. 851–862. Springer, Heidelberg (2005)

A Fuzzy Rule-Based Approach to Design Game Rules in a Mission Planning and Evaluation System

D. Vijay Rao and Jasleen Kaur

Institute for Systems Studies and Analyses
Defence Research and Development Organisation,
Metcalfe House, Delhi 110054, India
doctor.rao.cs@gmail.com

Abstract. Simulations and wargames offer powerful representations to model the mechanics and psychology of military operations that are inherently complex. They offer mechanisms to predict and assess the effectiveness of the mission plans and operations in achieving the military objectives. In this paper, we present a new approach to design the games rules of wargames using fuzzy rule bases, for quantitatively evaluating the effectiveness of air tasking missions. We determine the comparative damage relative to intended damage for a target, taking into account the effects of operational characteristics to compute possibilistic damage to the target as opposed to the probability of damage to the target. The cookie-cutter method to compute the damage is modeled as a fuzzy variable. Effectiveness of the mission is obtained by comparing the damage to targets with the cost and significance of the target in meeting the mission objectives. Damage assessment computation to targets using fuzzy rule bases gave more realistic results when used in field training and deployment of the system.

Keywords: Military simulation, Mission effectiveness, Rule-based systems, Damage Assessment Modeling, Fuzzy logic.

1 Introduction

Simulation has been applied extensively and successfully to a wide range of military problems, including wargaming, acquisition, logistics, and communications. It has been used as a decision support tool to evaluate how a battle force should be constituted, how to plan the force strength and structures, how it might be deployed, and how the weapon systems should be acquired and maintained. Wargames constitute an important class of applications that have proved to be an important tool for military system analyses and an inexpensive alternate to live training exercises. However, modeling the complexity and battle dynamics of the real world, assessing and predicting the outcomes of missions plans quantitatively and accurately is a very difficult endeavor [1],[3]. Air Wargame Simulation System (AWGSS) is a wargaming software that has been developed for planning and evaluating air tasking operations (missions) [3]. Given a military objective, the air component commander plans missions, minimizes the risk while targeting, chooses appropriate weapons for

H. Papadopoulos, A.S. Andreou, and M. Bramer (Eds.): AIAI 2010, IFIP AICT 339, pp. 53–61, 2010.

targeting while ensuring the intended damage is caused to the target. This is done considering the enemy forces' deployment, weather and terrain encountered en-route and suitability of aircraft, weapons for achieving the objectives. Traditionally, the occurrence of various events has been represented using probability distributions, a set of game rules that represent various situations and classical logic to compute the damage assessment. Damage assessments for targets were implemented using crisp cookie-cutter function that gives the probability of damage of a target. These methods are found to be inadequate in the present war scenario, to generate a realistic assessment of the mission plans due to uncertainty and vagueness of the factors considered. In this paper, we present a novel approach to design the games rules of wargames using fuzzy rule bases, for quantitatively evaluating the effectiveness of air tasking missions.

2 Fuzzy Rule-Based Representation of Game Rules

Traditional approaches to wargame simulations use classical logic for damage assessment. Classical two-valued logic system, crisp set theory and crisp probability on which the damage assessment is based, are inadequate and insufficient for dealing with real-life war scenario that involves complexity and different sources of uncertainty. Damage assessment for a target done using cookie-cutter function gives the probability of damage of a target. Crisp cookie-cutter function states that a target is damaged inside a circle of specified radius r, and no damage occurs outside it [3].

$$d(x, y) = \begin{cases} 1, & x^2 + y^2 \leq r^2 \\ 0, & \text{otherwise} \end{cases} \tag{1}$$

where, $d(x, y)$ is the damage function of the point target by a weapon whose point of impact is (x, y). The target is assumed to be completely damaged within the circle of radius r and no damage occurs outside r. The notion of probability stems from, and depends on, the idea of repeated trials. Under identical and repeatable laboratory conditions conducted on simple models, this probabilistic notion readily applies; but, in real-world systems, experiments are rarely identical and repeatable. Therefore, for the subjective assessment of complex military systems, probability has its limitations. Game rules in conventional mission processing are limited and simplistic, taking into account limited number of parameters and based on statistical distributions. They do not take into consideration, effects of weather conditions, operational environment, target attributes, significance attached with the target, visibility of target at time of target acquisition, and effects of terrain while assessing damage. Variables such as these play a vital role in mission planning, mission processing and damage assessment. Different commanders deciding on the result of a planned mission are usually subjective based on their experiences. In order to overcome the limitations of statistically steady state results for game rules, we propose a fuzzy rule based approach for the design of the game rules. What is gained through fuzzification is greater generality, higher expressive power, an enhanced ability to model real-life problems, a methodology for exploiting the tolerance for imprecision and which serves to achieve tractability and robustness [2],[4]. Fuzzy Logic is the logic behind approximate reasoning instead of exact reasoning. As knowledge acquisition in wargames design and development is obtained from pilots and defence analysts, it is usually true that facts and rules are neither totally certain nor totally consistent due to the varied experience sets of the pilots. This leads to the reasoning processes used by experts in certain situations as approximate. In this paper, the theory of fuzzy sets is used to help assess uncertain information derived from this approximate reasoning process. Features of Fuzzy Logic that give it a realistic implementation than classical logic are:-

1. Fuzzy Logic offers far greater resources for managing complexity and controlling computational cost. More complex the problem involved, the greater the superiority of fuzzy methods.
2. Fuzzy Logic has greater expressive power. It has the capability to capture and deal with meanings of sentences expressed in natural language. This capability allows to deal in mathematical terms with problems that require the use of natural language.
3. Fuzzy Logic has capability to capture human common-sense reasoning, decision making and other aspects of human cognition.
4. Data which are based on graded distinctions among states of relevant variables are usually called fuzzy data. When fuzzy data are processed, their intrinsic uncertainties are processed as well, and the results obtained are more meaningful, than their counterparts obtained by processing the usual crisp data.

In any given language the values of a linguistic variable are words, phrases, or sentences. For example, structural damage can be considered as a linguistic variable with values such as "severely damaged," or "moderately damaged." These are meaningful classifications but not clearly defined. With the use of fuzzy sets, however, we can quantify such terminology and apply it in a meaningful way to help solve a complex problem. An evident advantage of the fuzzy set approach is the possibility of representing numeric and linguistic variables in a uniform way and of using a formalized calculus to manipulate these variables [3],[4].

3 Identification of Factors for Damage Assessment

The metric "*comparative damage relative to intended damage*" gives a measure of damage caused to the target relative to the damage that was intended to be caused, when the weapon was delivered. It uses various factors as its input variables such as Offset, Weapon Suitability, Weapon Delivery Mode, Target Identification factor and Terrain. Meetings with the pilots and analysts helped in identifying following parameters to be considered while assessing damage.

(1) *Offset* is the measure of deviation of actual point of impact from desired point of impact. It is calculated using actual altitude, actual vertical flight path angle, actual wind speed and observed altitude, observed vertical flight path angle, observed wind speed by the weapon system trajectory calculation module and the aircraft speed as the input variables. Offset is a measure of Altitude induced error, wind induced error, vertical flight path angle induced error. [3],[5],[6],[7].

(2) *Weapon Suitability factor* takes into account weapon characteristics, weapon's suitability in the operational conditions and weapon-target matching. Apart from the geo-political and threat environment factors such as weapon/target matching, affordability and maintainability are also considered. The weapon is chosen for attack considers parameters such as attack conditions, significance attached with the target versus cost of weapon, platform for weapon delivery, target-weapon match and collateral damage. Weapon Suitability factor is rated on the scale of 0 to 10 considering following factors:-

-*Operational Environment and Weather Conditions*: Operational environment and weather conditions play an important role in deciding the suitability of the weapon for a target. For instance, poor weather and conflict induced environmental conditions such as smoke from bombing etc. may degrade or block the IR, EO or laser targeting sensors required for delivery of guidance ordnance. LGBs cannot be used in poor weather where target illumination cannot be seen, or where it is not possible to get target designator near the target. However, poor visibility does not affect satellite guided bombs. Only comparatively inaccurately unguided bombs could be delivered in poor weather.

-*Cost versus Target Significance*: The expenditure incurred in using a weapon against a target should be justified taking into account significance attached with the target. For instance, there are high costs associated with the usage of guided munitions; they must not be neglected in cost-benefit analyses and the deliberation processes that are intrinsic to capital equipment purchases. The comparatively high cost of guided bombs and resulting smaller inventories is also an important factor in weapon suitability factor.

-*Aircraft Compatibility*: Compatibility of the aircraft for weapon delivery is an important factor. Weapons must be compatible for carriage and release with the combat aircraft types.

-*Weapon-Target matching*: Target size and target hardness are factors that need to be considered for deciding the suitability of a weapon against a target. For large targets, use of unguided ordnance is appropriate. LGB capable aircrafts could deliver penetrating LGBs. LGBs are particularly useful against hardened, reinforced, buried reinforced targets. The aim of matching weapons to targets is to achieve the desired level of damage for the least number of sorties while minimizing the risk to the weapons delivery platform. Weapons employed against certain targets such as ships must be inherently maneuverable so they can accommodate target velocity after launch.

-*Collateral Damage*: Although advances in weapon aiming systems have improved the accuracy of unguided air-delivered munitions, collateral damage cannot be discounted. PGMs must be employed against military objectives where there is a possibility of collateral damage to civilians or civilian objects.

(3) *Weapon Delivery Mode* is another factor considered for calculating damage. It takes into account aircraft for weapon delivery, weapon chosen and target characteristics. A value is assigned on the scale of 0 to 10. Three basic delivery modes considered are Roll, Toss and Dive. Weapons can be either direct attack (freefall, accurate, or precision) or standoff. Direct attack weapons require an aircraft to maneuver to a release point. Stand-off weapons enhance an aircraft's survivability by

allowing it to deliver ordnance at increased distance from the enemy's defenses. Planners and crews prefer launch-and-leave weapons to those requiring line of sight with the target. These tend to be much more expensive and fewer in number than other weapons.

(4) *Target Identification factor*: Rapid and certain target detection and identification are the dominant factors in the success of all air-to-ground attacks. The ability of tactical fighters to penetrate enemy defenses and to acquire and identify ground targets successfully is a keystone of success in a mission. Enemy defenses have necessitated higher speeds and higher altitudes. The employment of standoff weapons demands acquisition and identification of targets at even greater ranges. It has been observed that aerial observers respond to targets in a manner indicating that detection / identification represents a continuum rather than discrete phenomena. At one extreme the response is based on the ability to merely discriminate the existence of a military object among non-military objects (detection). At the other extreme the observer can describe the object in precise detail (identification). Factors considered for computing the *Target Identification factor* are target size, percent contrast, illumination, terrain, altitude and speed of the aircraft at time of target acquisition.

-*Target Size*: As target size increases, probability of correct target identification increases. It may vary from small to large tactical targets, including personnel, trucks, and tanks to big targets as bridges, runways and taxi-tracks.

-*Contrast*: Target/Ground Brightness Contrast is expressed as a percentage.

-*Illumination*: Detection performance increases as illumination increases. Effects of decreases in illumination occurring after sunset and before sunrise are very important and need to be considered.

-*Terrain*: Types of terrain have been defined in terms such as number of slope changes per unit area and average slope change. Four different terrain types have been defined--fairly smooth, moderately rough, rough, and very rough. As the roughness of terrain increases, percent terrain view decreases, and decrease in detection performance is observed.

-*Altitude*: The relationship between altitude and target detection/identification is normally one in which there is assumed to be an optimal altitude; above and below this optimum altitude, detection is reduced. As altitude increases, detection performance decreases. As altitude is increased beyond an optimal point, detection probability falls off rapidly.

Data on all these factors are collected from handbooks and experimental field trials and subjective decisions from experts and defense analysts (in questionnaire form) are collected and recorded. They are then represented as decision matrices and decision trees which form the basis to design the membership functions and rules. The rules are then executed in the mission processing module and defuzzified to obtain the damage to target. These results are then compared to the expected output and fine-tuned before storing in the rule base. A decision to include the new rule or not is provided to the commander. Missions and results of the missions are stored as a case-base for retrieval and reuse of missions plans in new situations. The following fuzzy linguistic variables used in the design of the game rules are as follows:

Terrain: 1- 100

{*Fairly Smooth [0 -22]; Moderately Rough [14 - 49]; Rough [45-81]; Very Rough [75 - 100]*}.

Target Size (in feet):

{*Very small: 0 -100; Small: 70 – 190; Medium sized: 160 – 300; Large: 270 – 400; Fairly Large: 360 – 500; Extremely Large: 450 – 900*}

Damage: Offset (in meters):

{*Very Less:[0-23]; Less:[16-36]; Medium: [34-57]; Large: [56-80]; Very Large [78-100]*}

Weapon Target Match: [0 to 10]

{*Poor: [0-3.6]; Average: [3.36 – 6.669]; Good: [6.73 – 14.2]*}

Target Identification Factor: [0 -10]

{*Very poorly identified: [0-1.19]; Poorly identified [0.96 – 2.43]; Average identification [2.34 – 5.61]; Good identification [5.43 – 7.55]; Excellent identification [7.35 – 10]*}

Relative Damage (Damage relative to intended damage)**: [0 - 100]**

{Mild: [0-18]; Moderate: [16-36]; Average: [34-57]; Severe: [56-80]; Fully Damaged: [78-96]}

These fuzzy variables to calculate the *target identification factor* are depicted in the Mamdani FIS as shown in Fig.1 and firing of the rules to compute the relative damage to the target are depicted in Fig. 2. Offset is calculated using Actual altitude, actual vertical flight path angle, actual wind speed and Observed altitude, observed altitude, observed vertical flight path angle, observed wind speed by the weapon system trajectory calculation module and the aircraft speed as the input variables (Table3). Offset is a measure of induced error, wind induced error, and vertical flight path angle induced error.

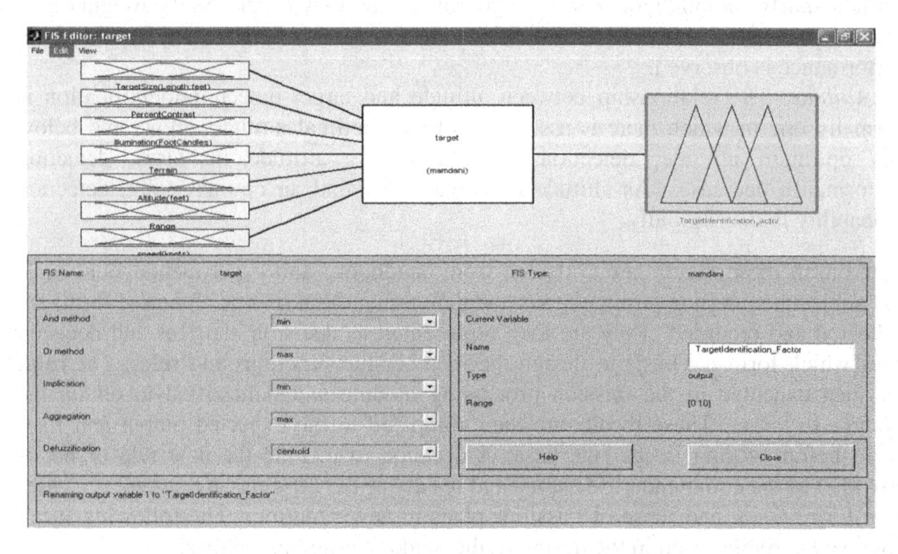

Fig. 1. Mamdami Fuzzy logic Inference System for computing *Target Identification factor*

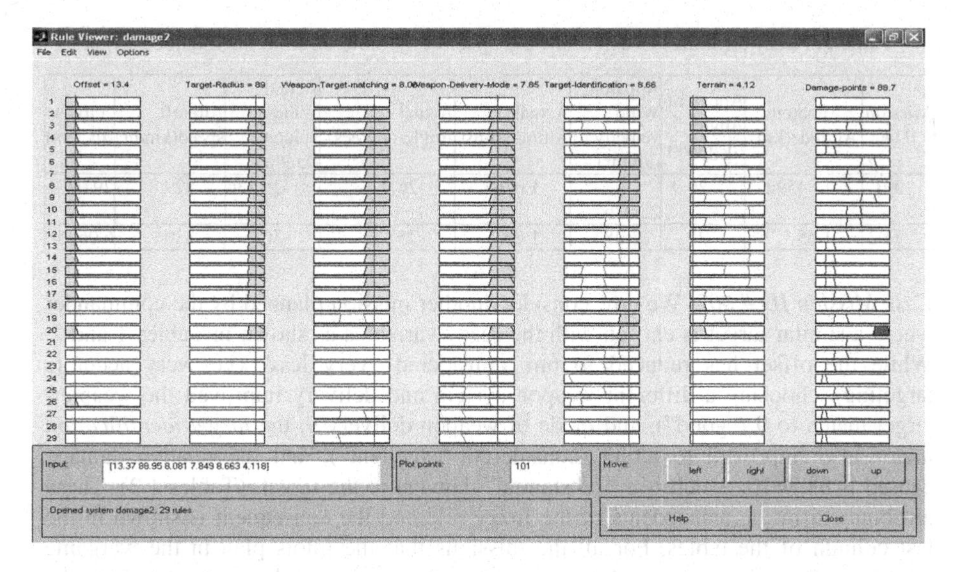

Fig. 2. Fuzzy Rules and their firing for the Mamdami FIS for finding *Damage caused* in relation to the *intended damage*

4 Results

Case Mission ID # 001: Consider a large area-target of size of 550 ft to be attacked, where the fuzzy variables *target-ground contrast* 80%, the *terrain*, rated 8, is fairly smooth, aircraft altitude is 900 ft, aircraft range is 5000 ft is flying at 100 knots speed.

Table 1. Fuzzy Rules to determine the Target Identification factor

MissionID	Target Size(ft)	Target-Ground Contrast%	Illumination (foot candles)	Terrain	Aircraft Altitude(feet)	Aircraft Range(feet)	Aircraft Speed (knots)	Target Identify Factor
#001	550	80	40	8	900	5000	100	7.3295
#002	550	80	60	7	750	4000	80	8.0333

Table 2. Fuzzy Rules to compute the *Relative damage to target*

MissionID	Offset (meters)	Target Radius (km)	Weapon-Target Match	Weapon Delivery Mode	Target Identify Factor	Relative damage
# 001	29.027	0.0900	6	6	7.3295	28.9187
# 002	6.0696	90	9	9	8.0333	88.7410

The *target identification factor* for this target is seen as "good" with value 7.3295. In this mission, on firing the rules for inference, the offset from the desired point of impact is 29m, considered "less"(i.e. fairly accurate targeting); *weapon-target match* is 6 (average), "good" *target identification factor* 7.3295, the relative damage caused is 28.9187 which is a "moderate" damage to the target.

Table 3. Fuzzy attributes to determine the *offset* of the weapon from the intended target

Mission ID #	Apparent Altitude(km)	Apparent Angle (degrees)	Apparent Wind velocity (km/hr)	Actual Altitude(km)	Actual Angle(degrees)	Actual Wind velocity (km/hr)	Aircraft speed(km/hr)	Offset (meters)
001	1.64592	-26.9	-25.236	1.6764	-26.9	-25.236	829.8	29.0271
002	1.65	-25	-28	1.65	-25	-30	830	6.0696

Case Mission ID # 002: We now consider another mission planned by the commander where a similar target is chosen with the fuzzy variables as shown in Tables 1 and 2. While the offset has reduced to 6m, considered "very less" (i.e. very accurate targeting), choosing a different weapon system and delivery improved the weapon-target match to 9 ("good"), and mode of weapon delivery 9, the *target identification factor* also improved to 8.033 (considered "excellent"), and the relative damage caused is 88.7410, which is a "substantial" damage to the target. (Tables 1,2). These attributes form the antecedents of the fuzzy rule and the consequent is shown in the last column of the tables. For all the missions that the pilots plan in the wargame exercises, these fuzzy game rules are used to infer the expected damage caused to the target. These missions form a part of a case-base which is used as part of the '*learning*' by the system for future instructional use.

5 Conclusions

We present a novel approach using a fuzzy rule-based system to design the game rules in a mission planning and evaluation system. The conventional crisp cookie cutter function used to compute the probabilistic damage caused to a target is replaced by a fuzzy cookie-cutter function, which takes into account many physical parameters before assessing the possibilistic damage caused to the target. This methodology of damage assessment computation of targets using fuzzy rule bases gave more realistic results when used in field training and deployment of the system.

Acknowledgments. The authors thank the Director, ISSA, for his guidance, support, encouragement and permission to present this work. We also thank Air Cmde S P Ojha, College of Air Warfare and Air Mshl. P K Mehra for the useful insights and discussions on air warfare tactics and game rules development; and the scientists of the air wargame team for their enthusiasm and support.

References

1. Banks, J. (ed.): Handbook of Simulation: Principles, Methodology, Advances, Applications, and Practice. John Wiley and Sons, New York (1998)
2. Cox, E.: The Fuzzy Systems Handbook, 2nd edn. Academic Press, New York (1999)
3. VijayRao, D.: The Design of Air Wargame System Simulation system. Technical report, Institute for Systems Studies and Analyses (2009)
4. Mendel, J.M.: Uncertain Rule-Based Fuzzy Logic Systems-Introduction and New Directions. Prentice Hall PTR, Upper Saddle River (2001)

5. Meitzler, T.J., Singh, H., Sohn, E.: Fuzzy Logic Technique to Determine Search Time and Probability of Detection for Targets of Interest in Background Scenes, TACOM Research Development and Engineering Center, Warren, MI (2002)
6. Banks, J. (ed.): Handbook of Simulation: Principles, Methodology, Advances, Applications, and Practice. John Wiley and Sons, New York (1998)
7. Meitzler, T., Toet, A.: Advances in Target Acquisition Modeling. Opt. Eng. 40(1756) (2001)

One-Dimensional Linear Local Prototypes for Effective Selection of Neuro-Fuzzy Sugeno Model Initial Structure

Jacek Kabziński

Institute of Automatic Control, Technical University of Lodz, Stefanowskiego 18/22, 90 924 Lodz, Poland
jacek.kabzinski@plodz.pl

Abstract. We consider a Takagi-Sugeno-Kang (TSK) fuzzy rule based system used to model a memory-less nonlinearity from numerical data. We develop a simple and effective technique allowing to remove irrelevant inputs, choose a number of membership functions for each input, propose well estimated starting values of membership functions and consequent parameters. All this will make the fuzzy model more concise and transparent. The final training procedure will be shorter and more effective.

Keywords: fuzzy modeling, neuro-fuzzy systems.

1 Introduction

We consider a Takagi-Sugeno-Kang (TSK) fuzzy rule based system used to model a memory-less nonlinearity from numerical data. It is typical for such application to perform a training of the system to improve modeling accuracy. One of the first (and very successful) approaches to the problem was Adaptive Neuro-Fuzzy Inference System (ANFIS) proposed by Jang [1]. The idea was to realize a fuzzy inference system as a neural network and to apply neural network learning techniques to train the fuzzy model. The ANFIS technique (where least squares optimization of consequent parameters is used and gradient decent error back propagation is applied to tune membership functions parameters) is offered as a standard tool of Matlab and is still very popular and employed to solve various modeling problems - for example [2,3]. Several other learning algorithms were developed to train fuzzy models including genetic, bacterial and PSO algorithms [4,5,6].

Before the training of parameters is started we have to propose initial structure of the fuzzy model. For a real-world problems we may consider a great number of potential inputs, but it is essential to select only the really important ones for system description. Also the number of membership functions for each input should be carefully chosen to compromise between model accuracy and complexity. The necessity of ANFIS input selection was noticed already by Jang [7] and this aspect of fuzzy model complexity reduction was investigated by many researchers [8], but the problem remains open. The most popular approaches to initial system architecture selection are based on several clustering techniques[8].

The purpose of this work is to develop a simple and effective technique allowing to remove irrelevant inputs, choose a number of membership functions for each input,

H. Papadopoulos, A.S. Andreou, and M. Bramer (Eds.): AIAI 2010, IFIP AICT 339, pp. 62–69, 2010.

propose well estimated starting values of membership functions and consequent parameters. All this will make the fuzzy model more concise and transparent. The resulting final training procedure will be shorter and more effective. The proposed approach is based on a concept of fuzzy projection of the data on each input space [9,10], combined with piece-wise linear approximation allowing to choose automatically a number of membership functions according to the required accuracy and to estimate starting values of all fuzzy system parameters. We demonstrate the effectiveness of the proposed technique investigating several examples and comparing results with popular clustering methods.

2 Neuro-Fuzzy Inference System

We consider a neural network realisation of TSK fuzzy model [1]. The network possesses N inputs $x_1,\ldots,x_N$, one output y, m_i membership functions μ_{i,j_i}, $j_i \in \{1,\ldots,m_i\}$ associated with the i-th input $i \in \{1,\ldots,N\}$, $N_R = \prod_{i=1}^{N} m_i$ rules of the form:

$$\begin{aligned} &IF \quad (x_1 \; IS \; \mu_{1,j_1}) \quad AND \ldots AND \quad (x_N \; IS \; \mu_{N,j_N}) \quad THEN \\ &y = p_{1,r}x_1 + \ldots + p_{N,r}x_N + q_r = f_r(x_1,\ldots,x_N) \end{aligned} \tag{1}$$

where $j_i \in \{1,\ldots,m_i\}$ and r is the rule number. We can choose any smooth membership function, for example generalized bell-shape function:

$$\mu_{i,j}(x) = \frac{1}{1+\left(\frac{x-c}{a}\right)^{2b}} \tag{2}$$

The rule firing strength is calculated as:

$$w_r = \prod_{i=1}^{N} \mu_{i,j_i}(x_i) \tag{3}$$

for the r-th rule and the normalised firing strength is:

$$\overline{w}_r = \frac{w_r}{\sum_{i=1}^{N_R} w_i} \; . \tag{4}$$

The final output is calculated as:

$$y = \sum_{r=1}^{N_R} \overline{w}_r f_r(x_1,\ldots,x_N) \; . \tag{5}$$

It is well known that this Sugeno fuzzy model may be realised as a five-layer neural network: the first layer calculates the membership functions values, the second layer - firing strengths, the third - normalised firing strengths, the fourth and the fifth calculates the final output [1]. The standard training procedure proposed for ANFIS combines gradient descent back-propagation method to tune membership functions parameters and least squares method to find optimal consequent parameters.

The system is supposed to model a memory-less nonlinearity given by input-output data with possible extraneous inputs. We assume that that the input-output data are given by

$$\left(x_{1,k},\cdots,x_{\tilde{N},k}\right)\to y_k \quad \tilde{N}\geq N, \quad k\in\{1,\ldots,m\}\,. \tag{6}$$

3 Initial Structure Selection Procedure

To test the significance of the i-th input, $i\in\{1,\ldots,\tilde{N}\}$ we consider a single-input fuzzy model (SIFM) described below:

- input - x_i, output - c_i,
- input linguistic categories: x_i *IS* $x_{i,k}$ $\;k\in\{1,\ldots,m\}$,
- membership functions:

$$\varphi_{i,k}(x)=\frac{1}{1+\left(\dfrac{x-x_{i,k}}{a}\right)^{2b}}, \tag{7}$$

- rules:

$$\begin{aligned}
&IF \quad x_i \quad IS \quad x_{i,k} \quad THEN \quad c_i = p_{i,k}x_i + q_{i,k}\\
&p_{i,k}=\frac{y_k}{x_{i,k}}, \quad q_{i,k}=0 \quad if\ x_{i,k}\neq 0\\
&p_{i,k}=0, \quad q_{i,k}=y_k \quad if\ x_{i,k}=0\\
&k\in\{1,\ldots,m\}.
\end{aligned} \tag{8}$$

The action curve given by the output c_i of this system for the input data - $x_{i,k}$ generalises information coded by $x_{i,k}\to y_k$. The degree of this generalisation depends on parameter a. As we want to cover the whole range of the *i-th* input by a few (say 3 or 5) membership functions, it is reasonable to take such a that the set in which a membership function is 'active', say $\{x: \varphi_{i,k}(x)>0.3\}$ covers 10-30% of the interval $\left[\min\limits_k(x_{i,k}),\ \max\limits_k(x_{i,k})\right]$. It may be proved that the shape of the action curve of SIFM is robust to single outliers in the measured data and to the measurement noise.

If the *i-th* input is inessential the curve generated by corresponding SIFM will be flat, if it is meaningful the curve will cover significant part of $\left[\min_k(y_k), \max_k(y_k)\right]$. Building SIFM for every input, plotting and testing it's action curve $A_i = \{(x_{i,k}, c_i(x_{i,k})) \mid k \in \{1,\cdots,m\}\}$ we are able to classify the importance of the input and to select N significant inputs with corresponding curves. We must remember that sometimes data symmetry may lead to false diagnosis that the input is irrelevant one. In this case other techniques for input selection must be used [8], or the modelling domain should be narrowed, or nonlinear data transformation may be applied.

Selection of membership functions for each input is based on piece-wise linear approximation of action curves derived above. Uniform or mean-square approach are both applicable. The input data have to be sorted in ascending order and repeated values must be removed – it is important especially for grid-type data. It is not necessary to obtain continuous piece-wise linear approximation. Mean square approximation means that starting from the left we are to approximate (by a linear function) maximal number of subsequent points $(x_{i,k}, c_i(x_{i,k}))$, under the condition that the mean-square error is limited by a given parameter δ. Uniform approximation consists in placing the longest possible line segments ('sticks') inside a δ-wide 'tube' surrounding the curve $A_i = \{(x_{i,k}, c_i(x_{i,k})) \mid k \in \{1,\cdots,m\}\}$ on the plane. Both approaches are easy to implement numerically and produces equivalent results. Of course design parameter δ influences the number of intervals and so the number of membership functions. As the result of piece-wise linear approximation for the i-th significant input we obtain m_i intervals

$$I_{i,j} = \left(x_{\min i,j}, x_{\max i,j}\right) \quad j \in \{1,2,\cdots,m_i\}\ . \tag{9}$$

and a linear polynomial

$$P_{i,j}(x_i) = p_{1\,i,j} x_i + p_{0\,i,j}, \quad x_i \in \left(x_{\min i,j}, x_{\max i,j}\right) \quad j \in \{1,2,\cdots,m_i\}\ . \tag{10}$$

for each interval.

The next step is to build a neuro-fuzzy (ANFIS) model to imitate curve $A_i = \{(x_{i,k}, c_i(x_{i,k})) \mid k \in \{1,\cdots,m\}\}$ with small number of membership functions. The proposed membership functions for the i-th significant input will be m_i generalized bell-shape functions $\mu_{i,j}(x)$ with parameters

$$a_{i,j} = \frac{1}{2}\left(x_{\max i,j} - x_{\min i,j}\right), \quad c_{i,j} = \frac{1}{2}\left(x_{\max i,j} + x_{\min i,j}\right), \quad j \in \{1,2,\cdots,m_i\}\ . \tag{11}$$

So the function $\mu_{i,j}(x)$ is spanned over $I_{i,j}$ and centred at the middle point of $I_{i,j}$. The choice of the third parameter $b_{i,j}$ is arbitrary – it is reasonable to start with $b_{i,j} = 1.5$ for all membership functions. The rules for the proposed SISO neuro-fuzzy model will be:

$$IF \quad x_i \quad IS \quad \mu_{i,j} \quad THEN \quad c_i = p_{1\,i,j} x_i + p_{0\,i,j} \quad , \quad j \in \{1,2,\ldots,m_i\}, \tag{12}$$

where starting values of parameters are taken from piecewise linear approximation results and are given by (10) and (11). Each rule describes a linear, local fuzzy prototype of fuzzy-filtered one-dimensional data.

Single input single output neuro-fuzzy models proposed for each significant input are trained using the data $(x_{i,k}, c_i(x_{i,k}))$, $k \in \{1, \cdots, m\}$. As the starting values of parameters (10, 11) are carefully chosen and as we train one-dimensional model, the training is fast and the results are accurate. Finally, after the training we get N one-dimensional neuro-fuzzy models, the i-th one is equipped with

- m_i membership functions $\mu_{i,j}(\cdot)$ with optimised parameters $(\hat{a}_{i,j}, \hat{b}_{i,j}, \hat{c}_{i,j})$,
- m_i rules *IF* x_i *IS* $\mu_{i,j}$ *THEN* $c_i = \hat{p}_{1\,i,j} x_i + \hat{p}_{0\,i,j}$ with optimised parameters $(\hat{p}_{1\,i,j}, \hat{p}_{0\,i,j})$.

The multi-input model, which will be a starting point for the final training is equipped with $N_R = \prod_{i=1}^{N} m_i$ rules:

$$IF\ (x_1\ IS\ \mu_{1,j_1})\ AND ... AND\ (x_N\ IS\ \mu_{N,j_N})\ THEN\ \ y = \hat{p}_{1\,1,j_1} x_1 + ... + \hat{p}_{1\,N,j_N} x_N \cdot$$
$$q_r = \frac{1}{N}\left(\hat{p}_{0\,1,j_1} + ... + \hat{p}_{0\,N,j_N}\right), \qquad j_i \in \{1, 2, \cdots, m_i\} \tag{13}$$

and membership functions parameters are $(\hat{a}_{i,j}, \hat{b}_{i,j}, \hat{c}_{i,j})$. The above neuro-fuzzy inference system is trained by a selected procedure to obtain the final model.

4 Examples

The results of the proposed approach were compared with three standard methods of generation initial structure of fuzzy inference system from numerical data applied in Matlab [11]:

- *genfis1* - generates a FIS structure from a training data set, using a grid partition on the data (no clustering), number and type of membership functions is given by the user;
- *genfis2* - generates Fuzzy Inference System structure from data using subtractive clustering, cluster radii is given by the user;
- *genfis3* - generates Fuzzy Inference System structure from data using FCM clustering, number of clusters is given by the user.

User-defined parameters were chosen to obtain the same number of membership functions as generated automatically by the proposed method.

We are to model two 2-dimmensional functions:

a) $y = x_1^2 + x_2^2 - \cos(18x_1) - \cos(18x_2)$ (Riastragin function),

b) $y = (x_1^2 + \sin(6x_2))^2$

over [0,1]×[0,1]. We generate 400 points of grid data in [0,1]×[0,1] and calculate corresponding values of y. We make a false assumption that the third input influences the function – we consider the triples (x_1, x_2, x_3) where value of x_3 is generated on random from [0,1]. As we see in both cases the curve A_3. is flat - we can eliminate input x_3 as an inessential one. Piece-wise linear mean square approximation was applied and for $\delta = 0.001$ numbers of membership functions was: for problem a) $[m_1, m_2] = [6,6]$, for problem b) $[m_1, m_2] = [2,3]$. Because of parameters generated by the linear, local prototype approach, the initial position of one-dimensional FIS is close to the desired one and the training is fast and effective. Finally the initial FIS was generated according to (13) and compared with those generated by *genfis1-3*.

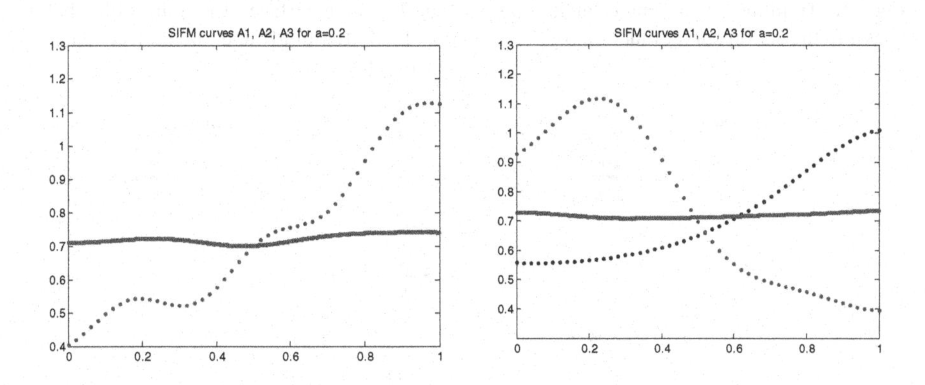

Fig. 1. Single-input fuzzy model curves *left*: function a), right: function b)

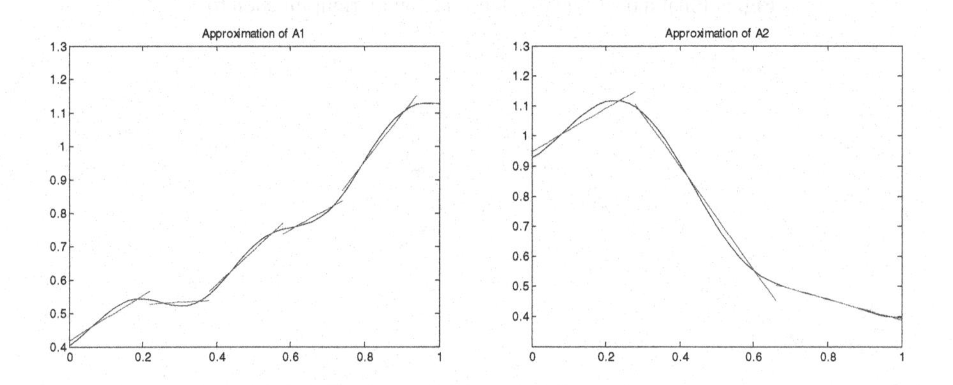

Fig. 2. Approximation of single-input fuzzy model curves *left*: function a), *right*: function b).

For both functions the starting error as well as the whole learning performance was superior in case of the proposed method. The design procedure is illustrated in fig. 1-5.

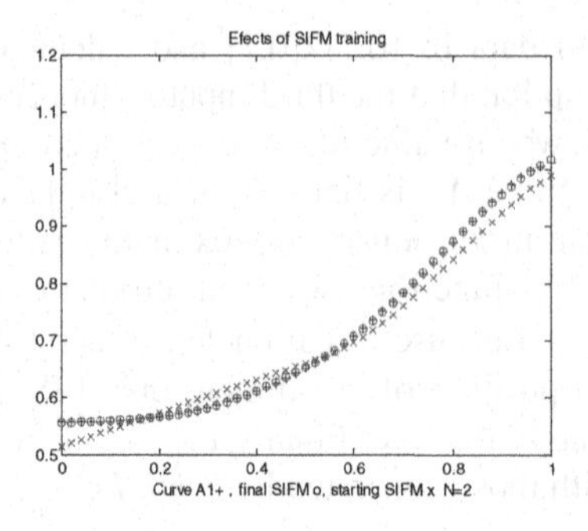

Fig. 3. Training of a single-input fuzzy model to the action curve, function b).

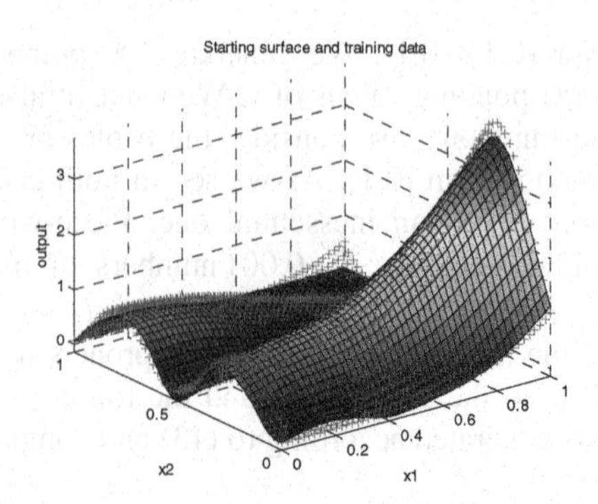

Fig. 4. Action surface of an initial FIS (before the final training) and numerical data +, function b).

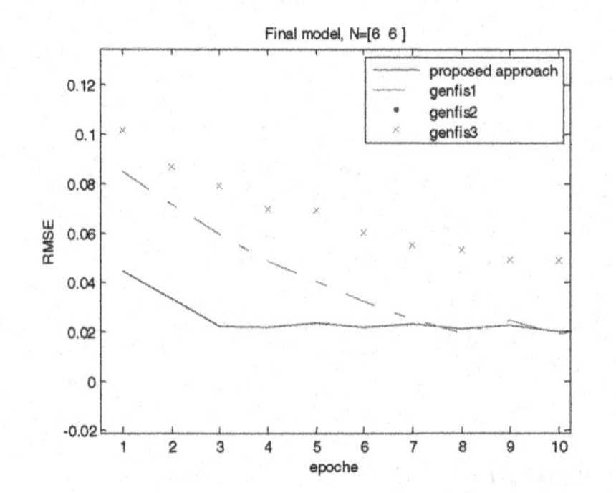

Fig. 5. Final model training: left function a), right function b).

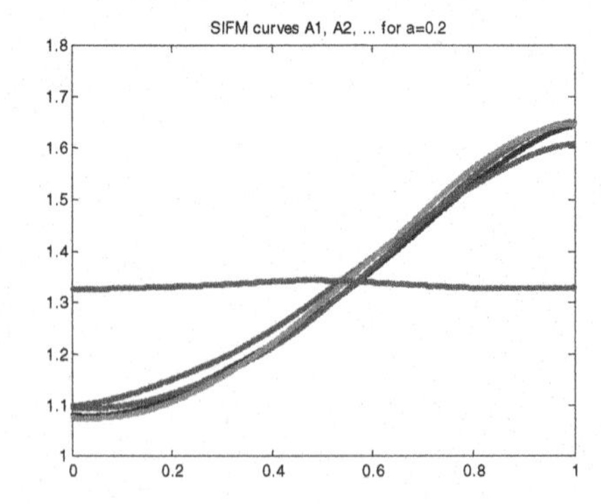

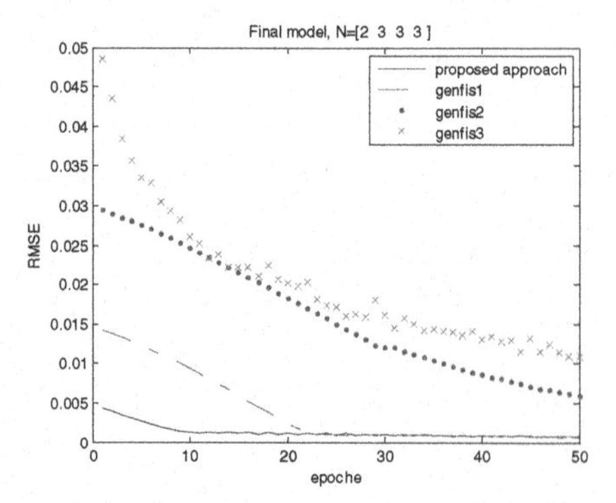

Fig. 6. Modeling a multidimensional ball: *left* - single-input fuzzy curves, *right* - final model training

The next task was modeling a section of a multidimensional ball with one 'dumb' input. 1000 points were generated on random in the unit multi-dimensional cube. The results are presented in fig. 6. Again the proposed method is better than the standard clustering procedures.

5 Conclusions

We propose simple and effective procedure for creating neuro-fuzzy TSK models of complex systems from the numerical input-output data. Significant inputs, number of membership functions for each input and initial values of all system parameters are set automatically based on accuracy requirements. Because the initial structure and parameters are set properly, we need a few training iterations for the neural network representation of our model to converge. The generated FIS may be used as a starting point to any training algorithms – also those incorporating reduction of fuzzy rules what is highly recommended, although was not discussed here. The proposed technique was compared with clustering methods implemented in Matlab. Results obtained from the proposed method provided several times smaller starting training error, leading to simpler models, smaller errors and reduction of necessary training time. Several important problems were not discussed here, for example elimination of relevant but dependent inputs.

References

1. Jang, J.R.: ANFIS: Adaptive-network-based fuzzy inference system. IEEE Trans. Syst. Man Cybern. 23, 665–684 (1993)
2. Buyukbingola, E., Sismanb, A., Akyildizc, M., Alparslanb, N.F., Adejared, A.: Adaptive neuro-fuzzy inference system (ANFIS): A new approach to predictive modeling in QSAR applications: A study of neuro-fuzzy modeling of PCP-based NMDA receptor antagonists. Bioorganic & Medicinal Chemistry 15, 4265–4282 (2007)
3. Kovac, N., Bauk, S.: The ANFIS-based route preference estimation in sea navigation. Journal of Maritime Research, III, 69–86 (2006)
4. Alcalá, R., Alcalá-Fdez, J., Casillas, J., Cordón, O., Herrera, F.: Local Identification of Prototypes for Genetic Learning of Accurate TSK Fuzzy Rule-Based Systems. International Journal Of Intelligent Systems 22, 909–941 (2007)
5. Lin, C.-J., Hong, S.-J.: The design of neuro-fuzzy networks using particle swarm optimization and recursive singular value decomposition. Neurocomputing 71, 297–310 (2007)
6. Gal, L., Botzheim, J., Koczy, L.T.: Modified bacterial memetic algorithm used for fuzzy rule base extraction. In: Proceedings of the 5th International Conference on Soft Computing as Transdisciplinary Science and Technology, pp. 425–431 (2008)
7. Jang, J.R.: Input selection for ANFIS learning. In: IEEE Int. Conf. Fuzzy Systems, vol. 2, pp. 1493–1499 (1996)
8. Hu, C., Wan, F.: Input selection in learning systems: a brief review of some important issues and recent developments. In: Proceedings of the 18th International Conference on Fuzzy Systems, pp. 530–535 (2009)
9. Lin, Y., Cunningham III, G.A., Coggeshal, S.V.: Using fuzzy partitions to create fuzzy systems from input-output data and set the initial weights in a fuzzy neural network. IEEE Trans. on Fuzzy Systems 5, 614–621 (1997)
10. Kabziński, J., Woźniak, P., Kuźmiński, K.: Effective selection of neuro-fuzzy Sugeno model architecture. In: Proc. of 17-th IASTED International Conf. Modelling, Identification & Control, pp. 97–100 (1998)
11. The MathWorks, Inc., http://www.mathworks.com/access/helpdesk/help/toolbox/fuzzy/genfis1.html,genfis2.html,genfis3.html

Lasso: Linkage Analysis of Serious Sexual Offences

A Decision Support System for Crime Analysts and Investigators

Don Casey[1,2] and Phillip Burrell[1]

[1] London South Bank University, Borough Rd, London SE1 0AA, U.K
caseyd@lsbu.ac.uk
[2] Metropolitan Police Service, London, U.K
phillb@lsbu.ac.uk

Abstract. One of the most important considerations when investigating a serious sexual offence is to find if it can be linked to other offences. If this can be done then there is a considerable dividend in terms additional evidence and new lines of enquiry. The central problem is the construction of a satisfactory typology of these crimes, but little progress has been made. It is the authors' contention that difficulties arise from the inadequacy of the adoption of the classical or 'crisp set' paradigm. Complex events like crimes cannot be described satisfactorily in this way and it is proposed that fuzzy set theory offers a powerful framework within which crime can be portrayed in a sensitive and perceptive manner that can enhance the search for associations between offences.

Keywords: fuzzy systems, decision support, crime linkage.

1 Introduction

The most influential crime classification system has been that proposed in the Crime Classification Manual [1], which is the work of senior F.B.I agents and advances the notion of an organized / disorganized dichotomy in serious offences and was developed from interviews with offenders [2]. The basis of this approach is that crimes can be differentiated by the level of planning associated with them. The authors extend this to assert that the dichotomy can be applied to the offender so that organized and disorganized crimes are committed by individuals who can be divided into discrete groups with distinct characteristics.

Very serious objections have been made to the methodology employed by the F.B.I: only 36 offenders were interviewed, no attempt was made to ensure this group was representative and the interviews conducted were not structured or consistent. An evaluation of the typology by distinguished psychologists working in the field [3] applied to 100 serial murderers provided no support for it.

In the most comprehensive research programme into the linkage of serious sexual offences by Grubin et al [4], the authors propose.

Our starting premise is that rape attacks can be organized into distinct types

H. Papadopoulos, A.S. Andreou, and M. Bramer (Eds.): AIAI 2010, IFIP AICT 339, pp. 70–77, 2010.

It is certainly the aim of investigation into any field to initially classify the objects contained within it but it is the contention of this paper that that although rapes can be organized into types that these will be far from 'distinct'. And that the attempt to discriminate between crimes in this way is likely to be not only barren but actively misleading in that they will be forced in to mutually exclusive groups that will misrepresent their complexity; a view arrived at after many years of research by Canter[5] one of the area's foremost investigators.

The actions of any individual criminal may therefore be thought of as a subset of all the possible activities of all criminals. Some of this subset overlaps with the subsets of many other criminals, and some with relatively few. It therefore follows that assigning criminals or crimes to one of a limited number of 'types' will always be a gross oversimplification.

Canter and his associates who are identified with Investigative Psychology have published numerous studies [6], [7] on sexual assault, homicide and other serious crimes but have been unable in any of them to construct a satisfactory typology with even the most relaxed rules of assignment [8], [9].

Grubin is forced to propose a highly redundant 256 element taxonomy of serious sexual assault. A classification system in which many, if not most, elements will never occur cannot be satisfactory. The assumption of the crisp set paradigm in this research appears to be the cause of the problems relating to these difficulties. This can be illustrated by a simple description of a crime such as 'a very violent assault on a middle-aged woman by a young man' which cannot be properly expressed in terms of crisp sets. It can lead to either the misallocation of fundamentally different offences to the same place or to crimes that bear strong resemblances to each other being regarded as entirely unconnected, a phenomenon referred to as linkage blindness [4] of which researchers are fully aware but have been unable to address.

2 Applicability of Fuzzy Systems

2.1 Geographic Profiling as a Fuzzy System

There has been only one area of research into crime clustering that has been widely acknowledged to have been successful in linking crimes and it is instructive that this can be regarded as relying on fuzzy sets, although this has not as yet been acknowledged.

There are various schools of geographic profiling [10] [11] but all are based on the idea proposed by environmental criminology [12] relating to 'mental maps', i.e. that individuals, including criminals, are much more likely to conduct their activities in areas known to them. In terms of offenders this means that they are likely to live close to, or have some other association with the locations of their crimes. This has been adapted to construct a variety of systems so that the co-ordinates of a number of offences, known or believed to be linked can be input to a function that returns a 'jeopardy surface'. This not only returns information regarding the offender's lifestyle but presents an area in which further crimes committed by him may be discovered although they are presently unlinked. Most importantly it indicates those areas where the offender is likely to live or have some other strong association with such as employment or previous address. Slightly different techniques are employed to differentiate the 'most likely' from the 'less likely' areas.

We can regard this, the most successful approach in the area as a fuzzy system in that a set of discrete values, geographical coordinates, are input to an algorithm that assigns a degree of membership of the fuzzy set 'offender has an association with', or something similar, to other co-ordinates in the region of the offences. Typically the result will be a map that resembles a series of concentric circles or clusters that are strongly reminiscent of a family of fuzzy sets.

2.2 Crime and Fuzzy Sets

Fuzzy set theory [13] allows us to represent crimes and criminals as highly descriptive objects in the concept space and to undertake experimental procedures to discover what the most significant differentiating features are, using mathematically and logically sound methods. We have been fortunate in being successful in obtaining data on 574 serious sexual offences from the Serious Crimes Analysis Section of the U.K National Policing Improvement Agency. We have excluded those offences that do not relate to serial stranger rapes, by which we mean a set of rapes committed by a single individual, unknown to the victim. This results in a much narrower dataset (n = 110, development set n = 83, test = 27). The development set which has provided the results in this paper consists of 28 series of average length 2.96.

2.2.1 Fuzzy Similarity

We can define the universe of crimes as a data set (X) of n elements

$$X = \{ x_1, x_2, x_3, \dots x_n \} \tag{1}$$

Where each crime (x_i) is defined by j features

$$x_i = \{ x_{i1}, x_{i2}, x_{i3}, \dots x_{ij} \} \tag{2}$$

In this case they could be the characteristics identified as significant by Grubin. We can then regard a crime as a datapoint in j dimensional space. If we introduce an index crime (x_m) into this space for comparison with another crime x_k we can define a fuzzy relation that captures the concept 'x_k is close to x_m', by using a membership function that measures the Euclidean distance between crimes and divides it by some value 'c', where c is a positive real number whose value is chosen as a reasonable representation of the concept 'close' in that application.

We have used c = d / 2 where d = the average distance between crimes. As a result the introduced crime becomes the centre of its own cluster. We can then define a number of crisp subsets around this centre by restricting membership of these sets to those elements that have a degree of membership which is greater than or equal to some value α in [0, 1]. This results in a crisp subset ${}^{\alpha}A$ of the fuzzy set A which is itself defined on the universal set X. This crisp subset is known as an α-cut of X:

$${}^{\alpha}A = \{ x \in X \mid A(x) \geq \alpha) \} \tag{3}$$

Here the crisp set $^{\alpha}A$ contains all the elements of the universal set X whose membership degrees are greater than or equal to the value of α. In this case we would generate a number of nested sets around the index crime, membership of which would reflect their 'closeness' to it. This illustrates very graphically the search strategy for crime analysts and investigators when one of these very serious crimes occurs and it is required to look for other crimes committed by the same offender.

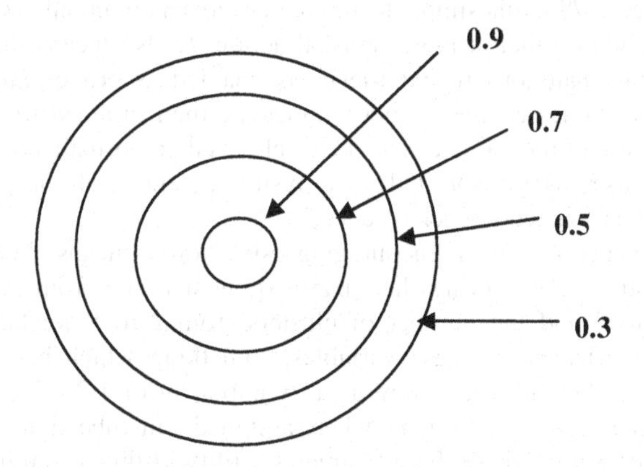

Fig. 1. Closeness to the index crime

2.2.2 Fuzzy c-Means Clustering

Fuzzy c-means clustering [14] is the most widely used fuzzy clustering strategy and effectively addresses the problem raised by Canter of exclusive types by defining a family of fuzzy sets on the universe X so that the sum of degrees of membership of all the classes of any datapoint is unity, there will be no empty classes and no class that contains all the datapoints. This is an iterative optimisation technique of the objective function below where a degree of fuzziness $1 \leq m < \infty$ is specified and elements assigned degrees of membership of the clusters until some termination criterion has been reached.

$$J_m = \sum_{i=1}^{N} \sum_{j=1}^{C} u_{ij}^{m} \left\| x_i - c_j \right\|^2 \tag{4}$$

u_{ij}^{m} is the degree of membership of x_i in cluster j and c_j is the cluster centre.

2.2.3 Initial Results

This data is extremely rich and comprises over 370 fields with dichotomous values relating to every feature of the crime. By employing only those variables that inform the domains identified as significant by Grubin and excluding ill-defined or poorly recorded data we have reduced this to 41: sex(11), escape (11) and control (19). These reflect the offender / victim transaction that is at the heart of this serious offence. A problem that arose was that these concepts, unlike those usually identified with fuzzy membership functions like age, height etc are cumulatively or hierarchically scaled.

This makes the use of a conventional membership function difficult. In order to overcome this we have proposed that the *amount* of these activities can be measured, i.e. the number of separate sexual, controlling or escape-centred actions.

The membership function we have employed derives closely from the techniques used by Canter [9] and Investigative Psychology which emphasizes the frequency of variables and their co-occurrence within crimes. Each variable is assigned a value c / n where c is the number of times it occurs in the dataset n, so a variable occurring 14 times has a value 14/83, this simple technique ensures a commonly occurring variable has a lower weighting than a more unusual action. It also means that a very simple form of learning analogous to experience is enabled as crimes are added and the distribution of variables changes. These values are then summed for all the variables for each dimension of the index crime. This value is then normalized using the highest sum in the dataset as divisor and as a result a degree of membership of each dimension can be assigned to each offence.

Our initial results have been encouraging using both c-means clustering and fuzzy similarity measures .By varying the fuzzy exponent (m) from the crispest fuzzy value of 1.25 to 3 and the number of clusters from 2 to 5 we have been able to evaluate the significance of these variables, something which has not before been done in this field. For instance using a fuzzy exponent of 1.25, 3 dimensions and 3 clusters, 15 of the 28 series of crimes where assigned a membership value ≥0.9 in one cluster. However it is unlikely that the greatest utility of this technique will be at this low level of fuzziness. Table 1 indicates an example of the very high level of consistency across the dimensions that can be achieved in linked series (bold underline) and the resulting strong level of clustering at m = 1.25 in the clusters we identify as A,B and C.

This differs from previous attempts to classify crimes in that it is as empty as possible of psychological precepts, as is reflected in the cluster names. As a result the assignment of crimes to classes is as a result of their positions in the concept space rather than their perceived association with wider psychological principles. If clustering is successful then there undoubtedly are psychologically meaningful alternatives for these labels but this would lie within the remit of criminal psychologists.

Table 1.

	Dimensions			Membership		
Crime	Control	Sex	Escape	A	B	C
x53	44	16	6	0.99	0	0.01
x54	44	2	6	0.95	0.01	0.04
x50	87	2	32	0.12	0.05	0.83
x51	59	2	32	0.29	0.04	0.67
x52	87	2	32	0.12	0.05	0.83

Overall where m=1.25 88% of crimes are assigned to a set with > 0.80 degree of membership: type A = 12%, type B = 20%. Type C = 53%. 15 of the 28 series are assigned to a single set.

In terms of the closeness of crimes we have looked at the overall number of possible comparisons between the 83 offences and then at the relative differences in closeness between the linked and unlinked crimes, i.e. those committed by different individuals and those committed by the same. The relatively small number of comparisons of linked offences is as a result of the high number of series of two crimes.

Table 2.

number of comparisons between crimes

	All	Unlinked	Linked
total	3403	3393	110

The very marked difference in the degrees of closeness between linked and unlinked crimes is encouraging, particularly as the gap between the two widens very significantly as degree of closeness increases. So at the average degree of closeness of 0.23 a linked crime is a little less than twice as likely to appear while at the highest degree of closeness, > 0.9, a linked crime is seven times as likely to figure.

The implication in terms of assisting crime analysts and investigators is clear: at the highest levels of closeness the greatest possibility of a match occurs and this declines in line with closeness. It provides the clearest and most effective search strategy for analysts in order to maximise their chances of a positive 'hit'.

Table 3.

degree of closeness between crimes

	All	Unlinked	Linked
above ave	1387	1309	78
> 0.23	40.76%	38.58%	70.91%
above 0.5	674	632	45
	19.81%	18.63%	40.91%
above 0.6	464	427	37
	14%	12.58%	33.64%
above 0.7	286	256	30
	8.40%	7.54%	27.27%
above 0.8	124	110	14
	3.60%	3.24%	12.73%
above 0.9	36	29	7
	1.00%	0.85%	6.36%

3 Conclusion

The problem of rigid typology that has hampered this area of research is precisely the one that fuzzy sets avoids. Because of the nature of the area under investigation any crisp classification method is bound to fail. Either a large number of crimes elude classification as in Investigative Psychology or an enormous system that specifies 256 type of stranger rape, which is itself a small subset of rape has to be proposed. These crimes must be placed somewhere if research is to be fruitful. The answer may be that instead of belonging nowhere or in a tiny compartment of a huge structure, that they belong in several places at the same time to differing degrees.

The success of geographic profiling in modelling criminal conduct is illuminating. An algorithm that can be regarded as a membership function uses a number of geographic locations, of linked crimes, as input in order to assign degrees of membership to a larger set of geographic points and thereby construct a fuzzy set. And in so doing effectively assist crime investigators in highlighting areas in which to find the offender. In this case longitude and latitude are the relevant dimensions on which the system operates. If one generalizes from this and is able to identify the pertinent dimensions that describe a landscape of actions rather than the physical landscape associated with crime then the achievements of geographic profiling may be possible. There is also an interesting symmetry in that the input to geo-profiling systems is a set of linked crimes and the desired output of LASSO is also a set of linked offences.

The 'set of meaningful numbers' called for in the earliest days of research [15] into this area can be achieved by using fuzzy set theory and in thus allow empirical research rather than the experiential and anecdotal or hypothetical approaches that have so dominated the field for so long and so unproductively.

References

1. Douglas, J.E., Burgess, A.W., et al.: Crime classification manual: A standard system for investigating and classifying violent crime. Simon and Schuster, New York (1992)
2. Ressler, R.K., Douglas, J.E.: Crime Scene and Profile characteristics of organized and disorganized murderers. FBI Law Enforcement Bulletin 54(8), 18–25 (1985)
3. Canter, D.V., Alison, L.J., et al.: The Organized/Disorganized Typology of Serial Murder. Myth or Model? Psychology, Public Policy and Law 10(3), 293–320 (2004)
4. Grubin, D., Kelly, P., et al.: Linking Serious Sexual Assault through Behaviour Home Office Research Study 215, London (2000)
5. Canter, D.: Offender profiling and criminal differentiation. Legal and Criminological Psychology 5, 23–46 (2000)
6. Santilla, A., Hakkanen, H., et al.: Inferring the Crime Scene Characterstics of an Arsonist. Interrnational Journal of Police Science and Management 5(1) (2003)
7. Hakkanen, H., Lindof, P., et al.: Crime Scene Actions and offender characteristics in a sample of Finnish stranger rapes. Journal of Investigative Psychology and Offender Profiling 1(2), 153–167 (2004)
8. Salfati, E.C., Canter, D.V.: Differentiating Stranger Murders: profiling offender characteristics. Behavioural Sciences and Law 17(3) (1999)

9. Canter, D.V., Bennell, C., et al.: Differentiating Sex Offences. Behavioural Sciences and Law 21 (2003)
10. Rossmo, D.K.: Geographic Profiling. Boca Raton Fl. CRC Press, Boca Raton (2000)
11. Canter, D., Coffey, T., et al.: Predicting serial killers' home base using a decision support system. Journal of Quantitative Criminology 16(4), 457–478 (2000)
12. Brantingham, P.J., Brantingham, P.L.: Environmental Criminology. Prospect Heights, IL, Waveland Press (1981)
13. Zadeh, L.: Fuzzy Sets. Information and Control (8), 228–353 (1965)
14. Bezdek, J.C.: Pattern Recognition with Fuzzy Objective Function Algorithms. Plenum, New York (1981)
15. Canter, D.: Facet theory: approaches to social research. Springer, New York (1985)

Forecasting Euro – United States Dollar Exchange Rate with Gene Expression Programming

Maria A. Antoniou[1], Efstratios F. Georgopoulos[1,2], Konstantinos A. Theofilatos[1], and Spiridon D. Likothanassis[1]

[1] Pattern Recognition Laboratory, Dept. of Computer Engineering & Informatics, University of Patras, 26500, Patras, Greece
[2] Technological Educational Institute of Kalamata, 24100, Kalamata, Greece
antonium@ceid.upatras.gr, sfg@teikal.gr, theofilk@ceid.upatras.gr, likothan@cti.gr

Abstract. In the current paper we present the application of our Gene Expression Programming Environment in forecasting Euro-United States Dollar exchange rate. Specifically, using the GEP Environment we tried to forecast the value of the exchange rate using its previous values. The data for the EURO-USD exchange rate are online available from the European Central Bank (ECB). The environment was developed using the JAVA programming language, and is an implementation of a variation of Gene Expression Programming. Gene Expression Programming (GEP) is a new evolutionary algorithm that evolves computer programs (they can take many forms: mathematical expressions, neural networks, decision trees, polynomial constructs, logical expressions, and so on). The computer programs of GEP, irrespective of their complexity, are all encoded in linear chromosomes. Then the linear chromosomes are expressed or translated into expression trees (branched structures). Thus, in GEP, the genotype (the linear chromosomes) and the phenotype (the expression trees) are different entities (both structurally and functionally). This is the main difference between GEP and classical tree based Genetic Programming techniques.

Keywords: Gene Expression Programming, Genetic Programming, Evolutionary Algorithms, System Modeling, time series, Euro-Dollar Exchange rate.

1 Introduction

The problem of discovering a mathematical expression that describes the operation of a physical or artificial system using empirically observed variables or measurements is a very common and important problem in many scientific areas. Usually, the observed data are noisy and sometimes missing. Also, it is very common, that there is no known mathematical way to express the relation using a formal mathematical way. These kinds of problems are called modeling problems, symbolic system identification problems, black box problems, or data mining problems [2].

H. Papadopoulos, A.S. Andreou, and M. Bramer (Eds.): AIAI 2010, IFIP AICT 339, pp. 78–85, 2010.

Most data-driven system modeling or system identification techniques assume an a-priori known model structure and focus mainly to the calculation of the model parameters' values. But what can be done when there is no a-priori knowledge about the model's structure?

Gene Expression Programming (GEP) is a domain-independent problem-solving technique in which computer programs are evolved to solve, or approximately solve, problems [3],[9]. GEP is a member of a broad family of techniques called Evolutionary Algorithms. All these techniques are based on the Darwinian principle of reproduction and survival of the fittest and are similar to the biological genetic operations such as crossover and mutation [9]. GEP addresses one of the central goals of computer science, namely automatic programming; which is to create, in an automated way, a computer program that enables a computer to solve a problem [1].

In GEP the evolution operates on a population of computer programs of varying sizes and shapes. GEP starts with an initial population of thousands or millions of randomly generated computer programs, composed of the available programmatic ingredients and then applies the principles of biological evolution to create a new (and often improved) population of programs. The generation of this new population is done in a domain-independent way using the Darwinian principle of survival of the fittest, an analogue of the naturally-occurring genetic operation of sexual recombination (crossover), and mutation. The fundamental difference between other Evolutionary Algorithms and GEP is that, in GEP there is a distinct discrimination between the genotype and the phenotype of an individual. So, in GEP the individuals are symbolic strings of fixed length representing an organism's genome (chromosomes/genotype), but these simple entities are encoded as non-linear entities of different sizes and shapes, determining an organism's fitness (expression trees/phenotype). GEP is a new evolutionary technique and its applications so far are quiet limited. However, it has been successfully applied in some real life problems [4], [5], [6].

In the current paper we present an integrated GEP environment with a graphical user interface (GUI), called jGEPModeling. The jGEPModeling environment was developed using the JAVA programming language, and is an implementation of the steady-state gene expression programming algorithm. In order to evaluate the performance of the jGEPModeling we tested it in the task of forecasting the next trading period of EUR/USD exchange rate based on the exchange rates of the past N periods, where N is a user-defined variable.

2 The Gene Expression Programming Algorithm

Gene Expression Programming is a new Evolutionary Algorithm proposed by Ferreira (2001) as an alternative method to overcome the drawbacks of Genetic Algorithms (GAs) and Genetic Programming (GP) [1], [3], [7], [9] . Similar to GA and GP, GEP follows the Darwinian principles of natural selection and survival of the fittest individual [8]. The fundamental difference between the three algorithms is that, in GEP there is a distinct discrimination between the genotype and the phenotype of an individual. This difference resides in the nature of the individuals, namely in the way the individuals are represented: in GAs the individuals are symbolic strings of fixed length (chromosomes); in GP the individuals are non-linear entities of different sizes and shapes (parse trees); and in GEP the individuals are also symbolic strings of fixed length representing an organism's genome (chromosomes/genotype), but these simple entities are encoded as non-linear entities of different sizes and shapes, determining an organism's fitness (expression trees/phenotype) [3].

GEP chromosomes are usually composed of more than one gene of equal length. Each gene is composed of a head and a tail. The head contains symbols that represent both functions and terminals, whereas the tail contains only terminals. The set of functions usually includes any mathematical or Boolean function that the user believes is appropriate to solve the problem. The set of terminals is composed of the constants and the independent variables of the problem. The head length (denoted h) is chosen by the user, whereas the tail length (denoted t) is evaluated by:

$$t = (n-1)h+1, \tag{1}$$

where n is the number of arguments of the function with most arguments. Despite its fixed length, each gene has the potential to code for Expression Trees (ETs) of different sizes and shapes, being the simplest composed of only one node (when the first element of a gene is a terminal) and the largest composed of as many nodes as the length of the gene (when all the elements of the head are functions with maximum arity). One of the advantages of GEP is that the chromosomes will always produce valid expression trees, regardless of modification, and this means that no time needs to be spent on rejecting invalid organisms, as in case of GP [9].

In GEP, each gene encodes an ET. In the case of multigenic chromosomes, each gene codes for a sub-ET and the sub-ETs interact with one another using a linking function (any mathematical or Boolean function with more than one argument) in order to fully express the individual. Every gene has a coding region known as an Open Reading Frame (ORF) that, after being decoded, is expressed as an ET, representing a candidate solution for the problem. While the start point of the ORF is always the first position of the gene, the termination point does not always coincide with the last position of a gene.

The flowchart of the GEP algorithm is shown in Figure 1. The process begins with the random generation of the linear chromosomes (or individuals) of the initial population. Then the chromosomes are expressed as ETs and the fitness of each individual is evaluated. After that, the individuals are selected according to their fitness in order to be modified by genetic operators and reproduce the new population. The individuals of this new population are, in their turn, subjected to the same developmental process: expression of the chromosomes, evaluation, selection according to fitness and reproduction with modification. The process is repeated for a certain number of generations or until a good solution has been found.

Next, follows a full description of the algorithm's steps:

1. Creation of initial population: The initialization in GEP is a very trivial task· in fact is the random creation of the chromosomal structure of the individuals. According to the nature of the problem, we must choose the symbols used to create the chromosomes, that is, the set of functions and terminals we believe to be appropriate to solve the problem. We must also choose the length of each gene, the number of genes per chromosome and how the products of their expression interact with one another.

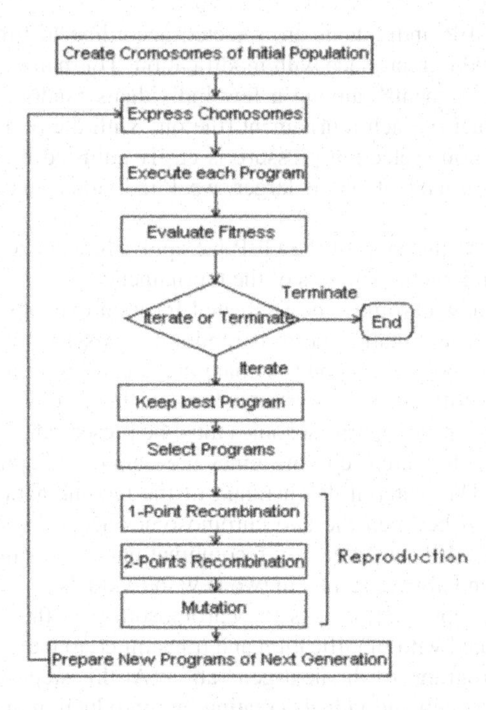

Fig. 1. Flowchart of gene expression algorithm

2. Express chromosomes: The second step is the expression of the chromosome of each individual as an ET. This process is also very simple and straightforward. For the complete expression, the rules governing the spatial distribution of functions and terminals must be followed. First, the start of a gene corresponds to the root of the ET, forming this node the first line of the ET. Second, depending on the number of arguments of each element (functions may have a different number of arguments, whereas terminals have an arity of zero), in the next line are placed as many nodes as there are arguments to the elements in the previous line. Third, from left to right, the new nodes are filled, in the same order, with the elements of the gene. This process is repeated until a line containing only terminals is formed.

3. Execute each program: Having expressed each individual to an ET is now easy to find and compute the mathematical (or Boolean) expression it codes. We implement this by a post order traversal on the ET.

4. Evaluate fitness: One crucial step in GEP is finding a function that performs well for all fitness cases within a certain error of the correct value. In the design of the fitness function, the goal must be clearly and correctly defined in order to make the system evolve in the intended direction.

5. Keep best Program: A feature that plays a significant role in GEP is the elitism. Elitism is the cloning of the best chromosome(s)/individual(s) to the next population (also called generation). By elitism, we guarantee that at least one descendant will be viable in the next generation, keeping at the same time the best trait during the process of adaptation.

6. Selection: In GEP, individuals are selected according to fitness by the tournament selection method to reproduce with modification. Tournament selection involves running several "tournaments" among a few individuals randomly chosen from the population. The winner of each tournament (the one with the best fitness) is selected for genetic modification. Selection pressure is easily adjusted by changing the tournament size. If the tournament size is larger, weak individuals have a smaller chance to be selected.

7. Reproduction: At this step of the GEP we apply the genetic operators of mutation and recombination on the winners of the tournaments.

a. Mutation: one new chromosome is created from an existing one, by mutating a single symbol. The mutation point is randomly chosen within the chromosome and then the symbol at this point is changing according to a mutation probability. In GEP, mutations can occur anywhere in the chromosome. However, the structural organization of chromosomes must be preserved.
b. Recombination: the parent chromosomes are paired and split up at exactly the same point(s). The material downstream of the recombination point(s) is afterwards exchanged between the two chromosomes according to a recombination probability. In GEP, an event of recombination always involves two parent chromosomes and always results in two new individuals.

Note that, during reproduction it is the chromosomes of the individuals, not the ETs that are reproduced with modification and transmitted to the next generation.

8. Prepare new programs of the next generation: At this step, we replace the tournament losers with the new individuals creating by reproduction in the population.

9. Termination criterion: We check if the termination criterion is fulfilled, if it is not we return to step 2. As a termination criterion it was used the maximum number of 500.000 generations that GEP was left to run.

10. Results: As a result we return the best individual ever found during the evolution process.

3 Modeling Experiments

In this section we present the performance of GEP environment in a real series forecasting problem. Particularly, we apply the GEP environment to an one- day- ahead forecasting and trading task of the Euro/Dollar (EUR/USD) exchange using the Euro Central fixing series.

The European Central Bank (ECB) publishes a daily fixing for selected EUR exchange rates: these reference mid-rates are based on a daily concertation procedure between central banks within and outside the European System of Central Banks, which normally takes place at 2.15 p.m. ECB time. The reference exchange rates are published both by electronic market information providers and on the ECB's website shortly after the concertation procedure has been completed. Although only a reference rate, many financial institutions are ready to trade at the EUR fixing and it is therefore possible to leave orders with a bank for business to be transacted at this level.

The ECB daily fixing of the EUR/USD is therefore a tradable level which makes using it a realistic approach.

Table 1. Dataset description

Name of period	Trading days	Beginning	End
Total dataset	2721	6 January 1999	23 October 2009
Training dataset	2178	6 January 1999	30 August 2007
Out-of-sample dataset [Test set]	543	31 August 2007	23 October 2009

Based on the exchange rates of the past N periods, where N is a user-defined variable, we forecast the next trading period EUR/USD exchange rate. The difficulty in this modeling problem is to identifυ and the capturε all the discontinuous, the nonlinearities and the high frequency multipolynomial components which characterize the financial series today.

After a certain number of experiments, we are in position to define standard values for some running parameters of GEP. Table 2 defines all these defaults parameters we use for our experiments.

Table 2. Default parameters derived by experimentation

Parameter	Value
Number of Generations	300.000
Function Set	{+, -, *, /, ^, $\sqrt{}$,abs, cos, sin, ln, exp, tan, min, max}
Constants Range	[-3, 3]
Mutation Probability	0.9
Population Size	1500
Tournament Size	4
Head Size	30
Type of Recombination	two points recombination

Next, we present our experiments with various values of past exchange rates used for our prediction.

Table 3. Results taken by experimenting with models from GEP which use various past exchange rates for prediction

Past Exchange Rates	Mean Square Train Error	Mean Square Test Error
1	3.528493651301604e-005	7.038129152437989e-005
2	3.525297196338534e-005	7.002301886682884e-005
3	3.529949324406443e-005	6.967481960607426e-005
4	3.510071391816327e-005	6.979235421769149e-005

We observed that using only three past values of past exchange rates for the prediction models, gives as the best performance. Increasing the number of previous values to be used as inputs did not enhanced our models and this reveals short term phenomena in the time series of EURO/USD exchange rate. The best model found has mean square test error 6.959490069577218E-5.

For comparison reasons, we experimented using our dataset with 3 other methods that have been previously used in this forecasting problem [11], [12]. These methods are the naïve method, the moving average predictor and Artificial Neural Networks. Specifically:

- Naïve strategy assumes that the most recent period change is the best predictor of the future [11], [12].
- Moving Average method uses a weighted average of past observations to smooth short-term fluctuations. In this approach 35 previous values were used as in [11], [12].
- Neural Network models are universal approximators capable of approximating any continuous function [13]. In this approach we used a feed-forward Multi-Layer Perceptron(MLP) using 4 previous values as inputs in order to make a fair comparison with our method. For the neural network training we used back propagation algorithm.

The performance of each model in the out of sample period is presented in Table 4. Because of the stochastic nature of the MLP method we experimented 100 times with it and we show the mean results. We observe that gene expression programming outperforms the three classic methods in our forecasting problem.

Table 4. Experimental results of three benchmark methods in the out of sample period

Method	Mean Square Test Error
Naïve strategy	1.3475e-004
Moving Average	7.5000e-005
MLP	7.4551e-005

4 Conclusions

The reported results taken by the application of our GEP implementation in the task of forecasting the next trading period of EUR/USD exchange rate based on the exchange rates of the past N periods, confirm our intuition that Gene Expression Programming is a very effective technique in system modeling and timeseries prediction. So, we conclude that our GEP environment can be used in a variety of problems in different scientific areas. Also, it is important to note that the rapid development of very fast computer hardware encourages the use of techniques like GEP, and is expected that these kinds of techniques will be used more and more for solving difficult problems in the future.

Now, concerning some future directions of the presented GEP environment these could be:

- The application of our GEP implementation in other system modeling tasks in various scientific areas, such as system identification, timeseries prediction, e.t.c..
- The further improvement of the tool in terms of speed, memory management and parallel implementation.

- The implementation of more sophisticated genetic operators, initialization techniques and selection techniques in the basic Gene Expression Programming technique.

Thus, our environment has been proved to be a powerful GEP tool with great expansion capabilities that we intend to perform in the near future.

References

1. Koza, J.R.: Genetic programming: on the programming of computers by means of natural selection. MIT Press, Cambridge (1992)
2. Winkler, S., Affenzeller, M., Wagner, S.: Identifying Nonlinear Model Structures Using Genetic Programming Techniques. In: Cybernetics and Systems 2004, pp. 689–694. Austrian Society for Cybernetic Studies (2004)
3. Ferreira, C.: Gene Expression Programming: A New Adaptive Algorithm for Solving Problems. Complex Systems 13(2), 87–129 (2001)
4. Lopez, H.S., Weinert, W.R.: EGIPSYS: An Enhanced Gene Expression Programming Approach for Symbolic Regression Problems. International Journal of Applied Mathematics in Computer Science 14(3), 375–384 (2004)
5. Margny, M.H., El-Semman, I.E.: Extracting Logical Classification Rules with Gene Expression Programming: Micro array case study. In: AIML 2005 Conference, Cairo, Egypt, December 19-21 (2005)
6. Dehuri, S., Cho, S.B.: Multi-Objective Classification Rule Mining Using Gene Expression Programming. In: Third International Conference on Convergence and Hybrid Information Technology (2008)
7. Goldberg, D.E.: Genetic Algorithms in Search, Optimization, and Machine Learning. Addison-Wesley, Reading (1989)
8. Darwin, C.: On the Origin of Species (1859)
9. Ferreira, C.: Gene Expression Programming: Mathematical Modeling by an Artificial Intelligence, 2nd edn. Springer, Heidelberg (2006)
10. Mitchell, M.: An Introduction to Genetic Algorithms. MIT Press, Cambridge (1996)
11. Dunis, C., Williams, M.: Modelling and Trading the EURO/USD Exchange Rate: Do Neural Network Models Perform Better?, Derivatives Use, Tradingand Regulation (2002)
12. Dunis, C., Laws, J., Sermpinis, G.: Modelling and trading the EUR/USD exchange rate at the ECB fixing. The European Journal of Finance (2009)
13. Haykin, S.: Neural Networks: A comprehensive foundation, 2nd edn. Prentice Hall, Upper Saddle River (1999)

Automatically Designing Robot Controllers and Sensor Morphology with Genetic Programming

Bert Bonte* and Bart Wyns

Ghent University, Deptartment of Electrical Energy, Systems and Automation,
Technologiepark 913, 9052 Zwijnaarde, Belgium
Bart.Wyns@UGent.be, Bert.Bonte@gmail.com
http://www.UGent.be

Abstract. Genetic programming provides an automated design strategy to evolve complex controllers based on evolution in nature. In this contribution we use genetic programming to automatically evolve efficient robot controllers for a corridor following task. Based on tests executed in a simulation environment we show that very robust and efficient controllers can be obtained. Also, we stress that it is important to provide sufficiently diverse fitness cases, offering a sound basis for learning more complex behaviour. The evolved controller is successfully applied to real environments as well. Finally, controller and sensor morphology are co-evolved, clearly resulting in an improved sensor configuration.

Keywords: Genetic Programming, Evolutionary Robotics, Corridor Following, Sensor Morphology, EyeBot.

1 Introduction

Robots are mainly used for relatively easy and repetitive tasks. Fully autonomous robots in realistic applications still are exceptions [16]. This is mainly caused by the highly complex design of such systems. More specifically, the development of robust controllers for real mobile robots is challenging.

In nature however, very powerful processes that are able to combine basic elements in a robust way resulting in very efficient and sometimes ingenious solutions exist. Using the concept of survival of the fittest, robot controllers and robot systems can be evolved just like organisms in nature. In this way, using natural processes, robust robot controllers can be automatically designed without complexity burden for human developers and even without programming.

The main objective of this contribution is developing robust controllers for corridor following using genetic programming [7]. The robot must navigate in a corridor system from start to end as efficiently as possible and without collisions. Unlike some other work in literature [17], we will not use a highly simplified simulation environment. The evolved mobile robot controllers must be robust enough to navigate successfully in corridor systems on which the robot was

* Presenting author.

H. Papadopoulos, A.S. Andreou, and M. Bramer (Eds.): AIAI 2010, IFIP AICT 339, pp. 86–93, 2010.

trained during the evolutionary process as well as new and unseen environments. Furthermore, the controller evolved in simulation must be transferable to the real robot preserving its behaviour learned in simulation. Finally, we also assess simultaneous evolution of controllers and sensor morphology using GP.

The remainder of this paper is as follows. Section 2 gives a brief overview of selected related work in literature. Then, Section 3 provides an introduction to GP, the standard parameters and the most important techniques used in the experiments. After that, in Section 4 the experimental setup is set out, including the simulation environment and the robot platform. Next, Section 5 discusses the evolution of robot controllers in simulation and the transfer to reality. Finally, Section 6 handles the combined evolution of controller and morphology.

2 Literature Overview

Evolutionary robotics in general [12] is a relative young discipline within the robotics research community, providing room for further improvement and advances. In this section we will give a brief literature overview closely related to the goals of this paper.

Most of the experiments are conducted in simplified simulation environments. The resulting controllers are limited to specific simulation parameters and conditions. Therefore, a lot of work in literature is devoted on the use of simulation environments or real robots [1,4,6,11,12,17,18]. Only using simple simulation environments without verifying on real robots afterwards is criticized a number of times [11,12]. Indeed, there are significant differences between controllers performing well in simulation versus reality.

An example of a basic simulation environment is found in [8]. GP is used for evolving wall following-behaviour, which is part of many higher level robot skills. Successful programs were evolved, clearly exhibiting the desired behaviour. The simulation consisted of a grid environment were the "robot" is on one of the cells. The robot can directly move to neighbouring cells. Sensors simply indicate whether a certain cell contains an obstacle or not. Similar experiments provide some basic proof for using GP in robotics but their practical use is questionable.

In [17] a simplified but noisy simulation environment for evolving corridor following behaviour with steady-state GP was used. In this simulation, the robot moves forward with a constant speed, the controller only determines the direction. Moreover, the corridors are just small enough to let the robot pass, so turning in the wrong direction is impossible. Mostly short corridor systems were used. By adding noise to sensors and actuators Reynolds evolved more or less robust controllers.

In [3], a vision-based line-following controller was evolved in simulation by incrementally improving the visual model in the simulation. In simulation, successfull controllers using mostly basic functions were evolved. Mainly caused by some oversimplifications in the simulator, the authors were not able to successfully transfer this behaviour to the physical robot.

Some experiments were conducted directly on real robots. An example is in [14], where an obstacle avoidance controller for the *Khepera*-robot is evolved, again using steady-state GP. The population size was only 50 individuals. Two different training environments were tested. Using simple operations, in 200-300 generations the desired behaviour was achieved in both environments. The resulting program was robust, as it was successful in other environments as well.

3 Genetic Programming

Genetic programming uses a population of computer programs encoding a robot controller and was introduced by John Koza in the early nineties [7]. Each generation, only the best performing programs are selected for breading and have their offspring added to the population of the next generation. The selection of the best programs is based on the fitness function, which is defined to model how well the desired behaviour is matched. A detailed explanation on the fitness function used in this study is given in Section 3.2.

3.1 Standard Parameters

GP requires various parameters to be set. Our GP uses a population size equal to 500 and is allowed to run for 100 or 200 generations. Crossover (90%) and reproduction (10%) are used. As selection algorithm, tournament selection with seven individuals is used. The function set contains two functions: *IfLess* (arity 4), and *PROGN2* (arity 2), both well known. Terminals to move forward and backward over a distance of 10 cm are also included. Turn left and right makes the robot turn 15 degrees in place. Three infrared distance sensors are used: front, left and right, each perpendicular on the robot. Finally, three threshold values are available: low, medium and high, respectively 75, 150 and 300 mm.

3.2 Fitness Function

At each generation all individuals are tested in three environments and can perform 500 movements in each environment. The fitness function is averaged over all three tested environments and consists of three components:

- As a first component, the distance in bird's eye perspective the robot has covered so far is measured. This distance can be easily calculated by applying the rule of Pythagoras to the coordinates of the starting and ending position of the robot. This basic component mainly differentiates between controllers in initial generations.
- The second component punishes every collision detected using either sensor on each side of the robot. A collision occurs when the robot's sensor return a reading smaller than 50mm. The penalty consists of substracting a fixed value (3) to the number of movements so far.

- Finally, a bonus component is added when the robot reaches the end of the corridor system (robot's position equals top right corner of the simulation environment). This consists of a fixed part (3) and a variable part. The variable part is relative to the number of spare movements and thus rewards efficient controllers. The scalar used in this study equals to 4. This component mainly differentiates between better controllers in later generations.

While evaluating fitness on a single environment most likely leads to brittle strategies, averaging over multiple fitness cases results in more general solutions. Here, three different enviroments are used for evaluation. To select these three environments, we use a variant of the layered learning approach [5]. Two environments belong to one category, while the third belongs to a more difficult category. As the number of generations increases, the overall difficulty level of these three fitness cases also increases. However, changing the composition of the set of fitness cases at each generation has detrimental effects on the evolved controller quality. Therefore, some time (i.e. generations) is left for the evolutionary process to optimize the individuals towards the presented set of environments.

4 Experimental Setup

We use the *EyeBot*-platform for our experiments [1]. The evolutionary process is carried out entirely in the EyeSim, the simulation environment of the platform. Main advantage is that programs for the EyeSim can be transferred immediately for execution on the real EyeBot. Hence we realise a major speedup and cost reduction by evolving the controllers in simulation. To gain realism, we add our own additive noise component (Gaussian distributed noise), which can be manipulated easier. The standard deviation we use is 3 for sensor noise and 2 for motor noise. Combined with this noise, the simulation environment facilitates the evolution of controllers for real applications, which is impracticable in simplified and naive simulation environments.

The GP process was handled by ECJ [1], a research platform supporting multiple evolutionary algorithms in Java. ECJ constructs controllers of which the fitness is evaluated in the EyeSim and returned back to ECJ, which performs all evolutionary computations.

Since GP is a probabilistic method, we consider three different runs of each experiment. This relatively low number is justified because we are interested in the best controller, not some mean value. Moreover, when considering too many runs, the computational cost becomes too high.

5 Evolving Robot Controllers

As stated before in literature, noise can improve simulation results and lead to more robust controllers. Main argument is that with noise, evolution is no

[1] http://www.cs.gmu.edu/ eclab/projects/ecj/

longer able to exploit coincidences in fitness cases only valid in simulation and therefore leads to more robust controllers. Next to the noisy sensor values, noise is added to the steering mechanism as well. Two experiments with a population of 500 individuals were run, resulting in very high mean fitness values. The first one, using 100 generations lead to an extremely robust solution, performing well on five verification environments (not seen during training). Significant further improvement is possible since the entropy of 2.08 is extremely high. However, the second experiment using 200 generations, was slightly more efficient and fluent in turns. Though, this controller was not as robust as the first, presumably caused by premature convergence, in turn caused by too long training on preliminary categories. An example trail of both controllers is depicted in Figure 1. Further real life experiments will only use the first controller.

An interesting remark is that the underlying strategy of nearly all successful controllers of the experiments is wall following. Mostly, the left wall is followed, navigating to the end of the corridor system without turning and proceeding in the wrong direction. This general strategy is a nice example of evolutionary processes to come up with simple yet general and efficient solutions.

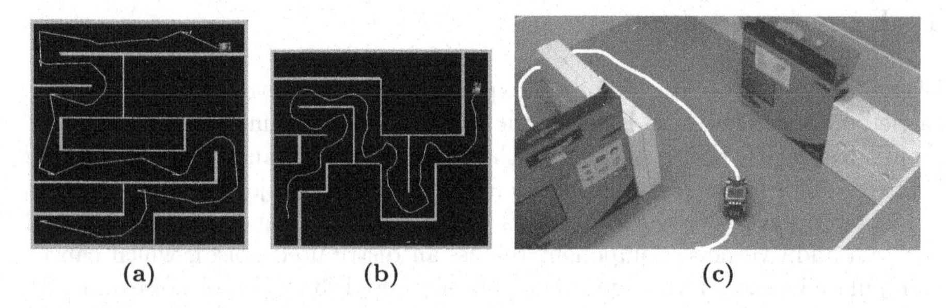

Fig. 1. Some example navigation trails. (a) The robot trail from the controller evolved in 100 generations. (b) The robot trail from the controller evolved in 200 generations. (c) The EyeBot with controller from (a) in a test environment. The white line denotes the robot trajectory.

When transferring the best controller to reality, performance slightly decreased. This was mainly caused by the fact that the real PSD sensors return increasing values when approaching a wall from a certain distance. After increasing the lowest threshold from 75 to 100 (the distance under which the sensor values become unreliable), this problem was solved. The robot was able to navigate efficiently through previously unseen environments. The robot successfully drives straight ahead, adjusts where necessary and most curves are taken smoothly. Nevertheless we noted slightly more collisions than in simulation. Figure 1(c) shows the real robot in a test environment. Remark that the left wall following is illustrated by omitting some right walls, yet resulting in a successful navigation.

6 Simultaneous Evolution of Controller and Sensor Morphology

6.1 Situation

In nature, both the controller and the body of an organism are evolved simultaneously. The analogy can be made to evolve controllers (software) and robot morphology (hardware), hopefully resulting in better suited robots and controllers for a given task. In this work, we will co-evolve the controller and sensor morphology [2,17,9,10]. Again in simulation and with noise added to the sensor values. The standard simulation environment was extended to support the use of an arbitrary number of sensors.

Because sensors are the main source of information on many robots, investigating their placement and quantity is an interesting task. Moreover, sensors can be moved easily on existing robots. However, the most important reason for examining sensor placement was found in preliminary experiments. Some – otherwise robust – controllers struggled with the detection of vertical walls. Since the frontal PSD-sensor measured parallel with the vertical wall, the latter was not detected and the robot repeatedly caused collisions. Therefore, the controller can be made more robust and general by better configuring sensor placement. Evolving sensor morphology can thus be advantageous.

6.2 Experimental Results

An experiment with 8 possible sensor locations, each addressed by a unique terminal leads to both a good controller and sensor configuration. In order to reach a minimal sensor suite, the fitness function awarded every successful controller according to the number of unused sensors. Note that the evolutionary process didn't reduce the number of sensors to evolve a simpler and more usable sensor morphology without this stimulus. The evolutionary process clearly omits the sensors one by one during evolution, starting with those that are least useful for the task. This is illustrated in Figure 2. In this case, these are the sensors that are placed orthogonal on the robot. Indeed, they only offer limited coverage (e.g. walls parallel with the sensor placement). Therefore the evolutionary process selects almost exclusively non perpendicular sensors, especially at the front of the robot. This is an important design issue that was not taken in account when designing the real robot, which only has perpendicular sensors.

A similar experiment with 35 sensors resulted in a good sensor configuration as well, again with non perpendicular sensors on the front. However, the resulting controller was not robust enough. Here we reached a complexity barrier for the evolutionary process; too many terminals lead to an explosion of the search space resulting in a problem that is too hard to solve with a standard general purpose PC in an acceptable time span. This is a concern that is shared by other authors in the field as well [15].

Another approach with one generic function for all possible sensor positions accepting three parameters (x and y position and angle on the robot) defining

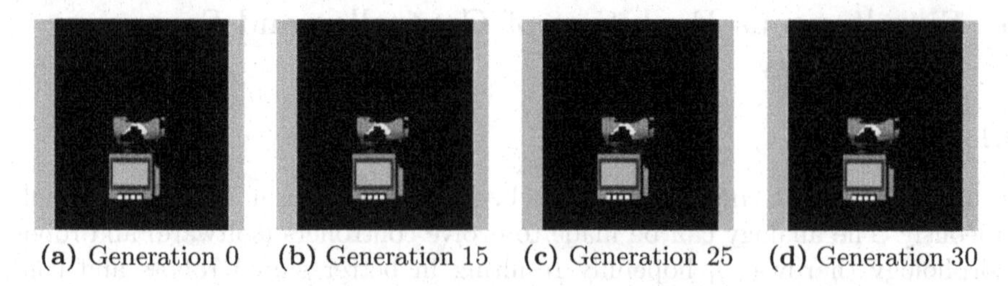

(a) Generation 0 (b) Generation 15 (c) Generation 25 (d) Generation 30

Fig. 2. The sensor usage during the evolutionary process. Sensors with limited ranges in terms of the measured surface are omitted during initial generations.

the sensor placement was investigated as well. Again we found that the fitness function must reward the limited use of sensors, otherwise too many sensors are used. This results in unnecessary complex controllers. However, more research is needed for quantifying the difference in result and computation time between a number of strictly defined terminals or a more generic function which is more flexible.

7 Conclusions

We demonstrated that, even with a basic PC and limited computation time, GP is able to evolve controllers for corridor following in a simulation environment by using a gradual form of layered learning. Moreover, this controller was transferred successfully to reality. Secondly, we also co-evolved controller and sensor morphology of the robot. This resulted in a significantly improved sensor configuration.

References

1. Bräunl, T.: Embedded Robotics: Mobile Robot Design and Applications With Embedded Systems, 2nd edn. Springer, Heidelberg (2006)
2. Buason, G., Bergfeldt, N., Ziemke, T.: Brains, bodies, and beyond: Competitive co-evolution of robot controllers, morphologies and environments. Genetic Programming and Evolvable Machines 6(1), 25–51 (2005)
3. Dupuis, J., Parizeau, M.: Evolving a Vision-Based Line-Following Robot Controller. In: Proceedings of the the 3rd Canadian Conference on Computer and Robot Vision, p. 75. IEEE Computer Society, Washington (2006)
4. Goosen, T., Brule, R., Janssen, J., Haselager, W.: Interleaving Simulated and Physical Environments Improves Evolution of Robot Control Structures. In: Dastani, M., de Jong, E. (eds.) BNAIC 2007: Proceedings of the 19th Belgium-Netherlands Artificial Intelligence Conference, November 5-6. BNAIC, vol. 19, pp. 135–142. Utrecht University, Utrecht (2007)
5. Gustafson, S., Hsu, W.: Layered Learning in Genetic Programming for a Cooperative Robot Soccer Problem. In: Miller, J., Tomassini, M., Lanzi, P.L., Ryan, C., Tetamanzi, A.G.B., Langdon, W.B. (eds.) EuroGP 2001. LNCS, vol. 2038, pp. 291–301. Springer, Heidelberg (2001)

6. Jakobi, N., Husbands, P., Harvey, I.: Noise and the reality gap: The use of simulation in evolutionary robotics. In: Morán, F., Merelo, J.J., Moreno, A., Chacon, P. (eds.) ECAL 1995. LNCS, vol. 929, pp. 704–720. Springer, Heidelberg (1995)
7. Koza, J.: Genetic Programming: On the programming of computers by Means of natural selection. MIT Press, Cambridge (1992)
8. Lazarus, C., Hu, H.: Using Genetic Programming to Evolve Robot Behaviours. In: Proceedings of the 3rd British Conference on Autonomous Mobile Robotics & Autonomous Systems (2001)
9. Lee, W.: Evolving Autonomous Robot: From Controller to Morphology. IEICE Transactions on Information and Systems 83(2), 200–210 (2000)
10. Lund, H., Hallam, J., Lee, W.: Evolving robot morphology. In: IEEE International Conference on Evolutionary Computation, pp. 197–202 (1997)
11. Nelson, A., Grant, E., Galeotti, J., Rhody, S.: Maze exploration behaviors using an integrated evolutionary robotics environment. Robotics and Autonomous Systems 46(3), 159–173 (2004)
12. Nolfi, S., Floreano, D.: Evolutionary Robotics: The Biology, Intelligence, and Technology of Self-Organizing Machines. MIT Press, Cambridge (2000)
13. Nordin, P., Banzhaf, W.: Complexity compression and evolution. In: Eshelman, L. (ed.) Proceedings of the Sixth International Conference on Genetic Algorithms, pp. 240–245. Morgan Kauffman, San Francisco (1995a)
14. Nordin, P., Banzhaf, W.: Genetic programming controlling a miniature robot. In: Working Notes for the AAAI Symposium on Genetic Programming, pp. 61–67 (1995b)
15. Ok, S., Miyashita, K., Nishihara, S.: Improving performance of gp by adaptive terminal selection. In: Mizoguchi, R., Slaney, J.K. (eds.) PRICAI 2000. LNCS, vol. 1886, pp. 435–445. Springer, Heidelberg (2000)
16. Pollack, J., Lipson, H., Ficici, S., Funes, P., Hornby, G., Watson, R.: Evolutionary techniques in physical robotics. In: Miller, J.F., Thompson, A., Thompson, P., Fogarty, T.C. (eds.) ICES 2000. LNCS, vol. 1801, pp. 175–186. Springer, Heidelberg (2000)
17. Reynolds, C.: Evolution of corridor following behavior in a noisy world. In: Cliff, D., Husbands, P., Meyer, J.-A., Wilson, S. (eds.) From Animals to Animats 3: Proceedings of the third International Conference on Simulation of Adaptive Behavior, pp. 402–410. MIT Press, Cambridge (1994)
18. Urzelai, J., Floreano, D.: Evolutionary Robotics: Coping with Environmental Change. In: Proceedings of the Genetic and Evolutionary Computation Conference (2000)

Multiple Criteria Performance Analysis of Non-dominated Sets Obtained by Multi-objective Evolutionary Algorithms for Optimisation

Gerrit K. Janssens[1] and José Maria Pangilinan[2]

[1] Hasselt University, Transportation Research Institute (IMOB),
Wetenschapspark Building 5
B-3590 Diepenbeek, Belgium
gerrit.janssens@uhasselt.be
[2] Saint-Louis University, Dept. of Computer Science
Baguio, Philippines
joey.pangilinan@slu.edu.ph

Abstract. The paper shows the importance of a multi-criteria performance analysis in evaluating the quality of non-dominated sets. The sets are generated by the use of evolutionary algorithms, more specifically through SPEA2 or NSGA-II. Problem examples from different problem domains are analyzed on four criteria of quality. These four criteria namely *cardinality* of the non-dominated set, *spread* of the solutions, *hyper-volume*, and *set coverage* do not favour any algorithm along the problem examples. In the Multiple Shortest Path Problem (MSPP) examples, the *spread* of solutions is the decisive factor for the 2S|1M configuration, and the *cardinality* and *set coverage* for the 3S configuration. The differences in *set coverage* values between SPEA2 and NSGA-II in the MSPP are small since both algorithms have almost identical non-dominated solutions. In the Decision Tree examples, the decisive factors are *set coverage* and *hyper-volume*. The computations show that the decisive criterion or criteria vary in all examples except for the *set coverage* criterion. This shows the importance of a binary measure in evaluating the quality of non-dominated sets, as the measure itself tests for dominance. The various criteria are confronted by means of a multi-criteria decision tool.

Keywords: evolutionary algorithms, multi-objective optimization, multi-criteria analysis.

1 Introduction

Many real-world optimization problems are multi-objective by nature and with objectives that are in conflict. Mathematical techniques are available to find best-compromise solutions by aggregating multiple objectives into a single function [9]. They have their drawbacks as they have difficulty dealing with concave and discontinuous Pareto fronts. Stochastic local or global search algorithms or population-based algorithms are often used when exact methods are infeasible to be applied and are

H. Papadopoulos, A.S. Andreou, and M. Bramer (Eds.): AIAI 2010, IFIP AICT 339, pp. 94–103, 2010.

used to solve difficult optimization problems. Among the many types of these algorithms, Evolutionary Algorithms (EA) seem particularly suitable to solve multi-objective optimization problems. An increasing number of research papers report comparative findings of several evolutionary algorithms in terms of computing speed and Pareto optimality as tested on various multi-objective problem instances or applications with known Pareto sets. In most practical or experimental cases however the Pareto sets are unknown.

The objective of the study is to describe the performance of evolutionary algorithms in terms of stability, computational complexity, diversity and optimality of solutions in different multi-objective optimization problems (MOOP). Multi-objective evolutionary algorithm (MOEA) experiments generate a variety of non-dominated sets in each problem domain. Comparisons of the quality of the non-dominated sets have not yet been presented in each of the cases. The stability of an algorithm is concerned with the sensitivity of the results to changes in the MOEA parameters settings. Computational complexity refers to the solution run-time complexity in terms of the size of the problem. Diversity measures the spread of solutions in the non-dominated set in order to provide the decision maker a true picture of trade-off solutions. Optimality measures the proximity of the best non-dominated set to the Pareto-optimal set.

Bosman and Thierens [1] argue that the quest for finding the components that result in the best EAs for multi-objective optimization is not likely to converge to a single, specific MOEA. They stated that the trade-off between the goals of proximity and diversity preservation plays an important role in the exploitation and exploration phases of any MOEA.

A comprehensive discussion of multi-objective evolutionary algorithms (MOEA) can be found in [7]. In addition, Coello [6] gives a summary of current approaches in MOEA and emphasizes the importance of new approaches in exploiting the capabilities of evolutionary algorithms in multi-objective optimization. Zitzler and Thiele [17] performed a comparative analysis of existing evolutionary algorithms in multi-objective optimization by means of well-defined quantitative performance measures. In this research a comparison is made using two well-known techniques: SPEA2 and NSGA-II.

Zitzler et al. [18] introduced the Strength Pareto Evolutionary Algorithm 2 (SPEA2), which is an extension and improvement of the original work by Zitzler and Thiele [17]. SPEA2 integrates a fitness assignment strategy, which considers the number of individuals that an individual dominates and the number of its dominators. It uses a nearest-neighbor density estimation technique that guides the search more efficiently and avoids the formation of new solutions in only a few clusters in the search space. SPEA2 has a truncation procedure that preserves the best solutions when the number of non-dominated individuals exceeds the external population size.

Deb et al. [8] introduced an elitist non-dominated genetic algorithm (NSGA-II) that uses not only an elite-preserving strategy but also an explicit-diversity preserving mechanism. Initially, NSGA-II creates a random parent population, sorts the population based on non-domination and assigns each solution a fitness value equal to its non-domination level.

Consider first a case where an optimal algorithm exists to obtain the Pareto set. For example, in the Competitive Facility Location Problem (CFLP) [11], the quality of the non-dominated sets generated by the MOEA can be calculated and compared to the Pareto-optimal set generated by an algorithm by Carrizosa and Plastria [5]. The *error ratio* metric, which measures the closeness of the non-dominated set to the Pareto front in terms of set membership, can be used to measure the quality of solutions in the CFLP. However, if such an algorithm does not exist or is not available, like for two other problems under study here, the Multi-Objective Shortest Path Problem (MSPP) and the Decision Tree (DT) experiments, only approximations to their Pareto-optimal sets are available for performance analysis.

The next sections present some performance metrics that are useful in measuring the quality of non-dominated sets when the Pareto-optimal set is unknown, and utilizes a multi-criteria tool to determine the best non-dominated set for MSPP, and DT problems based on the performance metrics. Several performance metrics exist in literature, and several comparative studies have been conducted that evaluate them.

The studies presented above show a variety of results and no single MOEA performs better in the different performance metrics but most studies compare their algorithms with either NSGA-II or SPEA2 or both. The studies above mostly evaluate the performance of the selection operators of each MOEA without investigating the effect of the parameter settings on its performance. The current study investigates the performance of NSGA-II and SPEA2 in selected multi-objective optimization problems by means of a multi-criteria method in which various performance metrics of the Pareto front are presented to the decision-maker.

2 Performance Metrics

Deb [7] states that there are two orthogonal goals for any multi-objective optimisation algorithm: (1) to identify solutions as close as possible to the true Pareto-optimal set and (2) to identify a diverse sets of solutions distributed evenly across the entire Pareto-optimal surface. This has led to several metrics that characterise either closeness, or diversity, or both. Examples of metrics which measure the closeness to the Pareto surface are the Error ratio [14] and the Set coverage [16]. Examples of metrics which measure the diversity across the Pareto surface are the Spacing [13] and the Spread [8]. A measure like the Hypervolume measures both closeness and diversity [17]. Most measures are unary quality measures, i.e. the measure assigns to each Pareto set approximation a number that reflects a certain quality aspect.

Hypervolume

The hypervolume metric [17] calculates the volume in the objective space covered by the members of the non-dominated set **Q**. For each solution $i \in \mathbf{Q}$, a hypercube $\mathbf{v}_i$ is computed from a reference point and the solution i as the diagonal corners of the hypercube, The reference point can be found by constructing a vector of worst objective function values. The hypervolume (HV) is calculated as:

$$HV = volume\left(\bigcup_{i=1}^{|Q|} v_i\right). \tag{1}$$

The hypervolume metric is interesting because it is a single metric which is sensitive both to the overall advancement of the non-dominated set and to the distribution of individual points across the set. The placement of the reference point is critical and determines the sense and the magnitude of the hypervolume. Problems may appear if objectives have dissimilar scales or if some objectives are unbounded.

Spacing
Schott [13] introduced a metric, which is a measure of the relative distances between consecutive solutions in the non-dominated set $\boldsymbol{Q}$ is calculated as:

$$S = \sqrt{\frac{\sum_{i=1}^{|Q|}\left(d_i - \bar{d}\right)}{|Q|}} \tag{2}$$

$$d_i = min_{k \in Q \wedge k \neq i} \sum_{m=1}^{M} \left| f_m^i - f_m^k \right| \tag{3}$$

$$\bar{d} = \frac{\sum_{i-1}^{|Q|} d_i}{|Q|} \tag{4}$$

Schott's metric measures the diversity of a non-dominated set.

Set coverage metric
This metric is based on Zitzler [16]. The metric computes the relative spread of solutions between two non-dominated sets $\boldsymbol{A}$ and $\boldsymbol{B}$. The set coverage metric $C(\boldsymbol{A}, \boldsymbol{B})$ calculates the proportion of solutions in $\boldsymbol{B}$ that are weakly dominated by solutions of $\boldsymbol{A}$:

$$C(\boldsymbol{A},\boldsymbol{B}) = \frac{\left|\left\{\boldsymbol{b} \in \boldsymbol{B} \,\middle|\, \exists \boldsymbol{a} \in \boldsymbol{A} : \boldsymbol{a} \preceq \boldsymbol{b}\right\}\right|}{|\boldsymbol{B}|} \tag{5}$$

The metric value $C(\boldsymbol{A}, \boldsymbol{B}) = 1$ means all members of $\boldsymbol{B}$ are weakly dominated by $\boldsymbol{A}$. On the other hand, $C(\boldsymbol{A}, \boldsymbol{B}) = 0$ means that no member of $\boldsymbol{B}$ is weakly dominated by $\boldsymbol{A}$. This operator is not symmetric, thus it is necessary to calculate $C(\boldsymbol{B}, \boldsymbol{A})$. The set coverage metric measures convergence based on the concept of dominance relations.

Cardinality
This metric counts the number of solutions in the non-dominated set. It measures neither diversity nor convergence.

In order not to limit the description of the quality of a non-dominated set by using only a single metric, a multi-criteria evaluation seems appropriate. Hence, the computations in this paper evaluate the quality of non-dominated sets according to four criteria, which are mentioned above. The multi-criteria tool employed in the computations is Decision Lab [15]. The Decision Lab software is a multi-criteria decision making software, which is based on the Preference Ranking Organization Method for Enrichment Evaluations (PROMETHEE) and the Graphical Analysis for Interactive Assistance (GAIA). The details of the PROMETHEE method are found in Brans et al. [2]. GAIA, which makes use of principal component analysis, is a descriptive complement to the PROMETHEE methods [3].

3 Computations and Results

Table 1 shows the computed values for each performance criterion in the different multi-objective optimization problems. There are four criteria. A smaller set cardinality is preferred, especially in the case of a continuous decision space. A spread that has smaller value means that the solutions on the non-dominated front are uniformly spaced therefore this criterion is minimized. Hypervolume and set coverage are maximized. Two non-dominated sets are compared in each MOOP. One set is generated by NSGA-II and the other by SPEA2. Decision Lab can rank more non-dominated sets but since the set coverage is a binary quality measure, only two non-dominated sets can be evaluated each time.

The Multi-objective Shortest Path Problem (MSPP) is an extension of the traditional shortest path problem and is concerned with finding a set of efficient paths with respect to two or more objectives that are usually in conflict. A variety of algorithms and methods such as dynamic programming, label selecting, label correcting, interactive methods, and approximation algorithms have been implemented and investigated with respect to the MSPP [10]. The problem is known to be NP-complete. For fitness and selection, two objective configurations are considered for finding efficient paths: (3-S) and (2-S|1-M). S-type objectives are sum problems that are to be minimised and M-type objectives are max-min problems that are to be minimised. A 50-node of 10% density is the basis for the computed values in both MSPP (Multi-objective Shortest Path Problem) configurations. The *hypervolume* values for the MSPP are blank since they cannot be computed. This reduces the number of MSPP criteria to three.

Table 1. Computed criteria values of non-dominated sets

Nondominated Set	Criterion			
	Cardinality (Minimize)	Spread (Minimize)	Hyper-volume (Maximize)	Set Coverage (Maximize)
MSPP 2S|1M				
NSGA-II	7	0.8		0
SPEA2	8	0.51		0
MSPP 3S				
NSGA-II	8	0.57		0
SPEA2	5	1.14		0.38
Housing dataset				
NSGA-II	5	0.43	5.47	0.17
SPEA2	6	0.38	6.98	0.6
Optical-digits				
NSGA-II	5	0.36	5.96	0
SPEA2	6	0.63	7.15	0.6

An evolutionary algorithm is a promising technique to build oblique decision trees (see [4] for a list of advantages). Benchmarking in this field is done by finding oblique partitions on a variety of datasets from the UCI (University of California at Irvine) machine learning repository. The Housing and the Optical Digits datasets are used as the non-dominated sets for the Decision Tree problem.

Decision Lab has a visualization tool that shows the relation between the criteria and the non-dominated set, and shows a preferred solution if it exists. Figure 1 shows a GAIA diagram for the MSPP that shows how each criterion relates to each action. The GAIA plane corresponds to the first principal components of the data, which ensures that a maximum quantity of information is available on the plane. An action in this case refers to a non-dominated set, generated by a specific algorithm. The orientation of the criteria axes indicates which criteria are in agreement with each other. The orientation of the position of an action indicates its strong features. The length of the axis correspond to a criterion's observed deviations between actions, the longer the axis the higher the deviation.

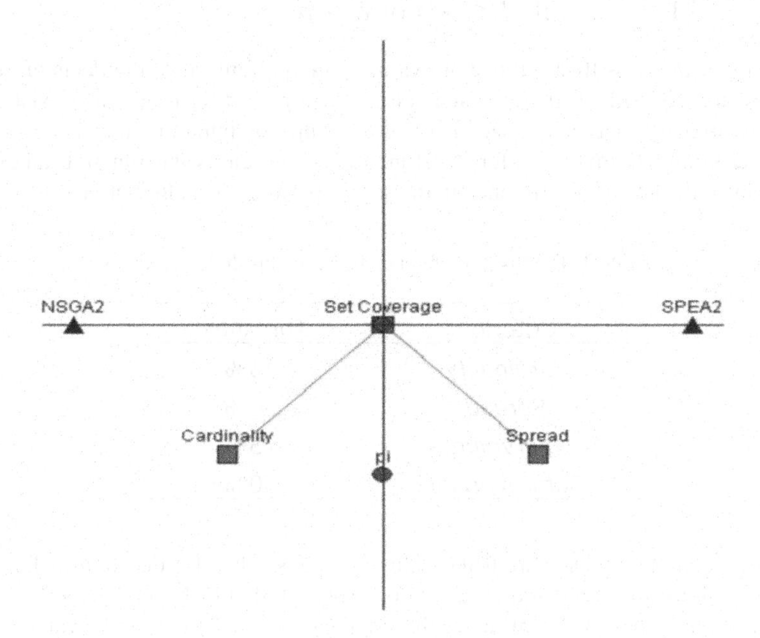

Fig. 1. GAIA diagram for a 2S|1M configuration using the 'usual criterion'

The orientation of the *pi*-vector on the figure, which is the decision axis, points to the preferred action or solution considering all the criteria. In this example, *pi* does not point towards any action, which means that there is no compromise solution. This due to the condition that NSGA-II has a strong feature on cardinality, SPEA2 has a strong feature on spread, and there is no difference regarding set coverage. All the criteria have the same weight. Adding different weights to each criterion obviously changes the orientation of the *pi* decision axis.

The criterion weight is independent from the scale of the criterion which means the larger the value the more important the criterion. In order to compare the different criteria independently from their measurement units, the PROMETHEE method provides six preference functions. Five out of six preference functions need some parameters specified by the decision-maker (only the 'usual criterion' does not have parameters). The parameters are used to computer the 'level of preference' of one action over another.

The preference function translates the deviation d between the values of two actions on a single criterion in terms of a preference degree. The preference degree is an increasing function of the deviation, defined on an interval [0,1]. A value 0 has to be interpreted an indifference, while a value of 1 has to be interpreted as strict preference. This research makes use of the 'linear preference function', which is defined by Brans et al. [2] as follows:

$$H(d) = \begin{cases} d/p & \text{if } -p \leq d \leq p, \\ 1 & \text{if } d < -p \text{ or } d > p \end{cases} \tag{6}$$

As long as d is less than a threshold value p, the preference of the decision maker increases linearly with d. If d becomes larger than p, a strict preference exists. The linear preference function is used in the following computations and is associated with all criteria. The linear preference function is chosen as the function takes into account even the smallest difference in the scores between two alternatives.

Table 2. The linear preference function threshold values

Criterion	Threshold
Cardinality	75%
Spread	50%
Hypervolume	75%
Set Coverage	50%

Table 2 contains some parameter values to be set by the decision-maker. The threshold values are expressed in a relative way. It should be understood that the choice of these values might influence the final decision. That means the values are to be set either on an objective consensus, or on a subjective but also consensus basis, or – in case these options are not available – are values which should be subject to sensitivity analysis.

Figures 1 and 2 are used, for the MSPP, to show which non-dominated set has better quality under two preference functions: the 'usual criterion' and the 'linear preference function'. They show the plane for the 2S|1M configuration. The number of criteria has reduced to three as mentioned previously. The values for *set coverage* in both sets are zero, which means that there are no weakly dominated solutions from each set or that their solutions are similar. Figure 2 shows that SPEA2 is the preferred solution after incorporating the thresholds from Table 2, and the decisive criterion should be interpreted as the 'spread' of solutions. The set coverage criterion is not a factor since both sets do not cover any weakly dominated solutions between them.

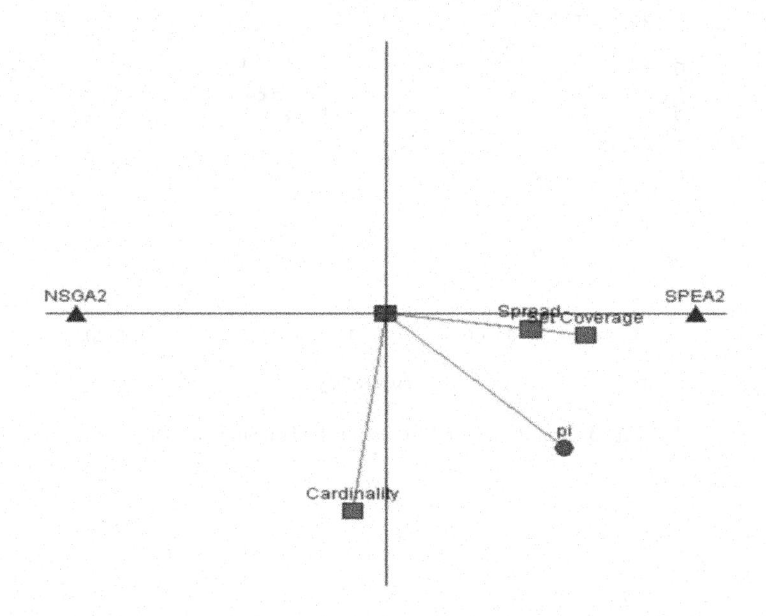

Fig. 2. GAIA diagram for a 2S|1M configuration with preferences

With respect to the application, decision trees or classifiers, the fronts shown in Figure 3 are the non-dominated sets from the ensembles of trees generated for one of the UCI datasets, the Housing Set. The Housing set deals with data regarding housing values in the suburbs of Boston (12 attributes, 506 observations). The OC1 (Oblique Classifier 1 [12]) solutions are dominated by the solutions of either SPEA2, or NSGA-II, or AP (axis-parallel tests). The non-dominated solutions of the AP classifier are dominated by either SPEA2 or NSGA-II. Most of the non-dominated solutions in NSGA are dominated by solutions in SPEA2. It seems that SPEA2 produces the better non-dominated set through the projection of their non-dominated sets but needs to be validated using the PROMETHEE method. The non-dominated sets of AP and OC1 need not be tested for performance quality as their solutions are dominated by both MOEAs solutions.

Figure 4 shows the GAIA plane of the Housing dataset options, and validates that SPEA2 is the preferred solution. The factors that favor SPEA2 are the hypervolume, the spread, and the set coverage. The result does not change when preference thresholds are added. In fact, the *pi* decision axis leans more to the direction of SPEA2 when preferences are added than it does when without any preferences.

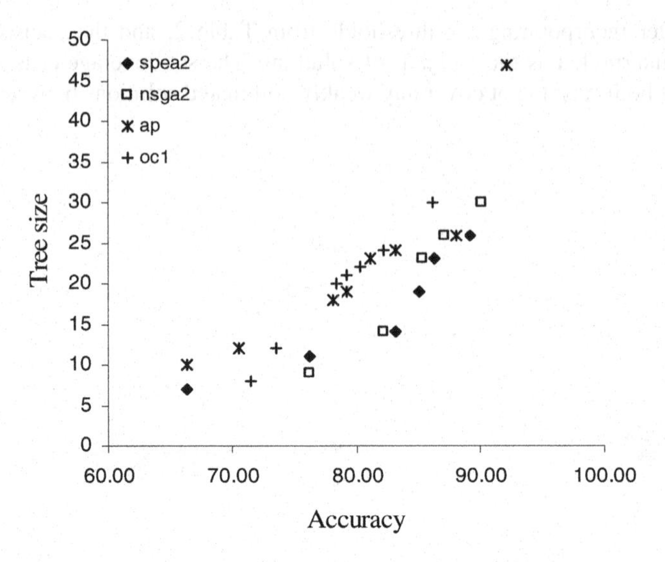

Fig. 3. Non-dominated fronts for the Housing dataset

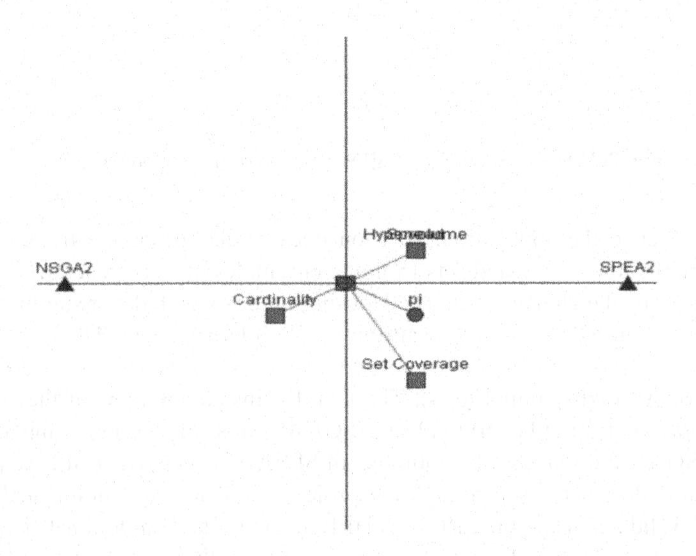

Fig. 4. GAIA plane for the Housing dataset using the usual criterion

4 Conclusions

Evolutionary algorithms have a great power of generating non-dominated sets in multi-objective optimization problems. Various strategies deliver various non-dominated sets. In order to evaluate the power of such a strategy, several aspects are to be taken into consideration like diversity and convergence to optimality. In literature various metrics have been proposed to measure both performance characteristics.

Some metrics measure only one of both, some measure a mix of both. To decide which strategy is better reduces to a multi-criteria problem. In this paper it is illustrated how a plain and clear method, the Promethee method, can be used to decide on the overall quality of an EA strategy. The illustration is made for a combinatorial problem, called the Multi-Objective Shortest Path Problem, and a data mining problem, called the Decision Tree problem.

References

1. Bosman, P.A., Thierens, D.: A balance between proximity and diversity in multiobjective evolutionary algorithms. IEEE Transactions on Evolutionary Computation 7(2), 174–188 (2003)
2. Brans, J.P., Vincke, P., Mareschal, B.: How to select and how to rank projects: the PROMETHEE method. European Journal of Operational Research 24, 228–238 (1986)
3. Brans, J.P., Mareschal, B.: The PROMCALC and GAIA Decision Support System for Multicriteria Decision Aid. Decision Support Systems 12, 297–310 (1994)
4. Cantu-Paz, E., Kamath, C.: Inducint oblique decision trees with evolutionary algorithms. IEEE Transactions on Evolutionary Computation 7(1), 54–68 (2003)
5. Carrizosa, E., Plastria, F.: On minquantile and maxcovering optimization. Mathematical Programming 71, 101–112 (1995)
6. Coello, C.: An updated survey of GA-based multiobjective optimization techniques. ACM Computing Surveys 32(2), 109–143 (2000)
7. Deb, K.: MultiObjective Optimization Using Evolutionary Algorithms. Wiley, Chichester (2001)
8. Deb, K., Agrawal, S., Pratap, A., Meyarivan, T.: A fast elitist multiobjective genetic algorithm: NSGA-II. IEEE Transactions on Evolutionary Computation 6(2), 182–197 (2002)
9. Ehrgott, M.: Multicriteria Optimization, 2nd edn. Springer, Berlin (2005)
10. Ehrgott, M., Gandibleux, X.: A survey and annotated bibliography of multiobjective combinatorial optimization. OR Spectrum 22(4), 425–460 (2000)
11. Eiselt, H.A., Laporte, G., Thisse, J.-F.: Competitive location models: a framework and bibliography. Transportation Science 27, 44–54 (1993)
12. Murty, S.K., Kasif, S., Salzberg, S.: A system for induction of oblique decision trees. Journal of Artificial Intelligence Research 2(1), 1–32 (1994)
13. Schott, J.: Fault tolerant design using single and multicriteria genetic algorithm optimization. Master thesis. Massachusetts Institute of Technology, Cambridge (1995)
14. Van Veldhuizen, D.A.: Multiobjective Evolutionary Algorithms: Classifications, Analyses and New Innovations. Ph.D. thesis, Graduate School of Engineering of the Air Force Institute of Technology, Dayton, Ohio (1999)
15. Visual Decision: Decision Lab and Decision Lab 2000. Visual Decision Inc., Montreal (2003)
16. Zitzler, E.: Evolutionary Algorithms for Multiobjective Optimization: Methods and Applications. Ph.D. thesis, Shaker Verlag, Aachen (1999)
17. Zitzler, E., Thiele, L.: Multiobjective evolutionary algorithms: a comparative study and the strength Pareto approach. IEEE Transactions on Evolutionary Computation 3(4), 257–271 (1999)
18. Zitzler, E., Laumanns, M., Thiele, L.: SPEA2: Improving the strength Pareto evolutionary algorithm for multiobjective optimization. In: Giannakoglou, K., Tsahalis, D., Periaux, J., Papailou, K., Fogarty, T. (eds.) Evolutionary Methods for Design, Optimization and Control, pp. 19–26. CIMNE, Barcelona (2002)

Efficiency and Robustness of Three Metaheuristics in the Framework of Structural Optimization

Nikos D. Lagaros[1] and Dimos C. Charmpis[2]

[1] Institute of Structural Analysis & Seismic Research,
National Technical University Athens
Zografou Campus, Athens 15780, Greece
nlagaros@central.ntua.gr

[2] Department of Civil and Environmental Engineering,
University of Cyprus,
75 Kallipoleos Str., P.O. Box 20537, 1678 Nicosia, Cyprus
charmpis@ucy.ac.cy

Abstract. Due to the technological advances in computer hardware and software tools, structural optimization has been gaining continuously increasing interest over the last two decades. The purpose of the present work is to quantitatively compare three metaheuristic optimization algorithms, namely the Differential Evolution, Harmony Search and Particle Swarm Optimization methods, in the framework of structural optimization. The comparison of the optimizers is performed with reference to their efficiency (overall computing demands) and robustness (capability to detect near-optimal solutions). The optimum design of a real-world overhead traveling crane is used as the test bed application for conducting optimization test runs.

Keywords: Structural Optimization, Metaheuristics, Differential Evolution, Harmony Search, Particle Swarm Optimization.

1 Introduction

Structural optimization typically aims in detecting the optimum design by minimizing the cost (or weight) of a structure subject to certain behavioral constraints imposed by relevant design codes. Structural optimization applications are associated with multiple local minima, large and non-convex search spaces and several (usually conflicting) constraints to be satisfied. As a result, such applications do not favor the use of Mathematical Programming (gradient-based) algorithms, which exhibit a satisfactory local rate of convergence to the nearest optimum, but they cannot assure that the global optimum can be found when confronted with complex optimization problems. Therefore, various heuristic probabilistic-based search algorithms, which present a better global behavior and are less vulnerable to local optima, have been applied during the last decades to meet the high demands of structural optimization problems.

Several structural optimization applications have been reported, which utilize algorithms inspired by natural phenomena, such as Evolutionary Programming [1],

H. Papadopoulos, A.S. Andreou, and M. Bramer (Eds.): AIAI 2010, IFIP AICT 339, pp. 104–111, 2010.

Genetic Algorithms [2] and Evolution Strategies [3], among others. More recently, new metaheuristic optimizers have been developed based on the simulation of social interactions among members of a specific species looking for food or resources in general. One such method is Particle Swarm Optimization (PSO) [4], which is based on the behavior reflected in flocks of birds, bees and fish that adjust their physical movements to avoid predators and seek for food. Two other metaheuristics with growing interest within the structural optimization community are Differential Evolution (DE) [5,6] and Harmony Search (HS) [7,8].

The purpose of the present work is to provide quantitative comparison results on the performance of the three aforementioned metaheuristics (DE, HS and PSO) in the framework of structural optimization. A typical overhead traveling crane (the crane bridge span, the trolley, the end carriages and the runway beam span) was selected from the catalogues of GH Cranes SA (a manufacturer of elevation systems) as the test bed problem for conducting optimization test runs. A 3D CAD model was developed for the crane and its cross-sectional dimensions have been optimized using DE, HS and PSO subject to constraints imposed by Eurocode Standards.

2 The Metaheuristic Algorithms

This section provides a short description of the three metaheuristic optimization algorithms (DE, HS and PSO) assessed in the present work. All three algorithms are applied to an optimization problem, which can be formulated as follows:

$$\begin{aligned} &\min_{x \in \mathcal{F}} f(\boldsymbol{x}) \\ &\textit{subject to } g(\boldsymbol{x}) \geq 0.0 \end{aligned} \tag{1}$$

where f denotes the objective function, $\boldsymbol{x}$ is the vector of design (decision) variables, F represents the design (search) space and g are the constraint functions of the problem.

2.1 Differential Evolution (DE)

In 1995, Storn and Price [6] proposed a new floating point evolutionary algorithm for global optimization and named it Differential Evolution (DE). DE has a special kind of differential operator, which is invoked to create new offspring from parent chromosomes instead of the classical crossover or mutation. DE is a novel parallel direct search method which utilizes NP parameter vectors $\boldsymbol{x}_{i,G}$ (i=1,..,NP) as a population for each generation G. The initial population is generated randomly. As a rule, a uniform probability distribution for all random decisions is assumed. In case a preliminary solution is available, the initial population is often generated by adding normally distributed random deviations to the nominal solution $\boldsymbol{x}_{\text{nom},0}$. The crucial idea behind DE is a new scheme for generating new vectors. DE generates new vectors by adding the weighted difference vector between two population members to a third member. If the resulting vector yields a lower objective function value than a predetermined population member, then the newly generated vector replaces the vector with which it was compared. The comparison vector can - but needs not to be - part of the generation process mentioned above. The best parameter vector $\boldsymbol{x}_{best,G}$ is evaluated for every generation G in order to keep track of the progress made during the optimization process. The extraction of distance and direction information from the population to generate random deviations results in an adaptive scheme with excellent convergence properties. Several variants of DE have been proposed so far, the two most popular of which are presented below as schemes DE1 and DE2.

Scheme DE1. The first variant of DE works as follows: for each vector $\boldsymbol{x}_{i,G}$, *a trial vector* $\boldsymbol{v}$ *is generated according to:*

$$\boldsymbol{v} = \boldsymbol{x}_{r1,G} + F \cdot (\boldsymbol{x}_{r2,G} - \boldsymbol{x}_{r3,G}) \tag{2}$$

The integers $r1$, $r2$ and $r3$ are chosen randomly from the interval $[1,NP]$ and are different from the running index i. F is a real and constant factor, which controls the amplification of the differential variation $(\boldsymbol{x}_{r2,G} - \boldsymbol{x}_{r3,G})$. In order to increase the diversity of the parameter vectors, the vector $\boldsymbol{u} = [u_1, u_2, \ldots, u_D]^T$ is defined as follows:

$$u_j = \begin{Bmatrix} v_j \; for \; j = \langle n \rangle_D, \langle n+1 \rangle_D \ldots, \langle n+L-1 \rangle_D \\ (x_{i,G})_j \qquad\qquad otherwise \end{Bmatrix} \tag{3}$$

where the acute brackets $\langle \; \rangle_D$ denote the modulo function with modulus D.

Scheme DE2. In the second variant of DE, the trial vector $\boldsymbol{v}$ *for each vector* $\boldsymbol{x}_{i,G}$ *is generated through the expression:*

$$\boldsymbol{v} = \boldsymbol{x}_{i,G} + \lambda \times (\boldsymbol{x}_{best,G} - \boldsymbol{x}_{i,G}) + F \times \left(\boldsymbol{x}_{r2,G} - \boldsymbol{x}_{r3,G}\right) \tag{4}$$

which introduces an additional control variable λ. The idea behind λ is to provide a means to enhance the greediness of the scheme by involving the current best vector $\boldsymbol{x}_{best,G}$. This feature can be useful for non-critical objective functions. The construction of $\boldsymbol{u}$ from $\boldsymbol{v}$ and $\mathbf{x}_{i,G}$, as well as the decision process, are identical to those of DE1.

2.2 Harmony Search (HS)

The Harmony Search (HS) algorithm was originally inspired by the improvisation process of Jazz musicians. Each musician corresponds to a decision variable; musical instrument's pitch range corresponds to decision variable's value range; musical harmony at certain time corresponds to the solution vector at certain iteration; and audience's aesthetics corresponds to the objective function. Just like musical harmony is improved as time passes, the solution vector is improved iteration by iteration.

HS utilizes a number of parameters: harmony memory size (HMS), harmony memory considering rate (HMCR), pitch adjusting rate (PAR), maximum improvisation (MI) and fret width (FW). HMS is the number of solution vectors simultaneously handled by the algorithm. HMCR is the probability of choosing one value from the musician's memory ($0 \leq$ HMCR ≤ 1). Thus, (1-HMCR) is the rate with which HS picks one value randomly from the total value range. PAR is the rate with which HS tweaks the value originally picked from memory ($0 \leq$ RAR ≤ 1). Thus, (1-PAR) is the rate with which HS keeps the original value obtained from memory. MI is the number of iterations. HS improvises one harmony (= vector) at each iteration. Finally, FW refers to the term fret, which is the metallic ridge on the neck of a string instrument dividing the neck into fixed segments; each fret represents one semitone. In the context of the HS algorithm, frets are arbitrary points which divide the total value range into fixed segments and fret width (FW) is the length between two neighboring frets. Uniform FW is normally used in HS. Mahdavi et al. [7] suggested that PAR increases linearly and FW decreases exponentially with iterations:

$$PAR(I) = PAR_{min} + (PAR_{max} - PAR_{min}) \times \frac{1}{MI} \tag{5}$$

$$FW(I) = FW_{max} \exp\left[\ln\left(\frac{FW_{min}}{FW_{max}}\right)\frac{I}{MI}\right] \tag{6}$$

Once the optimization problem is formulated and the parameter values are set, a random tuning process is performed. The HS algorithm initially improvises many random harmonies. The number of random harmonies is at least HMS (this number can be more than HMS, such as two or three times HMS [8]). Then, top-HMS harmonies are selected as starting vectors. Musician's harmony memory (HM) can be considered as a matrix.

$$HM = \left[\begin{array}{cccc|c} x_1^1 & x_2^1 & x_3^1 \dots & x_v^1 & f(x)^1 \\ x_1^2 & x_2^2 & x_3^2 \dots & x_v^2 & f(x)^2 \\ \dots & \dots & \dots & \dots & \dots\dots \\ x_1^{HMS} & x_2^{HMS} & x_3^{HMS} \dots & x_v^{HMS} & f(x)^{HMS} \end{array}\right] \tag{7}$$

Once the HM is prepared, a new vector (=harmony) is improvised based on the following operators:

Memory Consideration: When HS determines the new value $\boldsymbol{x}_i$, it randomly picks the j-th value from HM = $\{x_i^1 \dots x_i^{HMS}\}$ with probability HMCR. The appendix j is taken from a uniform distribution U (0, 1):

$$j \leftarrow \text{int}(U(0,1) \,.\, HMS) + 1 \tag{8}$$

Pitch Adjustment: Once the value x_i^{new} is randomly picked from HM in the above memory consideration process, it can be further adjusted to neighboring values by adding a certain amount to the value, with probability of PAR. For a discrete variable, if $x_i(k) = x_i^{new}$, the pitch-adjusted value becomes $x_i(k+m)$, where $m \in \{-1, 1\}$ normally;

and for a continuous variable, the pitch-adjusted value becomes $x_i^{new} + \Delta$, where $\Delta = U(0,1) * FW(i)$ normally. The above-mentioned three basic operations (random selection, memory consideration and pitch adjustment) can be expressed as follows:

$$\text{xj} \leftarrow \left\{ \begin{array}{l} \left. \begin{Bmatrix} xi \in \{xi(1),.....xi(k)....,xi(Ki)\} \\ xu \in \left[xi^L, xi^U \right] \end{Bmatrix} \right\} w.p.(1-HMCR) \\ xi \in HM = \{xi^1, xi^2,...,xi^{HMS}\} w.p.HMCR(1-PAR) \\ \left. \begin{Bmatrix} xi(k+m) if..xi(k) \in HM \\ xi + \Delta...if..xi \in HM \end{Bmatrix} \right\} w.p.HMCR * PAR \end{array} \right\} \quad (9)$$

2.3 Particle Swarm Optimization (PSO)

In Particle Swarm Optimization (PSO), multiple candidate solutions coexist and collaborate simultaneously. Each solution is called a 'particle' that has a position and a velocity in the multidimensional design space. A particle 'flies' in the problem search space looking for the optimal position. As 'time' passes during its quest, a particle adjusts its velocity and position according to its own 'experience', as well as the experience of other (neighbouring) particles. Particle's experience is built by tracking and memorizing the best position encountered. As every particle remembers the best position it has visited during its 'flight', the PSO possesses a memory. A PSO system combines local search method (through self experience) with global search method (through neighbouring experience), attempting to balance exploration and exploitation.

Each particle maintains two basic characteristics, velocity and position, in the multi-dimensional search space that are updated as follows:

$$\boldsymbol{v}^j(t+1) = w\boldsymbol{v}^j(t) + c_1\boldsymbol{r}_1 \circ \left(\boldsymbol{x}^{\text{Pb},j} - \boldsymbol{x}^j(t)\right) + c_2\boldsymbol{r}_2 \circ \left(\boldsymbol{x}^{\text{Gb}} - \boldsymbol{x}^j(t)\right) \quad (10)$$

$$\boldsymbol{x}^j(t+1) = \boldsymbol{x}^j(t) + \boldsymbol{v}^j(t+1) \quad (11)$$

where w is the inertia weight parameter, $\boldsymbol{v}^j(t)$ denotes the velocity vector of particle j at time t, $\boldsymbol{x}_j(t)$ represents the position vector of particle j at time t, vector $\boldsymbol{x}^{\text{Pb},j}$ is the personal best ever position of the j-th particle, and vector $\boldsymbol{x}^{\text{Gb}}$ is the global best location found by the entire swarm. The acceleration coefficients c_1 and c_2 indicate the degree of confidence in the best solution found by the individual particle ($c1$ - cognitive parameter) and by the whole swarm (c_2 - social parameter), respectively, while r_1 and r_2 are two random vectors uniformly distributed in the interval [0,1]. The symbol "$\circ$" of Eq. (10) denotes the Hadamard product, i.e. the element-wise vector or matrix multiplication.

Fig. 1 depicts a particle's movement in a two-dimensional design space. The particle's current position $\boldsymbol{x}^j(t)$ at time t is represented by the dotted circle at the lower left of the drawing, while the new position $\boldsymbol{x}^j(t+1)$ at time $t+1$ is represented by the dotted bold circle at the upper right hand of the drawing. The figure shows how the particle's movement is affected by: (i) it's velocity $\boldsymbol{v}^j(t)$; (ii) the personal best ever position of the particle, $\boldsymbol{x}^{\mathrm{Pb},j}$, at the right of the figure and (iii) the global best location found by the entire swarm, $\boldsymbol{x}^{\mathrm{Gb}}$, at the upper left of the figure.

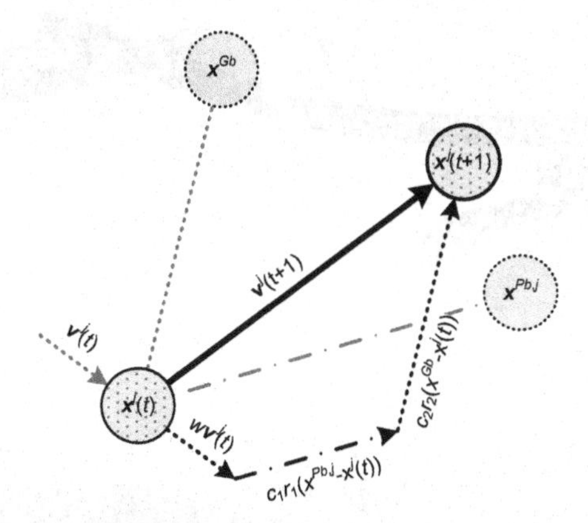

Fig. 1. PSO: Visualization of the particle's movement in a two-dimensional design space

3 Test Case

The three metaheuristic algorithms described in the previous section have been applied for the sizing optimization of a travelling Overhead Crane (Fig. 2). The design variables (11 in total) are web, top and bottom flange sizes both of the box girder and the runway beams and the thickness (Fig. 3). The design variables of the box girder are the breadth of the top flange (DV1), the distance of the webs (DV2), the height of the girder (DV3) and the thicknesses of top flange, bottom flange, webs (DV4, DV5 and DV6, respectively). The design variables of the runway beams are the breadth of the bottom flange (DV7), the height of the beams (DV8) and the thicknesses of top flange, bottom flange, web (DV9, DV10 and DV11, respectively). The objective function of the optimization problem corresponds to the total mass of the crane system. The constrains of the problem are imposed based on the Structural Eurocodes (Chapter 2): (i) the von Misses equivalent stress must not exceed the prescribed stress limit state of the selected material used for the crane (S235) divided by the coefficient 1.1; (ii) the vertical deflection must not exceed 15mm (=L/600, where L is the span of the bridge); (iii) the vertical deflection must not exceed 15mm (=L/600). Constraints violation is taken into account by penalizing the objective function.

Each of algorithms DE (scheme DE2), HS and PSO was executed 1,000 times. The performance results obtained are summarized in Tables 1 and 2 (CoV is defined as the standard deviation divided by the mean).

DE was found to give the minimum mass for the test example considered, while PSO also yielded almost the same minimum result. In general, DE and PSO appear to be rather robust algorithms, since they yield solutions close to the best result achieved with low variance when compared between several test runs. HS provides solutions of lower quality and with larger variation between several test runs.

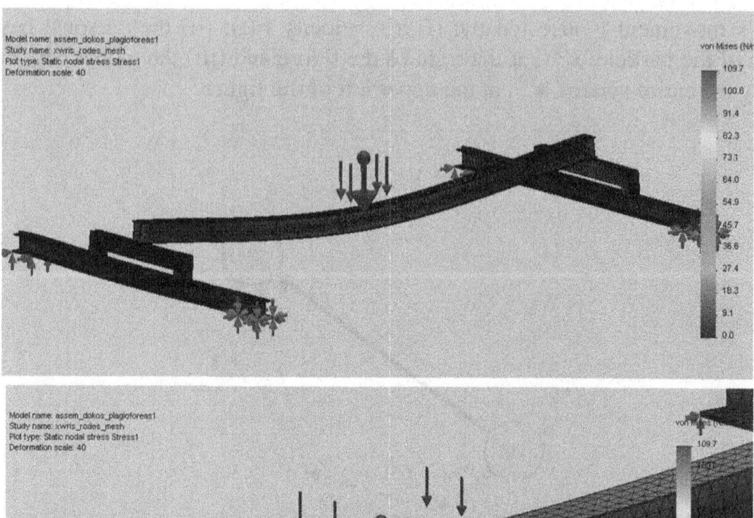

Fig. 2. Views of the crane and its finite element mesh

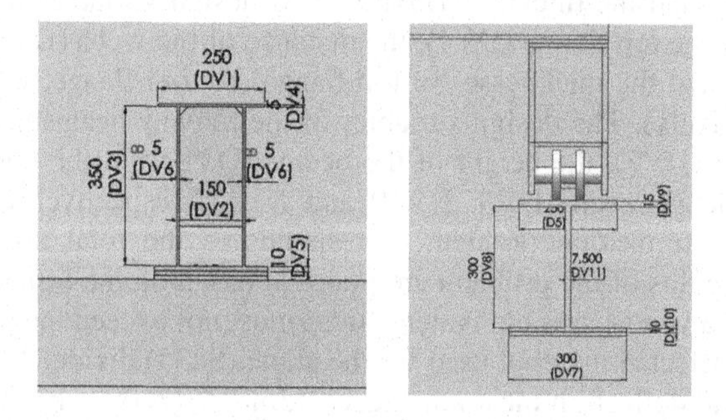

Fig. 3. The 11 design variables

Table 1. Minimum objective function value achieved by the three metaheuristics

Optimization Method	Minimum mass achieved (Kg) mean (CoV)	min	max
PSO	2007.3 (0.33%)	1999.8	2071.7
DE	2001.1 (0.41%)	1998.1	2084.1
HS	2286.2 (2.43%)	2076.8	2322.3

Table 2. Computational cost induced by the three metaheuristics (a time-consuming structural analysis of the crane is performed at each iteration of the optimization algorithms)

Optimization	Number of iterations		
Method	mean (CoV)	min	max
PSO	60912.0 (33.8%)	33000	157500
DE	95613.8 (80.5)	5000	371000
HS	5477.5 (72.6)	4422	20100

In terms of computing demands, HS seems to be the most inexpensive of the three meatheuristics. DE appears to be the computationally most demanding algorithm, while PSO performance lies between the ones of HS and DE.

4 Conclusions

Differential Evolution (DE) is the optimization method that was found to give the best design vector, but required the largest amount of iterations for convergence. Harmony Search (HS) is the optimization method that provided the worst design vector, but converged faster than both DE and PSO. PSO yielded solutions of almost the same quality as DE, but with less iteration demands.

In general, it can be stated that efficiency is inversely proportional to robustness. A fast algorithm (HS) cannot provide the solution quality of a robust yet computationally expensive algorithm (DE). PSO seems to be a good compromise between HS and DE, although it exhibits substantially larger computational demands that HS.

References

1. Fogel, L.J., Owens, A.J., Walsh, M.J.: Artificial Intelligence through Simulated Evolution. John Wiley, Chichester (1966)
2. Goldberg, D.E.: Genetic Algorithms in Search Optimization and Machine Learning. Addison Wesley, Reading (1989)
3. Rechenberg, I.: Evolutionsstrategie – Optimierung technischer Systeme nach Prinzipien der biologischen Evolution (1971); Reprinted by Fromman-Holzboog (1973)
4. Kennedy, J., Eberhart, R.: Particle Swarm Optimization. In: Proceedings of IEEE International Conference on Neural Networks IV (1995)
5. Price, K., Storn, R., Lampinen, J.: Differential Evolution - A Practical Approach to Global Optimization. Springer, Berlin (2005)
6. Storn, R., Price, K.: Differential Evolution - a Simple and Efficient Adaptive Scheme for Global Optimization over Continuous Spaces. Technical Report TR-95-012, International Computer Science Institute (1995)
7. Mahdavi, M., Fesanghary, M., Damangir, E.: An improved harmony search algorithm for solving optimization problems. Applied Mathematics and Computation 188, 1567–1579 (2007)
8. Degertekin, S.: Optimum design of steel frames using harmony search algorithm. Structural and Multidisciplinary Optimization 36, 393–401 (2008)

A Fuzzy Non-linear Similarity Measure for Case-Based Reasoning Systems for Radiotherapy Treatment Planning

Rupa Jagannathan[1], Sanja Petrovic[1], Angela McKenna[2], and Louise Newton[2]

[1] Automated Scheduling, Optimisation and Planning Research Group
School of Computer Science, University of Nottingham, Nottingham, UK
{rxj,sxp}@cs.nott.ac.uk
[2] Department of Medical Physics
Nottingham University Hospitals NHS Trust Nottingham, UK
{angela.mckenna,louise.newton}@nuh.nhs.uk

Abstract. This paper presents a decision support system for treatment planning in brain cancer radiotherapy. The aim of a radiotherapy treatment plan is to apply radiation in a way that destroys tumour cells but minimizes the damage to healthy tissue and organs at risk. Treatment planning for brain cancer patients is a complex decision-making process that relies heavily on the subjective experience and expert domain knowledge of clinicians. We propose to capture this experience by using case-based reasoning. Central to the working of our case-based reasoning system is a novel similarity measure that takes into account the non-linear effect of the individual case attributes on the similarity measure. The similarity measure employs fuzzy sets. Experiments, which were carried out to evaluate the similarity measure using real brain cancer patient cases show promising results.

Keywords: Case-based Reasoning, Fuzzy Logic, Radiotherapy Treatment Planning, Decision-Support Systems.

1 Introduction

Radiotherapy is a treatment method that uses ionizing radiation in cancer control. The aim of radiotherapy treatment planning (RTP) is to deliver a tumouricidal radiation dose over the tumour region while minimizing the radiation received by healthy tissue and critical organs in the vicinity of the tumour. To realize this goal, a detailed treatment plan is created for each patient that describes exactly how a patient will be irradiated to achieve the required dose distribution. Parameters of RTP include the number of beams, the gantry angle of beams (the angle by which the apparatus applying the radiation beam is moved around the patient's body) and wedges that shape the beam. Oncologists, together with medical physicists, use their subjective experience and expert clinical knowledge to generate treatment plans. The RTP decision support system, under development in collaboration with the Nottingham University Hospitals, City Hospital campus, uses case-based reasoning to capture this experience and aid oncologists in the computation of plan parameters.

H. Papadopoulos, A.S. Andreou, and M. Bramer (Eds.): AIAI 2010, IFIP AICT 339, pp. 112–119, 2010.

Case-based reasoning (CBR) is based on the concept that the solution of a problem can be derived from the solutions of similar problems [1]. The main advantage of CBR in RTP over numerical optimisation [2] and rule-based methods [3] is its capability to utilise non-quantifiable or subjective knowledge and benefit from the successes, errors and failures of previous treatment plans. In addition, since treatment plan generation is based on existing plans, new plans can be generated quickly and in line with the institution's preferences. Although, CBR has been widely applied in healthcare systems in the diagnosis and treatment of disease [4, 5], the research done on CBR in radiotherapy treatment planning has been limited. Case based reasoning has been used in dose planning for prostate cancer by Song et al [6] and Mishra et al [12]. The CBR system Roentgen, designed by Berger [7], aids radiotherapy planning for thorax cancer and retrieves cases from its case archive based on the similarity between patient geometric descriptors. However, no implementation details or experimental evaluation of the method are furnished. The work presented in this paper builds up on Berger's idea of using the geometry of the patient as input to the CBR system and implements this concept using a fuzzy non-linear similarity measure.

The choice of the similarity measure, which calculates how similar cases in the case base are to the target case, is fundamental. The popular nearest neighbour method [8] matches each attribute in the target case to its corresponding attribute in the archive case. The aggregate similarity is given by the weighted sum of the individual similarities. However, if the similarity values with respect to individual attributes are very different in terms of their distribution or range they can not be directly compared or summed up to give an accurate representation of the aggregate similarity between two cases. Another common assumption of many CBR systems is that the similarity with respect to a single case attribute influences the aggregate similarity linearly over its range. In reality, however, the attribute similarity between two cases often behaves in a non-linear fashion [9, 10]. That is, a high similarity value with respect to a non-linear attribute indicates a high suitability of the case solution to the target case but also a very low similarity of non-linear attribute has a detrimental effect on the suitability of the case. To overcome these limitations, our CBR system uses a novel fuzzy non-linear similarity measure. The performance of the proposed similarity measure is evaluated using real brain cancer patient cases obtained from the Nottingham City Hospital.

The paper is organized as follows. Section 2 describes the architecture of the CBR system and the non-linear fuzzy similarity measure. Section 3 presents some results obtained by comparing the performance of the standard weighted nearest neighbour and the fuzzy non-linear similarity measure. Section 4 discusses future research directions and concludes our work.

2 Architecture of the CBR System

Each case in the developed CBR system contains a patient description and a suitable treatment plan detailing the RTP beam configuration. The patient description consists of spatial information regarding the location of the planning target volume containing the tumour (PTV) and the organs at risk (OAR). The OAR commonly include the spinal cord, the eyes, the lens, the optic nerve and chiasm, the brainstem and the pituitary gland. The cases are sorted according to the OAR present in the vicinity of the PTV. Since procuring clinical patient data is a slow and difficult process, the case base currently contains only 24 brain cancer patient cases. However, we are continuously adding more cases to our case base as they become available.

2.1 Case Attributes

Case attributes identified to be relevant for similarity between two patients describe the location of the tumour and the spatial relationship between the tumour and the OAR. These attributes determine the geometry of a patient and are computed using 3D coordinates representing the PTV and OAR structures, which are extracted from the CT DICOM [11] image header files created for each patient. The following geometric descriptors are used.

- Volume of PTV, denoted by *V*.
- Distance between tumour and OAR, denoted by *E*, is defined as the minimum edge-to-edge distance between the tumour and the OAR.
- Angle between tumour and OAR, denoted by *A*, is defined as the angle between the line connecting the origin of the image patient coordinate system and the tumour centroid and the line connecting the origin and the OAR centroid.

2.2 A Fuzzy Similarity Measure

A good definition of a similarity measure between two cases is of crucial importance for the retrieval process, which should retrieve from the case base a case with a treatment plan suitable for the target patient. A suitable treatment plan is one whose plan parameters achieve the desired radiation dose distribution in the target case patient. After consultation with medical physicists, we have arrived at the assumption that patients with similar geometric descriptors have similar treatment plans. The CBR system Roentgen is based on this assumption as well [7]. In the retrieval process, the similarity with respect to the geometric descriptors is calculated between the target case and those cases in the case base that contain the same OAR.

The values of the three geometric descriptors are normalized to take values from the interval [0, 1]. Let C_T be the target case and C_C be a case from the case-base. The distance d_l between attribute value v_T in target case C_T and attribute value v_C in case C_C with respect to attribute l, where $l = A, E$ and V, is calculated as follows:

$$d_l = v_T - v_C \,. \tag{1}$$

Then, the similarity between C_T and C_C with respect to attribute l is:

$$s_l = 1 - d_l \,. \tag{2}$$

The similarity values with respect to each attribute have to be combined into an aggregate similarity value between two cases. In the nearest neighbour method, the aggregate similarity is generally defined as the weighted sum of the individual attribute similarities. This method however does not take into account the variations in the distribution of the similarity values of different attributes. In order to obtain an idea of the distribution of attribute similarities, we calculate the similarity between each case and every other case (with the same OAR) in our case base in a leave-one out fashion considering one attribute a time. Figure 1 shows the similarity calculated between all possible pairs of patient cases in the case base (arranged in decreasing order of similarity) for each attribute. The data suggests a trend in the distribution of the attribute similarity values between cases. We can see that the distribution and range of the similarity with respect to attributes *A* and *E* is comparable. However, it is different for attribute *V*. Hence, numerical similarity values with respect to attribute *A* or *E* do not necessarily have the same meaning as an equal numerical similarity value with respect to attribute *V*. To illustrate this point, consider the case pair C_{10} and C_{23}. In figure 1, we can see that the similarity between C_{10} and C_{23} with respect to the volume *V* is '0.72', which in linguistic terms can be considered as an 'average' volume similarity. However, if the similarity between C_{10} and C_{23} with respect to attribute angle was '0.72', we observe that this value is considered as a relatively low angle similarity compared to all other similarity values for that attribute and we conclude that the two cases are not very similar with respect to the attribute angle. That is, the same numerical similarity value has different meanings for different attributes. Thus, we can not compare directly the numerical values of the similarity and a simple sum of individual attribute similarities does not necessarily give an accurate representation of the similarity between two cases. In order to sensibly compare attribute similarities and generate an aggregate similarity measure, the CBR system has to understand what numerical similarity value actually constitutes a "high", "average" or "low" similarity for each attribute. We therefore propose the use of fuzzy sets. We define the fuzzy sets *High*, *Avg* and *Low*, which denote low similarity, average similarity and high similarity, respectively, for each attribute. The fuzzy membership functions of the three sets are defined for each attribute based on the minimum, maximum and average of the corresponding similarity values found across the case base. They, therefore, give a realistic indication of what constitutes a relatively "high similarity", "average similarity" or "low similarity" for an attribute.

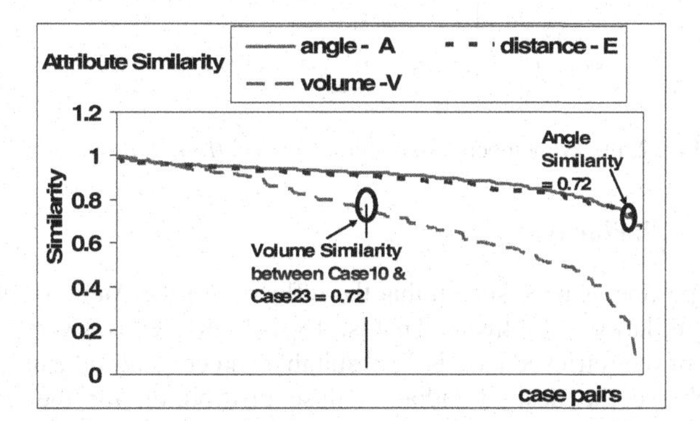

Fig. 1. The graph shows the similarity values obtained between all possible pairs of cases

Figure 2 presents the fuzzy membership functions defined for attributes *A*, *E* and *V*. The membership degree $\mu_{l,s}$ of similarity s_l with respect to attribute *l* is computed for each fuzzy set *s*, *High*, *Avg* and *Low*. The aggregate similarity consists of the *High*, *Avg* and *Low* component M_s defined as the sum of the membership degrees of the attribute similarities to the fuzzy sets *High, Avg* and *Low* of the corresponding attributes, shown in expression 3.

$$\mathrm{M}_s = \sum_{l=A,E,V}^{3} w_l \mu_{l,s} \,. \tag{3}$$

where w_l denotes the weight of attribute l , $l = A, E, V$ and $\mu_{l,s}$, $s = High, Avg, Low,$ is the membership degree of the attribute similarity to the fuzzy sets *High, Avg* and *Low.* A large value of component M_{High} indicates a high aggregate similarity between two cases, a large value of component M_{Avg} indicates an average similarity between two cases, while a large value of component M_{Low} indicates a low similarity between two cases. That is, M_{High} displays a net positive effect while M_{Avg} and M_{Low} display a net negative effect on the aggregate similarity. The aggregate similarity s_T between two cases is defined as the net contribution M_{High}, M_{Avg} and M_{Low} as shown in expression 4.

$$s_T = w_{High} M_{High} - w_{Avg} M_{Avg} - w_{Low} M_{Low} \,. \tag{4}$$

The weights w_{High}, w_{Avg} and w_{Low} determine the importance of the fuzzy sets *High, Avg* and *Low* respectively.

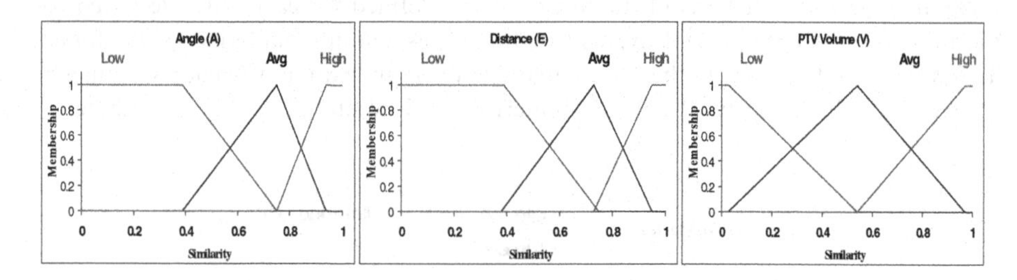

Fig. 2. Fuzzy membership functions for sets *High, Low* and *Avg*

2.3 Non-linear Similarity

Preliminary experiments have shown that the influence of the Angle similarity (s_A) on the aggregate similarity is not linear. That is, if s_A is high, then it is very likely that the treatment plan of the retrieved case is very suitable and contains a beam configuration that would achieve the desired radiation dose distribution for the target patient. However, if s_A is low, then the corresponding treatment plan is very likely to be rather unsuitable for the target case. In other words, to retrieve cases whose treatment plans are very suitable for the target case, the aggregate similarity between cases has to be increased if the value s_A is high and reduced or penalised if s_A is low. This effect is not covered by the standard method of multiplying weights with the similarity value assigned to attributes, since a high weight for attribute *A*, would increase the aggregate similarity value not only if s_A is high but also if s_A is low.

For the similarity with respect to *V*, s_V, the effect is non-linear too, but in an opposite fashion. If s_V is high, it does not substantially increase the likelihood that the treatment plan of the retrieved case is suitable. Similarly, if s_V is low, it does not necessarily mean that the corresponding treatment plan is highly unsuitable i.e. the influence of both high and low similarity due to attribute *V* is reduced or smoothed.

Expression 5 presents the aggregate similarity measure, which takes into account the non-linear effect of attributes, where k_l is the non-linearity factor of attribute l.

$$s_{TNL} = w_{High} \sum_{l=A,E,V}^{3} w_l \left(\mu_{l,High}\right)^{k_l} - w_{Avg} \sum_{l=A,E,V}^{3} w_l \left(\mu_{l,Avg}\right)^{k_l} - w_{Low} \sum_{l=A,E,V}^{3} w_l \left(\mu_{l,Low}\right)^{k_l} \quad (5)$$

To achieve a stretching non-linear effect, *k* should be smaller than 1, to achieve a smoothing non-linear effect, *k* should be greater than 1 and k should be equal to 1 if no non-linear effect is desired. After initial experiments, we have used $k = 1/3$, for attribute *A*; $k = 3$, for attribute *V;* and $k = 1$, for attribute *E*.

3 Experimental Results

We tested the performance of the fuzzy non-linear similarity measure using brain cancer patient cases. Since all cases contain the treatment plan used in that case, we can evaluate the similarity measure by comparing the treatment plan of the retrieved case with the actual treatment plan. In the experiments, we looked mainly at the beam configuration of the plan, in particular the beam angles. The beam angles consider the position of the gantry, the patient couch and the beams. The difference in the beam angle values in the plan of the retrieved case and the actual treatment plan of the target case gives the error. The error is averaged over all cases to give an indication of the quality of the similarity measure. A commonly employed evaluation strategy in CBR is the leave-one out cross validation method. Each case in the case base is consecutively made the target case and the most similar case among the remaining cases in the case base is retrieved. In the future when more cases are available, the case base can also be divided into training and test cases. Figure 3a shows the average errors in angle degrees per beam across all cases obtained when using the weighted nearest neighbour similarity measure (wNN), the fuzzy similarity measure (wFuzzy) given in expression (4) and the fuzzy non-linear similarity measure (wNL) given in expression (5). Following discussion with hospital staff, attribute *A* is the most important, followed by *E* and *V*. Therefore, the attribute weights are set at $w_A = 0.6$, $w_E = 0.3$ and $w_V = 0.1$ for all similarity measures. The weights for the fuzzy sets *High*, *Low* and *Avg* are set at $wHigh = 3$, $wLow = 2$ and $wAvg = 1$.

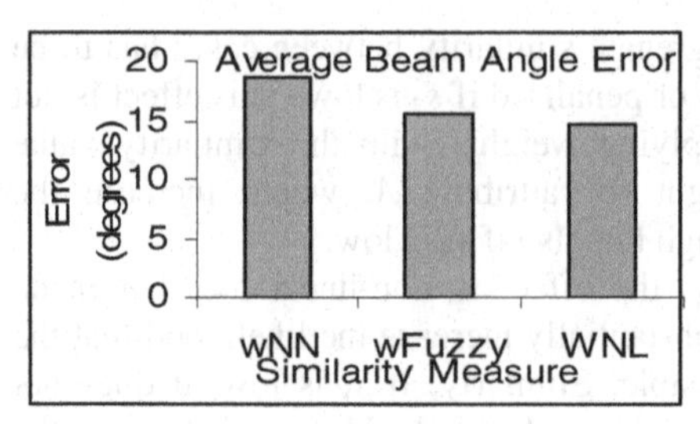

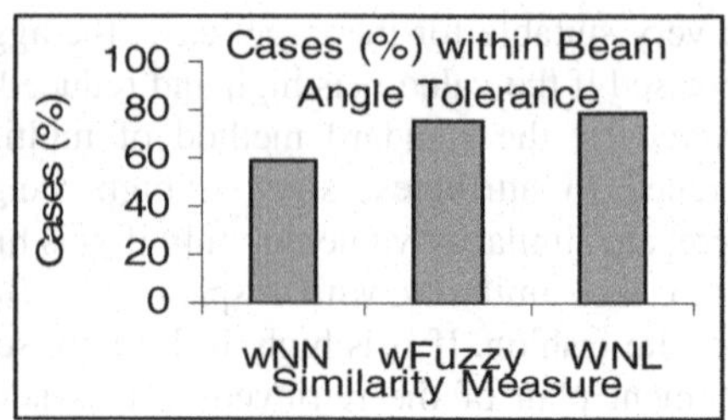

Fig. 3. a) The average error between the beam angles of the treatment plan of the retrieved case and the actual treatment plan. b) The percentage of cases, for which the similarity measure retrieves a treatment plan with acceptable beam angle difference.

We can see that the fuzzy similarity measure, wFuzzy, has a lower average error (18.6°) than the nearest neighbour method, wNN (15.5°). The error is further reduced with the fuzzy non-linear similarity measure, wNL (14.7°). Following consultation with hospital staff, an angle error of 20° per beam is acceptable for most cases. Figure 3b, shows the number of cases for which the similarity measure retrieves a case, in which the average error per beam is less than 20°. We can see that the percentage of cases within this beam angle tolerance is 58% for wNN, 75% for wFuzzy and 79% for wNL similarity measure.

4 Conclusion

In this paper, we described a case-based reasoning system (CBR) that is under development to aid oncologists with radiotherapy treatment planning in brain cancer.

The results confirm the assumption that similar treatment plans can be derived from similar geometric patient descriptors, as long as these descriptors are combined in an appropriate fashion in the similarity measure. Therefore, a key characteristic of our system is the fuzzy non-linear similarity measure that computes the similarity between cases in order to retrieve the case with the most suitable treatment plan. The fuzzy nature of the similarity measure allows aggregation of the similarity values with respect to single attributes to provide an accurate representation of the total similarity while taking into account the non-linear behaviour of attributes. The method was compared to the commonly used nearest neighbour method. The experimental results using real brain cancer patient cases show an improvement with the fuzzy non-linear similarity measure. The attribute weights and the non-linearity factor k have been set empirically, but the performance of the similarity measure can be improved by optimising the weights. A detailed weight analysis to identify the optimum values of the attribute weights and k is currently in process. We are aiming to find a set of weights that return the lowest average error with the leave-one-out cross validation method. Preliminary results show a significant reduction in beam angle error with optimized attribute weights.

In CBR systems the performance of the case retrieval mechanism and its similarity measure is crucial to the working of a CBR system. However, frequently, differences do exist between the target case and the retrieved case. Once the most similar case has been retrieved, its solution generally has to be adapted to the specific needs of the target case. Adaptation can be done by adjusting the beam configuration according to the geometric displacement in the location of the tumour and OAR structures of the target case compared to the retrieved case. Another method evaluates the plan using a treatment planning system to identify dose violations of OAR and confirm tumour coverage. The adaptation module of the CBR system then has to tweak the plan parameters of the retrieved case to resolve the violations. Adaptation can be carried out using if-then rules or by using another case-based reasoning inference mechanism.

Acknowledgements. The authors would like to thank the Engineering and Physics Science Research Council (EPSRC), UK, (Ref No. EP/C549511/1) and the City Hospital, Nottingham University Hospitals NHS Trust (UK) for their support.

References

1. Kolodner, J.: Case-Based Reasoning. Morgan Kaufmann, San Francisco (1993)
2. Oldham, M., Khoo, V.S., Rowbottom, C., Bedford, J.L., Webb, S.: A Case Study Comparing the Relative Benefit of Optimizing Beam Weights, Wedge Angles, Beam Orientations and Tomotherapy in Stereotactic Radiotherapy of the Brain. Phys. Med. Biol. 43, 2123–2146 (1998)
3. Haas, O.C.L., Bumham, K.J., Mills, J.A.: Hybrid Optimisation Technique for Radiotherapy Treatment Planning. In: Proceedings of the 1998 IEEE International Conference on Control Application, Trieste, Italy (1998)
4. Holt, A., Bichindaritz, I., Schmidt, R., Perner, P.: Medical Applications in Case-Based Reasoning. The Knowledge Engineering Review 20(3), 289–292 (2006)
5. Schmidt, R., Montani, S., Bellazzi, R., Portinale, L., Gierl, L.: Cased-Based Reasoning for Medical Knowledge-Based Systems. International Journal of Medical Informatics 64, 355–367 (2001)
6. Song, X., Petrovic, S., Sundar, S.: A Case-Based Reasoning Approach to Dose Planning in Radiotherapy. In: Wilson, D. (ed.) ICCBR 2007. LNCS (LNAI), vol. 4626, pp. 348–357. Springer, Heidelberg (2007)
7. Berger, J.: Roentgen: Radiation Therapy and Case-Based Reasoning. In: Proceedings of the Tenth Conference on Artificial Intelligence for Applications, pp. 171–177 (1994)
8. Cover, T.M., Hart, P.E.: Nearest Neighbor Pattern Classification. IEEE Transactions on Information Theory 13(1), 21–27 (1967)
9. Wang, R., Zeng, Y.: Nonlinear Nearest-Neighbour Matching and its Application in Legal Precedent Retrieval. In: Information Technology and Applications, ICITA 2005, vol. 1, pp. 341–346 (July 2005)
10. Cheng, C.B.: A Fuzzy Inference System for Similarity Assessment in Case-Based Reasoning Systems: An Application to Product Design. Mathematical and Computer Modelling 38, 385–394 (2003)
11. N.E.M.A.: DICOM - Digital Imaging and Communications in Medicine, `http://Medical.Nema.Org/`
12. Mishra, N., Petrovic, S., Sundar, S.: A Knowledge-light Nonlinear Case-based Reasoning Approach to Radiotherapy Planning. In: Proceedings of the 21st International Conference on Tools with Artificial Intelligence, ICTAI, pp. 776–783 (2009)

A Soft Computing Approach for Osteoporosis Risk Factor Estimation

Dimitrios Mantzaris[1], George Anastassopoulos[2], Lazaros Iliadis[3], Konstantinos Kazakos[4], and Harris Papadopoulos[5]

[1] Informatics Laboratory, Department of Nursing, Technological Educational Institute of Kavala, GR-68300, Didymoteicho, Greece
dmantzar@med.duth.gr
[2] Medical Informatics Laboratory, Democritus University of Thrace, GR-68100, Alexandroupolis, Greece
anasta@med.duth.gr
[3] Department of Forestry & Management of the Environment and Natural Resources, Democritus University of Thrace, GR-68200, Orestiada, Hellas
liliadis@fmenr.duth.gr
[4] Department of Orthopedics, University Hospital of Alexandroupolis, Medical School, Democritus University of Thrace, GR-68100
kazakosk@yahoo.gr
[5] Department of Computer Science and Engineering, Frederick University, 7 Y. Frederickou St., Palouriotisa, Nicosia 1036, Cyprus
h.papadopoulos@frederick.ac.cy

Abstract. This research effort deals with the application of Artificial Neural Networks (ANNs) in order to help the diagnosis of cases with an orthopaedic disease, namely osteoporosis. Probabilistic Neural Networks (PNNs) and Learning Vector Quantization (LVQ) ANNs, were developed for the estimation of osteoporosis risk. PNNs and LVQ ANNs are both feed-forward networks; however they are diversified in terms of their architecture, structure and optimization approach. The obtained results of successful prognosis over pathological cases lead to the conclusion that in this case the PNNs (96.58%) outperform LVQ (96.03%) networks, thus they provide an effective potential soft computing technique for the evaluation of osteoporosis risk. The ANN with the best performance was used for the contribution assessment of each risk feature towards the prediction of this medical disease. Moreover, the available data underwent statistical processing using the Receiver Operating Characteristic (ROC) analysis in order to determine the most significant factors for the estimation of osteoporosis risk. The results of the PNN model are in accordance with the ROC analysis and identify age as the most significant factor.

Keywords: Artificial Neural Networks, Probabilistic Neural Networks, Learning Vector Quantization Neural Networks, ROC analysis, Osteoporosis.

1 Introduction

Artificial Neural Networks, Genetic Algorithms and Fuzzy Logic are modern subfields of the emerging field of Computational Intelligence and are used in order to

H. Papadopoulos, A.S. Andreou, and M. Bramer (Eds.): AIAI 2010, IFIP AICT 339, pp. 120–127, 2010.

solve problems without prior knowledge and symbolic representation of their rules. Methods and algorithms derived from these subfields exhibit certain advantages and drawbacks when applied to real-life problems.

The huge mass of applications, in which ANNs can be used with promising results [1] is the essential element of their growth. ANNs have been widely implemented over the last years in many scientific areas, like the industry [2], image processing [3], communications [4] and medicine [5-7].

Physicians use medical protocols in order to diagnose diseases. Medical diagnosis is the accurate decision of upon the nature of a patient's disease, the prediction of its likely evolution and the chances of recovery by a physician, based on a set of clinical and laboratorial criteria, applicable to a particular case.

ANNs are suitable for disease prognosis since there is no need to provide a diagnostic rule to identify the disease, but a set of examples that represent the variations of the disease. ANNs have been used in many medical areas successfully, such as cardiology [8], oncology [9], orthopaedics [5], urology [9], surgery [6] etc.

The determination of the suitable ANN architecture is a critical point in some applications. An ANN with a small number of neurons involves deficient training, whereas a large number of processing elements limits its generalization ability. A typical approach for the proper design of an ANN is trial and error.

This study employees best-performance Probabilistic Neural Networks (PNN) [5] and also Learning Vector Quantization (LVQ) ones, which do not implemented in similar research efforts for osteoporosis estimation in the past [5], [10-13]. Moreover, the implemented ANNs are based on a vast data set comprising of 3426 records corresponding to an equal number of patients, whereas previous studies [10-13] were based on data sets with fewer cases.

The PNN with the best performance, as well as the Receiver Operating Characteristic (ROC) analysis were used for the investigation of the essential diagnostic factors of osteoporosis.

2 Osteoporosis and Data

Osteoporosis is a common bone disease that leads to an increased risk of fracture. In osteoporosis, the Bone Mineral Density (BMD) is reduced; the bone microarchitecture is disrupted whereas the concentration and the variety of proteins in bones are altered. Apart from the direct physical implications of a fracture, such as pain and inconvenience, osteoporotic fractures (involving the hip or the spine) are a major cause of morbidity and mortality.

Osteoporosis is a most frequent disease for women after menopause, when it is called *postmenopausal osteoporosis*, but it may also be developed in men. It may occur in anyone in the presence of particular hormonal disorders and other chronic diseases or as a result of medications, specifically glucocorticoids. In this case the disease is called *steroid or glucocorticoid induced osteoporosis*. It is a fact that a percentage as high as 75% of the women with osteoporosis disregards this disorder.

The diagnosis of osteoporosis is based on measuring the BMD of the patient. The most popular laboratorial method for doing this is the Dual Energy X-ray Absorptiometry (DEXA) scan. This examination, applied to specific bones, is costly and based on radiation absorption from the patient, so it is not recommended in all orthopedic cases.

The osteoporosis data, which were used for the design of the PNN and LVQ models, were obtained from the Orthopedic Clinical Information System of Alexandroupolis' University Hospital in Greece. This information system consists of a main database (MDB) where all medical records and statistical data is kept. For secure and real-time connection to the database, the information has been distributed, designing a replication scheme of the related database MDB. Although, the data belong to different instances, they seem to be included in a single database [14].

The healthcare specialists access the medical data via a friendly and ergonomic web interface. The users of the Orthopedic Clinical Information System have remote access to the MDB either by their desktop personal computers or by Personal Digital Assistants (PDAs) wireless connected to database.

Data related to four diagnostic factors were stored for each patient, namely: age, sex, height and weight. The diagnostic factors for osteoporosis risk prediction as well as their values are presented in Table 1 and described in detail in [5].

Table 1. Osteoporosis diagnostic factors coding

NN inputs	1	2	3	4
Variables	Age	Sex	Height	Weight
Coding	Numeric value (years)	1 (Female) 2 (Male)	Numeric value (cm)	Numeric value (kg)

The estimation of the osteoporosis risk factor was based on a T-score value, which is the patient's bone density compared to the normally expected in a healthy young adult of the specific sex. A T-score value less than or equal to -2.5 implies the development of osteoporosis, whereas a T-score greater than -2.5 is related to a normal person. These values of bone densitometry were divided into 2 classes, as presented in Table 2.

As it has already been mentioned, the present study is based on a data set comprising of 3426 records. Actually, the vast majority 3346 cases (97.7%) are related to women and only 80 cases (2.3%) are related to men with osteoporosis symptoms. The 1083 persons are pathological cases, as their T-score was less than or equal to -2.5.

Table 2. T-score values coding

T-score Value	Classification
$\leq$ -2.5	1
> -2.5	2

This data set was divided into a set of 2426 records for the training of both PNNs and LVQ ANN and another set of 1000 records for the evaluation and testing of their classification efficiency.

3 PNNs and LVQ Models in Medical Diagnosis Prediction

In this study, two osteoporosis risk prediction models based on a non-symbolic learning approach are presented. These proposed models use ANNs, particularly Probabilistic Neural Networks (PNNs) and Learning Vector Quantization (LVQ) ones.

PNNs are a variant of Radial Basis Function Networks and approximate Bayesian statistical techniques. The modus operandi of PNNs is familiar to the human decision making approach [8]. The patterns' classification by PNNs is based on Parzen's Probabilistic Density Function (PDF) estimator [8]. A PNN is a feed-forward neural network, consisting of two layers. The first layer, which consists of radial basis function units, computes the distances between each input vector and the training input vectors [15]. The obtained vector's values indicate the propinquity of the input vector to a training input. The second layer, which is a competitive layer, sums these contributions for each class of inputs to produce a vector of probabilities as output. These probabilities indicate the likelihood of the input vector to be classified in each of the available classes. The transfer function of the second layer detects the maximum probability and classifies the input vector to the corresponding class.

More details about the specific implementation of the used PNNs can be found in [5]. The PNN learns with exposure to training patterns with a single pass without the necessity of extended training. This feature implies that the PNN may manage large quantities of data much faster that other ANN architectures. The maximum accuracy requires the determination of a "smoothing factor", which represents the width of the calculated Gaussian curve for each probability density function.

The LVQ neural networks were first proposed by Kohonen. A LVQ network is a two-layer feed-forward network, consisting of a competitive layer and a linear layer. The first layer learns to classify the input vectors in the same way as the competitive layer of PNNs. The second layer transforms the competitive layer's classes into desired classifications defined by the designer of LVQ network. The classes of the competitive layer are called subclasses while the classes of the linear layer are called target classes [16]. Both of the competitive and linear layers have one neuron per class. The number of neurons for the hidden layer is always larger than the number of output neurons.

The LVQ method is used in training ANNs for pattern classification, where each output represents a particular class. Each class is referred by a vector of weights that sequentially, represents the centers of the classes. The training data set is used several times during the training phase in a random order. The training of LVQ ANNs is terminated when classes remain stable or a specific number of iterations has been carried out. A trained LVQ neural network is a vector comparator. When a new vector is presented to the input layer of a LVQ ANN it will be classified to a class with the closest center.

4 Experimental Results

The MATLAB Neural Network Toolbox was used for the construction and the assessment of the classification efficiency of the developed PNN and LVQ neural networks.

The effectiveness as well as the user-friendly interface conduced to the selection of the specific software for the purpose of this study.

The implemented PNN and LVQ architectures, the number of neurons in the input, hidden and output layers, as well as the transfer functions used in the hidden and the output layers for each network type, are summarized in Table 3. The spread of the radial basis function, representing the width of the Gaussian curve, varied during the experimental design phase of the PNNs. The three level architecture of the optimal PNN is 4-2426-2, whereas the spread value offering the best performance was equal to 0.6.

The three level architecture of the optimal LVQ neural network was 4-4-2. A statistical analysis over osteoporosis data set indicated that the data is grouped into four subclasses. As it was mentioned, the number of hidden neurons of an LVQ neural network equals to the number of subclasses, so the implemented LVQ has four neurons for its hidden layer.

Table 3. The optimal ANNs for the osteoporosis risk factor prediction

ANNs	Input Layer - Neurons	Hidden Layer		Output Layer	
		Transfer Function	Neurons	Function	Neurons
PNN	4	Radial Basis (spread = 0.6)	2426	Competitive (compet)	2
LVQ	4	Competitive	4	Linear	2

The obtained results from the optimal PNNs and LVQ ANNs for the osteoporosis risk prediction are summarized in Table 4. The performances of the proposed ANN models are depicted in columns two to seven. In this study, the classification efficiency of PNNs and LVQ ANNs was based on two criteria. The first criterion was the percentage (numbers in parenthesis correspond to real cases) of correct classified cases over testing (1000 cases), training (2426 cases) and overall data set (3426 cases). These percentages are recorded in columns two to four of Table 4. An important criterion for the evaluation of the ANN's generalization ability is the proper classification of the pathological cases. The implemented ANNs have to recognize patients with increased osteoporosis risk factor, so the fields five to seven store the percentage of pathological cases that have been categorized correctly, for testing, training and overall data, respectively.

Table 4. Experimental results using PNN and LVQ architectures

ANN model	Percentage of Successful Prognosis			Percentage of Successful Prognosis Over Pathological Situations		
	Testing Set	Training Set	Overall Set	Testing Set	Training Set	Overall Set
PNN	83.10 (831)	84.95 (2061)	84.41 (2892)	95.90 (304)	96.87 (742)	96.58 (1046)
LVQ	82.60 (826)	82.11 (1992)	82.25 (2818)	94.95 (301)	96.48 (739)	96.03 (1040)

It is shown clearly that PNN outperforms LVQ neural network, as the percentages of successful prognosis for overall and pathological cases for testing, training and for the entire data set are higher. Therefore, the PNN has been characterized as the optimal choice of ANN for being potentially used towards osteoporosis risk prediction.

The PNN was used for the significance estimation of the diagnostic risk factors and it was trained with 4 pruned data sets. Each of these sets used values related to 3 diagnostic factors instead of the total 4. In other words, one different diagnostic factor was omitted in each data set. The PNN was constructed and tested with the pruned data sets, and the obtained results are recorded in Table 5. The first column presents the diagnostic factor which was not considered during the PNN's development. The 2nd to 7th columns depict the performance of PNN for each of pruned data sets. The results of the 2nd to 4th columns are the percentages of successful prognosis over testing (1000 cases), training (2426 cases) and overall pruned data sets (3426 cases). The 5th to 7th columns record the percentage of successful prognosis of pathological cases for testing, training and overall pruned data sets.

Table 5. Experimental results using Pruned PNN

Omitted Factor	Percentage of Successful Prognosis			Percentage of Successful Prognosis Over Pathological Situations		
	Testing Set	Training Set	Overall Set	Testing Set	Training Set	Overall Set
Age	82.90	83.02	82.98	94.95	96.01	95.75
Sex	83.10	84.87	84.35	95.58	96.87	96.49
Height	83.70	82.98	83.19	99.68	99.22	99.35
Weight	83.00	82.85	82.89	99.68	99.87	99.82

The results of the pruned PNN were compared to the ones related to the full-sized PNN. The performance of PNN dropped dramatically in the case of the omission of the "*age*" feature. The PNN's effectiveness remained stable when the "*sex*" diagnostic factor was omitted, as the vast majority is women with osteoporosis symptoms. The absence of "*height*" improved the PNN's performance.

The omission of "*weight*" did not improve the classification efficiency of the pruned PNN according to the percentages of successful prognosis for testing, training and overall data set, however the percentages of successful prognosis over pathological situations are better than ones of the full-sized data set PNN. The results of Table 5 prove that the height and weight diagnostic factors reduce the generalization ability of PNN.

5 ROC Analysis

The available data set of osteoporosis records underwent statistical processing using Receiver Operating Characteristic (ROC) analysis for the determination of the contribution of each diagnostic factor to osteoporosis risk prediction [17].

The essential statistic from ROC is the Area Under Curve (AUC) for each diagnostic factor. The mathematical expressions for AUC, sensitivity and specificity are explained in [17]. The obtained results are summarized in Table 6. The 1st line presents the osteoporosis' diagnostic factors, while the AUC of ROC for each factor is recorded in the 2nd line of the Table 6.

As mentioned above, the AUC is the value of visual depiction of ROC. An AUC value greater that 0.5 indicates the importance of the contribution of a specific diagnostic factor. According to results of Table 6, the most important diagnostic factor is Age. The Sex's AUC is equal to 0.5, so this diagnostic factor seems not to have important contribution for osteoporosis risk estimation. This result was expected as the used data set consists of 3346 women, while the number of men is negligible.

Table 6. Area of diagnostic factors' curves

Diagnostic Factor	Age	Sex	Height	Weight
AUC	0.646	0.503	0.560	0.641

The results of pruned PNN and ROC analysis are convergence in terms of proposed diagnostic factors for osteoporosis risk prediction. Consequently, the Age factor is strongly recommended to be recorded for each patient.

6 Conclusions

Despite the fact that osteoporosis is an enormous public health problem with huge recourses required to deal with the immediate and long-term effects of fractures, no investigation has been performed on the effectiveness and weightiness of the various diagnostic factors in the clinical evaluation of patients. This paper evaluates the performance of two ANN topologies for osteoporosis risk prediction. The PNN outperformed the LVQ neural network in terms of successful prognosis of osteoporosis.

The PNN, which had the best performance, was used for further processing of the osteoporosis prediction data set. The PNN was designed, implemented and tested with pruned versions of the data set, for the evaluation of the most significant diagnostic factor towards osteoporosis risk. It was concluded that "age" is the most significant factor towards osteoporosis prediction.

Moreover, the available data were processed using the ROC analysis approach. The results of the optimal PNN were in accordance with the output of the ROC analysis.

In future work, it would be preferable to make the same experiments in an extended data set consisting of more male patients and more factors (such as smoking), which affect the presence of osteoporosis. The limited number of male data records used was an obstacle in the effort to reach generalized conclusions related to the "*sex*" factor. The additional diagnostic factors would contribute to an extensive study in terms of the osteoporosis disease. Also the scoring of the T-code values can be implemented by the use of fuzzy logic, offering a more comprehensive approach.

References

1. Dayhoff, J., DeLeo, J.: Artificial Neural Networks Opening the Black Box. Cancer Supplement 91, 1615–1635 (2001)
2. Iliadis, L.: An Intelligent Artificial Neural Network Evaluation System Using Fuzzy Set Hedges: Application in Wood Industry. In: The Annual IEEE International Conference on Tools with Artificial Intelligence (2007)

3. Hisashi, A., Tsuyoshi, O., Takahashi, N., Tanaka, M.: Sigma-Delta Cellular Neural Network for 2D Modulation. Neural Networks 21, 349–357 (2008)
4. Haralambous, H., Papadopoulos, H.: 24-hour Neural Network Congestion Models for Highfrequency Broadcast Users. IEEE Transactions on Broadcasting 55, 145–154 (2009)
5. Mantzaris, D., Anastassopoulos, G., Lymperopoulos, K.: Medical Disease Prediction Using Artificial Neural Networks. In: 8th IEEE International Conference on BioInformatics and BioEngineering (2008)
6. Papadopoulos, H., Gammerman, A., Vovk, V.: Confidence Predictions for the Diagnosis of Acute Abdominal Pain. In: AIAI 2009, pp. 175–184 (2009)
7. Economou, G.-P.K., Mariatos, E., Economopoulos, N., Lymberopoulos, D., Goutis, C.: FPGA Implementation of Artificial Neural Networks: An Application on Medical Expert Systems. In: 4th Int. Conf. on Microelectronics for Neural Networks and Fuzzy Systems, pp. 287–293 (1994)
8. Orr, R.: Use of a Probabilistic Neural Network to Estimate the Risk of Mortality after Cardiac Surgery. J. Medical Decision Making 17, 178–185 (1997)
9. Anagnostou, T., Remzi, M., Djavan, B.: Artificial Neural Networks for Decision-Making in Urologic Oncology. Reviews in Urology 5, 15–21 (2003)
10. Lemineur, G., Harba, R., Kilic, N., Ucan, O., Osman, O., Benhamou, L.: Efficient Estimation of Osteoporosis Using Artificial Neural Networks. In: 33rd Annual Conf. of IEEE Industrial Electronics Society (IECON), pp. 3039–3044 (2007)
11. Chiu, J., Li, Y., Yu, F., Wang, Y.: Applying an Artificial Neural Network to Predict Osteoporosis in the Elderly. Studies in Health Technology and Informatics 124, 609–614 (2006)
12. Mohamed, E., Maiolo, C., Linder, R., Pöppl, S., De Lorenzo, A.: Artificial Neural Network Analysis: A Novel Application For Predicting Site-Specific Bone Mineral Density. Acta Diabetologica 40, 19–22 (2003)
13. Rae, S., Wang, W., Partridge, D.: Artificial Neural Networks: A Potential Role in Osteoporosis. J. of the Royal Society of Medicine 92, 119–122 (1999)
14. Anastassopoulos, G., Kolovou, L., Lymperopoulos, D.: A Spatial Distributed Approach for Electronic Medical Record Administration. In: Recent Advances in Communications and Computer Science. Electrical and Computer Engineering Series, A series of Reference Books and Textbooks, WSEAS, pp. 407–412 (2003)
15. Iliadis, L.: Intelligent Information Systems and applications in risk estimation. Stamoulis Publishing, Thessaloniki (2007)
16. Gray, R.: Vector Quantization. IEEE Acoustic, Speech, and Signal Processing Magazine 1, 4–29 (1984)
17. Streiner, D., Cairney, J.: What's Under the ROC? An Introduction to Receiver Operating Characteristics Curves. The Canadian Journal of Psychiatry 52, 121–128 (2007)

Protein Secondary Structure Prediction with Bidirectional Recurrent Neural Nets: Can Weight Updating for Each Residue Enhance Performance?

Michalis Agathocleous[1], Georgia Christodoulou[1], Vasilis Promponas[2], Chris Christodoulou[1,*], Vassilis Vassiliades[1], and Antonis Antoniou[1]

[1] Dept. of Computer Science
[2] Dept. of Biological Sciences,
University of Cyprus,
P.O. Box 20537, 1678 Nicosia, Cyprus
michalis.agathocleous.09@ucl.ac.uk, cs06gc1@cs.ucy.ac.cy,
vprobon@ucy.ac.cy, cchrist@cs.ucy.ac.cy
v.vassiliades@cs.ucy.ac.cy, a.antoniou@cs.ucy.ac.cy

Abstract. Successful protein secondary structure prediction is an important step towards modelling protein 3D structure, with several practical applications. Even though in the last four decades several PSSP algorithms have been proposed, we are far from being accurate. The Bidirectional Recurrent Neural Network (BRNN) architecture of Baldi et al. [1] is currently considered as one of the optimal computational neural network type architectures for addressing the problem. In this paper, we implement the same BRNN architecture, but we use a modified training procedure. More specifically, our aim is to identify the effect of the contribution of local versus global information, by varying the length of the segment on which the Recurrent Neural Networks operate for each residue position considered. For training the network, the backpropagation learning algorithm with an online training procedure is used, where the weight updates occur for every amino acid, as opposed to Baldi et al. [1], where the weight updates are applied after the presentation of the entire protein. Our results with a single BRNN are better than Baldi et al. [1] by three percentage points (Q3) and comparable to results of [1] when they use an ensemble of 6 BRNNs. In addition, our results improve even further when sequence-to-structure output is filtered in a post-processing step, with a novel Hidden Markov Model-based approach.

Keywords: Protein Secondary Structure Prediction, Bidirectional Recurrent Neural Networks, Bioinformatics and Computational Biology.

1 Introduction

Proteins are linear polymers of amino acids, and their complement in any living cell provides a vast repertoire of functions required for maintaining life. Their

* Corresponding Author.

H. Papadopoulos, A.S. Andreou, and M. Bramer (Eds.): AIAI 2010, IFIP AICT 339, pp. 128–137, 2010.

functionality is determined by the detailed three dimensional arrangement of their constituent atoms, which specifies the shape of the molecule and the ways it can potentially interact with other biomolecules. Biochemical methods for experimental determination of protein tertiary structure are expensive, time consuming and frequently ineffective. On the contrary, modern high-throughput molecular biology techniques provide us daily with an increasing number of (putative) amino acid sequences, corresponding to proteins of unknown structure and function. It is widely accepted that protein sequence determines its 3D structure [2]; however, the exact mapping from the 1D to the 3D space has not yet been elucidated. Nevertheless, knowledge of local regular arrangements of amino acid residues (secondary structural elements), may indicate the structure adopted by a protein chain and provide useful constraints for further structural modelling.

The protein secondary structure prediction (PSSP) problem has been addressed with different computational approaches during the last four decades. Mainly due to the lack of necessary volumes of structural data, only relatively recently were machine learning methods (including artificial neural networks - NNs) introduced to this field. Such methods have resulted in considerably higher predictive performance compared to preceding empirical or statistical methods, with per residue accuracies (Q3 [3]) ranging from 63% to 76% [4], [5]. In particular, a fully connected feed forward NN and a single hidden layer, with a local input window of 13 amino acids using orthogonal (or one-hot) encoding has been applied by Qian and Sejnowski [6] reaching accuracy performance of Q3 = 64.3%. Based upon this work, Rost and Sander developed the so–called PHD method [7], [8], which used various machine learning methods including early stopping and ensemble of averages of different NNs. Their most substantial improvement was, however, the use of multiple sequence alignments (MSA) that enrich the input with evolutionary information [8]. Through their work the accuracy performance reached to Q3 = 74%. Another interesting approach, similar to PHD, is the work of Salamov and Soloveyev [9] that introduced a variant of the nearest-neighbor approach that could achieve similar accuracy (Q3 = 73.5%) using a single sequence as input. Cuff and Barton [10] built another MSA-enabled feed forward NN (of a larger size however) with comparable accuracy.

A major breakthrough in the field of PSSP is considered to be the work of Baldi and colleagues [1], who introduced a Bidirectional Recurrent Neural Network (BRNN). The main motivation of this work was the inability of feed forward NNs with a fixed-width sliding window to capture long-range dependencies. The aforementioned BRNN takes as input a fixed-width window centered each time at the residue of interest. However, it attempts to predict the secondary structure for the target residue by considering not only the local context (i.e., neighboring residues in the window) but also residues located on the left and right of the target residue [1]. These bidirectional dynamics proved to be able to capture both upstream and downstream information. When the BRNN processed evolutionary information in the form of MSAs, the predictive performance reached up to Q3 = 73.6% [1], which was increased to 75.1% on average, when an ensemble of

6 predictors was used. Along the same lines Chen and Chaudhari [11] developed a cascaded architecture, consisting of two BRNNs, where the second network filters the output of the first. As they report, their system achieves 74.38% (Q3) accuracy with a Segment Overlap (SOV [12]) score of approximately 66%.

In the introductory work of BRNNs for PSSP [1], for each position in a sequence presented to the network an input window is formed centered around this position. During the training phase, the protein is processed in its entirety before the weight updates are made. In this paper, our aim is to investigate how the prediction accuracy could be improved by: (i) updating the weights at every residue, which in a way constitutes a form of dynamic training and is more context-sensitive, and (ii) using different filtering approaches (a novel method based on Hidden Markov Models and a cascaded feed forward Artifical NN).

The rest of the paper is organised as follows: section 2 presents the methodology followed by the results and discussion in section 3; section 4 gives the conclusions and future work.

2 Methods

2.1 Data Collection and Preprocessing

Sequence Similarity-Based Redundancy Reduction: In order to train and validate the BRNN, we need a set of high quality data consisting of proteins with experimentally determined 3D structures deposited in the RSCB Protein Data Bank (PDB) (URL: http://www.pdb.org/, accessed 20 April 2009). Moreover, the resulting dataset should be maximal (in order to capture the knowledge we currently have available on protein structures) but also non-redundant, to avoid poor generalisation of the BRNN. For this purpose we utilised the PDB-Select25 dataset (URL: http://bioinfo.tg.fh-giessen.de/pdbselect/, accessed 20 April 2009), which is regularly produced by analysis of the PDB with the algorithm described in [13]. This dataset contained 4019 polypeptide chains that shared less than 25% overall pair-wise sequence identity.

Data Selection Criteria: It is of great importance to choose a suitable data set so that the BRNN can be properly trained, and several criteria should be fulfilled for the selection process:

- We retained only entries determined by X-ray diffraction, for which we can apply the resolution as a quantitative selection measure. In particular, a 3.0 Å threshold was used to discard structures of insufficient/questionable quality.
- We discarded entries with physical chain breaks, as empirically identified by at least one pair of successive C_α atoms with a distance longer than 4.0 Å, as well as proteins with large segments of undefined secondary structure.
- Chains with a length of less than 30 amino acids were discarded.
- DSSP [14] (see below), should provide a valid output file for any chain retained in the dataset.

We started with a dataset containing a total of 4019 protein chains, of which 2656 corresponded to structures determined by X-ray crystallography. Following the above procedure we ended up with 612 protein chains.

Secondary Structure Assignment: We have relied for secondary structure (SS) assignments on the widely used DSSP program [14]. More specifically, we reduce class assignments from the eight secondary structure (SS) types provided by DSSP (i.e. α-helix (H) , 3_{10}-helix (G) , π-helix (I) , β-strand (E) , β-bridge (B) , β-turn (T), bend (S) and 'other' ('.')) into three SS states (Helical: H, G; Extended: E, B; Random coil/Loop: I, T, S, '.'). From here onwards, we refer to these states as H, E and C respectively. DSSP results were fetched from the DSSP website (URL: http://swift.cmbi.kun.nl/gv/dssp/, accessed 20 April 2009) and transformed to the 3-state representation by an in-house parser. Since several protein chains contain segments of disordered regions where DSSP does not produce any output, for the purposes of this work we have decided to exclude any such entries.

Multiple Sequence Alignment Preprocessing: MSAs have been shown to significantly increase protein secondary structure prediction accuracy in recent applications [8]. This is because structure is considered to be more conserved than sequence [8]. Every position within an alignment contains an evolutionary record. We encode each input residue with a 20-dimensional vector, where ordinates correspond to the frequencies of the different 20 amino acid residues at the respective column of the MSA. Apparently, encoding for single sequences at the input reduces to the orthogonal encoding scheme. For the polypeptide chains collected with the previously described procedure, we have utilised unweighted profiles available from the HSSP database [15].

2.2 Modified Training Procedure for the BRNN Architecture

A NN must accept the amino acid at its input with all the necessary information that it needs in order to produce the right output. Taking into account that the formation of different secondary structural elements depends on the interaction between neighboring-in-space (not necessarily in sequence) amino acid residues, we chose the BRNN architecture for its ability to encapsulate information included in the amino acid residues that are coming before and after the residue at the examined position t; where t denotes the discrete time index in $[1, T]$, with T being the total length of the protein chain.

The BRNN architecture consists of two Recurrent Neural Networks (RNNs) and a Feed Forward Neural Network (FFNN). The RNNs are used for taking into account the information contained in a local segment of length L_s centered around position t. The Forward Recurrent Neural Network (FRNN) processes $j = \frac{L_s - 1}{2}$ amino acid residues located on the left side of the segment, computing iteratively from the far left side of the window (i.e., in position $t - j$) and moving towards the right until position t (inclusive) by taking into account a sliding sub-segment of length S_l. The Backward Recurrent Neural Network (BWRNN) processes the amino acids located on the right side of t, in a similar symmetric way.

During the recurrent network processing, a kind of memory is being formed since the NN correlates each sequence separately and holds an internal temporary knowledge [16]. The output from the two recurrent NNs and the output from the FFNN are correlated and predict the secondary structure state for residue t as indicated in equation 1.

$$O_t = \eta(F_t, B_t, I_t) \tag{1}$$

where $\eta(.)$ is realised by the FFNN, F_t is the forward (upstream) context, B_t is the backward (downstream) context and I_t is the input vector at time (sequence position) t. In the current work, we use an input vector encoding a single residue (corresponding to a window size of unity in Baldi's [1] implementation).

The contextual information from the protein is gathered into a pair of vectors F_t and B_t. Only after the F_t and B_t are computed, the algorithm can predict the state (as in [1]). In order for the amino acids to be examined, two learnable non-linear state transition functions $\phi(\cdot)$ and $\beta(\cdot)$ are applied. Algorithmically this is shown below:

1: **for** every sequence **do**
2:   **for** $t = 1$ to T **do**
3:     **for** $i_f = t - \frac{L_s-1}{2}$ to t **do**
4:       **for** $i_f^{'} = i_f - S_l$ to $i_f - 1$ **do**
5:         **if** $i_f^{'} < t - \frac{L_s-1}{2}$ **then**
6:           $F_{i_f^{'}} \leftarrow 0$
7:         **end if**
8:       **end for**
9:       $F_{i_f} \leftarrow \phi_t(\gamma\ F_{i_f-S_l}, \gamma\ F_{i_f-S_l+1},\ \ldots\ , \gamma\ F_{i_f-1}, I_{i_f})$
10:     **end for**
11:     **for** $i_b = t + \frac{L_s-1}{2}$ to t step -1 **do**
12:       **for** $i_b^{'} = i_b + S_l$ to $i_b + 1$ step -1 **do**
13:         **if** $i_b^{'} > t + \frac{L_s-1}{2}$ **then**
14:           $B_{i_b^{'}} \leftarrow 0$
15:         **end if**
16:       **end for**
17:       $B_{i_b} \leftarrow \beta_t(\gamma^{-1}\ B_{i_b+S_l}, \gamma^{-1}\ B_{i_b+S_l-1},\ \ldots\ , \gamma^{-1}\ B_{i_b+1}, I_{i_b})$
18:     **end for**
19:     $O_t = \eta_t(F_t, B_t, I_t)$
20:     *update* ϕ_t, β_t, η_t
21:   **end for**
22: **end for**

where $\gamma \in (0, 1]$ is a modified shift operator, which in effect adds a constant weight based on the importance given to the outputs of the FRNN and BWRNN. Intuitively, we chose $\gamma < 1$ (thus $\gamma^{-1} > 1$) to reflect the fact that protein chains are synthesised from the N-terminal to the C-terminal (i.e., from the left to the right side of the sequence respectively) with some secondary structural elements forming co-translationally [17].

Once the data located within the input vector enters the BRNN, the Mean Square Error function is applied and is used by the Backpropagation algorithm [18] for the BRNN to be trained. Training is performed based on two alternative output encoding schemes: (i) an orthogonal, and (ii) a 'winner-take-all' (WTA). The former scheme has three output units with binary values giving eight possible combinations, three of which are assigned to the three reduced SS states, and the rest are arbitrarily considered to be classified as random coil. The WTA encoding scheme, has three output units as well (corresponding to the reduced SS states) and assigns the SS state of the winning neuron as the prediction for the examined residue. For both schemes, once the error is calculated, the delta rule is applied to update the network weights.

3 Results and Discussion

A set of optimal NN parameters were empirically found following experimentation. For training the network, the dataset of 612 polypeptide chains was randomly split to 513 proteins for training and 99 proteins for testing.

Firstly, we explored BRNN architectures with two hidden layers. More specifically, the FFNN was composed of two fully interconnected hidden layers with 12 neurons each, whereas the BWRNN and FRNN consist of two hidden layers each, with the first hidden layer having 13 and the second 12 neurons.

Our first experiments, utilising single sequences at the input, achieved a highest result of 66.59% (Q3) with optimal parameters: $\gamma = 0.7$, learning rate $\alpha = 0.8$, momentum $m = 0.0$, $L_s = 15$, $S_l = 3$, and orthogonal output encoding.

The next set of experiments was performed using the MSA profiles as input. For 11 out of the 612 protein chains of the initial data set an HSSP profile was not available, thus the data set was slightly reduced to a total of 601 chains, which was again randomly split into a training and a test set of 504 and 97 protein chains respectively. Initially, the predictive accuracy value reached on average 70.82% (Q3, with the aforementioned optimal parameters), almost 4% higher than when training with single sequences. In order to improve the performance, we decided to apply a randomisation procedure on the data at every iteration (i.e., when all the training data is passed through the network). We consider randomisation to be equivalent to the insertion of noise during training, which could theoretically improve the results. As predicted, randomisation does indeed improve the prediction accuracy, as illustrated in Table 1. It has to also be noted that large L_s values gave better results than smaller ones. We experimented with L_s values up to 60 (where data sets were modified accordingly to exclude sequences shorter than L_s), and from the results it was obvious that the optimal L_s was 31 (which concurs with the input window size of [1]). As it can be seen from Table 1 the best result is 73.92% (Q3) accuracy; the corresponding SOV measure is 63.02. Experimentation with varying the number of neurons in all the layers of the constituent networks of the BRNN did not give any significant improvement in the results reported above.

Secondly, we decided to reduce the complexity of the BRNN architecture by using a single hidden layer. We also changed the output encoding to WTA. In

Table 1. Prediction accuracy results with a BRNN architecture with two hidden layers for different sizes of the local context window L_s. MSA profiles were used as input (for all parameter values, see text).

L_s	15	17	19	21	23	29	31
Q3%	72.94	73.14	72.79	71.80	73.09	73.41	73.92

addition, we experimented with different hidden layer sizes and the optimal results (shown in Table 2, columns 1 and 2) were obtained with: (i) a FFNN of 15 hidden neurons combined with RNNs of 17 hidden neurons (rows 1-4), and (ii) a FFNN of 51 hidden neurons combined with RNNs of 41 hidden neurons (rows 5-8). In order to assess whether these predictions are significantly different, we performed four repetitions of each configuration (see table 2), and descriptive statistics were calculated (data not shown). For the Q3s resulting for the two configurations no statistically significant difference could be observed, with a non-parametric Mann-Whitney test (p-value=0.57). As we can see from the results shown in table 2 there was a significant improvement of the Q3 prediction accuracy measure (up to 76.07%) and a SOV of up to 65.36 compared to architectures with two hidden layers.

Table 2. Prediction accuracy results with a BRNN architecture with one hidden layer, randomized input, WTA output encoding and MSA profiles as input. The first four rows correspond to a BRNN architecture with a FFNN of 15 hidden neurons and RNNs of 17 hidden neurons and the last four rows to a BRNN architecture with a FFNN of 51 hidden neurons and RNNs of 41 hidden neurons. The rightmost columns correspond to the performance metrics after filtering with an HMM and a feed forward ANN.

Q3%	SOV	Q3% HMM	SOV HMM	Q3% ANN	SOV ANN
76.07	64.32	76.57	70.32	76.60	71.90
75.32	64.66	75.17	67.67	75.47	72.90
75.14	62.12	76.38	68.13	75.59	63.05
74.81	65.36	74.69	67.99	75.11	69.88
75.26	64.53	75.75	69.55	76.04	70.40
75.49	64.45	75.90	69.51	76.33	72.91
76.07	62.43	76.84	69.15	76.44	71.19
75.26	65.21	75.90	69.09	73.61	71.32

Knowing that filtering the initial sequence-to-structure network outputs improves the prediction accuracy (see [11] and references therein), we decided to explore this possibility on our BRNN architectures which gave the best results (i.e., with one hidden layer). Two filtering approaches were employed, one based on a novel Hidden Markov Model (HMM) and one based on a cascaded Artificial NN (ANN, similar to the structure-to-structure network of [7]) for comparison. According to the results (shown in table 2, columns 3-6), both filtering approaches improve the results, in particular there is a significant increase in the SOV values (in the order of up to 10%). More specifically, for both configurations the HMM filtering was producing significantly higher SOV values (Mann-Whitney test: p-value = 0.03 for both cases) compared to the unfiltered results. In addition, the ANN filtering produced significantly higher SOV values only for the second configuration (Mann-Whitney test: p-value =0.03) but not for the first one (p-value=0.2). The increase in SOV was expected, since both filtering procedures are known to eliminate biologically irrelevant predictions, especially in cases where isolated residues are predicted in a SS state.

For our best achieved results, further analysis regarding the details of the prediction in different SS states has been conducted. More specifically, we counted all possible instances of observed versus predicted outcomes, through which confusion matrices were created, as shown in Table 3.

Table 3. Confusion matrices, showing the distribution of predictions for the three secondary structure states, for the best performing configurations: (a) for the best configuration listed in table 1 giving a Q3 of 73.92%, (b) best results of the unfiltered configuration shown in table 2 giving a Q3 of 76.07%, (c) best results of the configuration shown in table 2 filtered with a HMM giving a Q3 of 76.57%, and (d) best results of the configuration shown in table 2 filtered with an ANN giving a Q3 of 76.60%. Note that the displayed values correspond to the fraction of predicted residues with a given observed state, expressed as a percentage.

Obs *vs* Pred	H	E	C	H	E	C
		(a)			(b)	
H	62.63	3.39	33.98	71.65	5.46	22.89
E	7.22	45.03	47.74	7.50	59.85	32.65
C	6.55	4.31	89.12	8.01	7.20	84.79
		(c)			(d)	
H	73.24	4.39	22.37	72.74	4.16	23.37
E	5.77	60.62	33.61	0.00	57.99	42.01
C	8.75	7.24	84.01	0.00	14.21	85.79

Three are the main observations from the confusion matrices of Table 3:

1. Helices (H) and loops (C) are pretty accurately predicted.
2. Extended structures (E) suffer from under-prediction. However, the novel HMM-based filtering method seems to partially overcome this problem.
3. Most of the incorrect predictions involve the loop secondary structure state.

Observation 1 was expected since (i) most other works on PSSP report similar trends, and (ii) H and C are the most populated states in our datasets as opposed to strands (observation 2). For observation 3, we believe this could be an artifact of our output encoding scheme (see Section 2.2). It is worth pointing out that the ANN filtering method (see Table 3d) completely eliminates false predictions in the 'H' class.

4 Conclusions

In an attempt to tackle the PSSP problem, our modified training procedure for the BRNN, where the main modification is the updating of the weights for each residue, gives a prediction accuracy of Q3=76.07% with respective SOV = 64.32, which is better than the state-of-the-art results of [1] (Q3=73.6%) and clearly comparable to the results obtained in [1] when an ensemble of 6 BRNN-based predictors was used (i.e., Q3=75.1% on average). Even though we have a slight computational overhead with our approach, where weight updates occur for each residue presented to the BRNN, in practice our BRNNs train (in the worst case) within a couple of hours, and certainly we assume that (for most network parameter sets) our single BRNN would be computationally cheaper than the ensemble of 6 BRNNs reported in [1]. Training time is not an issue for this type of applications, since it not anticipated that the BRNNs will have to be trained very frequently; the most important issue in this application is the prediction accuracy. Certainly, an exact comparison with [1] cannot be made, unless the same training and testing sets are used, but even with that the optimal connectivity of Baldi's architecture is not available (personal communication with one of the authors of [1], G. Pollastri).

In the work of Baldi et al. [1], the BRNN has a short input window of residues centered at the prediction site. Although this small window aims to avoid overfitting, it does not capture variable long-range information, which is overcome by unfolding the RNNs throughout the sequence. Despite the fact that we minimise the input vector in order to contain information for a single residue, the novelty of our approach lies firstly on the fact that we use a more elaborate computation within the recurrent context windows (as described in section 2.2), and secondly on updating network weights at every amino acid residue. With the latter, even though we are not able to capture long-range dependencies, we manage to more accurately take into account all available local information and this seems to be justified from our results.

Improved prediction results were obtained when sequence-to-structure output was filtered, in a post-processing step, in order to take higher order SS correlations into account. In particular, our novel HMM-based filtering approach not only improved the unfiltered results, but it was shown to be on average marginally better than a standard feed forward ANN-based filtering approach and much better than the BRNN-based filtering results reported in [11].

We believe that if we were to use a ensemble of BRNN-based predictors with our training scheme and our novel filtering procedures, our results would be even better.

Acknowledgments

We gratefully acknowledge the support of the Cyprus Research Promotion Foundation, as well as the European Structural Funds for grant TPE/ORIZO/0308 (FR)/05.

References

1. Baldi, P., Brunak, S., Frasconi, P., Soda, G., Pollastri, G.: Exploiting the past and the future in protein secondary structure prediction. Bioinformatics 15, 937–946 (1999)
2. Anfinsen, C.: Studies on the principles that govern the folding of protein chains. Les Prix Nobel en 1972 1, 103–119 (1973)
3. Richards, F., Kundrot, C.: Identification of structural motifs from protein coordinate data: Secondary structure and first-level supersecondary structure. Proteins 3, 71–84 (1988)
4. Pollastri, G., Przybylski, D., Rost, B., Baldi, P.: Improving the prediction of protein secondary structure in three and eight classes using recurrent neural networks and profiles. Proteins 47, 228–235 (2002)
5. Rost, B., Eyrich, V.: EVA: large-scale analysis of secondary structure prediction. Proteins 5, 192–199 (2001)
6. Qian, N., Sejnowski, T.: Predicting the secondary structure of globular proteins using neural network models. Journal of Molecular Biology 202, 865–884 (1988)
7. Rost, B., Sander, C.: Improved prediction of protein secondary structure by use of sequence profiles and neural networks. Proc. Natl. Acad. Sci. USA 90(16), 7558–7562 (1993)
8. Rost, B., Sander, C.: Combining evolutionary information and neural networks to predict protein secondary structure. Proteins 19, 55–72 (1994)
9. Salamov, A., Soloveyev, V.: Protein secondary structure prediction using local alignments. Journal of Molecular Biology 268, 31–36 (1997)
10. Cuff, J., Barton, G.: Evaluation and improvement of multiple sequence methods for protein secondary structure prediction. Proteins 34, 508–519 (1999)
11. Chen, J., Chaudhari, N.S.: Cascaded bidirectional recurrent neural networks for protein secondary structure prediction. IEEE/ACM Trans. Comput. Biology Bioinform. 4(4), 572–582 (2007)
12. Zemla, A., Venclovas, C., Fidelis, K., Rost, B.: A modified definition of Sov, a segment-based measure for protein secondary structure prediction assessment. Proteins 34(2), 220–223 (1999)
13. Hobohm, U., Scharf, M., Schneider, R., Sander, C.: Selection of representative protein data sets. Protein Science 1, 409–417 (1992)
14. Kabsch, W., Sander, C.: Dictionary of protein secondary structure: pattern recognition of hydrogen-bonded and geometrical features. Biopolymers 22, 2577–2637 (1983)
15. Schneider, R., Sander, C.: The HSSP database of protein structure-sequence alignments. Nucleic Acids Research 24, 201–205 (1996)
16. Elman, J.: Finding structure in time. Cognitive Science 14, 179–211 (1990)
17. Baram, D., Yonath, A.: From peptide-bond formation to cotranslational folding: dynamic, regulatory and evolutionary aspects. FEBS Lett. 579(4), 948–954 (2005)
18. Rumelhart, D., Hinton, G., Williams, R.: Learning representations by back-propagating errors. Nature 323, 533–536 (1986)

Contourlet Transform for Texture Representation of Ultrasound Thyroid Images

Stamos Katsigiannis, Eystratios G. Keramidas, and Dimitris Maroulis

University of Athens, Department of Informatics and Telecommunications,
Panepistimioupoli, Ilisia, 15784, Athens, Greece
{stamos,e.keramidas,dmarou}@di.uoa.gr

Abstract. Texture representation of ultrasound (US) images is currently considered a major issue in medical image analysis. This paper investigates the texture representation of thyroid tissue via features based on the Contourlet Transform (CT) using different types of filter banks. A variety of statistical texture features based on CT coefficients, have been considered through a selection schema. The Sequential Float Feature Selection (SFFS) algorithm with a *k*-NN classifier has been applied in order to investigate the most representative set of CT features. For the experimental evaluation a set of normal and nodular ultrasound thyroid textures have been utilized. The maximum classification accuracy was 93%, showing that CT based texture features can be successfully applied for the representation of different types of texture in US thyroid images.

Keywords: contourlet transform, ultrasound images, feature extraction, thyroid, feature selection.

1 Introduction

Modern medical ultrasonography (US) presents a unique set of advantages including real-time data acquisition, low cost, absence of any side effects and high resolution imaging. Thus, US has become the most common imaging modality for certain types of medical examinations, including that of the thyroid gland. Currently a challenge for the image analysis community is the application of automated or semi-automated computational methods on US images, for the computer aided diagnosis (CAD) of different types of diseases. A key issue for any CAD approach remains the texture representation method utilized.

Texture representation is a fundamental issue in image analysis and computer vision. It has been extensively investigated in the literature for more than three decades. Numerous approaches have been proposed dealing with textural features extraction which can be divided into four main categories [1] statistical, signal processing, model-based and geometrical. Signal processing approaches have drawn much attention, resulting in the proposal of a variety of texture representation methods, including the power spectral methods using the Fourier spectrum, the Discrete Cosine Transform (DCT), the Discrete Hartley Transform (DHT) [2], and

H. Papadopoulos, A.S. Andreou, and M. Bramer (Eds.): AIAI 2010, IFIP AICT 339, pp. 138–145, 2010.

more recently, the Gabor filters [3], the Haar [4] basis functions, the Discrete Wavelet Transform (DWT) and the Contourlet Transform (CT) [5][6].

Although the Discrete Wavelet Transform has been successfully applied for a wide range of image analysis problems, for two dimensions it tends to ignore the smoothness along contours [5]. In addition, the DWT provides only limited directional information which is an important aspect of multidimensional signals [6]. These limitations have been partially addressed by the CT which can efficiently approximate a smooth contour at multiple resolutions. Additionally in the frequency domain, the CT offers a multiscale and directional decomposition, providing anisotropy and directionality, features missing from the DWT [5][6][7]. The CT has been successfully used in a variety of texture analysis applications, including SAR and natural image classification [8], content-based image retrieval [9], image denoising [10], despeckling of images [11], image compression [12], etc.

The aim of this study is to investigate the performance of the Contourlet Transform (CT) for the representation of medical ultrasound (US) textures of the thyroid gland. A set of statistical features calculated from the CT coefficients are evaluated through a supervised classification schema on real thyroid US images. Additionally, a feature selection phase has been applied through the Sequential Float Feature Selection (SFFS) algorithm, for the extraction of the most representative set of CT features.

The rest of this paper is organized in three sections. Section 2 presents the proposed methodology including the CT and the above mentioned feature extraction schema. This methodology is evaluated through an experimental study on real thyroid ultrasound images presented in section 3, whereas conclusions are presented in section 4.

2 Methodology

2.1 The Contourlet Transform

The CT is a directional multiresolution image representation scheme proposed by Do and Vetterli [5] which is effective in representing smooth contours in different directions of an image [6], thus providing directionality and anisotropy [5][9]. The method utilizes a double filter bank, in order to obtain a sparse expansion of typical images containing smooth contours. In this filter bank, first the Laplacian Pyramid (LP) is used to detect the point discontinuities of the image and then a Directional Filter Bank (DFB) to link point discontinuities into linear structures. This scheme results in an image expansion that uses basic elements like contour segments and thus it is named CT. The separation of directional and multiscale decomposition stages provides a fast and flexible transform with computational complexity $O(N)$ for N-pixel images when using Finite Impulse Response (FIR) filters [6][8].

The LP, introduced in [13], provides a way to obtain multiscale decomposition. In each decomposition level, it creates a downsampled lowpass version of the original image and a bandpass image. A coarse image with the lower frequencies and a more detailed image with the supplementary high frequencies containing the point discontinuities are obtained. This scheme can be iterated continuously in the lowpass image, restricted only from the size of the original image.

The next step consists of the DFB which is a 2-D directional filter bank proposed by Bamberger and Smith, that can be maximally decimated while achieving perfect reconstruction [6][14]. The original DFB is efficiently implemented via an l-level binary tree leading to 2^l subbands with wedge-shaped frequency partitioning. However, in the CT, a new method is applied [6] which avoids modulating the input image and uses a simpler rule for expanding the decomposition tree [15]. The simplified DFB consists of two stages. First, a two-channel quincunx filter bank [16] that divides the 2-D spectrum into vertical and horizontal directions, and then a shearing operator that just reorders the samples.

Bandpass images from the LP decomposition are fed into a DFB in order to obtain the directional information. This scheme can be iterated on the coarse image and the combined result is a double iterated filter bank, named contourlet filter bank [8], which decomposes images into directional subbands at multiple scales.

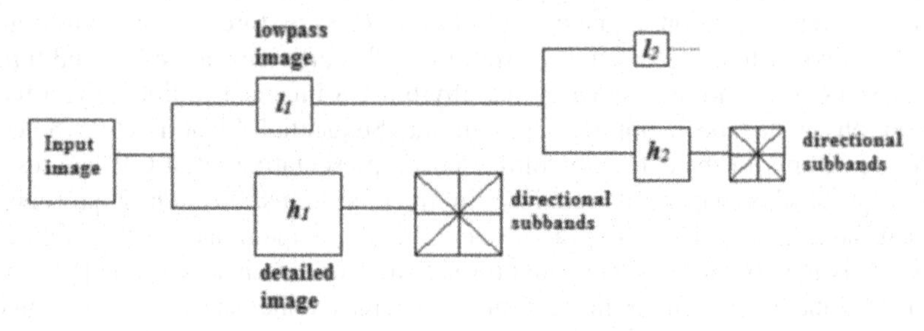

Fig. 1. The contourlet filter bank. Images l_i and h_i are the lowpass and detailed image, respectively, obtained from the ith level of decomposition with the LP. In every level of LP decomposition, h_i is fed into the DFB.

2.2 Feature Extraction

The set of statistical texture features in the contourlet domain evaluated in this study, have been proposed by Liu [8]. This set consists of the first order statistical measures that follow:

i) Energy. The energy E_{jk} of a subband image I_{jk} from the CT decomposition is defined as

$$E_{jk} = \frac{1}{M_{jk} \cdot N_{jk}} \sum_{n=1}^{N_{jk}} \sum_{m=1}^{M_{jk}} [I_{jk}(m,n)]^2 \quad (1)$$

Where I_{jk} is the subband image of the kth direction in the jth level. M_{jk} is the row size and N_{jk} the column size of the subband image I_{jk}.

ii) Standard deviation. Provides the means to capture the scale of the diversity of the image. The standard deviation S_{jk} of the subband image I_{jk} is defined as

$$S_{jk} = \sqrt{\frac{1}{M_{jk} \cdot N_{jk}} \sum_{n=1}^{N_{jk}} \sum_{m=1}^{M_{jk}} (I_{jk}(m,n) - \bar{I}_{jk})^2} \ . \tag{2}$$

$$\bar{I}_{jk} = \frac{1}{M_{jk} \cdot N_{jk}} \sum_{n=1}^{N_{jk}} \sum_{m=1}^{M_{jk}} I_{jk}(m,n) \ . \tag{3}$$

iii) Information entropy. For texture images, it represents the complexity of the texture information. The information entropy H_{jk} of a subband image I_{jk} is defined as

$$H_{jk} = -\sum_{n=1}^{N_{jk}} \sum_{m=1}^{M_{jk}} p_{jk}(m,n) \cdot \log p_{jk}(m,n) \ . \tag{4}$$

$$p_{jk}(m,n) = \frac{\left| I_{jk}(m,n) \right|^2}{\sqrt{\sum_{n=1}^{N_{jk}} \sum_{m=1}^{M_{jk}} \left[I_{jk}(m,n) \right]^2}} \ . \tag{5}$$

Using the above features, the feature vector for the subband image of the kth direction in the jth level is defined as $f_{jk}=(E_{jk}, S_{jk}, H_{jk})$. If the CT is defined as J level and for the jth ($j=1,\ldots, J$) level a K_j bands DFB is applied, then a total number o K subband images is obtained, where

$$K = \sum_{j=1}^{J} K_j \ . \tag{6}$$

By combining and rearranging all the feature vectors of the K subband images, the feature vector $F=\{ f_i \}, i=1,2,\ldots,K$ of the input image is obtained and normalized as proposed in [9].

2.3 Feature Selection

Feature selection offers more than one significant advantages, including reduction of computational complexity, improved generalization ability and robustness against outliers. A widely adopted algorithm for feature selection is the sequential floating forward selection (SFFS) algorithm [17] which has been used in a broad range of applications [18]. The idea behind the SFFS algorithm consists of consecutive forward selection steps, followed by a number of backward steps as long as the resulting subsets are better than the previously evaluated ones at the same level. Starting from an initially empty set of features, at each forward step an additional feature for which the classification accuracy is maximized is selected. Respectively, at each backward step the maximum subset that results in an improved classification accuracy is being selected.

3 Experimental Evaluation

Supervised classification experiments have been carried out to evaluate the performance of the proposed texture representation approach. The image set consists of real medical ultrasound images of the thyroid gland. A total of 72 thyroid ultrasound images were obtained from examinations performed on 43 patients, using a Philips HDI 5000 sonographic imaging system, with a spatial resolution of 470×470 pixels and amplitude resolution of 8 bits. During these examinations the parameters of the sonograph were kept the same. This set of ultrasound images includes hypogenic thyroid nodules, as it has been diagnosed by expert physicians. From each ultrasound image, an equal number of healthy and nodular sample blocks have been selected. The total number of non overlapping 32x32 pixel blocks (Fig. 2) resulting from this process was 200.

Fig. 2. Sample images from experimental dataset. (a) Normal and (b) nodular thyroid tissue.

The classification task was implemented by means of the non parametric and generally effective k-NN classification approach [24]. The distance measure used by the k-NN was the Euclidean and parameter k was $\{3, 5\}$. For all experiments conducted, the classification accuracy for all experiments was estimated by 10-fold cross validation [19].

Due to the small size of thyroid nodules, the sample blocks selected are also small (32x32 pixels) and they did not allow decomposition of more than three levels with the LP and six levels with the DFB. The first LP decomposition level supports up to six levels of DFB decomposition, decreased by one for every extra LP decomposition level. The filters applied for the LP were the filters Burt, 5-3 and 9-7 and for the DFB the 5-3 and 9-7 filters. All possible combinations of filters and decomposition levels were tested.

Detailed results about the maximum classification accuracy obtained using the k-NN classifier and different filter combinations are shown in Fig. 3. From this figure it can be noticed that the maximum classification accuracy is 77%. This resulted from the filter 9-7 for the LP and the DFB, with two levels of LP decomposition, decomposed into sixteen and two directional subbands respectively, from finer to coarser scale.

The application of the SFFS selection algorithm led to improved classification results for every experimental setup as shown in Fig. 4. In this case the maximum classification accuracy was 93% using only 36 out of 205 features and the confusion matrix is presented in Table 1. This accuracy has been obtained via 5-3 filters for the

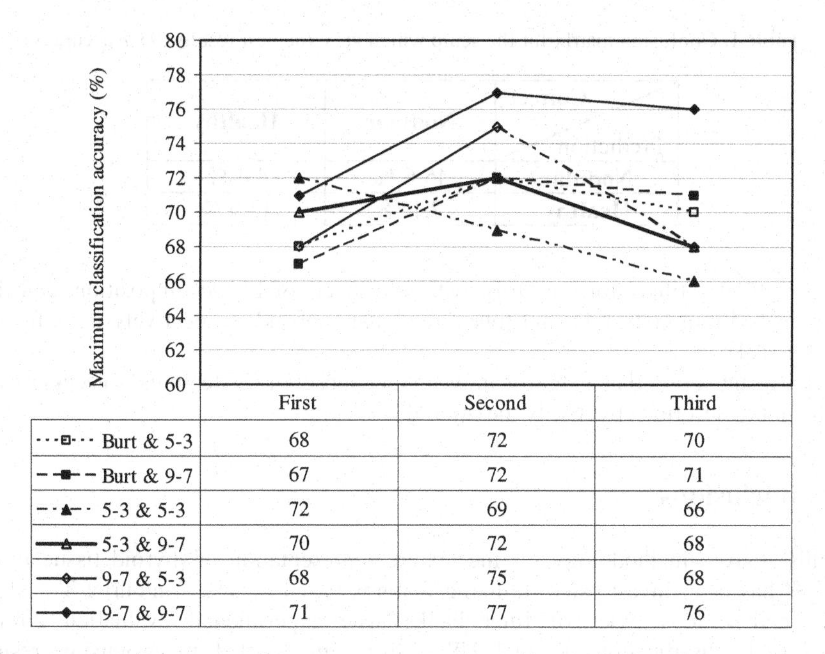

	First	Second	Third
Burt & 5-3	68	72	70
Burt & 9-7	67	72	71
5-3 & 5-3	72	69	66
5-3 & 9-7	70	72	68
9-7 & 5-3	68	75	68
9-7 & 9-7	71	77	76

Fig. 3. Maximum classification accuracy of the *k*-NN classifier for each filter combination at each level of LP decomposition (first, second, third). The first filter of the filter combinations refers to the one used for the LP and the second refers to the one used for the DFB.

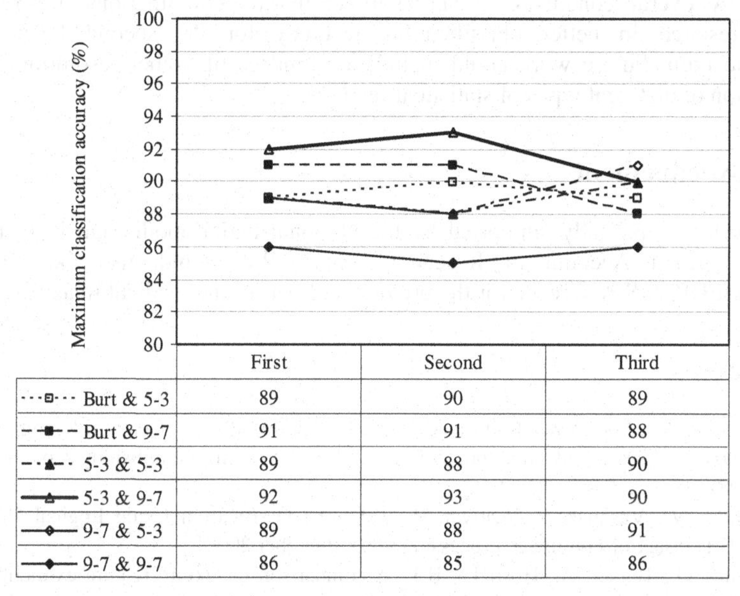

	First	Second	Third
Burt & 5-3	89	90	89
Burt & 9-7	91	91	88
5-3 & 5-3	89	88	90
5-3 & 9-7	92	93	90
9-7 & 5-3	89	88	91
9-7 & 9-7	86	85	86

Fig. 4. Maximum classification accuracy of the SFFS algorithm with the *k*-NN classifier, for each filter combination at each level of LP decomposition (first, second, third). The first filter of the filter combinations refers to the one used for the LP and the second refers to the one used for the DFB.

Table 1. Confusion matrix for the setup which gave the best results (93% accuracy)

Correct / Prediction	**Nodular**	**Healthy**
Nodular	45,6 %	4.4 %
Healthy	2.1 %	47.9 %

LP and the 9-7 filters for the DFB, with two levels of LP decomposition, and then decomposed into sixty four and four directional subbands respectively from finer to coarser scale.

This result shows that a significantly smaller subset of CT features is necessary and sufficient to describe effectively the thyroid US texture.

4 Conclusions

In this study a methodology for the texture representation of thyroid tissue in US images has been investigated, utilizing features based on the Contourlet Transform (CT) and various types of filter banks. The experimental evaluation through supervised classification on real US thyroid images led to promising results. Furthermore, through a feature selection phase the maximum classification accuracy reached 93% for a significantly smaller set of features. These results are considered to provide evidence for the effectiveness of CT texture representation of US thyroid images. As overall conclusion, it can be argued that the combinations of 5-3 and 9-7 filters resulted in better classification accuracy for the second level of LP decomposition. Future work could include US images of higher resolution and the evaluation of different types of statistical features.

Acknowledgement

This work was partially supported by the National and Kapodistrian University of Athens Special Account of Research Grants. We would also like to thank EUROMEDICA S.A. Greece for the provision of part of the medical images.

References

1. Tuceryan, M., Jain, A.K.: Texture Analysis. In: The Handbook of Pattern Recognition and Computer Vision, 2nd edn., pp. 207–248. World Scientific Publishing Co., Singapore (1998)
2. Manian, V., Vasquez, R., Katiyar, P.: Texture Classification Using Logical Operators. IEEE Transactions on image processing 9(10) (October 2000)
3. Pichler, O., Teuner, A., Hosticka, B.J.: A comparison of texture feature extraction using adaptive Gabor filtering, pyramidal and tree structured wavelet transforms. Pattern Recognition 29, 733–742 (1996)
4. Lonnestad, T.: A new set of texture features based on the Haar transform. In: 11th Int. Conf. Acoustics, Speech, Signal Processing, vol. 4, pp. 661–664 (1992)

5. Do, M.N., Vetterli, M.: Contourlets: A Directional Multiresolution Image Representation. In: Proc. of IEEE International Conference on Image Processing (ICIP), Rochester (2002)
6. Do, M.N., Vetterli, M.: The contourlet transform: an efficient directional multiresolution image representation. IEEE Trans. Im. on Proc. 14(12), 2091–2106 (2005)
7. Po, D.D.-Y., Do, M.: Directional multiscale modelling of images using the contourlet transform. IEEE Transactions on Image Processing 15(6), 1610–1620 (2006)
8. Liu, Z.: Minimum Distance Texture Classification of SAR Images in Contourlet Domain. In: 2008 International Conference on Computer Science and Software Engineering (2008)
9. Srinivasa rao, C., Srinivas kumar, S., Chatterji, B.N.: Content Based Image Retrieval using Contourlet Transform. ICGST-GVIP Journal 7(3) (2007)
10. Tsakanikas, P., Manolakos, E.S.: Improving 2-DE gel image denoising using contourlets. Proteomics 9(15), 3877–3888 (2009)
11. Varshney, L.R.: Despeckling Synthetic Aperture Radar Imagery using the Contourlet Transform. Application of Signal Processing (April 2004)
12. Karras, D.A., Karkanis, S.A., Mertzios, B.G.: Image Compression Using the Wavelet Transform on Textural Regions of Interest. In: 24th Euromicro Conf., vol. 2, pp. 633–639 (1998)
13. Burt, P.J., Adelson, E.H.: The Laplacian Pyramid as a Compact Image Code. IEEE Trans. on Communications, 532–540 (1983)
14. Bamberger, R.H., Smith, M.J.T.: A filter bank for the directional decomposition of images: Theory and design. IEEE Trans. Signal Proc. 40(4), 882–893 (1992)
15. Shapiro, J.M.: Embedded image coding using zerotrees of wavelet coefficients. IEEE Trans. on Sign. Proc., Wavelets and Signal Processing 41(12), 3445–3462 (1993)
16. Vetterli, M.: Multidimensional subband coding: Some theory and algorithms. Signal Proc. 6(2), 97–112 (1984)
17. Pudil, P., Novovicova, J., Blaha, S.: Statistical approach to pattern recognition: Theory and practical solution by means of PREDITAS system. Kyber. 27(1), 78 (1991)
18. Ververidis, D., Kotropoulos, C.: Fast Sequential Floating Forward Selection applied to emotional speech features estimated on DES and SUSAS data collections. In: Proc. European Signal Processing Conf. (EUSIPCO), Italy (2006)
19. Theodoridis, S., Koutroumbas, K.: Pattern Recognition, 3rd edn. Academic Press, London (2006)

Assessment of Stroke Risk Based on Morphological Ultrasound Image Analysis with Conformal Prediction

Antonis Lambrou[1], Harris Papadopoulos[2,1], Efthyvoulos Kyriacou[2,3], Constantinos S. Pattichis[3], Marios S. Pattichis[3,4], Alexander Gammerman[1], and Andrew Nicolaides[3,5]

[1] Computer Learning Research Centre, Royal Holloway, University of London, UK
{A.Lambrou,A.Gammerman}@cs.rhul.ac.uk
[2] Computer Science and Engineering Department, Frederick University, Cyprus
{H.Papadopoulos,E.Kyriacou}@frederick.ac.cy
[3] Computer Science Department, University of Cyprus, Nicosia, Cyprus
{anicolai,pattichi}@ucy.ac.cy
[4] Electrical and Computer Engineering Department, University of New Mexico
pattichis@ece.unm.edu
[5] Cyprus Institute of Neurology and Genetics

Abstract. Non-invasive ultrasound imaging of carotid plaques allows for the development of plaque image analysis in order to assess the risk of stroke. In our work, we provide reliable confidence measures for the assessment of stroke risk, using the Conformal Prediction framework. This framework provides a way for assigning valid confidence measures to predictions of classical machine learning algorithms. We conduct experiments on a dataset which contains morphological features derived from ultrasound images of atherosclerotic carotid plaques, and we evaluate the results of four different Conformal Predictors (CPs). The four CPs are based on Artificial Neural Networks (ANNs), Support Vector Machines (SVMs), Naive Bayes classification (NBC), and k-Nearest Neighbours (k-NN). The results given by all CPs demonstrate the reliability and usefulness of the obtained confidence measures on the problem of stroke risk assessment.

keywords: Conformal Prediction, stroke risk assessment, ultrasound images, confidence measures, carotid plaques.

1 Introduction

Visual classification of high-resolution ultrasound has made the non-invasive visualisation of the carotid bifurcation possible, and has thus been used in the study of arterial wall changes. Clinical applications of carotid bifurcation ultrasound include: i) identification and grading of stenosis of extracranial carotid artery disease often responsible for ischemic strokes, Transient Ischemic Attacks

H. Papadopoulos, A.S. Andreou, and M. Bramer (Eds.): AIAI 2010, IFIP AICT 339, pp. 146–153, 2010.

(TIAs) or Amaurosis Fugax (AF); ii) follow-up after carotid endarterectomy; iii) evaluation of pulsatile neck mass; iv) investigation of asymptomatic neck bruits where severe internal carotid artery stenosis is used as a predictive factor for future stroke; v) cardiovascular risk assessment where the presence of carotid bifurcation atherosclerotic plaques is associated with increased cardiovascular mortality. During the last 20 years, the introduction of computer aided methods and image standardisation has improved the objective assessment of carotid plaque echogenicity and heterogeneity [1], and has largely replaced subjective assessment that had been criticized for its poor reproducibility [2].

In this work, we propose the use of Conformal Prediction for assigning reliable confidence measures to the classification of plaques into symptomatic or asymptomatic, based on ultrasound images. To our knowledge, no other method which provides any type of confidence measures has been used before on this problem. Unlike most classification techniques, Conformal Predictors (CPs) can provide predictive regions which guarantee, under the i.i.d. assumption, that the error rate of the predictive regions will be bounded by a desirable significance level. Several machine learning classifiers can be incorporated into the Conformal Prediction framework. Work in [3,4,5,6,7,8] has been conducted for building and evaluating CPs using Support Vector Machines, k-Nearest Neighbours, Artificial Neural Networks, and Ridge Regression. Moreover, the Conformal Prediction framework has been applied to medical diagnostic problems with success for breast cancer [9], acute abdominal pain [10], ovarian cancer [11], and leukemia diagnosis [12].

We experiment on a real-world dataset which consists of morphological features derived from ultrasound images of atherosclerotic carotid plaques [13]. We apply the Conformal Prediction framework using four different classifiers: Artificial Neural Network (ANN); Support Vector Machine (SVM); Naive Bayes Classification (NBC); and k-Nearest Neighbours (k-NN). We compare the results and we show the reliability and practicality of the confidence measures obtained for the classification of atherosclerotic carotid plaques.

The rest of the paper is structured as follows. In section 2, we describe the data used, we give an overview of the Conformal Prediction framework, and explain how we have transformed four machine learning classifiers into CPs. In section 3, we describe our experimental settings and we give the results we have achieved. In section 4, we conclude and we outline our plans for future work.

2 Material and Methods

2.1 Atherosclerotic Carotid Plaque Data

A total of 274 carotid plaque ultrasound images associated with retinal or hemispheric symptoms (33 stroke, 60 TIA, and 44 AF) were used in this work. Patients with cardioembolic symptoms or distant symptoms (>6 months) were excluded from the study. Asymptomatic plaques were truly asymptomatic if they had never been associated with symptoms in the past, or symptomatic if they had been associated with retinal or hemispheric symptoms (Stroke, TIA or AF).

The ultrasound images were collected in the Irvine Laboratory for Cardiovascular Investigation and Research, Saint Mary's Hospital, UK, using an Advanced Technology Laboratories (ATL model HDI 3000 - Seattle, USA) duplex scanner with a linear broadband width 4-7 MHz (multifrequency) transducer, at a resolution of 20 pixels/mm. The gray scale images (gray levels 0-255) were normalized manually by adjusting the image linearly. The plaque identification and segmentation tasks are quite difficult and were carried out manually by a physician or vascular ultrasonographer who are experienced in scanning, both actions are described in [14]. The morphological features derived from the images are motivated from the need to study the structure of the plaque. In this work, we have used the group of L-images as described by the Multilevel binary morphological analysis in [13]. This group gave the best accuracy results.

2.2 Conformal Prediction

The Conformal Prediction framework provides a way for assigning reliable confidence measures to predictions, based on an underlying machine learning algorithm. Typically, we are given a training set of the form $\{(x_1, y_1), ..., (x_n, y_n)\}$, where x_i is a vector of real-valued attributes and $y_i \in \{Y_1, Y_2, ..., Y_c\}$ is a label given to the instance x_i. For a new instance x_{n+1}, we intend to predict the label y_{n+1} (i.e. the class of the instance). In order to make a prediction, we assume all possible classes $Y_h \in \{Y_1, Y_2, ..., Y_c\}$ for the new instance, and we test for each one how likely the prediction is of being correct. In order to test each assumption, we append the new instance x_{n+1} in our training set together with the assumed class Y_h, and we train the underlying machine learning algorithm on the extended training set

$$\{(x_1, y_1), ..., (x_{n+1}, Y_h)\}. \tag{1}$$

We then calculate a non-conformity score for each instance in (1). A non-conformity score indicates how different (or strange) an instance x_i is for its label y_i, compared to the other instances in (1). In section 2.3, we explain how we have modified four underlying algorithms in order to generate non-conformity scores. Once we have non-conformity scores for the instances in (1), we measure how likely the extended training set is of being i.i.d., using the p-value function

$$p(Y_h) = \frac{\#\{i = 1, ..., n+1 : a_i \geq a_{n+1}\}}{n+1}, \tag{2}$$

which compares the non-conformity score a_{n+1} of (x_{n+1}, Y_h) with all the other non-conformity scores. We call the output of this function the p-value of the class Y_h. For the wrong prediction Y_h, we expect that a_{n+1} will be relatively higher than most of the non-conformity scores. In such cases, we will get low p-values, whereas for the correct prediction we expect a higher p-value.

For the true label of x_{n+1}, the p-value function in (2) satisfies the following property for all probability distributions P, and for any significance level ϵ:

$$P\left(p(y_{n+1}) \leq \epsilon\right) \leq \epsilon. \tag{3}$$

Table 1. Example of a certain and an uncertain prediction at 95% confidence

Instance	x_1	x_2
$p(Y_1)$	0.8623	0.1920
$p(Y_2)$	0.0145	0.3768
Actual label	Y_1	Y_2
Predictive region for $\epsilon = 0.05$	$\{Y_1\}$	$\{Y_2, Y_1\}$

The property describes that when the given training set contains i.i.d. instances, the probability of the p-value of the training set to be less than or equal ϵ, is less than or equal ϵ. Consequently, we may output a set of possible predictions (i.e. a predictive region), which contains all the predictions with p-values greater than the significance level ϵ. Moreover, we always include the highest prediction in order to ensure that the predictive region will contain at least one prediction:

$$S = \{Y_h : p(Y_h) > \epsilon\} \cup \left\{ \arg \max_{h=1,\ldots,c} (p(Y_h)) \right\} . \tag{4}$$

Because of the property in (3), the probability of each set S not containing the correct prediction will be less than or equal to ϵ. As a result, the error of the predictive regions will be bounded to ϵ, and thus we can say that we have $1 - \epsilon$ confidence in our predictions. Alternatively, the CP may output a single prediction, which is the prediction with the highest p-value, complemented with a confidence measure, which is one minus the second highest p-value, and a credibility value which is the p-value of the prediction. The confidence measure shows how likely the output classification is of being correct, compared to all other possible classes. The credibility value gives an indication of how suitable the training set is for classifying the current instance (i.e. if the credibility value is very low, then the training set is not i.i.d. or the current instance is strange).

In Table 1, we give an example of a predictive region which contains a single label (certain prediction) and a predictive region which contains both labels for 95% confidence level. For instance x_1, the second p-value is 0.0145, which is less than the significance level of 0.05. Therefore, we can discard the second label at 95% confidence, and give a certain prediction which is the label that gives the highest p-value. In contrast, for instance x_2, the second largest p-value is 0.1920 and is greater than the significance level of 0.05. In this case, we cannot discard the second largest p-value at 95% confidence, and thus we have an uncertain predictive set, which contains both possible labels. Nevertheless, if we decrease confidence at 80.80% (or lower), we then have a certain prediction.

2.3 Non-conformity Measures

We describe how we derive non-conformity measures from four classical machine learning algorithms. Specifically, we give non-conformity measures for Artificial Neural Networks, Support Vector Machines, the Naive Bayes Classifier, and k-Nearest Neighbours.

Artificial Neural Networks. Artificial Neural Network (ANN) classifiers are usually trained on a training set to re-adjust the weights of the connections between the units inside the network. The output layer of a neural network has a unit o_j for each possible class, and given an instance x_i we predict the class Y_j corresponding to the unit which gives the highest value. We expect that, the more conforming an instance is for its class, the higher the corresponding o_j value would be. As proposed in [3], we can build a CP based on ANNs (ANN-CP), using the non-conformity measure

$$\alpha_i = 1 - o_t, \tag{5}$$

for any (x_i, y_i) where $y_i = Y_t$. Alternatively, we can use the following non-conformity measure which is again defined in [3]:

$$\alpha_i = \frac{\max_{j=1,\ldots,c:j\neq t} o_j}{o_t}. \tag{6}$$

That is, we use the maximum of the output units which do not correspond to the label of the given instance as the numerator, since a higher value from those units would also indicate a more strange instance, and would give a higher non-conformity score when divided by o_t.

Support Vector Machines. Support Vector Machines (SVMs) identify boundary instances for each class, and fix a separating hyperplane that maximises the margin between them. For the purpose of building a CP using SVM (SVM-CP), we use the distance of each instance from the separating hyperplane, and the class it belongs to, in order to produce non-conformity scores. For $Y = \{-1, 1\}$, we use the non-conformity measure

$$\alpha_i = -y_i h(x_i), \tag{7}$$

where $h(x_i)$ is the output of the SVM for the given instance x_i. The output of $h(x_i)$ is negative if the instance belongs to class -1, and positive if it belongs to class 1. If the prediction is correct, then the further the instance is from the hyperplane, the less the non-conformity score will be. In contrast, if the prediction is incorrect, the non-conformity score will increase as the distance from the hyperplane increases.

Naive Bayes Classifier. The Naive Bayes Classifier (NBC) is named after Bayes' theorem, and the "naive" assumption of attribute independence. The classifier multiplies the probabilities of the attributes given their class, and outputs the probability of label y_i given instance x_i. We can use the output probability to define a non-conformity measure and build a CP based on NBC (NBC-CP):

$$\alpha_i = 1 - P(y_i|x_i). \tag{8}$$

Nearest Neighbours. The k-Nearest Neighbours (k-NN) method computes the distance of a test instance from the other instances that are provided in the training set, and finds its k nearest instances. The prediction of the algorithm is the class which is the majority of the k instances. In the case of building a CP based on k-NN (k-NN-CP), we use the distances of the k nearest instances to define a non-conformity measure. The simplest approach is to calculate the total of distances of the k instances that belong to the class of instance x_i, since the nearer the instance is to its class, the less strange it is. Nonetheless, for a more accurate non-conformity measure we also take into consideration the distances of the k nearest instances that belong to other classes, since the nearer the instance x_i is to other classes the more strange it is. We build our k-NN-CP using the non-conformity measure defined in [4,8]:

$$\alpha_i = \frac{\sum_{j=1,\dots,k} s_{ij}}{\sum_{j=1,\dots,k} o_{ij}}, \tag{9}$$

where s_{ij} is the jth shortest distance of x_i from the instances of the same class, and o_{ij} is the jth shortest distance of x_i from the instances of other classes.

3 Experiments

We have applied Principal Component Analysis (PCA) on the dataset and selected its 6 features which accounted for 98% of its variance. For evaluation, we have applied the Leave-One-Out (LOO) method. Both these choices were made in order to be able to compare our results with [13]. The ANN-CP was structured with one hidden layer consisting of 3 units, and the output layer consisting of 2 units (one for each class). All units had a sigmoid activation function. We have used a learning rate of 0.3 and a momentum rate of 0.2. The ANN was trained for 500 epochs with 10% validation set, which was used to stop training when the performance on the validation set was deteriorating. For the SVM-CP, we used a Radial Basis Function (RBF) kernel mapping with a spread parameter of 0.1, and for the k-NN-CP we set the parameter $k = 10$.

3.1 Results

In Table 2, we compare the accuracy achieved by the four CPs with their corresponding machine learning algorithms. The difference of accuracy of each method with the corresponding CP is not significant, as expected. We would like to highlight that our aim is not to improve the accuracy level of the classification task, but rather to produce more informative predictions. On average all methods have an accuracy of 70.48%, while the SVM provides the best accuracy which is 73.72% (reported in [13]). In the left part of Table 3, we compare the certainty rates of the CPs, for given confidence levels 95%, 85%, and 75%. The certainty rates are calculated as the rate of predictive regions that contain only a single label. We do this in order to measure the quality of the p-values provided by each CP, and therefore the efficiency of the confidence measures. Although the SVM

Table 2. Accuracy comparison of 4 classical algorithms and the corresponding Conformal Predictors

Method	Classifier	CP
ANN	71.53%	71.90%
SVM	**73.72%**	73.36%
NBC	67.15%	66.79%
k-NN	69.34%	70.07%

Table 3. Certainty and error rates for 3 levels of confidence

Confidence level	95%	85%	75%	Confidence level	95%	85%	75%
ANN-CP Certainty	**33.9%**	**66.4%**	**87.9%**	**ANN-CP Error**	4.7%	13.9%	23.0%
SVM-CP Certainty	18.6%	54.3%	85.0%	**SVM-CP Error**	4.7%	14.9%	24.8%
NBC-CP Certainty	21.9%	58.7%	81.7%	**NBC-CP Error**	4.7%	14.9%	24.8%
k-NN-CP Certainty	28.1%	63.8%	86.8%	**k-NN-CP Error**	4.7%	14.9%	24.8%

has the best accuracy, it is the ANN-CP that provides the best certainty rates. The ANN-CP gives a 33.9% of certain predictive regions with 95% confidence. This is due to the difficulty of the classification task, which is reflected by the low 73.72% accuracy. Nevertheless, a 33.9% of patients will have a certain prediction and the error of such predictions will be at most 5%. Given the difficulty of the task, the 33.9% of certainty is arguably a useful result. Moreover, as we decrease the confidence level, the certainty rates increase dramatically. For example, at 85% confidence, the certainty rate given by the ANN-CP has increased to 66.4%.

In the right part of Table 3, we provide the error rates of all CPs for confidence levels 95%, 85%, and 75%. The error rates are calculated as the rate of the predictive regions that did not contain the true label. We confirm the validity of our confidence measures, as the error rates are below the significance level.

4 Conclusion

The classification of symptomatic and asymptomatic atherosclerotic plaques is a crucial task as it can be used to predict the risk of stroke. The accuracy levels are low since plaques with clear symptomatic features might have not given an event for reasons that need to be investigated. In this work, we have applied the Conformal Prediction framework on four machine learning algorithms in order to assign reliable confidence measures to the recognition of symptomatic or asymptomatic plaques; thus assess the risk of stroke. Our results demonstrate the validity of the produced confidence measures and their practicality. We believe that our contribution is suitable for the classification of plaques, as valid confidence measures may increase the quality of the decision-making process.

In the future, we aim to improve our confidence measures using meta-learning methods, such as boosting and ensemble algorithms. Moreover, we would like to experiment with different non-conformity measures for optimising further the accuracy and confidence measures of our methods.

References

1. Belgaro, G., Nicolaides, A., Laurora, G., Cesarone, M., Sanctis, M.D., Incandela, L., Barsotti, A.: Ultrasound morphology classification of the arterial wall and cardiovascular events in a 6-year follow-up study. Arteriosclerosis, Thrombosis, and Vascular Biology 16(7), 851–856 (1996)
2. Nicolaides, A., Shifrin, E., Bradbury, A., Dhanjil, S., Griffin, M., Belcaro, G., Williams, M.: Angiographic and duplex grading of internal carotid stenosis: can we overcome the confusion? Journal of Endovascular Therapy 3(2), 158–165 (1996)
3. Papadopoulos, H.: Inductive conformal prediction, theory and application to neural networks. In: Fritzsche, P. (ed.) Tools in Artificial Intelligence, I-Tech., Vienna, Austria, pp. 315–330 (2008), http://intechweb.org/downloadpdf.php?id=5294
4. Proedrou, K., Nouretdinov, I., Vovk, V., Gammerman, A.: Transductive confidence machines for pattern recognition. In: Elomaa, T., Mannila, H., Toivonen, H. (eds.) ECML 2002. LNCS (LNAI), vol. 2430, pp. 381–390. Springer, Heidelberg (2002)
5. Saunders, C., Gammerman, A., Vovk, V.: Transduction with confidence and credibility. In: Proceedings of the 16th International Joint Conference on Artificial Intelligence, Los Altos, CA, vol. 2, pp. 722–726. Morgan Kaufmann, San Francisco (1999)
6. Vovk, V., Gammerman, A., Saunders, C.: Machine-learning applications of algorithmic randomness. In: Proceedings of the 16th International Conference on Machine Learning (ICML 1999), pp. 444–453. Morgan Kaufmann, San Francisco (1999)
7. Papadopoulos, H., Proedrou, K., Vovk, V., Gammerman, A.: Inductive confidence machines for regression. In: Elomaa, T., Mannila, H., Toivonen, H. (eds.) ECML 2002. LNCS (LNAI), vol. 2430, pp. 345–356. Springer, Heidelberg (2002)
8. Papadopoulos, H., Vovk, V., Gammerman, A.: Qualified predictions for large data sets in the case of pattern recognition. In: Proceedings of the 2002 International Conference on Machine Learning and Applications (ICMLA 2002), pp. 159–163. CSREA Press (2002)
9. Lambrou, A., Papadopoulos, H., Gammerman, A.: Evolutionary conformal prediction for breast cancer diagnosis. In: 9th International Conference on Information Technology and Applications in Biomedicine (ITAB 2009). IEEE, Los Alamitos (2009)
10. Papadopoulos, H., Gammerman, A., Vovk, V.: Confidence predictions for the diagnosis of acute abdominal pain. In: Iliadis, L., Vlahavas, I., Bramer, M. (eds.) Artificial Intelligence Applications & Innovations III. IFIP International Federation for Information Processing, vol. 296, pp. 175–184. Springer, Heidelberg (2009)
11. Bellotti, T., Luo, Z., Gammerman, A., Delft, F.W.V., Saha, V.: Qualified predictions for microarray and proteomics pattern diagnostics with confidence machines. International Journal of Neural Systems 15(4), 247–258 (2005)
12. Bellotti, T., Luo, Z., Gammerman, A.: Reliable classification of childhood acute leukaemia from gene expression data using confidence machines. In: Proceedings of IEEE International Conference on Granular Computing (GRC 2006), pp. 148–153 (2006)
13. Kyriacou, E., Pattichis, M.S., Pattichis, C.S., Mavrommatis, A., Christodoulou, C.I., Kakkos, S., Nicolaides, A.: Classification of atherosclerotic carotid plaques using morphological analysis on ultrasound images. Applied Intelligence 30(1), 3–23 (2009)
14. Langsfeld, M., Gray-Weale, A.C., Lusby, R.J.: The role of plaque morphology and diameter reduction in the development of new symptoms in asymptomatic carotid arteries. J. Vasc. Surg. 9, 548–557 (1989)

Concept Based Representations as Complement of Bag of Words in Information Retrieval*

Maya Carrillo[1,2] and Aurelio López-López[1]

[1] Coordinación de Ciencias Computacionales, INAOE
Luis Enrique Erro 1, Santa Maria Tonantzintla, Puebla, México, C.P.72840
{cmaya,allopez}@inaoep.mx
[2] Facultad de Ciencias de la Computación, BUAP
Av. San Claudio y 14 Sur Ciudad Universitaria, 72570 Puebla, México

Abstract. Information Retrieval models, which do not represent texts merely as collections of the words they contain, but rather as collections of the concepts they contain through synonym sets or latent dimensions, are known as Bag-of-Concepts (BoC) representations. In this paper we use random indexing, which uses co-occurrence information among words to generate semantic context vectors and then represent the documents and queries as BoC. In addition, we use a novel representation, Holographic Reduced Representation, previously proposed in cognitive models, which can encode relations between words. We show that these representations can be successfully used in information retrieval, can associate terms, and when they are combined with the traditional vector space model, they improve effectiveness, in terms of mean average precision.

Keywords: Information Retrieval, Concept Based Representation, Vector Model, Random Indexing, Holographic Reduced Representation.

1 Introduction

Information Retrieval (IR) is a discipline involved with the representation, storage, organization, and access to information items [1]. IR systems are designed to provide, in response to a user query, references to documents which could contain the information desired by the user. To compare documents and queries, these have to be represented in an appropriate way to be processed. Sometimes, features are extracted from documents without performing any advanced processing; this produces what is known as Bag of Words representation (BoW), where the document attributes are words or word stems.

Merely considering the words of a document has shown not to be enough for representing content. For instance, consider two documents using the same set of words, but one discussing the topics in a positive sense, while the other refers to the same topic but in a negative sense. Then, a document would be more

* The first author was supported by Conacyt scholarships 208265, while the second author was partially supported by SNI, Mexico. We would also like to thank the CNRG at the University of Waterloo, Canada for the resources provided.

H. Papadopoulos, A.S. Andreou, and M. Bramer (Eds.): AIAI 2010, IFIP AICT 339, pp. 154–161, 2010.

appropriately represented if syntactic and semantic information was included. There is research that has sought to include such semantic aspects as Latent Semantic Analysis [9], which has included implicit context information in the indexing process. The information is obtained by grouping terms that have similar meaning using Singular Value Decomposition (SVD). However, this method is quite computationally expensive.

On the other hand, there have been other efforts to represent more precise concepts than only words. For instance, Mitra et al., Evans and Zhai [5],[6], among others have investigated the use of phrases as part of text representation since the early days of information retrieval. Their overall performance improvement has been only marginal, however. Recently Vilares, et al. in [22] have extracted binary dependencies (i.e. noun-modifier, subject-verb and verb-complement), their experiments have shown some improvement.

This paper, as an alternative to representing concepts, considers the use of Random Indexing (RI) to produce context vectors, which capture the implicit "semantics" of documents and queries without expensive reduction techniques as SVD. Thereafter, the context vectors are used to represent documents as Bag of Concepts (BoC) [7]. Besides this, we present the use of Holographic Reduced Representation (HRR) [2] to include syntactic relations between words. These techniques, to the best of our knowledge, have not been used in IR.

Nowadays, the traditional IR engines are able to retrieve the majority of relevant documents for most collections, but generally the ranking of the retrieved results leads to poor performance, in terms of precision. Therefore, we propose to use BoC and HRR to re-rank the results generated by the traditional vector space model (VSM) [4]. Our assumption was that the BoW could be enriched with information from a concept-based representation to improve its precision. Our results achieved with the English CLEF2005 collection for Adhoc track, have confirmed our hypothesis showing an improvement of over 16% in mean average precision (MAP).

The remainder of this paper is organized as follows: Section 2 provides a brief description of related work, particularly on including phrases in information retrieval. Section 3 presents Random Indexing, how it is used to create BoC representations, and related work. Section 4 introduces the concept of Holographic Reduced Representations (HRRs) and presents how to use them to represent documents. Section 5 explains the experimental setup and the results obtained. Finally, section 6 concludes the paper and gives some directions for further work.

2 Previous Work

There are several previous works, suggesting the use of phrases to index and retrieve documents. For instance, Evans & Zhai [5] present an approach to index noun phrases for IR. They describe a hybrid method to extract meaningful sub-compounds from complex noun phrases. Mitra et al. [6] present a study that compares the usefulness of phrase recognition by using linguistic and statistical methods. Croft et al. describe an approach where phrases identified in natural language queries are used to build structured queries for a probabilistic retrieval model [17]. Despite their many implementations, retrieval experiments with phrases have shown inconsistent results. Recently, a number of retrieval approaches have investigated the effectiveness of language modeling approach in modeling statistical phrases such as n-grams or proximity-based phrases, showing promising results [18], [19].

Our study differs from previous phrase-based approaches, in one aspect. We express phrases using a representation that reflects syntactic structure. This structure is then distributed across the document representation, rather than taking the phrases as new terms extending the space dimension.

3 Random Indexing

Random Indexing (RI) is a vector space methodology that accumulates context vectors for words based on co-occurrence data. The technique can be described via the following two steps:

1. A unique random representation known as index vector is assigned to each context (i.e. document). Index vectors are binary vectors with a small number of non-zero elements, which are either +1 or -1, with equal amounts of both. For example, if the index vectors have twenty non-zero elements in a 512- dimensional vector space, they will have ten +1s, ten -1s and 492 0s. Index vectors serve as indices or labels for contexts.
2. Index vectors are used to produce context vectors by scanning through the text and every time a given word occurs in a context; the index vector of the context is added to the context vector of the word [8].

The above steps can be exemplified as follows: Let's suppose we have an eight-dimensional space, two non-zero elements and two documents *D1: Regular Right Part Grammars and their Parsers*, whose index vector is [0,1,0,0,-1,0,0,0] and *D2: Boolean Matrix Methods for the Detection of Simple Precedence Grammars* with [0,1,0,0,0,-1,0,0] index vector. Then, the context vector for *Grammars* is the addition of both index vectors, since such word appears in them, producing the vector [0,2,0,0,-1,-1,0,0]. Word context vectors generated through this process are used to build document vectors as Bag of Concepts (BoC). Thus, a document vector is the sum of the context vectors of its words.

Random Indexing, in the same manner as Latent Semantic Analysis (LSA), attempts to capture implicit "semantic" relations, but RI has additional advantages as: a) it does not have to use reduction techniques like Singular Value Decomposition (SVD) to reduce the space dimensionality; b) It is an incremental method, which means that we do not have to process all the data before we can start using the context vectors [8].

There are several works that have validated the use of RI in text processing tasks: for example, Kanerva et al. in [20] used Random Indexing to solve the part of the TOEFL, in which given a word, the subject is asked to choose its synonym from a list of four alternatives. Sahlgren & Karlgren [21] demonstrated that Random Indexing can be applied to parallel texts for automatic bilingual lexicon acquisition. Sahlgren & Cöster [7] used Random Indexing to carry out text categorization.

4 A Representation from Cognitive Science

Distributed representation allows to integrate connectionism and cognitivism, where mental representations (symbols) are specified by distributed patterns of neural activities, while is possible to introduce formal algebraic operations on these distributed patterns, to mathematically model cognitive operations [10]. This approach, whose principles in cognitive science were formulated at the end of the 1960's [11], [12], [13], [14], [15], was culminated by the Holographic Reduced Representation (HRR) formulated by T. Plate [2]. The HRR, is a method for representing compositional structure in analogical reasoning. HRRs are vectors whose entries follow a normal distribution $N(0, 1/n)$. They allow us to express structure using a circular convolution operator to bind terms. Circular convolution operator ($\otimes$) binds two vectors $\mathbf{x} = (x_0, x_1, \ldots, x_{n-1})$ and $\mathbf{y} = (y_0, y_1, \ldots, y_{n-1})$ to give $\mathbf{z} = (z_0, z_1, \ldots, z_{n-1})$ where $\mathbf{z} = \mathbf{x} \otimes \mathbf{y}$ is defined as:

$$z_i = \sum_{k=0}^{n-1} x_k \; y_{i-k} \qquad i = 0 \; to \; n-1 (\text{subscripts are modulo}-n) \tag{1}$$

Illustrating the use of HRRs to represent simple predicates, we suppose an instantiated frame, which is the superposition of the frame name and the role-filler bindings (roles convolved with their respective fillers). For example, in a very simplified frame for *eating*, the vector for the frame name is *eating* and the vectors for the roles are eat_{agent} and eat_{object}. This frame can be instantiated with the fillers *Peter* and *the_fish* to produce "Peter eats the fish" as follows:

$$s_1 = eat + eat_{agent} \otimes peter + eat_{object} \otimes the\_fish \tag{2}$$

Accordingly, we adopt HRRs to build a text representation scheme in which two-word terms could be represented. Therefore, to define an HRR document representation, the following steps are done: a) Determine the index vectors for the vocabulary by adopting the random indexing method, as described earlier; b) Syntactically tag the documents using a natural language processing tool; c) Bind the *tf.idf*-weighted index vector of each word to its role. This "side" role is an HRR which serves to distinguish the right side from the left side of the two-word term. d) Add the resulting HRRs (with the two-word nouns encoded) to obtain a single HRR vector; e) Multiply the resulting HRR by an attenuating factor α; f) Normalize the HRR produced to obtain the vector which represents the document. In [23] and [24] there are some examples of this representation.

Fishbein, and Eliasmith have used the HRRs together with Random Indexing for text classification, having BoC as their baseline [3]. We have reported other result using this representation in [23] with smaller collections and adding the HRRs to a special representation named Index Vector Representation. In [24] we reported results using GEOCLEF collection from 2005 to 2008 and using HRRs to code prepositional location phrases. It is important to mention that up to now, we are not aware of other research that uses RI together with HRRs.

5 Experimentation

The proposed document representation was generated for the English document collection used in CLEF 2005 for the Adhoc track. This is composed of 56,472 news articles taken from the Glasgow Herald (British) 1995 and 113,005 from LA Times (American) 1994 for a total of 169,477. The queries used were 50 from number 251 to number 300. They contain title, a brief description, and a narrative. The experiments, described below, were done taking only the title and description fields.

The experiments were divided into two stages. The aim of the first was to obtain as many relevant documents as possible; this was carried out by Lemur[1], an open source system designed to facilitate research in information retrieval. The VSM using *tf.idf* weighting scheme and a cosine measure to determine vector similarity were configured in Lemur and the results generated were used as a baseline.

In the second stage, the list of documents generated by Lemur, with elements of three attributes *query-document-similarity* was re-ranked. In the re-ranking process the Lemur similarity list was combined (adding the similarity values) with two additional similarity lists. These lists were multiplied by 1/4 before the merging process to give more importance to the similarities obtained by Lemur. Of these additional lists, the former one was generated when documents and queries were represented as BoC and compared; the latter list was produced when the same documents and queries were represented as HRRs and compared. These lists were built taking only the first 1000 documents for each query to diminish the processing time. However, it should be mentioned that, on average, the time to build the BoC representation for a query and its associated documents was 2.9 min., and 4.6 min. to build the HRRs. In contrast, the time for comparing the query with the one thousand documents was only 0.124 seconds in both representations.

The HRR document representations were built as specified in section 4 where α was equal to 1/6, taking only two-word terms. MontyLingua [16] was used for parsing the documents and queries and extracting the two-word terms. Examples of these composed terms are: *civil war, serial killer, Russian force, military action.* Both BoCs and HRRs representations were weighted and compared using the same schemes used for the VSM. We carried out several previous experiments intended to assess the effects of dimensionality, limited vocabulary, vector density and context definition. In our experiments the vector dimensionality was 4096 and the density 20 non-zero elements. We removed stop words and used stemming in the same way as the traditional VSM. These parameters, which were determined after our preliminary experiments, produced suitable data for our proposal. However, defining their right value is an open research topic.

[1] http://www.lemurproject.org/

5.1 Evaluation

The results after re-ranking the documents were evaluated with two metrics: Mean Average Precision (MAP), which is defined as the average of all the *AvgP* obtained for each query. *AvgP* is defined as:

$$AvgP = \sum_{k=1}^{m} P(r) \text{ x } rel(r)/n \tag{3}$$

Where *P(r)* is the precision at *r* considered documents, *rel(r)* is a binary function which indicates if document *r* is relevant or not for a given query *q*; *n* is the number of relevant documents for *q*; *m* is the number of relevant documents retrieved for *q*.

The second metric is R-Precision (*R-Prec*), which is defined as the precision reached after *R* documents have been retrieved, where *R* is the number of relevant documents for the current query.

5.2 Results

Table 1 compares Lemur results, with those produced after adding to it, BoC and HRR similarity lists. This process thereby produces Lemur + BoC and Lemur + BoC + HRR lists. Notice how BoC improves MAP and R-Prec by always being above 15%. Although BoC gave a high improvement in MAP, HRR additionally improved it by 1.3%. In contrast, the R-Prec decreased 2% when the HRRs were added. Consequently the HRRs, as was expected, help to emphasize the representation of specific concepts as observed for the MAP increment, but they cause a loss of generality reflected in the R-Prec decrement. The results obtained by both Lemur + BoC and Lemur + BoC + HRR, performing a paired t-student test, were found to be statistically significant in a 99% confidence interval in terms of MAP.

Table 2 shows the same comparison, but now in terms of precision at the indicated number of documents. The proposed representations increased the

Table 1. MAP and R-Prec results for ADHOC CLEF 2005 English document collection

Metric	Lemur	Lemur+BoC	%Diff	Lemur+BoC+HRR	%Diff
MAP	0.2903	0.3392	**16.84**	0.3430	**18.15**
R-Prec	0.3103	0.3579	**15.34**	0.3514	**13.24**

Table 2. Precision at 5, 10, 15, 20, 30 and 100 documents

RI System	P@5	P@10	P@15	P@20	P@30	P@100
Lemur	0.4600	0.4160	0.3960	0.3610	0.3120	0.1854
Lemur + BoC	0.5280	0.4640	0.4293	0.3980	0.3540	0.1996
% Difference	**14.78**	**11.54**	**8.41**	**10.25**	**13.46**	**7.66**
Lemur + BoC + HRR	0.5360	0.4740	0.4387	0.3990	0.3527	0.1998
% Difference	**16.52**	**13.94**	**10.78**	**10.53**	**13.04**	**7.77**

precision in all cases, but it should be noted that the difference is higher at low recall levels. It is difficult to compare our results with those mentioned in related work, because the authors worked with different collections, metrics and environments. However, as mentioned in [25] "Sparck Jones has suggested that a difference in the scores between two runs that is greater than 5% is noticeable, and a difference that is greater than 10% is material".

6 Conclusions and Future Works

In this article, we have presented a proposal for representing documents and queries that according to the experiments, has shown itself to be feasible and able to capture "semantic relations" and encode two-word terms. BoC improved the initial ranker. HRRs produced a slight gain in precision. However, they have the potential to encode other relations between concepts (e.g. syntactical relations such as subject-verb, or additional information as identifying named entities).We think if more types of relations are considered it could lead to a higher improvement. Based on our results, it seems reasonable to conjecture that these new representations when combined with the VSM to re-rank the documents, increase precision. Our results showed an improvement above 16% in MAP. Additional study and experimentation will be necessary to quantify the usefulness of the proposed representations. Then, we will continue working with other collections that provide us with more specific contexts to be represented. This allows us to thoroughly explore the usefulness of the proposed representations to improve IR effectiveness.

References

1. Baeza-Yates, R., Ribeiro-Neto, B.: Modern Information Retrieval. Addison Wesley, Reading (1999)
2. Plate, T.A.: Holographic Reduced Representation: Distributed representation for cognitive structures. CSLI Publications, Stanford (2003)
3. Fishbein, J.M., Eliasmith, C.: Integrating structure and meaning: A new method for encoding structure for text classification. In: Macdonald, C., Ounis, I., Plachouras, V., Ruthven, I., White, R.W. (eds.) ECIR 2008. LNCS, vol. 4956, pp. 514–521. Springer, Heidelberg (2008)
4. Salton, G., Wong, A., Yang, C.S.: A vector space model for automatic indexing. Communications of the ACM 18(11), 613–620 (1975)
5. Evans, D., Zhai, C.: Noun-phrase Analysis in Unrestricted Text for Information Retrieval. In: Procs. of the 34th Annual Meeting on ACL, pp. 17–24 (1996)
6. Mitra, M., Buckley, C., Singhal, A., Cardie, C.: An Analysis of Statistical and Syntactic Phrases. In: Procs. of RIAO 1997, 5th International Conference, pp. 200–214 (1997)
7. Sahlgren, M., Cöster, R.: Using Bag-of-Concepts to Improve the Performance of Support Vector Machines in Text Categorization. In: Procs. of the 20th International Conference on Computational Linguistics, pp. 487–493 (2004)

8. Sahlgren, M.: An introduction to random indexing. In: Methods and Applications of Semantic Indexing Workshop at the 7th International Conference on Terminology and Knowledge Engineering, Copenhagen, Denmark (2005)
9. Deerwester, S., Dumais, S., Furnas, G., Landauer, T., Harshman, R.: Indexing by latent semantic analysis. Journal of the ASIS 41, 391–407 (1990)
10. Kvasnicka, Vladimir: Holographic Reduced Representation in Artificial Intelligence and Cognitive Science. In: Neural Network World, vol. 14, part 6, pp. 521–532 (2004)
11. Gabor, D.: Holographic model for temporal recall. Nature 217, 1288–1289 (1968)
12. Metcalfe Eich, J.: Levels of processing, encoding specificity, elaboration, and charm. Psychological Review 92, 1–38 (1985)
13. Murdock, B.B.: A theory for the storage and retrieval of item and associative information. Psychological Review, 316–338 (1982)
14. Slack, J.N.: The role of distributed memory in natural language processing. In: Advances in Artificial Intelligence: Procs. of the Sixth European Conference on Artificial Intelligence, ECAI 1984. Elsevier Science Publishers, New York (1984)
15. Willshaw, D.J., Buneman, O.P., Longuet-Higgins, H.C.: Non-holographic associative memory. Nature 222, 960–962 (1969)
16. Liu, Hugo: MontyLingua: An end-to-end natural language processor with common sense (2004), `http://web.media.mit.edu/hugo/montylingua`
17. Croft, W.B., Turtle, H.R., Lewis, D.D.: The use of phrases and structured queries in information retrieval. In: Procs. of the 14th Annual International ACM/SIGIR Conference, pp. 32–45 (1991)
18. Metzler, D., Croft, W.B.: A markov random field model for term dependencies. In: Procs. of SIGIR 2005, pp. 472–479 (2005)
19. Gao, J., Nie, J., Wu, G., Cao, G.: Dependence language model for information retrieval. In: Procs. of SIGIR 2004, pp. 170–177 (2004)
20. Kanerva, P., Kristoferson, J., Anders Holst, A.: Random indexing of text samples for latent semantic analysis. In: Procs. of the 22nd Annual Conf. of the Cognitive Sc. Society, pp. 103–106. Erlbaum, New Jersey (2000)
21. Sahlgren, M., Karlgren, J.: Automatic bilingual lexicon acquisition using Random Indexing of parallel corpora. Journal of Natural Language Engineering Special Issue on Parallel Texts 11(3), 327–341 (2005)
22. Vilares, J., Gómez-Rodríguez, C., Alonso, M.A.: Managing Syntactic Variation in Text Retrieval. In: Procs. of the ACM Symposium on Document Engineering. Bristol, United Kingdom, pp. 162–164. ACM Press, New York (2005)
23. Carrillo, M., Eliasmith, C., López-López, A.: Combining Text Vector Representations for Information Retrieval. In: Matoušek, V., Mautner, P. (eds.) TSD 2009. LNCS, vol. 5729, pp. 24–31. Springer, Heidelberg (2009)
24. Carrillo, M., Villatoro-Tello, E., López-López, A., Eliasmith, C., Montes-y-Gómez, M., Villaseñor-Pineda, L.: Representing Context Information for Document Retrieval. In: Procs. Flexible Query Answering Systems, 8th International Conference, Roskilde, Denmark, pp. 239–250 (2009)
25. Buckley, C., Voorhees, E.M.: Evaluating Evaluation Measure Stability. In: Procs. of ACM SIGIR 2000, pp. 33–40 (2000)

Information Fusion for Entity Matching in Unstructured Data

Omar Ali and Nello Cristianini

Intelligent Systems Laboratory, Bristol University
Merchant Venturers Building, Woodland Road, Bristol, BS8 1UB, United Kingdom
{Omar.Ali,Nello.Cristianini}@bristol.ac.uk
http://patterns.enm.bris.ac.uk

Abstract. Every day the global media system produces an abundance of news stories, all containing many references to people. An important task is to automatically generate reliable lists of people by analysing news content. We describe a system that leverages large amounts of data for this purpose. Lack of structure in this data gives rise to a large number of ways to refer to any particular person. Entity matching attempts to connect references that refer to the same person, usually employing some measure of similarity between references. We use information from multiple sources in order to produce a set of similarity measures with differing strengths and weaknesses. We show how their combination can improve precision without decreasing recall.

Keywords: Information fusion, entity matching, information extraction, social network, data cleaning, web mining.

1 Introduction

The global media system produces a vast amount of news articles which discuss named entities, such as people, places and organisations. Extracting these references to named entities can enable us to study interactions between them and the news stories that they appear in.

In this paper we describe a complete system for the extraction and matching of named entities from unstructured news data, where large numbers and multiple sources of information are used to improve the precision of the extracted output. Our aim is to investigate how the fusion of information can improve the performance of the system.

To this end, the system has been collecting on-line news articles from multiple sources over a period of nearly two years. These come from RSS feeds, which are visited many times per day. Articles are passed through a named entity extraction tool, Gate [1], which provides us with references to named entities in each article. These references do not have one-to-one correspondences with named entities as there are many ways to refer to the same entity.

Currently the system contains 9.3 million English articles, which contain references to 3.8 million people. Amongst these references are many sources of noise, such as rare misspellings of names and errors in the entity extraction process.

H. Papadopoulos, A.S. Andreou, and M. Bramer (Eds.): AIAI 2010, IFIP AICT 339, pp. 162–169, 2010.

Any study based on the people that appear in the media will be affected by the problem of multiple references, so these must first be resolved. Given the scale of the input this must be automated and we should be very careful not to merge similar-looking, yet different references, such as 'Tim Johnson' and 'Jim Johnson'.

We describe our approach to entity matching, fusing three sources of information to improve precision, without reducing recall and we test a method for the removal of ambiguous references based on their degree in a co-reference network.

2 Methods and Problem Formulation

In this section we formalise the problem of entity matching and discuss the similarity measures that will be used in our experiments.

Problem. We consider a set of references to named entities, $\{r_1, r_2, ...r_n\} \in R$. References may occur in many articles and pairs that appear in the same article are said to co-occur. The number of articles that a reference A appears in is its frequency, $f(A)$ and the number of times that any pair of references co-occur is their co-occurrence frequency, $f(A, B)$.

This definition allows us to construct a graph of cooccurrences, $G(V, E)$, which consists of references from set R as vertices and co-occurrences between references as edges. In this graph we wish to find pairs of references that represent the same person.

String Similarity. Most references to the same person share something in common. These range from multiple words, for example 'Hillary Clinton' and 'Hillary Rodham Clinton', to small differences in spelling like 'Hilary' or 'Hillary'. We use three measures of string similarity from the SimMetrics library [13]: **Jaro-Winkler**, **Q-Gram** and **Levenshtein**.

Titles, such as 'Mr', 'President', or 'Governor' should not be matched, therefore we maintain a list of titles and remove them prior to comparison. This leads us to rate 'President Barack Obama' and 'Barack Obama' very highly, whilst avoiding matches such as 'President Obama' and 'President Bush'.

Note that string similarity alone cannot resolve cases where similar names belong to different entities, as is the case for 'George W. Bush' and 'George H. W. Bush'. In order to resolve this ambiguity we require additional information.

Neighbourhood Similarity. Matching references are likely to be either directly linked, or to share a very similar set of neighbours. For this reason we employ measures of neighbourhood similarity in order to locate potential matches.

Finding references with similar neighbours could require us to compare all pairs, which is infeasible on large inputs. Fortunately we do not need to measure similarity between references that are more than two steps away from each other in the input graph, as we prove that their similarity must be zero.

Definition 1. *A graph $G(V, E)$ is composed of a set of vertices (references), V and a set of edges (co-occurring references), E. Each edge connects exactly two vertices. The* ***distance****, $d(u, v)$ between vertices u and v is the length of the* *shortest path between them. The* ***neighbours*** *of a vertex,* $N(v)$ *are the set of all vertices that are* ***adjacent*** *to* v*. If* u *and* v *are adjacent,* $d(u,v)=1$*.*

A ***neighbourhood similarity measure****,* $S_{nbr}(u,v)$ *describes the amount of overlap of the neighbourhoods of* u *and* v*. In general, if* $N(u)\cap N(v)=\emptyset$ *then* $S_{nbr}(u,v)=0$ *and if* $N(u)\cap N(v)=N(u)\cup N(v)\neq\emptyset$ *then* $S_{nbr}(u,v)=1$*.*

Proposition 1. *Any pair of vertices that have some neighbours in common,* $N(u)\cap N(v)\neq\emptyset$ *must be within two steps of one another, i.e.:*
$S_{nbr}(u,v)>0, \forall u,v\in V, \text{ s.t. } d(u,v)\leq 2$

Proof. Assume we have a pair of vertices, (u,v), where $S_{nbr}(u,v)>0$ and $d(u,v)>2$. If $S_{nbr}(u,v)>0$, then $N(u)\cap N(v)\neq\emptyset$, $\exists w\in V$ s.t. $w\in N(u), w\in N(v)$. Therefore $d(w,u)=1, d(w,v)=1$ and $d(u,v)\leq d(w,u)+d(w,v)=2$, which contradicts our assumption. □

We use the **Jaccard** and **Adar** neighbourhood similarity measures [3], which are defined in (1) and (2). Adar is similar to Jaccard with the addition of a uniqueness score for each reference.

$$S_{jaccard}(r_i,r_j)=\frac{|N(r_i)\cap N(r_j)|}{|N(r_i)\cup N(r_j)|} \tag{1}$$

$$S_{adar}(r_i,r_j)=\frac{\sum_{c\in(N(r_i)\cap N(r_j))}u(c)}{\sum_{c\in(N(r_i)\cup N(r_j))}u(c)} \text{ where } u(c)=\frac{1}{log(|N(c)|)} \tag{2}$$

Co-reference Similarity. Named entities are extracted using a named entity recognition tool that is part of the Gate distribution [1]. Gate also finds co-references [2], which are references to the same entity that occur within the same article. This information is used as another clue in our entity matching process. We treat this as a binary measure as shown in (3).

$$S_{coref}(r_i,r_j)=\begin{cases}1, & \text{if } f(r_i,r_j)>0\\ 0, & \text{otherwise}\end{cases} \tag{3}$$

Note that since this can only link references that appear in the same article it will not resolve systematic spelling differences from different sources, such as 'Gaddafi' and 'Kaddafi'. For these we can only rely on string similarity.

The degree a node in this network reveals information about its ambiguity and is used to filter out ambiguous references, as we show in the next section.

Combination of Measures. All measures are normalised to be in the range $[0,1]$ and we test weighted combinations of them as shown, for example, in (4).

$$S(r_1,r_2)=\alpha S_1(r_1,r_2)+\beta S_2(r_1,r_2)+\gamma S_3(r_1,r_2)$$
$$\text{where } \alpha,\beta,\gamma\in[0,1], \text{ s.t. } \alpha+\beta+\gamma=1 \tag{4}$$

3 Experimental Set-Up

In this section we describe how our experiments were set up, then present and discuss our findings. We also discuss the difficulties of measuring performance in an unstructured setting. This is a setting where references are not retrieved from a database where a certain degree of standardisation has been enforced, but are extracted from free text using statistical methods.

3.1 Measuring Performance

Matching entities that are automatically extracted from unstructured data is difficult due to the variability of references and errors in extraction and spelling. Freely available data sets for this task are difficult to find.

In order to be able to measure performance on this type of data we produced our own data set, consisting of 1417 pairs of marked references. This breaks down into two sets containing 901 correct and 516 incorrect pairs[1].

In web extraction applications it is more important to guarantee the quality of extracted information than it is to extract it all. As there is plenty of data available, we focus on high precision rather than high recall, in a similar vein to [15]. In addition we focus on relative changes in performance resulting from information fusion, rather than on absolute performance.

Precision and Recall. The input is a set of entity references, R from our entity database. All pairs of references could potentially be linked, $P = R \times R$.

Our algorithms produce a set of linked pairs, $P_L \subseteq P$. The ground-truth data consists of two sets of pairs, correct and incorrect, or $P_T \subseteq P$ and $P_F \subseteq P$, respectively. Precision and recall are calculated according to (5).

$$\text{Precision} = \frac{|P_T \cap P_L|}{|P_T \cap P_L| + |P_F \cap P_L|}, \quad \text{Recall} = \frac{|P_T \cap P_L|}{|P_T - P_L|} \tag{5}$$

3.2 Basic Set Up

For experimental purposes we focus on one week of references from the 20th to the 27th December 2009. During this time we observed 114,388 references and 2,008,225 co-occurring pairs, in N = 96,880 articles. The size of this graph is reduced by discarding edges that co-occur less than 5 times. This reduces the size of the vertex and edge sets to, $|V| = 8,264$ and $|E| = 57,361$.

χ-Square Test of Independence. To ensure that the remaining edges are statistically significant we use Pearson's χ-square test of independence [14]. In this setting we examine the contents of an article and make a pair of observations on the presence of references A and B. Our null hypothesis is that references A and B are statistically independent so their expected frequency of co-occurrence should equal $\frac{f(A)f(B)}{N}$, where N is the number of articles and $f(A)$ is the number of articles reference A is seen in.

[1] Data available at: `http://patterns.enm.bris.ac.uk/entity-matching-data`.

Our test makes multiple comparisons between reference pairs and the test statistic is calculated for all pairs of references which co-occur. In reality we implicitly compare $g = \frac{n(n+1)}{2} - n$ pairs, for n total references. Therefore we apply the Bonferroni correction and adjust the significance level to $1/g$.

For this network we have $g \approx 3.42 \times 10^7$ and an adjusted significance of 2.93×10^{-10}. Applying the χ-square test of independence at this significance level leaves us with $|V| = 8,230$ and $|E| = 53,462$.

3.3 Results

We now present results for string, neighbourhood and co-reference similarities, along with weighted combinations of the three.

Note that all measures are implicitly aided by neighbourhood information, as comparisons are made between references that are up to two steps apart in the co-occurrence network. This step ensures that we do not calculate similarities for all pairs of references. When we discuss neighbourhood similarity we are referring to *further* specific measures on the immediate neighbours of each reference.

We quote values of precision and recall which demonstrate the peak precision of each method. This excludes any results where precision becomes unstable when the number of output pairs becomes too small.

String Similarity. Table 1 shows results for string similarity measures, along with weighted combinations of them. All measures exhibit reasonably high precision but have subtly different types of reference-pairs that they can and cannot match. JaroWinkler is biased toward matching the beginning of a string, which leads to incorrect matches such as, 'Robert Casey' and 'Robert Costa'.

Q-gram is able to handle more variation in the input as it operates on sets of 3-grams, so order and positioning of characters is less important. This works in the case of 'Thomas James Leggs Jr.' and 'Thomas J. Leggs Jr.', but fails on pairs such as 'James Anderson' and 'Pamela Anderson'.

Levenshtein is a lot more selective, as each insertion, modification or deletion incurs a cost. Very similar strings like 'Tim Johnson' and 'Jim Johnson' are still a problem, however this is true for all of the measures we have described.

Neighbourhood Similarity. The addition of neighbourhood to string similarity did not lead to any improvement. We ran a number of weighted combinations of these measures and in all cases precision fell to around 50%.

This loss of precision was due to the inclusion of many false positives, which were typically strongly-connected pairs such as 'Warren Taylor' and his ex-wife 'Karen Taylor'. Neighbourhood measures have no way of discriminating references that are similar due to being related in some way from those which are the same. Similar to findings in [3], references which are similar but not matching are scored highly and incorrectly matched.

Co-Reference Similarity. Table 1 also shows precision and recall for co-reference similarity, followed by the addition of string and then neighbourhood similarity measures. Co-reference similarity alone shows strong performance, however we can improve this by adding in string measures.

Table 1. Precision and recall for measures, plus weighted combinations of them. Combining string measures seems to improve precision slightly, although Levenshtein performs best alone. Neighbourhood measures reduce performance (we show Levenshtein combined with both Jaccard and Adar — all other combinations perform comparably). Co-reference shows high precision and the addition of string similarity improves this, without a reduction in recall. The final line shows the effect of removing high-degree (ambiguous) references.

Method	Precision	Recall
Levenshtein	86.36	2.11
Q-Gram	80.77	2.33
JaroWinkler	79.17	2.11
$\frac{1}{4}$JaroWinkler + $\frac{3}{4}$Levenshtein	83.33	2.22
$\frac{1}{2}$JaroWinkler + $\frac{1}{2}$Levenshtein	84.21	1.78
$\frac{3}{4}$JaroWinkler + $\frac{1}{4}$Levenshtein	82.61	2.11
$\frac{1}{4}$JaroWinkler + $\frac{3}{4}$Q-Gram	71.43	2.77
$\frac{1}{2}$JaroWinkler + $\frac{1}{2}$Q-Gram	84.21	1.78
$\frac{3}{4}$JaroWinkler + $\frac{1}{4}$Q-Gram	82.35	1.55
$\frac{1}{4}$Q-Gram + $\frac{3}{4}$Levenshtein	83.33	1.66
$\frac{1}{2}$Q-Gram + $\frac{1}{2}$Levenshtein	86.36	2.11
$\frac{3}{4}$Q-Gram + $\frac{1}{4}$Levenshtein	85.00	1.89
$\frac{1}{2}$Levenshtein + $\frac{1}{2}$Jaccard	53.19	2.77
$\frac{1}{2}$Levenshtein + $\frac{1}{2}$Adar	48.78	2.22
CoRef	91.76	8.66
$\frac{1}{2}$Levenshtein + $\frac{1}{2}$CoRef	95.35	9.10
$\frac{1}{2}$Q-Gram + $\frac{1}{2}$CoRef	95.92	5.22
$\frac{1}{2}$JaroWinkler + $\frac{1}{2}$CoRef	94.87	8.21
$\frac{1}{3}$Levenshtein + $\frac{1}{3}$Jaccard + $\frac{1}{3}$CoRef	91.67	1.22
CoRef (ambiguous nodes removed)	91.49	19.09

Combining co-reference and string similarity gives some increase in performance. A co-reference error, which links 'Vicki Kennedy' to 'Ted Kennedy' is scored highly in terms of co-reference but relatively poorly by the Levenshtein measure, resulting in its removal. There is also the addition of pairs that cannot be found by co-reference alone, such as 'Mahmud Abbas' and 'Mahmoud Abbas'.

This combination results in increased precision without decreasing recall over the baseline string and co-reference measures. The co-reference network map also allow us to identify and remove references that are ambiguous (high-degree nodes).

4 Related Work

Existing methods [6,7] use on-line data, such as Wikipedia, to place references to entities into some context. An entity can then be disambiguated according to words or other entities that it appears alongside. Similarly, Bekkerman and

McCallum [5] obtain context from a list of contacts in a user's mailbox. They suggest that web pages related to a person will exist in a community of web pages related to that person's contacts.

Most methods [12] operate on structured data so do not have to handle noise associated with automatically extracting named entities from unstructured text.

Whilst state-of-the-art entity matching tools produce high accuracies, they do not scale well to large data sets. Rastogi [10] addresses this by defining a general framework for large scale entity matching. Dalvi [11] also discusses the more general problem of connecting related concepts.

Our system processes a high volume of unstructured data from the global media system, automatically extracting and storing named entities on a large scale. This introduces a lot of noise into our entity matching task and must be resolved with high precision in order that we may launch further studies with high confidence in our data.

5 Conclusions and Future Work

In this paper we have explored the use of information fusion for the task of information extraction from unstructured text. We have considered the problem of entity matching and have shown that combining multiple sources of information results in improved performance. Note that we have concentrated on relative changes in performance resulting from information fusion, rather than on absolute performance.

We looked at three similarity measures, all of which have different strengths and weaknesses. String similarity is a very effective method of comparison but is also very vulnerable to connecting similar-looking names. Co-reference information also gives high precision results and we have demonstrated that combining this with string similarity leads to increased precision, without reducing recall.

Neighbourhood similarity proved to be poor at this task as it assigns high scores to related references when we are looking for pairs which are the same. Whilst it does not prove useful as a similarity measure, the use of neighbourhood *information* is essential in order to reduce computation and produce a system which can scale.

In future work we will look to improve the recall of the system. The co-occurrence network is improved by considering greater volumes of news, for the simple reason that infrequent references can be disregarded, which greatly reduces noise in the input. We will also investigate further the performance improvements due to removing ambiguous (high-degree) nodes from the co-reference network. Examples include common forenames, the removal of which produces a much cleaner network. These noise-reduction steps will allow us to select lower thresholds to increase recall, without reducing precision.

On a more general level, our task is to find nodes in a network which are strongly related to one another. In this case the relation is one of identity, i.e., do these references refer to the same person? Other strong relations like family or work are regularly found and we wish to ignore these.

This also raises the issue of domain knowledge. One desirable property of an entity matching system is that it could be applied to other domains, as we also collect references to locations and organisations. It is very clear from our results that domain knowledge is very important and that a general system for matching entities of different types is a long way off. Increasing the amount of domain knowledge that we include should prove very effective at reducing the number of false positives that are produced.

Acknowledgements. We would like to give thanks to H. Cunningham, D. Maynard, K. Bontcheva and V. Tablan for creating and releasing Gate [1]. O. Ali is supported by a DTA grant, N. Cristianini is supported by a Royal Society Wolfson Merit Award and both are supported by the 'Pascal2' Network of Excellence.

References

1. Cunningham, H., Maynard, D., Bontcheva, K., Talban, V.: GATE: A framework and graphical development environment for robust NLP tools and applications. In: ACL (2002)
2. Dimitrov, M.: A light-weight approach to coreference resolution for named entities in text. MSc Thesis, University of Sofia (2002)
3. Bhattacharya, I., Getoor, L.: Collective entity resolution in relational data. In: TKDD (2007)
4. Newcombe, H., Kennedy, J., Axford, S., James, A.: Automatic linkage of vital records. Science (1959)
5. Bekkerman, R., McCallum, A.: Disambiguating web appearances of people in a social network. In: WWW (2005)
6. Han, X., Zhao, J.: Named entity disambiguation by leveraging Wikipedia semantic knowledge. In: CIKM (2009)
7. Cucerzan, S.: Large-scale named entity disambiguation based on Wikipedia data. In: EMNLP-CoNLL (2007)
8. Minkov, E., Cohen, W., Ng, A.: Contextual search and name disambiguation in email using graphs. In: SIGIR (2006)
9. Jijkoun, V., Khalid, M., Marx, M., De Rijke, M.: Named entity normalization in user generated content. In: AND (2008)
10. Rastogi, V., Dalvi, N., Garofalakis, M.: Large-scale collective entity matching. In: VLDB (2009)
11. Dalvi, N., Kumar, R., Pang, B., Ramakrishnan, R., Tomkins, A., Bohannon, P., Keerthi, S., Merugu, S.: A web of concepts. In: PODS (2009)
12. Köpcke, H., Rahm, E.: Frameworks for entity matching: A comparison. Data Knowledge Engineering (2009)
13. Chapman, S.: SimMetrics Library. NLP Group, University of Sheffield (2006), `http://www.dcs.shef.ac.uk/~sam/simmetrics.html`
14. Duda, R., Hart, P., Stork, D.: Pattern classification, 2nd edn. Wiley Interscience, Hoboken
15. Talukdar, P., Brants, T., Liberman, M., Pereira, F.: A context pattern induction method for named entity extraction. In: CoNLL (2006)

An Example-Tracing Tutor for Teaching NL to FOL Conversion

Themistoklis Chronopoulos, Isidoros Perikos, and Ioannis Hatzilygeroudis

Department of Computer Engineering & Informatics, School of Engineering
University of Patras, Greece
txronop@yahoo.gr, {perikos,ihatz}@ceid.upatras.gr

Abstract. In this paper we present an Example-tracing Tutor for the conversion of a sentence written in natural language (NL) to a sentence written in first order logic (FOL), which is a basic knowledge representation language. The tutor is based on the scripting of the process of the NL to FOL conversion and it has been authored using the Cognitive Tutoring Authoring Tool (CTAT) in which we have implemented a completed student interface and we also have created a Behavior Recorder graph for the above process.

Keywords: cognitive tutor, example-tracing tutor, knowledge representation tutor.

1 Introduction

Knowledge Representation & Reasoning (KR&R) is a fundamental topic of Artificial Intelligence (AI). A basic KR language is First-Order Logic (FOL), the main representative of logic-based representation languages, which is part of almost any introductory AI course and textbook [1, 2]. To make automated inferences, Clause Form (CF), a special form of FOL, is used. Teaching FOL as a knowledge representation and reasoning language includes many aspects. One of them is translating natural language (NL) sentences into FOL formulas. It is an ad-hoc process; there is no specific algorithm that can be automated within a computer. This is mainly due to the fact that NL has no clear semantics as FOL does. Also, most of existing textbooks do not pay the required attention to that. They simply provide the syntax of FOL and definitions of the logical symbols and terms [1, 2]. Even more specialized textbooks do the same [3]. At best, they provide a kind of more extended explanations and examples [4]. They do not provide any systematic guidance towards it. Given the above, students usually find difficulties in learning the task of formalizing NL sentences in FOL, which confronts to tutors' common experience.

In [5], we introduced a structured process for guiding students in translating a NL sentence into a FOL one, namely the SIP process. In [6], we presented a web-based system implementing the SIP process, i.e. helping students in learning how to convert NL sentences into FOL formulas. Having used the above system for some time, we resulted in the following findings: (a) At a first stage, students may not be necessary to work with sentences that produce formulas with more than three groups of atoms or with more than one group of formulas. (b) Tutors would like to use a graphical way of

H. Papadopoulos, A.S. Andreou, and M. Bramer (Eds.): AIAI 2010, IFIP AICT 339, pp. 170–178, 2010.

describing the SIP steps for each formula and a way of massively inserting them. Also, they would like related hints or feedback messages to be presented to the users in case of errors. (c) A stand-alone version of the system would be useful.

Example-tracing tutors [7] are a recent type of tutors that perform model tracing, provide context-sensitive instruction, in the form of hints and error feedback messages, and are flexible to multiple possible solution strategies and paths. Authoring for this flexibility is based on the explicit demonstration of alternative paths in each problem. No programming is required. CTAT (Cognitive Tutor Authoring Tools) is an authoring tool for creating example-tracing tutors [8, 9, 10]. So, CTAT is suitable for implementing the SIP process and satisfying the above requirements.

In this paper, we present a new lighter implementation of the SIP process in the form of an example-tracing tutor.

2 Related Work

There various systems that are used or have been created for teaching or helping in teaching some kind of logic and logic-based reasoning. Logic Tutor [11] is an intelligent tutoring system (ITS) for learning formal proofs in PL based on natural deduction. As an intelligent system, it adapts to the needs of the students via keeping user models. So, it provides context-sensitive feedback and exercises tailored to the users. Logic-ITA [12] is actually an extension to Logic Tutor, where a tutor part has been added, used as a teaching assistant system. P-Logic Tutor [13] is also a kind of intelligent tutoring system aiming at teaching students fundamental aspects of PL and theorem proving. To this end, it provides an interactive web-based interface. All the above systems, although deal with learning and/or teaching logic, they are not concerned with how to use predicate logic as a KR&R language. They do not deal with how to formalize a NL sentence into FOL.

As far as we are aware of, there is only one system that claims doing the latter. It is called KRRT (Knowledge Representation and Reasoning Tutor) [14]. It is a web-based system that aims at helping students to learn FOL as a KR&R language. It deals with both knowledge representation in and reasoning with FOL. The translation form NL to FOL takes place in its KR part. However, the only help provided to the students is at syntactic and logical equivalence levels. The student gives his/her FOL proposal sentence and the system checks its syntax and whether it is the correct one (here equivalent sentences are acceptable). However, it does not provide any guidance about how to make that translation or even what is the kind of error made. The system in [11] does the same as the system presented here, but there are also significant differences that concern (a) the user interface, (b) the way it works internally for student interaction checking, (c) the way hints/help are/is structured and (d) the way new sentences are inserted.

3 A Structured and Interactive Process for NL to FOL Conversion

One problem in converting natural language into first order logic has to do with the unclear semantics that natural language has. Natural language has no clear semantics as FOL has. However, the main difficulty comes from the lack of a systematic way of making the conversion. In a previous work [5], we introduced SIP (Structured and Interactive Process), a process that guides a student in translating/converting a NL sentence into a FOL one. SIP, which has been somewhat simplified here (to refer to simpler sentences, i.e. sentences that result in formulas with at most one group of formulas), is as follows:

1. Spot the verbs, the nouns and the adjectives in the sentence and specify the corresponding predicates or function symbols.
2. Specify the number, the types and the symbols of the arguments of the function symbols (first) and the predicates (next).
3. Specify the quantifiers of the variables.
4. Construct the atomic expressions (or atoms) corresponding to predicates.
5. Divide produced atoms in groups of the same level atoms.
6. Specify the connectives between atoms of each group and create corresponding logical formulas.
7. Form the group of formulas
8. Specify the connectives between formulas and create the next level formula.
9. Place quantifiers in the right points in the produced formula to create the final FOL formula

To demonstrate the steps of the above process, we present the conversion of the NL sentence "*All humans eat some food*" into a FOL formula in Fig. 1.

4 The Cognitive Tutor Authoring Tools (CTAT)

CTAT (Cognitive Tutor Authoring Tools) is an authoring tool for building tutors. Two types of tutors can be built using CTAT: example-tracing tutors and cognitive tutors [7, 8, 9, 10]. The first type is based on tracing specific pre-configured examples and requires no AI programming, whereas the second type requires AI programming and is based on a cognitive model, which is rule-based. Example-tracing tutors are easy to implement, but provide less flexibility, whereas cognitive tutors is quite more difficult to build, but can be quite more flexible. In this paper, we use CTAT for building an example-tracing tutor because, (a) it is easier to build, (b) we want to systematically analyze a large number of examples to extract possible cognitive patterns for building a cognitive tutor later, (c) we want to systematically analyze various types of hints or feedback needed.

Developing an example-tracing tutor in CTAT involves the following steps:

1. Creation of the graphical user interface (GUI).
2. Demonstration of alternative correct and incorrect solutions.
3. Annotation of the solutions steps with hint and feedback messages.

CTAT offers the *GUI Builder*, a tool for building the user interface of the tutor in the first step. GUI builder facilitates creating an interface in a graphical away without any programming, using a "recordable widget" palette added to Java NetBeans.

Additionally, CTAT offers *Behavior Recorder*, a tool for building "behavior graphs", which are graphs representing alternate correct and incorrect solutions to example problems, used in the second step above. For each problem, a corresponding behavior graph is created, which demonstrates student correct and incorrect problem solving behavior. Each such graph can be annotated with hints and feedback messages, in step three. Hints concern correct links, whether feedback messages concern incorrect links. The steps on those graphs are associated with corresponding items of the user interface built in the first step.

Step 1. specify predicates/functions
"humans" → predicate: *human*
"food" → predicate: *food*
"eat" → predícate: *eats*

Step 2. Number, types and symbols of arguments

Predicate	Arity	Types	Symbols
human	1	*variable*	*x*
food	1	*variable*	*y*
eats	2	*variable, variable*	*x, y*

Step 3. Quantifiers
x → ∀
y → ∃

Step 4. Atoms
Atom 1: *human(x)*
Atom 2: *food(y)*
Atom 3: *eats(x,y)*

Step 5. Groups of atoms of the same level
Alternate 1
GroupAtom1: {*human(x), food(y)*}
GroupAtom2: {*eats(x, y)*}

Alternate 2
GroupAtom1: {*human(x)*}
GroupAtom2: {*food(y), eats(x, y)*}

Step 6. Connectives and formulas of groups
Alternate 1
GroupAtom1 → Form1: *human(x)* ^ *food(y)*
GroupAtom2 → Form2: *eats(x, y)*

Alternate 2
GroupAtom1 → Form1: *human(x)*
GroupAtom2 → Form2: *food(y)* ⇒ *eats(x, y)*

Step 7. Connectives and formula of last group
Alternate 1
GroupForm1-1 → Form1-1:
(human(x) ^ *food(y))* ⇒ *eats(x, y)*

Alternate 2
GroupForm1-1 → Form1-1:
human(x) ⇒*(food(y))* ⇒ *eats(x, y))*

Step 8. Final formula
Alternate 1
(∀x) (∃y) (human(x) ^ food(y)) ⇒ eats(x, y)

Alternate 2
(∀x) (∃y) human(x) ⇒ (food(y)) ⇒ eats(x, y))

Fig. 1. Application of SIP process to "All humans eat some food"

At run time, CTAT's example-tracing engine implements the example-tracing function (or model tracing algorithm). This means that during the real-time use of the tutor, it maps the user problem solving behavior to the corresponding behavior graph and compares it with the one in the graph. Based on the results, the tutor provides either context-sensitive error feedback, when it matches an incorrect action link in the graph, or context-sensitive hints at student's request.

5 An Example-Traced Tutor for Converting a Sentence from NL to FOL

According to the process presented in Section 4, we first created the student interface of our system to reflect the NLtoFOL SIP process, as shown in Fig. 2. Actually a separate student interface was implemented for each step or group of steps of the process. All those student interfaces were integrated into one interface as different step tabs, through which a student can try to convert a NL sentence into a FOL formula following the NLtoFOL SIP process. Each tab, except first, corresponds to a step of the NLtoFOL SIP process. The first tab ("Atoms") corresponds to steps 1-4 of the process. In each problem solving cycle the student follows the NLtoFOL SIP process selecting one tab at a time, selecting on it the interface elements to work on and performing a problem solving action.

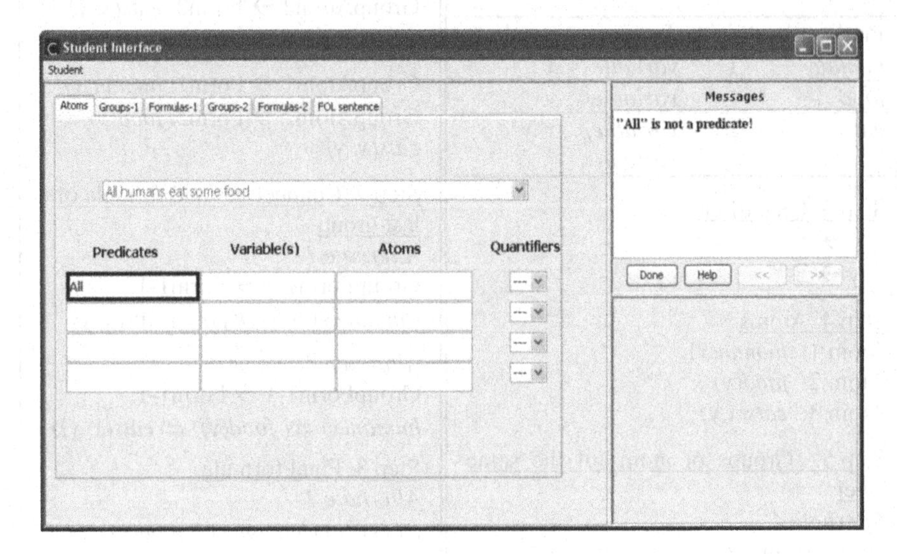

Fig. 2. Student graphical interface

Then, we selected a number of NL sentences and for each of them we demonstrated both correct and incorrect problem-solving behavior, recorded by the Behavior Recorder, and as many behavior graphs as the sentences were created. Also, alternatives solution paths for the conversions of the same sentences, where applicable, were recorded as alternative solutions paths (see Fig. 3). These cases are used as the basis for Example-Tracing Tutors to provide guidance to students.

6 Feedback from the Tutor

CTAT's Example-Tracing Engine uses the Behavior Graph to guide a student through a problem, comparing the student's problem-solving behavior against the graph. It provides positive feedback when the student's behavior matches steps in the graph, and negative feedback otherwise. If the student's input matches a link in the graph that was marked as an incorrect action, then any error feedback message attached to that link is presented to the student. When the student requests a hint, the hint messages attached to a link out of the current state in the graph are displayed. [15]

Tracing the student's step-by-step solution enables the tutor to provide individualized instruction in the problem solving context. Prototypically our tutor provides immediate feedback on each problem solving action: recognizably correct actions are accepted and unrecognized actions are rejected.

The FOL tutor provides feedback on each problem solving action, by accepting correct actions, which is shown to the student by green color and tagging errors instead of accepting them, which is shown to the student by red color.

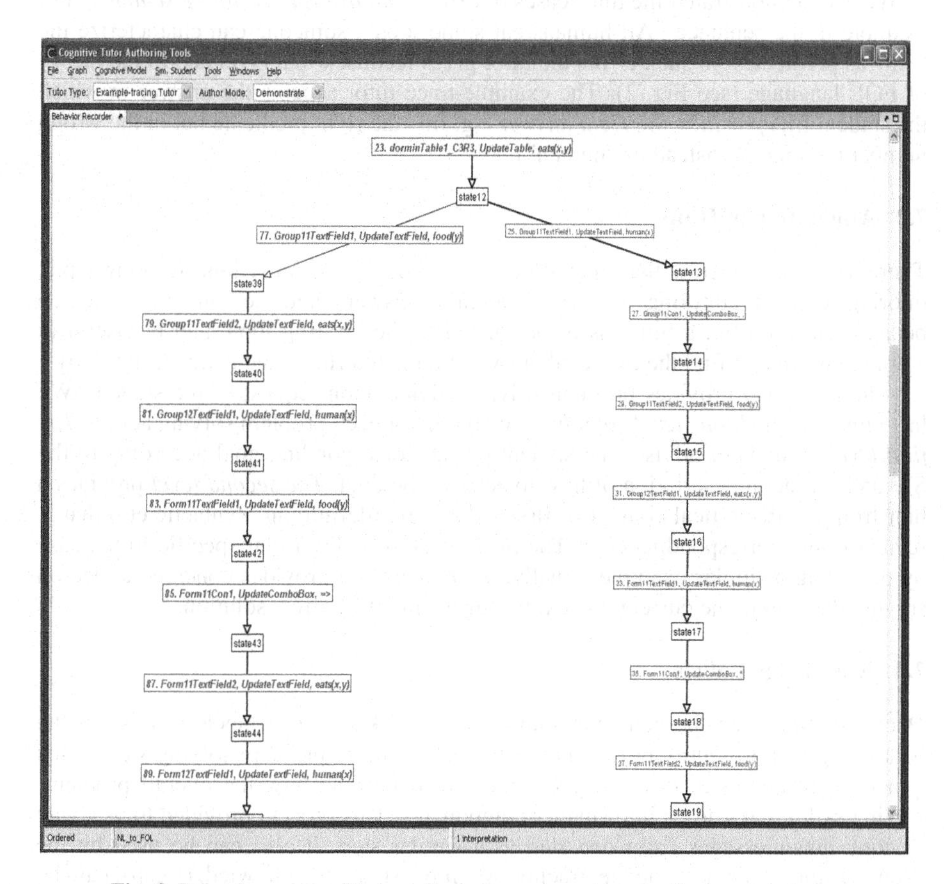

Fig. 3. Behavior graph with two alternative conversions for the same sentence

7 Annotation of Solution Steps

7.1 Incorrect Steps

In general any student input that is not recognized by the tutor is marked as incorrect, but *by defining incorrect steps* in the graph, the tutor will be able to provide a customized error feedback message for the specified input which. In each message we have included an example to demonstrate the correct use.

We focused to *common errors* that happen at the conversion of sentences from natural language to first order logic language, such as:

- Misuse of AND connective
- The order of quantifiers
- Use of function
- Grouping atoms of the same level
- Grouping formulas of the same level

We also demonstrated the tutor cases of *errors that are related to the sentence*. For example in the sentence "All humans eat some food", someone can characterize the "All" as predicates. In such errors the tutor gives feedback that is related to the theory of FOL language (see Fig. 2). The example-trace tutor accepts the answer in which the student fills *partially the right answer* e.g. fills the right predicate but in the wrong number ("humans" instead of "human").

7.2 Annotation of Hints

There are several factors that may affect the choice of a specific hint: tutoring topic, tutoring context, tutoring history, student's answer, and so on. First, to be pedagogically useful, a hint has to be related to the tutoring topic and be useful in helping the student find the expected answer. So the tutoring topic is important. [16]

The tutor also provides problem solving advice upon request of the student. We have implemented *four levels of advice* available for each problem solving action. *The first level* reminds or advises the student on the corresponding goal according to the SIP and a general description of how to achieve the goal. *The second level* provides a hint from the theoretical context of first order logic (definitions, syntactic etc) that is related to the corresponding step. The *third level* provides a hint specific to the case by providing a similar example. Finally, *the fourth level* provides concrete advice on solving the goal in the current context by suggesting the correct solution.

7.3 Knowledge Labels

Once we completed the behavior graph, we added knowledge labels to links in the behavior graph, to represent the knowledge behind those problem-solving steps. This is a form of cognitive task analysis, since we determine how the overall problem-solving skill breaks down into smaller components. This process provided to us a way to copy hint messages from one step to a similar step. It also can be used by the PseudoTutor to do knowledge tracing whereby students' knowledge gaps can be assessed and the tutor can select subsequent activities to address those gaps. [9]

We have added the following knowledge labels to links in the behavior graph: *FindPredicate*, *FindArgument*, *SpecifyQuantifier*, *ConstructAtom*, *FormulateGroup*, *SpecifyConnective*, *ConstructFormula*. At the same time, it is a way of planning the cognitive model, since we intend in future work, to create production rules corresponding to each identified skill. [9]

8 Conclusions

We have implemented an example-trace tutor for the conversion of sentences in NL to sentences in FOL, according to the SIP. We also have created worked-out examples of conversions for several sentences of the same level with the Behavior Recorder. The system has been preliminary used by a group of five students and the results are satisfactory. Four of them were satisfied in a large degree (>75%).

The examples can be used as the basis for Example-Tracing Tutors to provide guidance to students. The examples will also be used in future work, as planning cases and semi-automatic test cases for development of cognitive models, using the WME Editor and the Production Rule Editor, to create a production rule model.

References

1. Russell, S., Norvig, P.: Artificial Intelligence: a modern approach, 2nd edn. Prentice Hall, Upper Saddle River (2003)
2. Luger, G.F.: Artificial Intelligence: Structures and Strategies for Complex Problem Solving, 5th edn. Addison-Wesley, Reading (2004)
3. Brachman, R.J., Levesque, H.J.: Knowledge Representation and Reasoning. Elsevier, Amsterdam (2004)
4. Genesereth, M.R., Nilsson, N.J.: Logical Foundations of AI. Morgan Kaufmann, Palo Alto (1987)
5. Hatzilygeroudis, I.: Teaching NL to FOL and FOL to CL Conversions. In: Proceedings of the 20th International FLAIRS Conf., Key West, FL, pp. 309–314. AAAI Press, Menlo Park (May 2007)
6. Hatzilygeroudis, I., Perikos, I.: A Web-Based Interactive System for Learning NL to FOL Conversion. In: Damiani, E., Jeong, J., Howlett, R.J., Jain, L.C. (eds.) New Directions in Intelligent Interactive Multimedia Systems and Services – 2. SCI, vol. 266, pp. 297–307. Springer, Heidelberg (2009)
7. Aleven, V., McLaren, B.M., Sewall, J., Koedinger, K.R.: A new paradigm for intelligent tutoring systems: Example-tracing tutors. International Journal of Artificial Intelligence in Education 19(2), 105–154 (2009)
8. Aleven, V., McLaren, B.M., Sewall, J., Koedinger, K.R.: The Cognitive Tutor Authoring Tools (CTAT): Preliminary Evaluation of Efficiency Gains. In: Ikeda, M., Ashley, K.D., Chan, T.-W. (eds.) ITS 2006. LNCS, vol. 4053, pp. 61–70. Springer, Heidelberg (2006)
9. Koedinger, K., Aleven, V., Heffernan, N., McLaren, B., Hockenberry, M.: Opening the door to non-programmers: authoring intelligent tutor behavior bydemonstration. In: Lester, J.C., Vicari, R.M., Paraguaçu, F. (eds.) ITS 2004. LNCS, vol. 3220, pp. 162–174. Springer, Heidelberg (2004)
10. Koedinger, K.R., Aleven, V.: Toward a Rapid Development Environment for Cognitive Tutors. In: Hoppe, U., Verdejo, F., Kay, J. (eds.) Proceedings of the 11th International Conference on Artificial Intelligence in Education, AI-ED 2003, pp. 455–457 (2003)
11. Abraham, D., Crawford, L., Lesta, L., Merceron, A., Yacef, K.: The Logic Tutor: A multimedia presentation. Electronic Journal of Computer–Enhanced Learning (2001)
12. Abraham, D., Yacef, K.: Adaptation in the Web-Based Logic-ITA. In: De Bra, P., Brusilovsky, P., Conejo, R. (eds.) AH 2002. LNCS, vol. 2347, pp. 456–461. Springer, Heidelberg (2002)
13. Lukins, S., Levicki, A., Burg, J.: A tutorial program for propositional logic with human/computer interactive learning. In: SIGCSE 2002, pp. 381–385. ACM, NY (2002)

14. Alonso, J.A., Aranda, G.A., Martın-Matceos, F.J.: KRRT: Knowledge Representation and Reasoning Tutor. In: Moreno Díaz, R., Pichler, F., Quesada Arencibia, A. (eds.) EUROCAST 2007. LNCS, vol. 4739, pp. 400–407. Springer, Heidelberg (2007)
15. Aleven, V., Sewall, J., McLaren, B.M., Koedinger, K.R.: Rapid Authoring of Intelligent Tutors for Real-World and Experimental Use. In: Proceedings of 6th IEEE International Conference on Advanced Learning Technologies (ICALT 2006), pp. 847–851 (2006)
16. Zhou, Y., Freedman, R., Glass, M., Michael, J.A., Rovick, A.A., Evens, M.W.: Delivering Hints in a Dialogue-Based Intelligent Tutoring System. In: Proceedings of the Sixteenth National Conference on Artificial Intelligence (AAAI 1999), pp. 128–134 (1999)

Learning the Preferences of News Readers with SVM and Lasso Ranking

Elena Hensinger, Ilias Flaounas, and Nello Cristianini

Intelligent Systems Laboratory, University of Bristol, UK
{elena.hensinger,ilias.flaounas,nello.cristianini}@bristol.ac.uk

Abstract. We attack the task of predicting which news-stories are more appealing to a given audience by comparing 'most popular stories', gathered from various online news outlets, over a period of seven months, with stories that did not become popular despite appearing on the same page at the same time. We cast this as a learning-to-rank task, and train two different learning algorithms to reproduce the preferences of the readers, within each of the outlets. The first method is based on Support Vector Machines, the second on the Lasso. By just using words as features, SVM ranking can reach significant accuracy in correctly predicting the preference of readers for a given pair of articles. Furthermore, by exploiting the sparsity of the solutions found by the Lasso, we can also generate lists of keywords that are expected to trigger the attention of the outlets' readers.

Keywords: Learning to Rank, News Content Analysis, User Preferences, Support Vector Machines, Lasso.

1 Introduction

We are interested in understanding and predicting which news articles are most intriguing for readers of a given news outlet. This problem is of importance to both media analysts and news editors, as well as to engineers designing recommendation systems.

"What makes stories newsworthy?" has been the classical question in media studies [1]. A related question is "What makes stories popular?", and it can be addressed today due to the online availability of news media and the capability to capture preferences of readers by their clicks on news items. There are many possible answers to this 'Popularity question': it does not only depend on the textual content of the article, but it is also influenced by its position on the web page [2], on the images or videos associated with it and on its timing and according novelty [3].

The study of ranking user preferences has been recently approached by researchers in data mining, text processing and web analysis. Examples and applications in the area include the study of popularity of micro-blogging [3], advertisement selection [4] and search engine log analysis [5].

In our study, we focus on news media data with the goal to learn and predict preferences of news readers. We work on the content of news articles that readers

H. Papadopoulos, A.S. Andreou, and M. Bramer (Eds.): AIAI 2010, IFIP AICT 339, pp. 179–186, 2010.

will use in order to make their reading choices: the title and short description – the text snippet that is presented to readers before they click on a given link – from RSS (Real Simple Syndication) feeds of 10 news outlets with online presence, including broadcast media such as 'CBS' and newspapers such as 'LA Times'. We used two feeds per outlet: a) the 'Main' feed which contains the main stories of the day, as selected by the outlets' editors and b) the 'Most popular' feed formed by the readers' click choices. Can this information be sufficient to predict the interests and preferences of readers? To our knowledge, this question has not been addressed in literature.

We model the task as a problem of ordinal learning, or 'Learning to Rank': given a set of ordered data pairs, we learn a utility function that can be used to score every article, in a way that the scores can be used to rank them appropriately. The methods we use to infer linear utility functions are based on quadratic programming: Ranking SVMs (Support Vector Machines), as introduced by Joachims [5], and Lasso [6] that we applied for the ranking task. The motivation for the latter choice is to obtain sparse solutions, and hence extract the keywords that are most likely to predict article popularity and to trigger readers' attention.

In Sect. 2 we describe the two algorithms and present their relative merits in terms of generalisation performance and feature extraction ability. While Ranking SVM and Lasso originally solve different tasks, we show how Lasso can be applied to learn pairwise ranking.

In Sect. 3 we discuss the application of the two methods to news media datasets and report our results. For most outlets, the SVM-based ranking method can make predictions that are significantly stronger than random guessing. Under the posed experimental framework the Lasso-based ranking method performs less efficiently but still making significantly valid predictions and it has its actual strength in getting sparse solutions. Therefore we use the utility function inferred by the SVM to rank articles and the Lasso utility function to detect the strongest keywords that make articles become popular.

2 Task and Techniques

In this section we first present how the task of learning pairwise preference relationships can be transformed into a classification problem, secondly the ranking SVM and Lasso methods we used to solve it, and last their empirical comparison on synthetic data.

2.1 Pairwise Ranking as Classification

We aim to learn the ranking of items $x \in \mathbb{R}^n$, based on partial information about the preference relationship between pairs (x_i, x_j). We denote that item x_i is preferred to x_j by $x_i \succ x_j$. By defining a linear utility function $u : \mathbb{R}^n \rightarrow \mathbb{R}$ of the form $\langle w, x \rangle$, we can express this relationship as:

$$x_i \succ x_j \iff u(x_i) > u(x_j) \iff \langle w, (x_i - x_j) \rangle > 0 \qquad (1)$$

where $u(x_i)$ is the utility value of x_i. The task of learning the relationship between items x_i and x_j can now be expressed as a binary classification problem on the vector of their *difference.* which we will denote as $s_k \in S$, with S being the set of pair differences. We assign class label $y_k = +1$, if $\langle w, s_k \rangle \geq 0$, and $y_k = -1$ otherwise.

2.2 Ranking SVM

SVMs is a well studied method for binary classification [7,8]. The SVM formulation has been extended to learn pairwise preference relationships based on difference vectors of form s_k through the Ranking SVM [5]. Slack variables ξ_k allow to deal with non-linearly separable data, and the entire task for l training items of S is expressed as a quadratic optimisation problem:

$$\begin{aligned} \text{minimise}_{\xi,w} \quad & \langle w, w \rangle + C \textstyle\sum_{k=1}^{l} \xi_k \\ \text{subject to} \quad & y_k(\langle w, s_k \rangle) \geq 1 - \xi_k, \\ & \xi_k \geq 0 \quad \forall k = 1, \ldots, l \end{aligned} \tag{2}$$

The solution is a weight vector w that can not only compute the preference relationship between two items x_i and x_j, but can also give the rank, *i.e.* the utility value, of item x_i since $u(x_i) = \langle w, x_i \rangle$.

2.3 Lasso for Ranking

The second method we used is a special form of linear regression, the Lasso (Least Absolute Shrinkage and Selection Operator)[6]. In our setting rather than searching for a fit to real values we try to classify the pairwise differences s_k to their labels y_k. In order to produce a sparse solution, the regression problem is constrained on the L_1-norm, denoted as $||\cdot||_1$, of the computed weight vector w through a parameter $t \in \mathbb{R}$. The according quadratic optimisation problem is:

$$\begin{aligned} \text{minimise}_{w} \quad & \textstyle\sum_{k=1}^{l} (\langle w, s_k \rangle - y_k)^2 \\ \text{subject to} \quad & ||w||_1 \leq t \end{aligned} \tag{3}$$

The solution is a weight vector w which, as in the case of SVMs, is used for calculating the utility value of item x_i as $u(x_i) = \langle w, x_i \rangle$.

2.4 Empirical Comparison of Ranking SVM and Lasso

We compared Ranking SVM and Lasso in an experiment on small scale synthetic data to explore their potential on: a) correctly predicting class labels for the pairwise ranking task and b) handling sparse data and especially their ability to correctly identify relevant features, that is excluding features with zero value.

We synthetically created data items of 50 features with uniformly distributed values in $(0, 1)$. A weight value from the real interval $[0, 10]$ was randomly assigned to each feature, and 50% of features were randomly set to zero to simulate sparse data. We created difference vector pairs s_k and assigned class label $+1$ or -1, depending on the utility values of x_i and x_j. In all our computations we trained on 100 different data sets, ranging from 10 to 500 pairs of items, and we tested on 100 item pairs, averaging over 10 different data initialisations. We set parameters C in Ranking SVM and t in Lasso to their optimal values, as found for the described experimental setup.

To evaluate the two methods on preference prediction performance we used accuracy, defined as the percentage of correctly classified pairs. The resulting learning curves are illustrated in Fig. 1A. Both approaches improve their performance with increasing amount of training data, reaching an asymptotic bound on their performance. While Ranking SVM scores better in most cases, 'Lasso Ranking' can keep up successfully in this setting of predicting pairwise preferences.

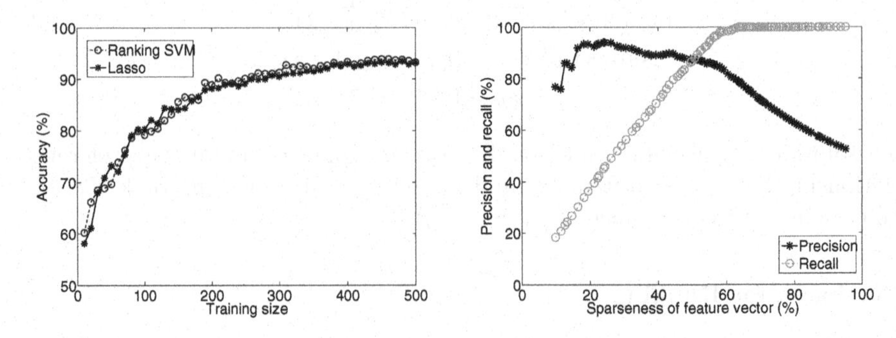

Fig. 1. A. Learning curves for Ranking SVM and Lasso. B. Lasso's ability to identify zero-valued features.

To evaluate performance on identifying zero valued features (ZF) we used precision and recall measures, defined as:

$$precision = \#\, of\ correctly\ identified\ ZF\ /\ \#\, total\ retrieved\ ZF \quad (4)$$

$$recall = \#\, of\ correctly\ identified\ ZF\ /\ \#\, total\ true\ ZF\ in\ features \quad (5)$$

Figure 1B presents precision and recall on detecting zero features using Lasso as a function of different levels of sparseness of the solution w for the 500 training pairs case. In our experimental setup, Ranking SVM produced dense solutions, even with the relaxation of counting features with absolute values less than 0.0005 as zero-valued.

Based on these results, we have an indication that Lasso can perform comparably with Ranking SVM in terms of prediction accuracy on pairwise difference data. Furthermore, it can identify relevant features in data, allowing to retrieve a sparse and meaningful solution.

3 Experimental Results on News-Outlets Data

We tested Ranking SVM and Lasso on real world data from 10 news-media outlets monitored from June 1st to December 31st, 2009. Since there are no publicly available datasets that provide annotated articles to use for learning readers' preferences we created our own system to gather this information from outlets that offer their content online [9].

Our system processes the content of media outlets that offer their content as RSS feeds, an easily machine readable format. Feeds contain for each article only limited information: a title, a summary and publication date. Most outlets serve multiple different RSS feeds of which we are interested in two specific ones, namely the 'Main' feed, and the 'Most Popular Stories' feed. The first one contains articles that comprise the leading stories of the outlet as chosen by their editors and roughly correspond to the front page of a newspaper. The second contains articles that readers found most interesting – these articles are not necessarily present in the 'Main' feed. We define feeds as 'Most Popular' if they carry the 'Most clicked' articles. For the outlets which miss such feeds, we adopted the 'Most emailed stories' or the 'Most discussed stories'. The article titles and summaries were gathered on a daily basis and processed with typical text mining methods such as stemming [10], stop words removal, indexing and TF-IDF (bag-of-words) representation of articles [11].

For each day of the period of study and for each outlet separately, we created pairs of articles from all combinations of articles that a) appeared in both 'Main' and 'Most Popular' feeds, and b) articles that appeared only in 'Main' feed and not in the 'Most Popular' one. The pairwise preference relationship is given by the fact that some articles of the 'Main' feed were preferred by the outlet's readers, while others were not. In total we formed 405,487 pairs – on average 5942 pairs per outlet per month. For 'Florida Times-Union' we managed to collect data for five months and for 'LA Times' for three months only.

We train both Ranking SVM and Lasso on one month per outlet and test on the following month's data. We have seven months of data resulting in six training/testing setups per outlet. For this set of experiments we used the SVM^{Rank} implementation of Ranking SVM [5]. We measured classification accuracy for three different values of C parameter, namely 1, 10 and 100, and we report the best results per outlet in Fig. 2.

In the case of Lasso, we are interested to discover the most interesting keywords that characterise popular articles, thus we adjusted the t parameter to achieve a sparse solution of roughly 100 words per outlet per month. Average month-to-month prediction results of readers' preferences are reported in Fig. 3. In both cases, we consider classifiers to have significant performance if they achieve an average accuracy over 50% for the 95% confidence intervals. Using SVMs, we get significant performance for seven outlets, reaching up to 73.14% for 'LA Times'. For 'Guardian' we had the least available data – only 482 pairs on average per month rather than thousands as in the rest of the outlets – and this explains the low performance. For 'CNN' and 'Time' we managed to get only marginally significant results, and we assume the diversity of their audience and

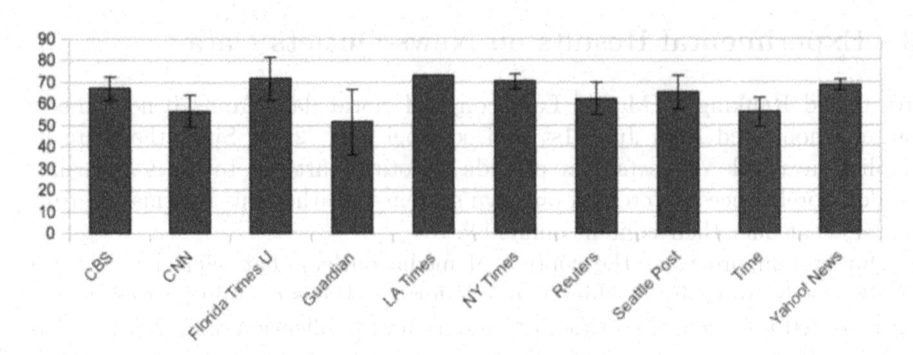

Fig. 2. Average month-to-month prediction accuracies per outlet using Ranking SVMs. Error bars are set to 95% confidence interval. Notice that for 'Guardian' we have limited amount of articles per day, and for 'LA Times' only three months of data resulting in a small confidence interval.

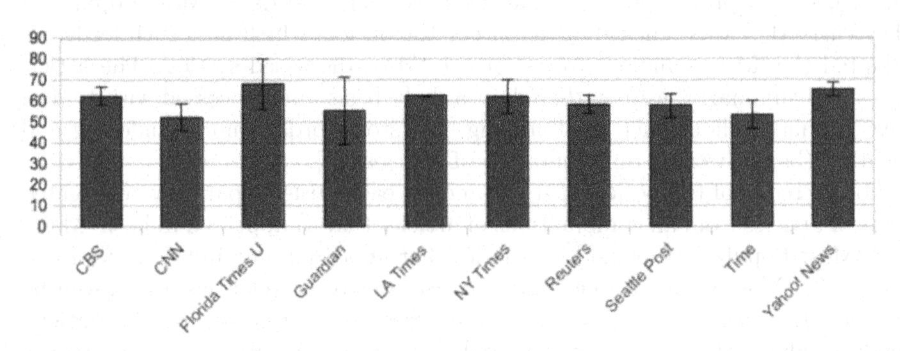

Fig. 3. Average month-to-month prediction accuracies per outlet using Lasso. Error bars are set to 95% confidence interval.

topic coverage to be the reason. As expected, we achieved on average 5.78% less accuracy using Lasso for the seven outlets for which we had significant results. However, it is the same outlets that failed to give significant results for SVMs which also fail for Lasso.

We used the SVM classifiers that predict article popularity with significant performance to detect the most popular stories per month for each outlet. To identify the most promising keywords that can trigger the audience's attention, we applied Lasso. As an example, we present the titles of the top-3 most popular articles per outlet and the top-10 keywords for December 2009 in Table 1. The stories we found reflect stories that were in the focus of media for that period of time. The stemmed keywords are similarly sensible and understandable. The different findings between outlets reflect a) differences of editors' choices on which stories to cover, and b) differences of what readers of different media outlets preferred to read.

Table 1. Titles of most popular articles per outlet as ranked using Ranking SVMs and keywords per outlet as identified using Lasso, for December 2009

Outlet	Titles of Top-3 Articles	Top-10 Keywords
CBS	Sources: Elin Done with Tiger - Tiger Woods Slapped with Ticket for Crash - Tiger Woods: I let my Family Down	defend, murphi, tiger, lake, mom, flight, bengal, man, goldman, unredact
Florida Times-Union	Pizza delivery woman killed on Westside - A family's search for justice, 15 years later - Rants & Raves: Napolitano unqualified	wade, middleburg, chest, send, charlton, petti, kill, augustin, shoot, levar
LA Times	Pacquiao to fight Mayweather in March - Bone marrow transplant 'gets rid of' sickle cell anemia - Disney toys get Pixar animation guru's touch	insul, immigr, disnei, futur, california, cash, studi, radiat, lo, selznick
NY Times	Poor Children Likelier to Get Antipsychotics - Surf's Up, Way Up, and Competitors Let Out a Big Mahalo - Grandma's Gifts Need Extra Reindeer	studi, visual, art, monei, gift, drug, journal, counterterror, water, korean
Reuters	Dubai says not responsible for Dubai World debt - Boeing Dreamliner touches down after first flight - Iran's Ahmadinejad mocks Obama, "TV series" nuke talks	secur, health, shot, landmark, ahmadinejad, exxon, oilfield, wood, jihad, test
Seattle Post	Hospital: Actress Brittany Murphy dies at age 32 - Actor Charlie Sheen arrested in Colorado - Charlie Sheen accused of using weapon in Aspen	patrolman, gainesvil, breakfast, girl, insur, salvat, hoax, ohio, sheen, utah
Yahoo! News	Yemen is growing front in al-Qaida battle - Report: US helped Yemen's strike against al-Qaida - AP source: Al-Qaida operative killed by US missile	care, storm, famili, white, sheen, slai, brittani, qaida, medicar, volcano

4 Conclusions

In this paper, we presented the feasibility of applying machine learning algorithms for the task of modelling readers' preferences on stories gathered from real world news-media outlets. We formulated the problem as a Learning to Rank task on preference-ordered pairs of items and applied two different methods, each with its own strengths. Ranking SVM allowed us to calculate a ranking of all articles and to retrieve the most popular articles per outlet. Lasso was applied to identify the keywords that relate to popularity of articles.

To rank articles we used only limited textual information, their titles and summaries, and we managed to achieve more than 70% accuracy in several cases. This is quite impressive given the limited data available and all the other factors that affect readers' choices, such as the placement of the article in the layout of the web-page, the presence of accompanying photos or videos, the potential sentiment content in the article, or the exact timestamp. Also the segmentation of the audience is very coarse: the readers of a given outlet are bundled into the same set, as we do not have more refined information about them (*e.g.* their location). That is we managed to quantify the contribution of content to the choices of readers, isolating this from other factors. Even more, in this research we focused only on the main page of each outlet – stories that the editors expect to be interesting and popular. We expect to be able to achieve a higher performance if we don't restrict the data to the first page only, but use the entire content of an outlet as a negative set of non-popular stories.

Future plans include research on higher level features to describe articles, such as topic or sentiment to improve performance in learning, predicting, and understanding what affects readers' preferences.

Acknowledgements. The authors would like to thank Mark Schmidt for his implementation of Lasso LARS algorithm. We used 'QPC – Quadratic Programming in C' Matlab Interface in the empirical comparing of SVM and Lasso section. I. Flaounas is supported by the A. S. Onassis Public Benefit Foundation; N. Cristianini is supported by a Royal Society Wolfson Merit Award. This research was partly supported by the "University of Bristol Bridging the Gaps Cross-Disciplinary Feasibility Account" (EP/H024786/1). Group activities are supported by PASCAL2 Network of Excellence.

References

1. Gans, H.J.: Deciding What's News: A Study of CBS Evening News, NBC Nightly News, Newsweek, and Time, 25th anniversary edition edn. Northwestern University Press (2004)
2. Wu, F., Huberman, B.A.: Popularity, novelty and attention. In: Proceedings 9th ACM Conference on Electronic Commerce (EC 2008), pp. 240–245 (2008)
3. Szabó, G., Huberman, B.A.: Predicting the popularity of online content. CoRR abs/0811.0405 (2008)
4. Ghose, A., Yang, S.: An empirical analysis of search engine advertising: Sponsored search in electronic markets. Management Science 55(10), 1605–1622 (2009)
5. Joachims, T.: Optimizing search engines using clickthrough data. In: Proceedings of the Eighth ACM SIGKDD International Conference on Knowledge Discovery and Data Mining (KDD), pp. 133–142 (2002)
6. Tibshirani, R.: Regression shrinkage and selection via the lasso. Journal of the Royal Statistical Society, Series B (Methodological) 58(1), 267–288 (1996)
7. Boser, B.E., Guyon, I., Vapnik, V.: A training algorithm for optimal margin classifiers. In: Proceedings of the 5th Conference on Computational Learning Theory (COLT), pp. 144–152 (1992)
8. Cristianini, N., Shawe-Taylor, J.: An introduction to support vector machines and other kernel-based learning methods. Cambridge University Press, Cambridge (2000)
9. Flaounas, I.N., Turchi, M., Bie, T.D., Cristianini, N.: Inference and Validation of Networks. In: Buntine, W., Grobelnik, M., Mladenić, D., Shawe-Taylor, J. (eds.) Machine Learning and Knowledge Discovery in Databases. LNCS, vol. 5781, pp. 344–358. Springer, Heidelberg (2009)
10. Porter, M.: An algorithm for suffix stripping. Program 14, 130–137 (1980)
11. Liu, B.: Web Data Mining, Exploring Hyperlinks, Contents, and Usage Data. Springer, Heidelberg (2007)

A Comparison of Two Ontology-Based Semantic Annotation Frameworks

Quratulain Rajput and Sajjad Haider

Artificial Intelligence Lab, Faculty of Computer Science
Institute of Business Administration
Karachi, Pakistan
{qrajput,sahaider}@iba.edu.pk

Abstract. The paper compares two semantic annotation frameworks that are designed for unstructured and ungrammatical domains. Both frameworks, namely ontoX (ontology-driven information Extraction) and BNOSA (Bayesian network and ontology based semantic annotation), extensively use ontologies during knowledge building, rule generation and data extraction phases. Both of them claim to be scalable as they allow a knowledge engineer, using either of these frameworks, to employ them for any other domain by simply plugging the corresponding ontology to the framework. They, however, differ in the ways conflicts are resolved and missing values are predicted. OntoX uses two heuristic measures, named level of evidence and level of confidence, for conflict resolution while the same task is performed by BNOSA with the aid of Bayesian networks. BNOSA also uses Bayesian networks to predict missing values. The paper compares the performance of both BNOSA and ontoX on the same data set and analyzes their strengths and weaknesses.

Keywords: Ontology, Information Extraction, Bayesian Network, Machine Learning, Semantic Annotation.

1 Introduction

A large amount of useful information over the web is available in unstructured or semi-structured format. This includes reports, scientific papers, reviews, product advertisements, news, emails, Wikipedia, etc [1]. Among this class of information sources, a significant percentage contains ungrammatical and incoherent contents where information is presented as a collection of words without following any grammatical rules. Many of the popular retail sites such as eBay[1] and craigslist[2] fall into this category. These websites are made up of user-provided data, called *posts,* which contains useful information but due to lack of semantics are not easily searchable. Moreover, the automatic extraction of required information from such posts is also a big challenge [2]. The semantic web technologies, such as OWL/RDF, provide a major advancement to handle this challenge [3] as they aid in the semantic annotation of new and existing data sources.

[1] www.ebay.com
[2] www.craigslist.org

H. Papadopoulos, A.S. Andreou, and M. Bramer (Eds.): AIAI 2010, IFIP AICT 339, pp. 187–194, 2010.

The semantic annotation process for existing data sources consists of many phases including information extraction, knowledge management, storage of extracted data in RDF/OWL, and user interfaces [4]. Much of the research, however, has been focused on extracting information from varying type of data sources. Leander et al. [5] provides a good survey of the techniques used in this area. Recently, a lot of research has been done on ontology based semantic annotation as it helps in making the application independent from domain knowledge and easily scalable. Few of the important contributions in this area are BYU [6], MnM [7], OntoX [8] and BNOSA [9] [10] which is an extension of our previous work OWIE [11] and E-OWIE [12].

This paper compares the performance of ontoX and BNOSA frameworks. These frameworks are selected due to their similarities in the way they exploit ontology for knowledge building, rule generation and information extraction. They, however, differ in the way conflict resolution is handled. ontoX uses heuristic measures while BNOSA uses Bayesian networks to resolve conflicts in case multiple values are extracted for a single attribute.

The rest of the paper is organized as follows. Section 2 describes ontoX and BNOSA while Section 3 compares the similarities and differences between them. The performance of both frameworks on the same data set is analyzed in Section 4. Finally, Section 5 concludes the paper and provides future research directions.

2 Selected Semantic Annotation Framework

2.1 ontoX: Ontology Driven Information Extraction

Burcu et al. [8] presented a framework to extract relevant data from existing web documents. The methodology suggests an ontology-driven information extraction process. It builds a domain specific ontology which is utilized during the information extraction phase. A tool, ontoX, was implemented based on the proposed scheme. The ontoX system consists of three main modules: i) Ontology Management Module, ii) Rule Generation Module and iii) Extraction Module.

Ontology Management Module (OMM): This module builds a domain-specific ontology to be used for information extraction. Different constructs of the ontology such as classes, object properties and data-type properties (also called attributes) as well as additional properties defined in the owl:AnnotationProperty element are utilized during the information extraction phase. The additional properties based on context keywords, constraints, quality properties and temporal properties are defined for each construct in the ontology.

- *Keywords:* The context keywords help in finding the location of relevant information in a corpus.
- *Constraints:* Constraints are used to narrow down the range of possible values belonging to an attribute. Standard data types are used to define such ranges. For example, *int* data type specifies that the value is of integer type, *float* data type specifies that the value is of float type, and so on. The only exception is the *string* data type which is specified as xsd:Name[3] with the aid of named-entity-probability heuristic.

[3] The value space of xsd:Name is a set of all strings that match the Name production of XML 1.0.

- *Quality properties:* Quality properties enrich the constructs in ontology by creating two additional properties. The first one is *confidence-level,* which takes values between [0, 1]. It indicates the level of confidence of the ontology engineer that the construct is relevant. It also helps in situations when the same value is assigned to two different attributes. If such a situation arises, the property with the higher confidence level is considered the winner. The other property is *relevance*, which takes one of the two values {true, false}. It tells the system that a user is not very interested in this property, but if found then extract it because the construct is part of the domain of interest.
- *Temporal properties:* Temporal properties provide two kinds of services: temporal extraction and change management. With temporal extraction property, a user can state if she wants her input data to be extracted and can suggest valid-time-begin and valid-time-end properties for every construct defined in the ontology. With change management property, users can provide suggestions regarding out-of-date concepts if they do not appear in the corpora anymore.

Rule Generation Module (RGM): This module is responsible for generating extraction rules for each data type defined in the ontology. The rules are used to identify possible values of each attribute.

Extraction Module (EM): This module applies the rules generated by RGM to extract values corresponding to each attribute. The extraction module consists of two main steps: pre-processing and extraction. The pre-processing step removes the stop words (such as "and", "it", "at", "to", etc.), and finds the location of the pre-defined data types. The extraction step assigns the located data to the corresponding attribute. If more than one keyword is assigned to an attribute, then data is searched/located in the neighborhood of all the keywords. If more than one value is assigned to an attribute, ontoX uses a heuristic, level of evidence, to resolve this conflict. In case of same value being assigned to more than one attributes, ontoX uses another heuristic, confidence level, to resolve this conflict.

2.2 BNOSA: Bayesian Network and Ontology Based Semantic Annotation

BNOSA (Bayesian Network and Ontology based Semantic Annotation) is a semantic annotation framework that is designed to extract relevant information from unstructured and ungrammatical domains such as craigslist [9]. It utilizes ontology as well as Bayesian networks to perform information extraction and semantic annotation tasks. The information extraction in BNOSA is conducted in two phases:

Phase-I: Similar to ontoX, this phase utilizes different constructs of ontology such as classes, object properties and data type properties for information extraction. It also stores additional information in owl:AnnotationProperty elements. The additional information consists of context keywords and value constraints. Unlike ontoX, BNOSA does not define Quality and Temporal properties.

- *Context Keywords:* The context keywords help in finding the location of relevant information in a corpus.
- *Constraints*: This feature is also similar to the corresponding feature in ontoX except the way *string* data types are handled. BNOSA applies simple pattern

matching rules for all instances of a string type attribute. All such instance values are stored in the comment section of the ontology.

Once information is specified in the ontology, this phase generates rules for each data type at run time. The context keywords are located first and then with the help of these rules, data within the neighbourhood of these context keywords is searched. The extracted data is then assigned to the corresponding attribute.

Phase-II: If more than one value is assigned to an attribute or no value is extracted at all then Bayesian networks are used in Phase-II for conflict resolution and missing value prediction. Phase-II is mainly divided into two modules: Bayesian network learning module and prediction module.

- *Learning Module:* This module first performs data cleaning and data pre-processing (discretization of continuous data and removal of anomalies), and then learns the probabilistic relationships that exist among the attributes by learning the structure and parameters of a BN.
- *Prediction Module:* If there are missing and/or conflicting values in the extracted data set then this module uses Bayesian inference mechanism to predict missing values and to resolve conflicts. In case of missing values, all the non-missing/non-conflicting values are considered as hard evidence and the posterior marginal probability of the missing attribute is computed. The value with the highest probability is considered the missing value if it satisfies a particular threshold value. In case of multiple values belonging to an attribute (conflict resolution), the attribute's value is considered as missing and all the non-missing/non-conflicting values are entered as hard evidences in the BN. The posterior marginal probability of this attribute is computed next. Among the multiple values, the value with the highest posterior probability is selected as the winner and is assigned to the corresponding attribute.

3 Comparative Analysis

This section identifies the similarity and dissimilarity of ontoX and BNOSA in terms of their knowledge specification mechanisms and information extraction processes.

3.1 Ontology Specification

Both approaches use ontology to define the domain knowledge as well as some additional information that is utilized during the information extraction phase.

The similarities in ontology specification are:

- The comment section of the owl:AnnotationProperty is used to define and store the context keywords. It must be stated, however, that finding/defining all relevant keywords requires manual analysis of documents collection which in itself is a labor intensive and time consuming process.
- Ontology constructs are used to define domain knowledge.
- The ranges of possible values of an attribute are constrained by defining its data type such as int, float, date, etc.

The differences in ontology specification are:

- ontoX uses owl:AnnotationProperty to define few additional properties as well. This includes quality and temporal properties. Such properties are not defined by BNOSA.
- To handle a string type attribute, ontoX defines its range through xsd:Name property if the attribute has large number of instances. In case of only few instances, it uses the enumeration construct. BNOSA, on the other hand, defines all possible instance values in the comment section without specifying context keywords.

3.2 Information Extraction

To extract values of each attribute, rules need to be generated. Both approaches define ranges of attributes' values as data types provided by OWL. The rules, thus, are generated according to the corresponding primitive data type. These rules, defined in the form of regular expressions, are used to extract values from the text.

The similarities in the extraction process are:

- Both approaches try to locate an attribute value within the neighborhood of the corresponding context keywords.
- Each attribute can have more than one keywords associated with it therefore more than one value can be extracted for an attribute.

The differences in the extraction methodology are:

- ontoX considers the neighborhood of a context keyword as the area ranging from its previous keyword to its next keyword. BNOSA, on the other hand, considers a fixed number of characters surrounding the keyword as its neighborhood.
- If one value is found to be relevant to more than one attribute than ontoX resolves it using confidence level, while BNOSA assigns this value to each of the attributes.
- To select one value from more than one extracted value, ontoX computes the level of evidence, while BNOSA resolves it using the Bayesian network.
- BNOSA also uses Bayesian networks to predict missing values which is not performed by ontoX.
- OntoX also has an ontology change detection mechanism which is not implemented in BNOSA yet.

4 Experimental Results

This section evaluates the extraction results of ontoX and BNOSA when they are applied on the same data set. To compare the results, the same data set has been selected as used by Burcu et al. [8], which is a collection of digital camera reviews available on a retail website[4]. A sample data is shown in Table 1. The evaluation is done on the basis of recall and precision values. The values are computed as:

[4] http://www.steves-digicams.com/

Table 1. Camera data available in reviews

Mega Pixel	Optical Zoom	Display	Storage Medium	Movie Format	Battery
5.0		2.5	SD	QuickTime	
5.0	12	1.8	SD	QuickTime	Lithiumion
	3	2.0	SD		AA
7.1	4	2.0			
6.0	12	2.0			
5		1.8	XD	AVI	lithium
5	3			VGA	
	4		FC		

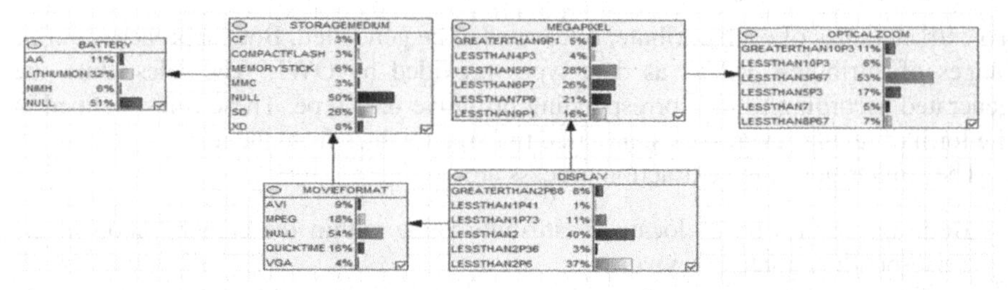

Fig. 1. Bayesian network learned model for digital camera

$$Recall = \frac{C}{N}, \quad Precision = \frac{C}{C + I}$$

where N is the number of values extracted from the document, C is the number of correctly extracted values, and I is the number of incorrectly extracted values.

As discussed earlier, BNOSA uses Bayesian network to predict missing values and to resolve conflicts. But before being used for this purpose, a Bayesian network needs to be learned from the available data. The first step in this process requires discretizing numerical values. The discretized data is then used to train the Bayesian network. Fig. 1 shows the Bayesian network learned from the discretized camera data set.

Once a BN has been learned, it is used for conflict resolution and for missing value prediction as explained in Section 2.2. ontoX, on the other hand, uses level of evidence for conflict resolution. Unlike BNOSA, there is no mechanism in ontoX to predict the missing values. To keep the results and the analysis of the experiment consistent, the prediction feature of BNOSA is not utilized during the experiments. Similarly, when there are a large number of possible instances belonging to a string type attribute, such as Model of a camera, ontoX models these instances through xds:Name property and uses a heuristic to resolve conflicts. The results reported in [8] show a good rate of recall and precision values for such string data types. BNOSA, on the other hand, stores all the possible values in the comments sections. Despite some obvious limitations, the approach used by ontoX - for large number of instances belonging to a string data type - is more generic and is superior in terms of its expressiveness. The method employed by BNOSA, however, would still generate better results as it stores all the instances in the comments section of the ontology. It is for these reasons; string data types with large number of instances are not compared in this analysis. String type attributes with small number of instances, however, are still being considered because both ontoX and BNOSA store them in the ontology albeit differently.

The results of the extraction process by ontoX and BNOSA are shown in Table 2 and Table 3, respectively. The corresponding graphical representation of the results is shown in Fig. 2. It can be seen from the tables and the figure that in general BNOSA performs better than ontoX and, in few cases, the difference in the recall and precision values is significantly higher. For example, the recall and precision values of ontoX for MegaPixel is within the range of 0.5. In contrast, BNOSA produces recall and precision values for the same attribute in the range of 0.9. Other entries can be read and interpret in a similar fashion.

This difference in the performance is primarily due to different heuristics/methods used by these approaches. For instance, when selecting the neighborhood area of a keyword, BNOSA considers a fixed number of characters on either side of the keyword. ontoX, on the other hand, uses the whole area between a keyword and its left and right neighboring keywords. This, in some cases, may produce more conflicts. Another reason might be the use of level of evidence in ontoX. Being a user-defined metric/heuristic, an inaccurate specification may degrade the performance of the whole ontoX framework. In contrast, BNOSA uses Bayesian networks which learn the probabilistic relationship from the available data set and can be considered more robust when compared to a user-defined metric. It must be stated, however, that BN learning requires the availability of a large data set to accurately reflect the probabilistic relationship that exist among the variables in a problem domain.

Table 2. Extraction results using ontoX

	Number of fact	Correctly identified facts	Incorrectly identified facts	Recall	Precision
Megapixel	137	70	63	0.51	0.52
Optical zoom	124	105	22	0.84	0.82
Display size	113	93	23	0.82	0.80
Storage medium	61	15	56	0.25	0.22
Movie format	56	41	59	0.73	0.41

Table 3. Extraction results using BNOSA

	Number of fact	Correctly identified facts	Incorrectly identified facts	Recall	Precision
Megapixel	137	121	7	0.88	0.95
Optical zoom	125	104	17	0.83	0.86
Display size	113	108	19	0.96	0.85
Storage medium	61	46	5	0.75	0.90
Movie format	55	50	52	0.91	0.49

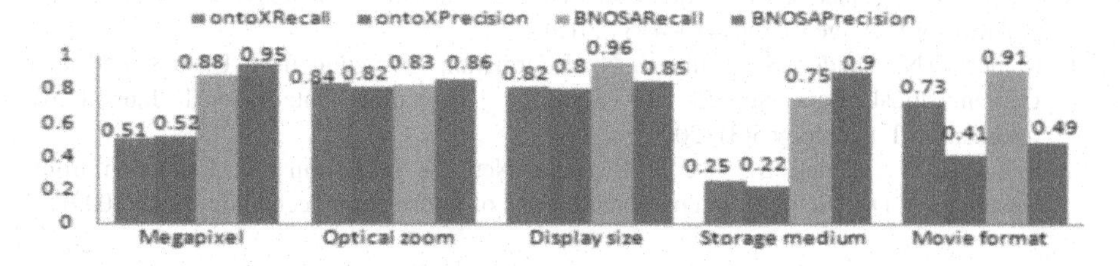

Fig. 2. Graphical view of extraction result

5 Conclusion

The paper compared the information extraction capabilities of two semantic annotation frameworks, namely BNOSA and ontoX. Both frameworks model and store knowledge about a problem domain in an ontology and uses it during the information extraction phase. They, however, differ in ways, information is located, conflicts are resolved and missing values are predicted. Experiments were conducted on a sample data set to analyze the strengths and weaknesses of both frameworks. The results suggest that BNOSA performs significantly better than ontoX on the selected data set. The statement, however, cannot be generalized unless thoroughly tested on many data sets and that too of a much larger size as compared to the one selected in this study. The comparison of BNOSA with ontoX and few other information extraction techniques on multiple large data sets is one of the future research directions.

Acknowledgements. The first author is extremely grateful to Dr. Burcu Yildiz for helping her in running ontoX as well as for sharing the camera data set which made the comparison reported in this paper possible.

References

1. Ittersum, R., Spalding, E.: Understanding the difference between structured and unstructured documents (2005)
2. Michelson, M., Knoblock, C.A.: Semantic annotation of unstructured and ungrammatical text. In: Proceedings of the 19th International Joint Conference on Artificial Intelligence, pp. 1091–1098. Morgan Kaufmann Publishers Inc., Edinburgh (2005)
3. Antoniou, G., Harmelen, F.V.: A semantic Web primer. MIT Press, Cambridge (2004)
4. Reeve, L., Han, H.: Survey of semantic annotation platforms. In: Proceedings of the 2005 ACM Symposium on Applied Computing, pp. 1634–1638. ACM, Santa Fe (2005)
5. Laender, A.H.F., Ribeiro-Neto, B.A., Silva, A.S.D., Teixeira, J.S.: A brief survey of web data extraction tools. Sigmod Record 31, 84–93 (2002)
6. Embley, D.W., Tao, C., Liddle, S.W.: Automating the extraction of data from HTML tables with unknown structure. Data and Knowledge Engineering 54, 3–28 (2005)
7. Vargas-Vera, M., Motta, E., Domingue, J., Lanzoni, M., Stutt, A., Ciravegna, F.: MnM: Ontology Driven Semi-Automatic and Automatic Support for Semantic Markup, pp. 379–391 (2002)
8. Yildiz, B., Miksch, S.: ontoX-A Method for Ontology-Driven Information Extraction. In: Gervasi, O., Gavrilova, M.L. (eds.) ICCSA 2007, Part III. LNCS, vol. 4707, pp. 660–673. Springer, Heidelberg (2007)
9. Rajput, Q.N., Haider, S.: BNOSA: A Bayesian Network and Ontology based Semantic Annotation Framework (submitted)
10. Rajput, Q.N.: Semantic Annotation Using Ontology and Bayesian Networks. In: Advances in Artificial Intelligence, pp. 416–418 (2010)
11. Rajput, Q.N., Haider, S., Touheed, N.: Information Extraction from Unstructured and Ungrammatical Data Sources for Semantic Annotation. International Journal of Information Technology 5(3) (2009)
12. Rajput, Q.N., Haider, S.: Use of Bayesian Network in Information Extraction from Unstructured Data Sources. International Journal of Information Technology 5(4) (2009)

A Tool for Automatic Creation of Rule-Based Expert Systems with CFs

Ioannis Hatzilygeroudis and Konstantinos Kovas

Department of Computer Engineering & Informatics, University of Patras
GR-26500 Patras, Greece
{ihatz,kobas}@ceid.upatras.gr

Abstract. This paper introduces a tool, namely ACRES (Automatic CReator of Expert Systems), which can automatically produce rule-based expert systems as CLIPS scripts from a dataset containing knowledge about a problem domain in the form of a large number of cases. The rules are created via a simple systematic approach and make use of certainty factors (CFs). CFs of same conclusions can be combined either using the MYCIN method or a generalization of MYCIN's method. This latter method requires calculation of some weights, based on a training dataset, via the use of a genetic algorithm. Creation of an expert system is outlined. Small scale experimental results comparing the above methods with each other and a neural network are finally presented.

Keywords: expert systems, certainty factors, unsupervised learning, MYCIN, genetic algorithm, CLIPS.

1 Introduction

A lot of datasets are available nowadays, containing known instances from various domains (e.g. UCI Machine Learning Repository). Effort is required to retrieve information from these datasets and use them as a knowledge base for predicting new instances. A tool is therefore needed that can automatically turn these raw data to rules that can classify new data accordingly using a model of uncertainty. Existing efforts at providing such automated tools are mostly commercial and thus targeted to professionals or knowledge engineers (e.g. [1]), making the process difficult for intermediate users or domain experts that want to make use of existing knowledge in datasets. The uncertainty models implemented are also not efficient in many cases or lack the ability to adjust to the requirements of each individual problem. The main goal of this work is to introduce a tool (ACRES), which can make use of datasets and other available sources of knowledge, in order to easily create expert systems that use that knowledge to cope with new instances of a problem. A generalized version of the MYCIN certainty factors (CFs) is used as the way to represent uncertainty, allowing optimization of the expert system. In section 2 we present existing tools for the creation of expert systems. In section 3 we present certainty factors as a way to represent uncertainty in expert systems and present the two methods we implemented. The way the automatic generation of expert systems from datasets can be achieved is described in Section 4. In section 5, an example creation of an expert system is outlined. In

H. Papadopoulos, A.S. Andreou, and M. Bramer (Eds.): AIAI 2010, IFIP AICT 339, pp. 195–202, 2010.

section 6, some experimental results are presented. Finally, section 7 concludes and discusses future works.

2 Related Work

One of the most well-known tools for developing rule-based expert systems is CLIPS (C Language Integrated Production Systems) environment from NASA [4]. JESS (Java Expert System Shell), its Java counterpart, has also received great interest [7]. Expert System Creator [9] is a commercial software tool for the development of expert system based applications, representing domain knowledge using rule sets, decision tables or classification trees. It is mainly targeted to professionals, domain experts or knowledge engineers and less for simple users. EXSYS [1], an also commercial Expert System Development Software, is easier to use and provides many different uncertainty models to choose from. Our proposed tool focuses on the automatic generation of expert system knowledge bases from existing datasets with the minimum required effort from the user. At the same time it offers the ability to optimize the output expert system taking into account specific requirements or priorities set by the developers-users.

3 Expert Systems and Certainty Factors

3.1 Combining Conclusions with Certainty Factors

MYCIN [2] was a medical expert system developed in Stanford University in the early 1970s. It was the first one that introduced Certainty Factors (CFs) as a way to represent uncertainty when a conclusion is made by a rule. Although CFs have some problems [8], they still remain a simple and useful way of dealing with uncertainty.

CFs are associated with a simple computational model that permits to estimate the confidence in conclusions being drawn. A CF is a number between –1 (definitely false) and +1 (definitely true), which measures the expert's belief (positive number) or disbelief (negative number) to a conclusion. However, usually CFs that are positive numbers between 0 (definitely false) and 1 (definitely true) are resulted, due to the nature of most problems. In these cases, any CF less than 0.5 indicates disbelief, whereas any CF equal to or greater than 0.5 indicates belief in a conclusion. Given that CFs are positive, when we have the following rules with the same conclusion and CF1, CF2 respectively and they are fired, the combined certainty CF for conclusion, according to MYCIN theory, is given by the formula:

$$CF = CF_1 + CF_2\,(1\text{-}CF_1) = CF_1 + CF_2 - CF_1 * CF_2 \tag{1}$$

3.2 Generalizing Certainty Factor Computation

The above formula didn't give satisfactory results in many cases. So in the expert system PASS [3], a generalized version of (1) was introduced:

$$CF = w_1 * CF_1 + w_2 * CF_2 + w * CF_1\,CF_2 \tag{2}$$

where w1, w2 and w are numeric weights that should satisfy the following equation:

$$w_1 + w_2 + w = 1 \tag{3}$$

to assure that 0 • CF • 1.

To use formula (2), however, the weights w_1, w_2, w should be first determined. In PASS, statistical data about the problem was used, as a training data set to determine the weights by hand. The tool presented in this paper computes the above weights automatically, utilizing a genetic algorithm.

4 Automatic CReation of Expert Systems (ACRES)

4.1 Rule and CF Generation

One of the main functions of the tool is to convert raw data in a dataset to a set of rules. We consider that each instance of the data set contains discrete values corresponding to attributes related to a problem domain. One of the attributes represents the class (or output) attribute. We consider binary class attributes, i.e. datasets referring to two classes (A and B). So, class attribute takes one of two values, e.g. yes-no or true-false. We distinguish between positive instances (those belonging to the class 'yes' or 'true') and negative instances (the rest ones). Rule and CF generation is then achieved by the following process:

1. Cluster instances in groups, so that each group contains instances that have identical values for all non-class attributes.
2. From each such group produce one rule with as conditions the attribute-value pairs of the instances and as conclusion the class attribute-value pair (where the value is the one of the positive instances).
3. Associate with each rule a CF defined as

   $$\mathrm{CF} = n_p/N$$

   where n_p is the number of the positive instances in the group and N the number of all instances. That is, a CF is defined as the frequency of the positive instances in the group.

4.2 Expert System Creation

ACRES can produce one CLIPS-based expert system (ES) from a given data set. The process is as follows:

1. Specify the attributes and the class attribute of the data set.
2. Divide attributes in two possibly (typically) overlapping groups.
3. Extract from the dataset two subsets, corresponding to the two attribute groups. Each subset includes instances having values for the attributes of the corresponding group and the class attribute.
4. Produce rules with CFs separately from the two subsets using the process in Section 4.1.
5. Produce the rule implementing computation of the combined CF, based on formula (2) above.
6. Produce the basic expert system
7. Determine the weights w_1, w_2 and w of formula (2).

The rationale behind the division of the attributes and the dataset in two groups/subsets is the need to have two stages of reasoning or two alternate ways of reasoning for the same conclusion, from which the need for CFs combination comes.

In step 7, if we want to produce a MYCIN-like expert system, the system sets $w_1 = w_2 = 1$ and $w = -1$. If not, the genetic algorithm is used.

4.3 Genetic Algorithm

We use a simple genetic algorithm to determine the best values for weights w_1, w_2 and w, since it is an optimization problem. An initial population of possible solutions, called *atoms* (or chromosomes or genomes), is created. Then the atoms of the population are evaluated via a fitness function and the best of them are selected to pass to the next generation after having been processed through two processes, i.e. crossover and mutation, which resemble corresponding genetic processes. The algorithm stops when any of the set termination criteria is met.

In such an algorithm, there are two crucial design decisions: the representation scheme of an atom and the fitness function. In our case, an atom is a representation of the values of weights w_1 and w_2, since w can be calculated as $w = 1 - (w_1 + w_2)$. The fitness function should evaluate how well the represented weight values in an atom affect the effectiveness of the expert system under creation. Usually, three metrics are used for evaluating classification-oriented expert systems: *accuracy* (abbr. *acc*), *sensitivity* (abbr. *sen*) and *specificity* (abbr. *spec*) [6, 7]. So, the fitness function should be based on them. Although accuracy is the most important, sensitivity and specificity should be taken into account too. Also, sometimes a system is required to give more emphasis to one of them. In our GA implementation, the fitness function is defined as

$$fitness = w_{ACC} * acc + w_{SEN} * sen + w_{SPE} * spec + w_{BAL} * bal \quad (3)$$

where w_{ACC}, w_{SEN}, w_{SPE} and w_{BAL} are user-defined factors and '*bal*', which is an abbreviation for '*balance*', is defined by the following expression:

$$bal = 1 - |sen - spec| \quad (4)$$

which represents how balanced sensitivity and specificity are.

With the above definition we can define a variety of fitness functions, depending on which metric and in what degree we consider as more significant for the under creation expert system.

To compute fitness for each atom of the population, we need to evaluate the expert system under construction. So, the CLIPS engine is embedded in our system and is called to run the constructed expert system for all instances of the dataset. Based on the results we compute the values of the three metrics and then the fitness for each instance.

4.4 Systems Comparison

As is stated above, ACRES can produce two types of expert systems with CFs, one of MYCIN type and one of WEIGHTED type. The second type uses the generalized way of CF combination. This is related to the *expert system creation mode* of the tool.

However, apart from that, ACRES give the possibility to compare the two types of systems. So, in its *method comparison mode*, the system creates two similar systems from the same dataset, one of MYCIN and one of WEIGTHED type, and compares them. Comparison process is as follows:

1. Split the dataset in two sets, a training and a test set.
2. Produce the basic expert system using the training set as the data set.
3. Produce the two expert systems (the MYCIN-type and the WEIGHTED-type).
4. Evaluate the two systems (i.e. compute *acc, sen* and *spec*) using the test set.
5. Repeat steps 1-4 *k* times (*k* is defined by the user).
   Repetition of steps 1-4 actually implements a *k*-fold validation process.

4.5 Implementation Issues

Both modes were implemented as console applications with C++, integrating the CLIPS library to load the expert systems created and the GALIB library for the genetic algorithm. The system uses a population size = 15, $p_c = 0.7$, $p_m = 0.05$ and terminates after 50 generations. This last parameter was chosen after having made a number of experiments. The graphical user interface was developed with Visual C++.

5 Example Use of the Tool

This section presents an example use of the presented tool, using a dataset from the UCI repository [8]. The breast cancer dataset was chosen, containing 286 instances of patient instances.

There are 10 variables in the dataset. The user should prepare two files. The one is the dataset, as provided in the UCI repository. In our case, it is the "breast.data" dataset file. The second file is the variables file ("VariablesNames.txt"), which contains the names of the attributes of the problem related to the dataset. The first variable is the class variable. In our case, it has two classes: "no-recurrence-events" and "recurrence-events".

In Fig. 1 the graphical user interface is shown, in the expert system creation mode. The user can browse for the above files. After loading them, he/she can choose the output/class variable. Next the user specifies two variable groups for which rules will be created. The way this is done is something that requires some knowledge about the domain (an expert's advice would help here). The groups can have common variables. Then, he/she can choose to create a MYCIN type system and/or a WEIGHTED type one. If the latter type is chosen, the GA parameters for the fitness function should be specified.

In the method comparison mode we can evaluate the two methods for combining conclusions. The user can specify the train/test datasets ratio and the cross validation folds. We've used 1/3 for the test/train ratio and 3-fold cross validation. After pressing the "Compare Methods" button the results of the evaluations of the two systems are depicted. Based on the results one can decide which of the methods is more appropriate for the problem and switch to the "expert system creation" mode. In our example, the WEIGHTED version of the expert systems does better.

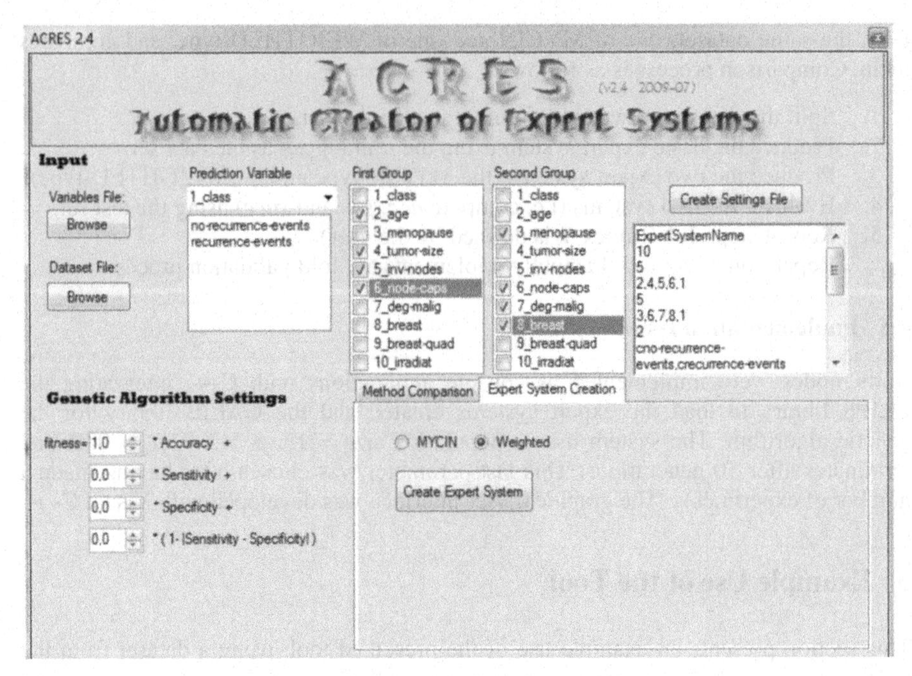

Fig. 1. ACRES Graphical User Interface

The expert system is created as a CLIPS program file. The program consists of four template definitions. The first template, named 'data', has as slots the variables of the dataset. The next two templates, named 'result1' and 'result2', respectively keep the results (classes and corresponding CFs) of the two reasoning stages for the two classes. The fourth template, named 'final' includes the final results from the combination of two different stage rules. The program also includes two groups of rules, one for each group/stage of variables/reasoning. Finally, it contains a rule that combines the results of two different stage rules. In Table 1, two rules, one from the first group and one from the second group of rules are presented, whereas Table 2 shows the final combination rule for the WEIGHTED version of the expert system (where w_1= 2.72175, w_2= 0.666972, w= -2.38872).

Table 1. Representative rules from the two rule groups

(defrule group1_1
 (data (age 40-49)
       (menopause premeno)
       (tumor-size 20-24))
 =>
 (assert (result1
       (no-recurrence-events
              0.823529)
         (recurrence-events
              0.176471))))
	(defrule group2_10
	 (data (inv-nodes 3-5)
	       (node-caps yes)
	       (deg-malig 2))
	 =>
	 (assert (result2
	       (no-recurrence-events
	              0.583333)
	       (recurrence-events
	              0.416667))))

Table 2. Rule for combination of results

```
(defrule ResultCombination
   (result1   (no-recurrence-events ?r1_0)
                        (recurrence-events ?r1_1))
   (result2   (no-recurrence-events ?r2_0)
                        (recurrence-events ?r2_1))
   ?y1<-(result1 (no-recurrence-events ?r1_0)
                        (recurrence-events ?r1_1))
   ?y2<-(result2 (no-recurrence-events ?r2_0)
                        (recurrence-events ?r2_1))
=>(bind ?r_0 (+   (* ?r1_0 2.72175) (* ?r2_0 0.666972)
                        (* ?r1_0 ?r2_0 -2.38872) ) )
  (bind ?r_1 (+   (* ?r1_1 2.72175) (* ?r2_1 0.666972)
                        (* ?r1_1 ?r2_1 -2.38872) ) )
  (assert (final (no-recurrence-events ?r_0)
                        (recurrence-events ?r_1)) )
  retract ?y1
  retract ?y2)
```

The CLIPS file can then be loaded in the CLIPS Expert System shell and make predictions about new instances of the problem.

6 Experimental Results

We used ACRES to produce two expert systems, one of MYCIN-type and the other of WEIGHTED-type) for predicting the success (or failure) of a technical high school student to the National exams in Greece. We used a real dataset consisting of 373 instances (student records). There were five attributes plus the class attribute. We also used WEKA [6] to produce a neural network for the same target. We used the same training and test sets for all systems and a 5-fold cross validation.

From Table 3, where results are presented, it is concluded that our ACRES weighted method of creating expert systems is comparable and in cases better than neural networks method. Additionally, construction of such systems in ACRES is much easier and controllable.

Table 3. Comparison of methods

METRIC	MYCIN	WEIGHTED	NN
acc	0.7334	0.8152	0.803
sen	0.7446	0.5639	0.630
spe	0.7291	0.9095	0.869

7 Conclusions and Future Work

We present a tool that can be used to easily create an expert system from existing knowledge stored as instances in a dataset. The tool introduces a method for creating rule-based expert systems with CFs in a CLIPS-based format. The method results in a two stages rule-based reasoning with CFs whose results are combined using a generalization of MYCIN's policy. The tool also allows for comparison of the two types of systems, the MYCIN-based one and the WEIGHTED one, which is based on the generalized formula for CF propagation. Experimental results are promising.

There are however a number of possible improvements that are opportunities for further work. For example, the way the variable groups are selected is currently something completely assigned to the user. A way to provide him with a tool that can detect dependency formations between the dataset variables would be helpful. Also, regarding the input dataset, the tool requires that its variables have discrete values. If they don't, the user must take care of discretizing them. So, a facility helping towards this target is desirable. A more advanced solution to this would be the use of fuzzy logic combined with CFs (in a FuzzyCLIPS style) in the produced rules.

References

1. Awad, E.: Building Knowledge Automation Expert Systems with Exsys Corvid. Exsys Inc. (2003)
2. Buchanan, B.G., Shortliffe, E.H.: Rule-Based Expert Systems. In: The MYCIN Experiments of the Stanford Heuristic Programming Project. Addison-Wesley, Reading (1984)
3. Hatzilygeroudis, I., Karatrantou, A., Pierrakeas, C.: PASS: an Expert System with Certainty Factors for Predicting Student Success. In: Negoita, M.G., Howlett, R.J., Jain, L.C. (eds.) KES 2004. LNCS (LNAI), vol. 3213, pp. 292–298. Springer, Heidelberg (2004)
4. Culbert, C., Riley, G., Donnell, B.: CLIPS Reference Manual, vols. 1-3. Johnson Space Center, NASA (1993)
5. Frank, A., Asuncion, A.: UCI Machine Learning Repository. University of California, School of Information and Computer Science, Irvine, CA (2010), http://archive.ics.uci.edu/ml
6. Hall, M., Frank, E., Holmes, G., Pfahringer, B., Reutemann, P., Witten, I.H.: The WEKA Data Mining Software: An Update. SIGKDD Explorations Newsletter 11(1), 10–18 (2009)
7. Hill, F.: Jess in action: rule-based systems in Java. Manning Publishing (2003)
8. Hackerman, D., Shortliffe, E.: From certainty factors to belief networks. AI in Medicine 4, 35–52 (1992)
9. Pop, D., Negru, V.: Knowledge Management in Expert System Creator. In: Scott, D. (ed.) AIMSA 2002. LNCS (LNAI), vol. 2443, pp. 233–242. Springer, Heidelberg (2002)

Non-standard Reasoning Services for the Verification of DAML+OIL Ontologies

Yingjie Song and Rong Chen

School of Information Science & Technology,
Dalian Maritime University, Dalian 116026, P.R. China
tiantian_yingjie@sina.com, tsmc.dmu@gmail.com

Abstract. Ontology has a pivot role in the development of Semantic Web which provides the understanding of various domains that can be communicated between people and applications. Motivated by J. S. Dong's work, we propose a new approach to interpreting DAML+OIL in a lightweight modeling language for software design, Alloy, which is used to provide a non-standard reasoning service for the verification of DAML+OIL ontologies. To do so, Jena is first used to parse ontology documents into classes, properties and statements, next we use algorithms to translate them into Alloy model, the Alloy Analyzer is then used to check and reason about such model. The experiments show that our method greatly improves J. S. Dong's work, and distinguishes from the traditional ontology reasoners in property checking and reasoning.

Keywords: Ontology Reasoning; DAML+OIL; Alloy; Semantic Web.

1 Introduction

A Semantic Web [4], as the next generation of the Web, provides well-defined notations and techniques for humans and applications to quickly and accurately access Web information and services. In the development of Semantic Web there is a pivot role of ontology, since ontology languages provide modeling primitives for converting notations in nature language into machine-readable logical formulas, from which autonomous software agents may infer and come to conclusions [1].

Ontology languages, such as DAML+OIL, enhance computer programs through structured organizational information and rules, with which it is able to understand the logical relationship between them. Ontology reasoning is crucial in that inconsistent ontology cannot be shared or used by autonomous software agents. A number of ontology inference engines, such as FaCT [3], RACER [2], and FaCT++ [10] have been developed with the advancement of ontology languages to facilitate ontology creation, management, verification, merging, etc.. However, the checking and reasoning of complex ontology-related properties cannot be done by them.

There is a role for software engineering techniques and tools that contribute to the Semantic Web development. J. S. Dong first proposes the use of Alloy [5] in checking and reasoning about the semantic relationship between web resources [1]. We propose a novel transformation from DAML+OIL ontology to Alloy, which greatly improves J. S. Dong's work; our approach scales up well and can work on a larger scope of property checking.

H. Papadopoulos, A.S. Andreou, and M. Bramer (Eds.): AIAI 2010, IFIP AICT 339, pp. 203–210, 2010.

The rest of the paper is organized as follows: Section 2 gives a brief introduction to DAML+OIL and Alloy. In the section 3, a simple ontology example is given described in DAML+OIL, and then the ontology document is analyzed by jena, the results of which will be used as the inputs of the algorithms, which are used to transferred the DAML+OIL into Alloy model.

2 Overview of DAML+OIL and Alloy

2.1 Logical Characteristic of DAML+OIL

DAML+OIL [7] is a successor language to DAML [8] and OIL [9] that builds on earlier W3C standards such as RDF, RDF Schema, and the language components of OIL. DAML+OIL layered on top of RDFS it inherited RDFS ontological primitives (subclass, range, domain). As a semantic Web ontology language, DAML+OIL provides users a richer set of modelling primitives (transitivity, cardinality, ...) that are commonly found in frame-based languages.

Although DAML+OIL is tightly integrated with RDFS, which provides the only specification of the language and its only serialization, DAML+OIL defines the semantic of the language to give a meaning to any ontologies that conform to the RDFS specification, including "strange" constructs such as slot constraints with multiple slots and classes. It contains richer modeling primitives than RDF. This is made easier by the fact that the semantics of DAML+OIL is directly defined in both a model-theoretic and an axiomatic form. Theoretically, DAML+OIL is undecidable, but its processor to detect the occurrence of constraints and warn the user of the consequences.

2.2 Alloy

Alloy [5] is a textual, declarative modelling language rooted in first order relational logic, which is widely accepted as micromodels of software in the software engineering community. For relationships between web resources are focus point in the Semantic Web, we believe that it will be a new application domain for Alloy. An Alloy model consists of *Signatures*, *Relations*, *Facts*, *Functions* and *Predicates*. *Signatures* represent the entities of a system and *Relations* are used to describe relations between such entities. *Facts* and *Predicates* introduce constraints over such *Signatures* and *Relations*. Whereas *Facts* are constraints to be always valid, *Predicates* are named parameterized contraints for depicting operations, *Functions* are named expression with parameters that return results.

Alloy comes with a tool, the Alloy Analyzer [6], which supports fully automated analysis of Alloy models through simulation and *Assertion* checking. While *Assertions* are assumptions about the model that can be checked. Simulation yields a random instance that is consistent with the model. Given a user specified scope on the model elements bounding the domain, the analyzer first translates an Alloy model into boolean formulas, and then invokes a SAT-solver to find an instance. If an instance that violates the assertion is found within the scope, the assertion is not valid and the instance is returned as a counterexample.

3 Description of the Approach

The specific process of our approach on ontology reasoning is shown in Fig.1: ontology documents are analyzed by Jena, and the results contain three parts: Classes C, Property P and Statements S. Next, the results are converted into Alloy model using daml2Alloy algorithm. And then we use Alloy Analyzer to check the model. In case an error, we check back the original ontology and correct it accordingly. The jena is used again to check the corrected ontology. These three steps are explained in more detail in the following.

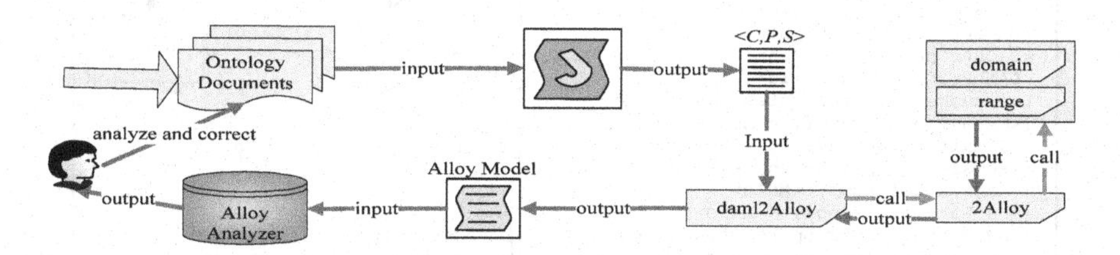

Fig. 1. The specific process of ontology reasoning

3.1 Parsing DAML+OIL Documents

To handle DAML+OIL ontologies, we adopt Jena [10] as a frontend of our framework to parse textual DAML+OIL documents. Jena provides APIs for maniplulating RDF graphs, abstracting from which it provides the ontology API for OWL and DAML ontologies. Our DAML+OIL parser, based on Jena parser, reads a DAML+OIL document of animal ontology (shown in Fig.2), which defined four classes: `Animal`, `Male`, `Man` and `Female`. While `Man` is subclass of `Male`, `Male` and `Female` are disjointed subclasses of `Animal`. `hasFather`, `hasParent` and `hasChild` are three properties such that `hasParent` and `hasChild` are inverse to each other and `hasFather` is subproperty of `hasParent`. We translate it into RDF triples, which are composed of *Classes*, *Properties* and *Statements*.

```
......
<rdfs:Class rdf:about="Animal">
  <rdfs:label>Animal</rdfs:label>
  <rdfs:comment>
    This class of animals is illustra-
tive of a number of ontological idioms.
  </rdfs:comment>
</rdfs:Class>
<rdfs:Class rdf:about="Male">
  <rdfs:subClassOf
rdf:resource="Animal"/>
</rdfs:Class>
<rdfs:Class rdf:about="Female">
  <rdfs:subClassOf
rdf:resource="Animal"/>
  <daml:disjointWith
rdf:resource="Male"/>
</rdfs:Class>
```

```
<rdfs:Class rdf:about="Man">
  <rdfs:subClassOf rdf:resource="Male"/>
</rdfs:Class>
......
<rdf:Property rdf:about=" hasParent">
  <rdfs:domain rdf:resource=" Animal"/>
  <rdfs:range rdf:resource=" Animal"/>
</rdf:Property>
<rdf:Property rdf:about=" hasFather">
  <rdfs:subPropertyOf     rdf:resource="
hasParent"/>
 <rdfs:range rdf:resource=" Male"/>
</rdf:Property>
<rdf:Property rdf:about=" hasChild">
<daml:inverseOf  rdf:resource="  hasPar-
ent"/>
</rdf:Property>
     ......
```

Fig. 2. A DAML+OIL document of animal ontology

We use a simple ontology about animal as a running example to show the outputs of Jena. It contains a sequence of *Classes*, *Properties* and *Statements*, each having a couterpart in the original document. Such *Classes*, *Properties* and *Statements* provide programmatic objects like DAMLClass, DAMLProperty and RDFTriples for our coversion algorithm in the next section. As illustrated in Fig.1, our DAML+OIL parser reads such a textual document and converts it into RFD triples as follows:

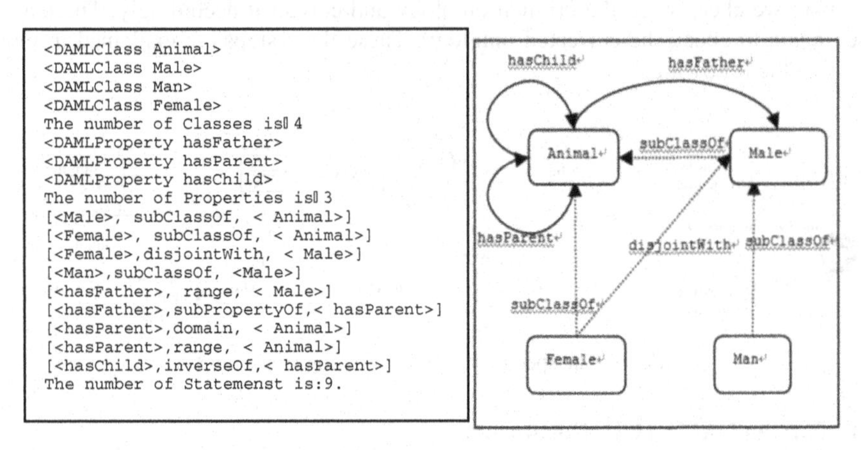

Fig. 3. Jena outputs of "Animal" ontology

3.2 Generation of the Alloy Model

Given RDFTriples and the related classes, we move on to generate the equivalent Alloy model, which contains Alloy classes and the relationship between them.

As shown below, our Algorithm daml2Alloy (C, P, S, Σ) converts the input *Classes* (denoted as C), *Properties* (denoted as P) and *Statements* (denoted as S) into a textual Alloy model Σ. To do so, we map each class into an Alloy signature, and the relationships between such classes are represented by Alloy primitives such as *Relations*, *Facts*, *Functions* and *Predicates*.

Algorithm 1: daml2Alloy (C, P, S, Σ)
Input: a set C of *Classes*, a set P of *Properties* and a set S of *Statements*
Output: a textual Alloy model Σ

1. $\Sigma \leftarrow \omega$;
2. **FOR** each $c \in C$
3. $\Sigma \leftarrow \Sigma +$ 2Alloy(c, S);
4. **FOR** each $s \in S$
5. **IF** *s.predicate.name* = "disjointWith"
6. **IF** *s.subject.subClassOf* ≠ *s.object.subClassOf*
7. $\Sigma \leftarrow \Sigma +$ "pred {no c1:" + *s.subjec.name* +", c_2:" + *s.object.name* +" | $c_1 = c_2$}";
8. **IF** *s.predicate.name* ="complementOf"
9. $\Sigma \leftarrow \Sigma +$ "pred {" + *s.subject.name* +"=" + $C -$ *s.object.name* +"}";
10. **IF** *s.predicate* meets one of Conditions in Table 3
11. the corresponding generated Alloy predicate is appended to Σ
12. **RETURN** Σ;

In our algorithm, the Alloy model Σ is initialized to be an empty string ω on line 1, each class c is converted into a signature by invoking 2Alloy (c, S) algorithm

Table 1. More cases for converting statements to Alloy

<table>
<tr><th>Case</th><th>Condition</th><th>Alloy predicates generated</th></tr>
<tr><td>1</td><td>s.predicate.name="subPropertyOf"</td><td>"pred subPropertyOf{all r:"+ s.subject.range +"| r in "+ s.object.range +"}"</td></tr>
<tr><td>2</td><td>s.predicate.name="samePropertyAs"</td><td>"pred samePropertyAs{"+s.subject+ "=" +s.object+"}"</td></tr>
<tr><td>3</td><td>s.predicate.name="inverseOf"</td><td>"pred inverseOf{"+s.subject+"=~"+s.object+"}"</td></tr>
<tr><td>4</td><td>s.predicate.name="TansitiveProperty"</td><td>"pred TansitivePropertyOf { a,b,c ∈" +s.subject+" | a.("+s.predicate+") = b && b.("+s.predicate+") =c ⇒ a. ("+ s.predicate+") = c}"</td></tr>
<tr><td>5</td><td>s.predicate.name="UniqueProperty"</td><td>"pred UniqueProperty{#("+s.predicate.range+")=1}"</td></tr>
<tr><td>6</td><td>s.predicate.name="UnambiguousProperty"</td><td>"pred UnambiguousProperty { # ("+ s.predicate.domain +") =1 }"</td></tr>
<tr><td>7</td><td>s.predicate.name="toClass"</td><td>"pred toClass{all (("+s.predicate.domain +").("+s.predicate+")) in ("+s.predicate.range+")}"</td></tr>
<tr><td>8</td><td>s.predicate.name=hasClass</td><td>"pred hasClass{ some((" + s.predicate.domain +").("+s.predicate+")) in ("+s.predicate.range+")}"</td></tr>
<tr><td>9</td><td>s.predicate.name=hasValue</td><td>"pred hasValue{#("+s.predicate.range+")=1}"</td></tr>
<tr><td>10</td><td>s.predicate.name=cardinality</td><td>"pred cardinality{#("+s.predicate.range+")=" + s.object"}"</td></tr>
<tr><td>11</td><td>s.predicate.name=maxCardinality</td><td>"pred maxCardinality{#("+s.predicate.range+")<= "+s.object "}"</td></tr>
<tr><td>12</td><td>s.predicate.name=minCardinality</td><td>"pred minCardinality{#("+ s.predicate.range+")>="+s.object "}"</td></tr>
<tr><td>13</td><td>s.predicate.name=hasClassQ</td><td>"pred hasClassQ{ some(("+s.predicate.domain+").("+s.predicate+")) in ("+s.predicate.range+")}"</td></tr>
<tr><td>14</td><td>s.predicate.name=cardinalityQ</td><td>"pred cardinalityQ{#("+s.predicate.range+")="+s.object+"}"</td></tr>
<tr><td>15</td><td>s.predicate.name=maxCardinalityQ</td><td>"pred maxCardinalityQ{#("+s.predicate.range+")<="+ s.object+" }"</td></tr>
<tr><td>16</td><td>s.predicate.name=minCardinalityQ</td><td>"pred minCardinalityQ{#("+s.predicate.range+")>="+s.object "}"</td></tr>
</table>

(see below), next follows the conversion of each statement into *Predicates* in a loop through lines 4~11, finally the produced model Σ is returned. Since each statement $s \in S$ contains RDF elements in the form of a RDF triple <*subject*, *predicate*, *object*>, we use *s.subject*, *s.predicate*, *s.object* to denote the three RDF elements respectively. Moreover, we further use *e.name* (*e.subClassOf*) to represent the name of a RDF element *e* (its parent's class) in that each RDF triple depicts the relationship between RDF elements. For instance, when it comes to two disjointed classes on lines 5~6, we should further consider whether they have the same parent class, if not, a new Alloy predicate is generated for depicting such a constraint[1]. Lines 8~9 handle a new case like lines 5~6, more generation on lines 10~11 is summarized in cases in Table 1, where a specific textual Alloy predicate is generated when some condition holds.

The next algorithm 2Alloy (*c*, *S*) is used to produce a signature for a class *c* with respect to a set *S* of *Statements*. The idea behind this conversion is as follows: let *c* be the input class, we first create an Alloy signature named as *c.name* on line 1. If *c* has parent class, i.e., *c.subClassOf* is not empty by checking the input *Statements*, we think *c* extends its parent class *c.subClassOf.name* on line 3. When it comes to a property on line 5, its domain and range is calculated before appending the resulting signature σ on line 7.

```
Algorithm 2: 2Alloy (c, S)
Input: a Class c, a set S of Statements
Output: a signature σ
7.   σ← "sig " + c.name;
8.   IF c.subClassOf≠ω
9.      σ← σ+ "extends" +

1.   σ← σ+ "{";
2.   FOR each p∈P
3.     IF domain(p, S) = c.name
4.       σ← σ+ p.name + " : " + range(p, S);
5.   σ← σ+ " }";
6.   RETURN σ ;
```

As shown in Algorithm 3, the domain of a property is calculated recursively; the domain is associated with an object on line 2 when a RDF triple satisfies such a

[1] No constraint is generated when two disjointed classes have the same parent, because Alloy 4.0 defaultly assumes classes are disjointed.

condition that its subject is a property and its predicate depicts a domain. Otherwise, the parent property is recursively checked until a qualified RDF triple is reached. We use *parent (property)* to represent the parent property of the parameter *property*. Note that the calculation of the range of a property with respect to *Statements* is quite similar to Algorithm 3, we omit it for obviousness.

Algorithm 3: domain (*p*, *S*)
Input: A property *p* of *P*, Statement *S*
Output: the domain of *p*
5. IF $\exists s \in S$, *s.subject* = *p* and *s.predicate* = "domain"
1. *p.domain* = *s.object.name*;
2. ELSE
3. *p.domain* = *domain*(*parent*(*p*), *S*);
4. RETURN *p.domain*;

Taking the "Animal" ontology as an example again, it is used to show how the conversion is achieved. The DAML+OIL document will be transferred into Alloy model as Fig.4.

```
sig Animal{
   hasParent:Animal,
   hasFather:Male,
   hasChild:Animal
}
sig Female extends Animal{}
sig Male extends Animal{}
sig Man extends Male{}
pred inverseOf{hasParent=~hasChild}
pred subPropertyOf{all
a:Animal|a.hasFather in a.hasParent}
```

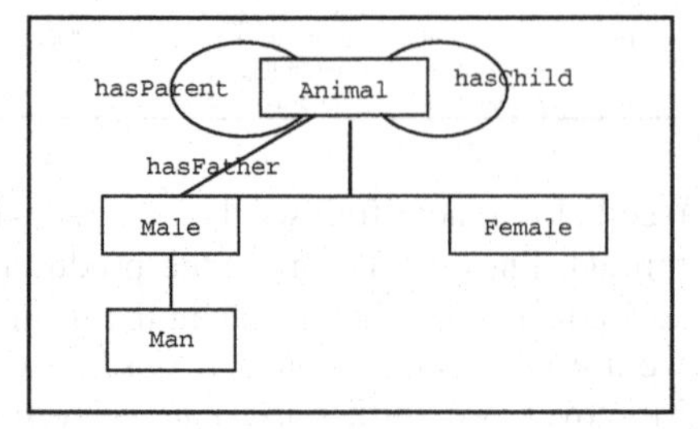

Fig. 4. The generated Alloy model of "Animal" ontology

3.3 Verifying Ontologies with the Alloy Analyzer

Semantic Web reasoning is one of key issues for ontology design, construction and maintenance, which contains the ontology consistency checking, subsumption reasoning, which task is to derive a class is another's parent, and implication relation checking. The correct ontology required to meet at least on instance. This is achieved through Alloy Analyzer to generate an instance of the model in given scope. As shown in Figure 5, there is an inconsistency occurred in the "Animal" model. This is because

```
New Open Reload Save Execute Show
module Animal

sig Animal{
   animalHasFather:Male
}
sig Female extends Animal{
   femaleHasFather:Woman
}
sig Male extends Animal{}
sig Woman extends Female{}

pred subPropertyOf{
   all a:Female.femaleHasFather| a in Animal.animalHasFather
}
 pred show{}
run show for 10
```

Fig. 5. Checking result of Alloy Analyzer

there is a `subProperty` constraint between `femalHasFather` and `animalHasFather`, `Woman` is `subClassOf Male` is implied since `Male` is the range of `animalHasFather` and `Woman` is the range of `femaleHasFather`.

Note that if Alloy Analyzer can't find a solution, it maybe for the too small scope. If there is something of inconsistent in reasoning with Alloy, assertion is an important criterion. When Alloy Analyzer can't find a counterexample, the assertion is reasonable, else Alloy Analyzer will generate a counterexample and some predicates and facts should be added to improve the model.

4 Comparison of Experimental Results

In J. S. Dong's article, classes and properties in semantic ontology were converted to subclasses of resource, and then predicates were used to establish the relationship between different resources. As a proof tool of program correctness, theorem machine prove and knowledge representation, the shortcoming of first order logic is that in reasoning it prone to "combinatorial explosion". Method we used is to make use of the characteristics of Alloy, we convert the properties of the semantic ontology to the properties of the signature in Alloy model. It has been greatly improved in the scale. Table 2 shows the operation results in the case of the same scope of the two methods.

Table 2. Comparison of Experimental Data

	scope	GenerateCM + GeneratePM				J. S. Dong			
		vars	primary vars	clauses	time	vars	primary vars	clauses	time
Family	(5)	1074	135	1674	32	3025	265	7875	47
	(10)	3168	520	5281	63	15530	1430	44475	156
	(30)	27468	4560	48201	391	257182	30090	806831	4828
	(40)	48618	8080	85861	578	570992	69320	1810146	14688
Course	(10)	5477	440	11868	63	11845	1440	31357	125
	(20)	17507	1680	41626	203	67632	9480	197133	859
	(30)	38737	3720	93696	453	204687	30120	612883	3578
	(40)	69887	6560	171406	813	460162	69360	1396873	8812
airportCode	(5)	1196	125	1910	16	2416	270	5573	31
	(20)	14716	1700	26300	156	67758	9480	193719	1109
	(35)	44656	5075	81200	688	44656	5075	81200	5781
	(50)	90796	10250	166250	1422	866928	13200	2627124	17422
Document	(10)	11458	1800	19691	125	31271	2070	87236	375
	(20)	37528	6000	67371	422	169783	10740	535897	2172
	(30)	80318	12600	146711	1125	501543	32010	1655392	6860
	(40)	139108	21600	256451	2235	1109903	71880	3750587	31016

As is shown in Table 2, we analyze and compare with several ontologies. Variables, primary variables, clauses and runtime are the four mainly elements for comparison. For each ontologies, we give four scopes to illustrate that with the expansion of the scope, our approach reflects the increasingly better performance comparing to the method of article [1].

The first ontology we used is the Family ontology, which contains three classes and five properties. Its Alloy model is composed of three signatures, five properties and one predicate. As is shown in Table 2, with the scope increasing, the advantages are more and more obvious. When the scope is equal to 40, the number of variables is 10 times larger than ours, the number of primary variables is more than 8 times and the runtime is 25 times. It is the similar with the other examples.

5 Conclusion

We propose an approach to convert the DAML+OIL ontology to Alloy model, and then using the Alloy Analyzer to automatically check and reason the generated model. First, we use Jena to analyze the ontology document and get classes, properties and statements of it. Next, we propose two algorithms to generate the Alloy model of the ontology. Finally, the tasks of checking and reasoning are executed by Alloy Analyzer. We have applied our approach to several ontologies, it can discover errors and inconsistencies, and when there are something wrong, we can efficiently correct the errors on the assistance of the counterexample given by Alloy Analyzer.

Acknowledgments

This work is supported by National Natural Science Foundation of China (60775028), the Major Projects of Technology Bureau of Dalian No.2007A14GXD42, and IT Industry Development of Jilin Province.

References

1. Jin, H.W., Dong, S., Sun, J.: Checking and reasoning about semantic web through alloy. In: Araki, K., Gnesi, S., Mandrioli, D. (eds.) FME 2003. LNCS, vol. 2805, pp. 796–814. Springer, Heidelberg (2003)
2. Haarslev, V., Möller, R.: RACER User's Guide and Reference Manaual: Version 1.7.6 (December 2002)
3. Horrocks, I.: The fact system. In: de Swart, H. (ed.) TABLEAUX 1998. LNCS (LNAI), vol. 1397, pp. 307–312. Springer, Heidelberg (1998)
4. Hendler, J., Berners-Lee, T., Lassila, O.: The semantic web. Scientific American (May 2001)
5. Jackson, D.: Alloy: a lightweight object modelling notation. ACM Trans. Softw. Eng. Methodol. 11(2), 256–290 (2002)
6. Jackson, D., Schechter, I., Shlyahter, H.: Alcoa: the alloy constraint analyzer. In: ICSE 2000: Proceedings of the 22nd International Conference on Software Engineering, pp. 730–733. ACM, New York (2000)
7. Horrocks, I., Patel-Schneider, P.F., van Harmelen, F.: Reviewing the design of daml+oil: an ontology language for the semantic web. In: Eighteenth National Conference on Artificial Intelligence, pp. 792–797. American Association for Artificial Intelligence, Menlo Park (2002)
8. van Harmelen, F., Patel-Schneider, P.F., Horrocks, I. (eds.): Reference description of the daml+oil ontology markup language (March 2001)
9. Broekstra, J., Klein, M., Decker, S., Fensel, D., Horrocks, I.: Adding formal semantics to the web building on top of rdf schema. In: Proc. of the ECDL 2000 Workshop on the Semantic Web (2000)
10. Tsarkov, D., Horrocks, I.: Fact++ description logic reasoner: System description. In: Furbach, U., Shankar, N. (eds.) IJCAR 2006. LNCS (LNAI), vol. 4130, pp. 292–297. Springer, Heidelberg (2006)
11. Lutz, C.: The complexity of reasoning with concrete domains revised version. Technical report (1999)

Algorithms for the Reconciliation of Ontologies in Open Environments

Yaqing Liu*, Rong Chen, and Hong Yang

School of Information Science & Technology, Dalian Maritime University,
116026 Dalian, China
liuyaqing234@yeah.net, tsmc.dmu@gmail.com

Abstract. The dynamic changing feature of Semantic Web determines that the ontology which is a part of Semantic Web needs constantly to be modified in order to adapt outer environment. In this paper we make a careful analysis of the ontology changes' complexity under open environment. The main contents discussed are as follow. At first we point out all possible relation types between any two ontology change sequences including directly conflict relation, indirectly conflict relation, dependent relation and compatible relation according to ontology change's definition. And then we propose a new algorithm named Algorithm of Searching Maximum and Sequential Ontology Change Sequence Set(ASMSOCSS) to find all maximum and sequential ontology change sequence subset in the prime ontology change sequence set and prove the independence of the result which may be got after running ASMSOCSS. At last we put forward the algorithm by using these maximum and sequential ontology change sequence sets to create new ontology versions according to the dependence relation between ontology change sequences.

Keywords: Ontology Changes Sequence, Maximum and Sequential Ontology Changes Sequence, Ontology Change.

1 Introduction

Ontology Evolution is the timely adaptation of an ontology to the arisen changes and the consistent propagation of these changes to dependent artifacts. Ontology change disposal, as a part of ontology evolution, focuses on exploring some ontology evolution's methods and technologies to modify ontology on the assumption with not breaking ontology consistency. By far, a lot of research work has been done on ontology change disposal and they may be classified into some based on logical reasoning[6] and others based on belief revision[8].But all of them mainly focus on ontology change disposal under centralized environment. The research on ontology change disposal under open environment is infrequent. [1] discussed ontology change disposal under open environment for the first time. Its main contribution is to define the

* The corresponding Author. This work was supported by National Natural Science Foundation of China (60775028), Dalian Science &Technology Program (2007A14GX042),Young Key Teachers Foundation Projects of Dalian Maritime University (2009QN034).

H. Papadopoulos, A.S. Andreou, and M. Bramer (Eds.): AIAI 2010, IFIP AICT 339, pp. 211–218, 2010.

mapping relation between any two ontology versions by analyzing these logs of editing ontology. But [1] is not all-inclusive because it can't provide some guide for ontology evolution's trend in semantic level.

Multiple ontology versions will be achieved under open environment. But these ontology versions are not all worthy. Moreover, too many ontology versions will make it more difficult to manage the base of ontology versions. Unfortunately, how to get worthy ontology versions under open environment is seldom concerned. In addition, the problem on ontology change of analysis and disposal is independent of the problem on merging multiple ontology versions. And it is well known that the work of merging ontology versions is very heavy. If the work of merging ontology versions can be integrated into the course of analysis and disposal of ontology changes it will greatly save the work of ontology evolution. In this paper, we propose a new method which may not only create a worthy ontology version but also avoid the work of merging ontologies through analyzing and disposing ontology change sequences.

This paper is organized as follow. The whole scheme of ontology change disposal under open environment is given in section 2. And then we make certain all possible relation types between ontology change sequences in section 3. We propose Algorithm of Searching Maximum and Sequential Ontology Change Sequence Set in section 4 and put forward the algorithm used to create new ontology versions in section 5. Related works are mentioned in section 6 and conclusion and the next work are arranged in the last section.

2 Description of the Approach

Our approach on ontology change in an open environment is composed of four steps as illustrated in Figure 1.

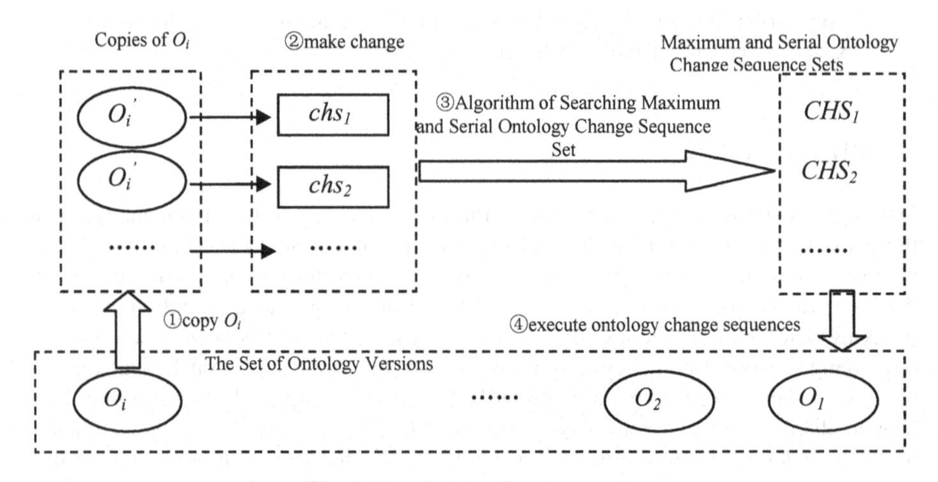

Fig. 1. Description of our approach

Step 1: Having accesses to a version O_i of an ontology, several users have separate copies O_i' of this ontology in their working spaces.

Step 2: Each user makes changes chs_j to his/her copy respectively.

Step 3: Given sequences of users' ontology changes, we algorithmically analyze the relevance between ontology change operations and search for the maximal consistent subset of the whole ontology change operations, which is called the maximal ontology changes set. We put such subsets together and denote them as a collection of ontology change sequences, i.e., $\{CHS_1,CHS_2,\ldots, CHS_k\}$, where each CHS_i $(1\leq i\leq k)$ represents a maximum sequential ontology changes.

Step 4: Several versions of ontologies may be derived from the original O_i by applying one of maximal ontology changes. Obviously, $\{CHS_1,CHS_2,\ldots, CHS_k\}$ will produce k distinguished ontologies.

The above-mentioned step 3 and step 4 are the crucial parts of our approach, next we will give their details in the remainder of this paper.

3 Formal Description of Ontology Change

Definition 1: An **ontology** O is defined as a 5-tuple:

$$O=\{C,R,H^c,Rel,A^o\}\ . \tag{1}$$

where:

- C is the set of ontology concepts.
- $H^C\subseteq C\times C$ is a set of taxonomic relationships between concepts.
- R is the set of non-taxonomic relationships. The function Rel: $R\rightarrow C\times C$ maps the relation identifiers to the actual relationships.
- A^o is a set of axioms, usually formalized into some logic language.

For brevity, O is short for ontology throughout this paper. Its instances are denoted by $O.I$, concept set by $O.C$, non-taxonomic relationships by $O.R$. We think of an ontology as a knowledge base, which contains not only the elements of an ontology, but also instances of the concepts C and relationships R.

Stojanovic categorizes all ontology changes into "Add" ontology changes and "Remove" ontology changes respectively. To highlight what type of changes is made to what object, we redefine ontology change as follows:

Definition 2: An **ontology change** ch is defined as:

$$ch=\{name,type,object,args\}. \tag{3}$$

where:

- $name$ is the identifier of this change ch.
- $type\in\{$"*Add*", "*Remove*"$\}$, is type of ch.
- $object\subset O.I\cup O.C\cup O.R$, are the elements which ch act on. O is an ontology. $O.C$ is the set of concepts of O and $O.R$ is the set of non-taxonomic relationships of O.
- $args\subset O.I\cup O.C\cup O.R$, is a list of one or more change arguments. There are changes with one, two or three arguments.

For example, ch={"*AddSubConcept*", "*Add*", {$subc_2$}, {$supc$}} denotes an ontology change that a concept $subc_2$ is added to an ontology as subconcept of $supc$. This can be pictured as Figure 2.

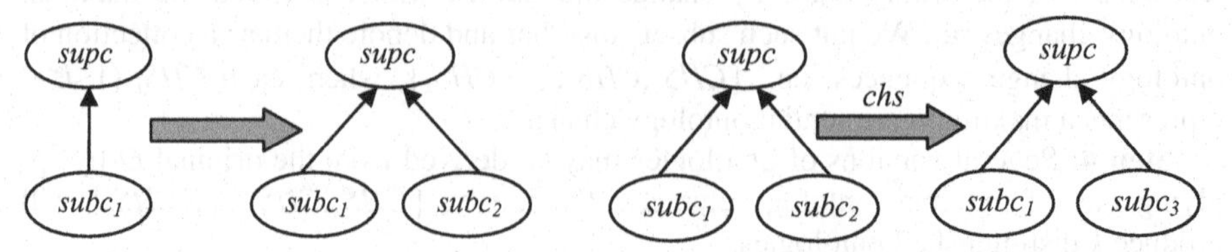

Fig. 2. An illustration of an ontology change

Fig. 3. An illustration of an ontology change sequence

Definition 3: An **ontology change sequence** chs is defined as

$$chs{=}{<}ch_1ch_2...ch_n{>},\ ch_i\ (1{\leq}i{\leq}n) \text{ is an ontology change} \quad (4)$$

if and only if n=1 or $\forall\, 1{\leq}i{\leq}n\text{-}1$, ch_i always is executed ahead of ch_{i+1}.

3.1 Argument Pool and Remove Pool of an Ontology Change Sequence

For an ontology change sequence $chs{=}{<}ch_1ch_2...ch_n{>}$, the **argument pool** of chs is used to enumerate all $ch_i.args$ when $ch_i.type$ is "Add" and the **remove pool** of chs is used to enumerate all $ch_j.object$ when $ch_j.type$ is "Remove". Further, the argument pool $chs.APool$ and the remove pool $chs.RPool$ of an ontology change sequence $chs{=}{<}ch_1ch_2...ch_n{>}$ can be obtained when we apply the *ACAPRP* algorithm to chs.

Given a simple example illustrated in Fig.3, we suppose:

- $chs{=}{<}ch_1ch_2{>}$
- ch_1={"*RemoveConcept*", "*Remove*",{$subc_2$},{ }}
- ch_2={"*AddSubConcept*", "*Add*",{$subc_3$},{$supc$}}

When *ACAPRP* algorithm is applied to chs the output is $chs.RPool$={$subc_2$} and $chs.APool$={$supc$}.

3.2 Relation Types between Two Ontology Change Sequences

Definition 4: An ontology change sequence chs_1 **depends on** another ontology change sequence chs_2 if and only if $chs_1.RPool \cap chs_2.APool \neq \varnothing \wedge chs_2.RPool \cap chs_1.APool = \varnothing$. $dependence(chs_1,chs_2)$ means that chs_1 depends on chs_2 and $\neg dependence(chs_1,chs_2)$ means that chs_1 doesn't depend on chs_2.

Definition 5: An ontology change sequence chs_1 **directly conflicts** with another ontology change sequence chs_2 if and only if $chs_1.RPool \cap chs_2.APool \neq \varnothing \wedge chs_2.RPool \cap chs_1.APool \neq \varnothing$. $directlyConflict(chs_1,chs_2)$ means that chs_1 and chs_2 directly conflict with each other and $\neg directlyConflict(chs_1,chs_2)$ means that chs_1 and chs_2 don't directly conflict with each other.

```
Algorithm 1:ACAPRP(chs)
Input: an ontology change sequence <ch1ch2......chn>
Output: chs.RPool, chs.APool
1.  chs.RPool←∅;   chs.APool←∅;          i←1;
2.  WHILE(i<=n)
3.     chs.APool←chs.APool∪chi.args;
4.     IF chi.type="Remove"
5.        chs.RPool←chs.RPool∪chi.object;
6.     IF chi.type="Add" and chi.object⊆ chs.RPool
7.        chs.RPool←chs.RPool-chi.object
8.     i++;
9.  Return chs.RPool and chs.APool
```

Definition 6: For a few of ontology change sequences chs_1, chs_2, …, chs_n, chs_1 **indirectly conflicts** with chs_n iff *dependence*(chs_1,chs_2), *dependence*(chs_2,chs_3),…, *dependence*(chs_n,chs_1) all are true, where *indirectlyConflict*(chs_1,chs_n) means that chs_1 and chs_n indirectly conflict with each other and ¬*indirectlyConflict*(chs_1,chs_n) means that chs_1 and chs_n don't indirectly conflict with each other.

Definition 7: An ontology change sequence chs_1 is **compatible** with another ontology change sequence chs_2 iff $chs_1.RPool \cap chs_2.APool=\varnothing \wedge chs_2.RPool \cap chs_1.APool=\varnothing$, where *compatible*($chs_1$,$chs_2$) means that chs_1 and chs_2 are compatible each other.

4 Algorithm of Searching Maximal Ontology Changes Set

According to the relation type between two ontology change sequences, four lemmas may be deduced.

Lemma 1: If an ontology change sequence chs_1 directly conflicts with another ontology change sequence chs_2, not all of chs_1 and chs_2 are executed no matter what the execution order may be.

Lemma 2: If an ontology change sequence chs_1 depends on another ontology change sequence chs_2, chs_1 and chs_2 can be all executed if and only if chs_2 is executed ahead of chs_1.

Lemma 3: If an ontology change sequence chs_1 indirectly conflicts with another ontology change sequence chs_n, not all of chs_1 and chs_n are executed no matter what the execution order may be.

Lemma 4: If an ontology change sequence chs_1 is compatible with another ontology change sequence chs_2, chs_1 and chs_2 may always be executed no matter what the execution order may be.

According lemma 1,2,3 and 4, if any two of a group of ontology change sequences are not *directlyConflict* or *indirectlyConflict* all these ontology change sequences may be executed. In order to find all maximum subset of all ontology change sequences that may be executed from a given group of ontology change sequences, we propose the Algorithm of Searching Conflict Set(ASCS) to be used to find all ontology change sequence pairs which indirectly conflict or directly conflict with each other and the Algorithm of Searching Maximum and Sequential Ontology Change Sequence Set (ASMSOCSS).

Definition 8: Given a group of ontology change sequences $CHS=\{chs_1, chs_2,......, chs_n\}$, $MSOCSS_{CHS}$ is defined as **Maximum and Sequential Ontology Change Sequence Set** if and only if $\neg\exists T\subseteq CHS$ makes all $T\supset MSOCSSM_{CHS}$ and $\forall t_1,t_2\in T$, $\neg directlyConflict(t_1,t_2)\wedge\neg indirectlyConflict(t_1,t_2)$ to be true.

In ASMSOCSS, step 2 means to traverse all elements of *CONF* in unknown order. It is puzzling whether the order of traversing elements can change the final MSOCSSSet or not. The problem can be explained by lemma 5.

Lemma 5: The final MSOCSSSet is identical no matter what the order of traversing elements may be.

Proof

$\forall$(e1,e2),(e3,e4)$\in$CONF, e1$\in$CHS$\wedge e_2\in$CHS$\wedge e_3\in$CHS$\wedge e_4\in$CHS is given.

Suppose that $e_1\neq e_3\wedge e_1\neq e_4\wedge e_2\neq e_3\wedge e_2\neq e_4\wedge e_1\neq e_2\wedge e_3\neq e_4$ is true.

If *(e1,e2)* is traversed ahead of *(e3,e4)* , $MSOCSSSet_0=\{CHS/\{e_1\},CHS/\{e_2\}\}$ is got. After *(e3,e4)* is traversed, $MSOCSSSet_1=\{CHS/\{e_1,e_3\}$, $CHS/\{e_1,e_4\}$, $CHS/\{e_2,e_3\}$, $CHS/\{e_2,e_4\}\}$ is got. In turn, if *(e3,e4)* is traversed ahead of *(e1,e2)* , $MSOCSSSet_0=\{CHS/\{e_3\},CHS/\{e_4\}\}$ is got. After *(e1,e2)* is traversed ,$MSOCSSSet_2=\{CHS/\{e_3,e_1\}$, $CHS/\{e_3,e_2\}$, $CHS/\{e_4,e_1\}$, $CHS/\{e_4,e_2\}\}$ is got. Obviously $MSOCSSSet_1=MSOCSSSet_2$.$MSOCSSSet_1=MSOCSSSet_2$ is easy proved when $e_1=e_3\wedge e_1\neq e_4\wedge e_2\neq e_3\wedge e_2\neq e_4\wedge e_1\neq e_2\wedge e_3\neq e_4$ is true.

$\forall$*(e1,e2),(e3,e4)*$\in$*CONF*, the same conclusion may be drew easily whichever of $e1\notin CHS\wedge e_2\in CHS\wedge e_3\in CHS\wedge e_4\in CHS$ or $e1\notin CHS\wedge e_2\notin CHS\wedge e_3\in CHS\wedge e_4\in CHS$ or $e1\notin CHS\wedge e_2\in CHS\wedge e_3\notin CHS\wedge e_4\in CHS$ or $e1\notin CHS\wedge e_2\notin CHS\wedge e_3\notin CHS\wedge e_4\notin CHS$ is given.

So we may know that the order of traversing any two elements of *CONF* is exchangeable. Further, Lemma 5 may be deduced easily according to related mathematical characteristics. □

5 Algorithm of Generating New Ontology Versions

According to ASMSOCSS, $\forall E\in MSOCSSSet$, $\forall(e_1,e_2)\in E$, e_1 must not directly conflict or indirectly conflict with e_2. So $\forall E\in MSOCSSSet$, all elements in E are executed once. When a $MSOCSS_{CHS}$ is applied to an ontology O, a new ontology version can be got. The algorithm of generating a new ontology version is AGNOV.

```
Algorithm 2 : ASCS(CHS)
Input: a group of ontology change sequences {chs1, chs2,……, chsn}
Output : a conflict set CONF
1.  CONF←Ø;        i1←1;  i2←1;
2.  WHILE(i1<=n)
3.    WHILE(i2>=i1 and i2<=n)
4.      IF(directConflict(chsi1,chsi2) or indirectConflict(chsi1,chsi2)) and i1 ≠ i2
5.        CONF←CONF∪{{chsi1,chsi2}};
6.      i2++;
7.    i1++;    i2←i1;
8.  RETURN CONF
```

```
Algorithm 3 : ASMSOCSS(CHS,CONF)
Input : a group of ontology change sequences {chs1, chs2,……, chsn} and
        ASCS({chs1, chs2,……, chsn})
Output : the set of all possible maximum and serial ontology change sequence
         set MSOCSSSet
1.  MSOCSSSet←{CHS};     i1←1;  i2←1;
2.  FOR EACH (chs1,chs2)∈CONF
3.    FOR EACH e∈MSOCSSSet
4.      IF chs1∈e and chs2∈e
5.        MSOCSSSet←MSOCSSSet-{e};
6.        e1←e-{chs1};  e2←e-{chs2};
7.        IF ¬ ∃e'∈CAND , e1⊂e'
8.          MSOCSSSet←CAND ∪ {e1 };
9.        IF ¬ ∃e'∈CAND , e2⊂e'
10.         MSOCSSSet←CAND ∪ { e2};
```

```
Algorithm 4 : AGNOV(O)
Input : an ontology O
Output : the set of all new ontology version OVs
1. OVs←Ø;
2. FOR EACH E∈MSOCSSSet
3.   O'←O;
4.   WHILE(E≠Ø)
5.     IF ∃ chsi∈E,∀ chsj∈E/{chsi }, ¬dependence(chsi, chsj)
6.       O'←chsi (O);  E←E/{chsi };
7.   OVs←OVs ∪ {O'};
8. RETURN OVs
```

6 Related Work

Ontology versioning[5] typically involves the storage of several ontology versions and identification issues, the relationship between different versions as well as compatibility information.[9] uses the term versioning to describe their approach of ontology change. They define ontology versioning as the ability to manage ontology changes and their effects by creating and maintaining different variants of the ontology. Adequate methods and tools must be used to distinguish and identify the versions. [4] presents a new ontology evolution approach. The approach keeps track of the different virtual versions of ontology concepts throughout their lifetime by combining the manual request for changes by the ontology engineer with an automatic change detection mechanism. [7] proposes a logic framework used to reason with multversion ontologies. In the framework such problems can be solved as querying log of semantic change, selecting a appropriate ontology version, etc.

7 Conclusion and Future Work

We propose a formal method used to analyze and dispose a group of ontology change sequences under open environment. But our discussion is specific to the same ontology subject to a group of ontology change sequences. In future, we will use graph theory to rephrase such a problem.

References

1. Klein, M., Proefschrift, A., Christiaan, M., Klein, A., Akkermans, J.M.: Change management for distributed ontologies. Technical report (2004)
2. Peter, Haase, P., Van Harmelen, F., Huang, Z.: A framework for handling inconsistency in Changing_Ontologies, pp. 353–367. Springer, Heidelberg (2005)
3. Kalyanpur, A., Parsia, B., Sirin, E.: Repairing unsatisfiable concepts in owl ontologies. In: Sure, Y., Domingue, J. (eds.) ESWC 2006. LNCS, vol. 4011, pp. 170–184. Springer, Heidelberg (2006)
4. Plessers, P., De Troyer, O.: Ontology change detection using a version log. In: Gil, Y., Motta, E., Benjamins, V.R., Musen, M.A. (eds.) ISWC 2005. LNCS, vol. 3729, pp. 578–592. Springer, Heidelberg (2005)
5. Noy, N.F., Chugh, A., Liu, W., Musen, M.A.: A framework for ontology evolution in collaborative environments. In: Cruz, I., Decker, S., Allemang, D., Preist, C., Schwabe, D., Mika, P., Uschold, M., Aroyo, L.M. (eds.) ISWC 2006. LNCS, vol. 4273, pp. 544–558. Springer, Heidelberg (2006)
6. Haase, P., Haase, P., Stojanovic, L.: Consistent evolution of owl ontologies, pp. 182–197. Springer, Heidelberg (2005)
7. Huang, Z., Stuckenschmidt, H.: Reasoning with multi-version ontologies: A temporal logic approach. In: Gil, Y., Motta, E., Benjamins, V.R., Musen, M.A. (eds.) ISWC 2005. LNCS, vol. 3729, pp. 398–412. Springer, Heidelberg (2005)
8. Flouris, G., Flouris, G., Plexousakis, D.: On belief change and ontology evolution. Technical report, University of Crete (2006)
9. Noy, N.F., Klein, M.: Ontology evolution: Not the same as schema evolution. Knowl. Inf. Syst. 6(4), 428–440 (2004)

Knowledge-Based Support for Software Engineering

Dencho Batanov

Department of Computer Science and Engineering
Frederick University, 7 Frederickou Str., Pallouriotisa,
Nicosia 1036, Cyprus
com.bd@fit.ac.cy

Abstract. The existing ambiguity of the notion of software engineering is mainly due to the fact that it is based on and depends on knowledge. The new definition of the term "software engineering", proposed in this paper, encounters that fact. The main subject of discussion in the paper is how three different types of knowledge, namely declarative explicit, declarative structured (ontologies) and tacit can be used for effective support of software engineering as both practice and academic subject. Illustrative examples are shown along with some trends for more intensive use of knowledge for support of software engineering.

Keywords: Knowledge, Software Engineering, Software Development Life Cycle (SDLC), Ontology, Expert System.

1 The Software Engineering Paradigm

Technology has two different but inseparable meanings – as tools and processes [1]. Although the technological tools (products) are more popular and attractive in our everyday life they could not be created and produced without respective processes. Good examples are engineering of any product in general and software engineering in particular. To understand the software engineering paradigm, which we are interested in, it would be helpful to briefly trace back the ambiguity and evolution of the meaning of the term "software engineering" since its inventing at the 1968 NATO Conference on Software Engineering [2]. Numerous definitions of the term have been given over time emphasizing diverse aspects of the notion and more than forty years still there are differences and disagreements. Most of those definitions are centered on the point that the development of software from the initial phase of requirements analysis and specification to the maintenance of software product is strongly linked to the notion of engineering as both academic discipline and profession.

Engineering is a mix of craft and sciences [3] with more dominating and increasing role of sciences in the last centuries because of the demands for more complex functionalities, higher qualities and greater quantities of the products as well as for more complicated management of the process. For some more specific products however crafts, which are characterized by learning by doing, idiosyncratic approach and production of handmade artifacts [4], are the basis of engineering process. Software is without any doubt quite specific product – unique, invisible, getting better over time, flexible and therefore easily modifiable, scaleable in large boundaries, etc., and it is

H. Papadopoulos, A.S. Andreou, and M. Bramer (Eds.): AIAI 2010, IFIP AICT 339, pp. 219–229, 2010.

not surprising that engineering of such products requires specific term – "software engineering". There are a lot of publications regarding the nature of this term with speculations, discussions, agreements and disagreements with existing definitions, the place of software engineering in science, practice and education and possible ways of its evolution. The objective of this paper is not to survey those publications but it is worth noting that they vary from emphasizing the absence of fundamental theory [5] through the need of a theory for software engineering [6] and the differences between software engineering and computer science [7] to even looking for some similarities and differences between fashion, politics and software engineering [8]. Central point in all these publications is that software engineering is not a rigorous discipline comparing to the core subjects constituting computer science as area of research, education and practice, such as data structures and algorithms, queuing theory, complexity, languages syntax and semantics, machine learning, etc. Analysis of the nature of those disciplines leads to an interesting thesis, expressed by Chuck Connell in [8]: *Software engineering will never be rigorous discipline with proven results, because it involves human activity.* And even more: *We should stop trying to prove fundamental results in software engineering and accept that the significant advance in this domain will be general guidelines.* Such a thesis is valid at least to some extent for al engineering disciplines but having in mind the specifics of software products and the existing practice of developing software systems of any size, it could be accepted in full for software engineering although, as the author states, the statements cannot be proved. The problem with this thesis is that the term "human activity" is too broad and vague that obviously cannot be used to clearly define the term of "software engineering". On the other hand the activities which distinguish distinctly the humans from the other forms of life are based on creating and continuous use of knowledge. There are different types of knowledge and different approaches to its classification, for example one is to classify knowledge as static, dynamic, declarative, procedural, heuristic, knowledge of methods and knowledge of equipment and tools; another is to separate knowledge in three different levels – surface, domain and deep; and as another option knowledge in classified in two large groups – explicit (objective) and tacit (subjective). No matter how the knowledge is classified the practice of software engineering shows that knowledge of different types is intensively used in all its phases. This can be expressed by the following definition of the term of software engineering:

Software engineering is a systematic approach to the management and development of software systems based on use of all kinds of knowledge, which is embedded in the final software product.

Fig. 1 shows the basic idea of the above definition.

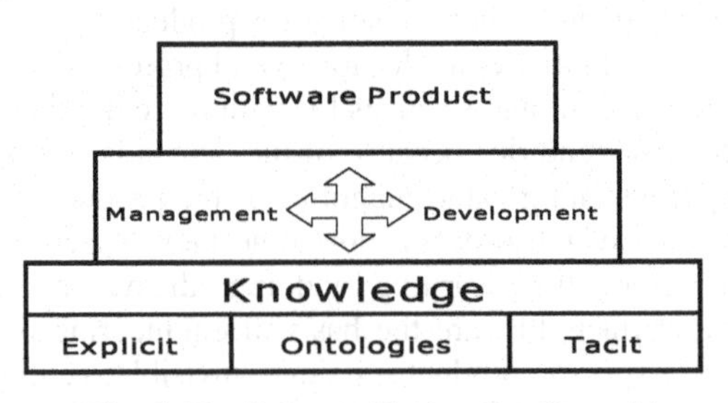

Fig. 1. The Software Engineering Pyramid

Central point of the definition is the focus on use of all kinds of knowledge, which in this particular case is classified in three groups: explicit (declarative knowledge), ontologies (declarative structured knowledge) and tacit (individual-related knowledge, which is result of accumulated experience and expressed mainly in a form of rules). It is worth noting at least three interesting properties of the definition: (a) it deals with the two meanings of technology – tools (software products) and processes (management and development); (b) it covers the two major activities in engineering of software systems – management (team organization, choosing the strategy, planning, scheduling, budgeting, cost estimation, maintenance, etc.) and development (use of models, methods, techniques and tools as elements of the chosen methodology), and (c) the quality of the tool (software system) strongly depends on the quality of the processes (management and development).

The software development life cycle (SDLC) consists of well defined phases that developers carry out during the process of software system development and which are subject of research and education. The above definition adds some new aspects of he development, research and education regarding: (a) the necessity of identifying the specific kind of knowledge, which is appropriate for a given phase of SDLS, and (b) use of existing or creating new methods, techniques and tools for gathering, representing and manipulating knowledge for support of the respective phase of SDLC.

The next sections of the paper represent illustrative examples of using the three types of knowledge from the definition for support of software engineering. The use of declarative explicit knowledge in well-known forms as description of models, methods, techniques and tools, manuals, documentation, standards, etc., is briefly mentioned in Section 2. Section 3 is dedicated to one promising way to use structured declarative knowledge in a form of ontology for supporting some of the phases of software development life cycle. In Section 4 an approach of using tacit knowledge for building expert system for supporting requirements analysis phase is described. Conclusions are outlined and some recommendations for further work are made in Section 5.

2 Use of Declarative Explicit Knowledge

Without doubt this is the kind of knowledge, which is most well known, popular and available. For more than forty years a lot of specialized knowledge has been created, accumulated and disseminated in form of collections of models, methods, techniques and tools, related books, papers, reports on good and bad practices, curricula, training and certification programs with respective teaching and learning materials, manuals, standards and so on. This valuable repository of knowledge is a great and widely used opportunity for practitioners, educators, students and scientists to learn and know more in the field of software engineering. This kind of knowledge, as it is shown in [4] is the basis of transition of the software engineering discipline from craft to profession.

Declarative explicit knowledge, although usually thematically classified, is non-structured in nature. To use such type of knowledge the users need guidance by experienced people to be able to navigate among the numerous sources of related information. That is why this knowledge is used mostly in education, including training and certification. Software engineering education is of primary importance for preparing software developers and is a subject of teaching in all academic institutions all over the world. A good example of accumulated knowledge for creating and applying software engineering curricula for universities is the model, proposed by the Joint Task Force on Computing Curricula of IEEE Computing Society and ACM [2]. The curricula seem to be developed for major of MSc in software engineering but the model can be definitely used for creating the content of related courses at bachelor degree level. We should not forget however that software engineering is based on knowledge and skills from a number of basic courses in Computer Science, such as data structures and algorithms, principles of programming languages, object-oriented programming, databases, etc., which constitute the necessary body of knowledge. In this regard the role of lecturers in software engineering is vital for advising students what to select and read. For example, is not enough to state in the beginning of the course that "software engineering is systematic and disciplined approach to software development" – it is absolutely necessary to support this statement through the entire course with relevant readings, case studies, analyses of good and bad practices and so on. Another good example is to include Software Engineering Project as a separate credited subject in the curriculum giving the students the opportunity to apply their knowledge and skills to development of real software systems. For this purpose they are required to find themselves a lot of additional sources of information about the methods, techniques and tools to be used, analytically compare them and finally to make decision what to be chosen. This is the only way to convert knowledge into practice.

Another useful example of relying on explicit declarative knowledge is the so called codified body of knowledge [4], represented by two significant projects – SEEKA (Software Engineering Education Knowledge Areas) and SWEBOK (SoftWare Engineering Body of Knowledge). SEEKA is more oriented to the knowledge areas that should be covered in an undergraduate curriculum in software engineering while SWEBOK concerns knowledge and practices, which can be applied to most projects most of the time. Although the slight differences both of them offer extremely helpful information about software engineering as a subject of learning and practice.

Software engineering in practice is different from software engineering in education. The differences are in the size and complexity of the projects, the number of people involved, the organization and management of the teams, the required quality of the software product and related compliance with standards, the time and budget constraints, etc. Accordingly, along with the traditional descriptions and manuals of models, methods, techniques and tools, there are additional sources of explicit declarative knowledge, which support the work of practitioners in software engineering. Examples of such sources are the large number of approved and working standards for software quality assurance, the regular publications of professional societies like IEEE and ACM and their special interest groups in software engineering, specialized journals and proceedings of conferences and workshops, analytical reviews of good and bad practices, project reports and so on. Another well known for specialists example is the so-called "Capability Maturity Model" (CMM) [4], created in the Software Engineering Institute at Carnegie Mellon University. The model helps evaluating the software products in a standardized way, which contributes to improving the working processes and the quality of the product as a whole.

In fact all necessary sources of explicit declarative knowledge are available somewhere in the world repository. Is it enough for more effective and efficient software product development? Not, of course, simply because the access to those resources is difficult and time consuming especially when the developers need the information online. Recent Web technologies however offer solution of this problem through Web services and currently emerging cloud computing. This can be considered as a challenging research and implementation topics for knowledge-based support of software engineering.

3 Use of Declarative Structured Knowledge (Ontologies)

Ontology here is defined as declarative structured knowledge because it can be derived from the structural representation of concepts (entities, objects, classes) linked through existing or established relationships in a given problem domain. This is another interpretation of the definition of ontology as a specification of a representational vocabulary for a shared domain of discourse: definitions of classes, relations, functions, and other objects [9] or, more generally, a specification of conceptualization [10]. In this section I will show as illustrative example how ontologies as form of knowledge can be used to support some of the most difficult phases of SDLC in object-oriented software engineering. The readers who would be interested in this example as complete representation can find more detailed description of respective models, methods and techniques in [11], [12].

The motto of classical object-oriented software development may be formulated in different ways, but its essence can be stated simply: "Identify and concentrate on objects in the problem domain description first. Think about the system function later." At the initial analysis phase, however, identifying the right objects, which are vital to the system's functionality, seems to be the most difficult task in the whole development process from both theoretical and practical point of view. Object-oriented software development is well supported by a huge number of working methods, techniques, and tools, except for this starting point - object identification and building the related system object model. Converting the text description of system problem domain and respective functional requirements into an object model is usually left to the intuition and experience of developers. One commonly accepted rule of thumb is, "If an object fits within the context of the system's responsibilities, then include it in the system." However, since the members of the development team are likely to have different views on many points, serious communication problems may occur during the later phases of the software development process. Here is the place where the knowledge represented by ontologies can help. It is worth noting that an ontology is either built already for a given problem domain or, if not, can be created using respective methods, techniques and tools (languages) specific for the field of ontology development, which is not objective of this paper. Fig. 2 illustrates the way in which ontology can be used as a supporting tool for the process of building the object model of the software system and converting its elements (objects) into abstract data types (ADTs). As far as the implementation of ADTs are classes and they are the basic building modules of object-oriented software, it becomes clear that actually the ontologies can help the entire analysis and design phases of SDLC.

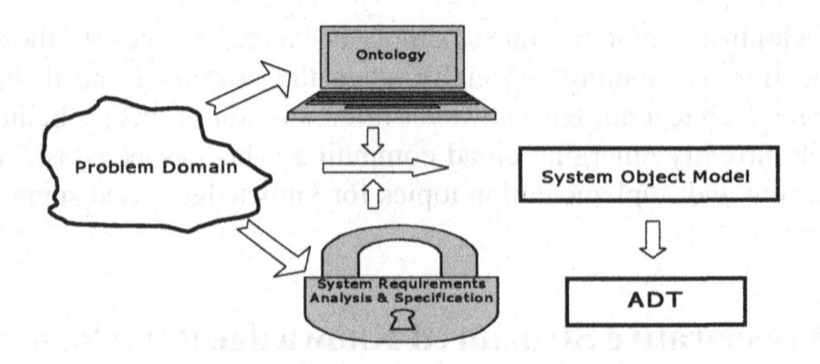

Fig. 2. Ontologies and Building Object Model (ADT)

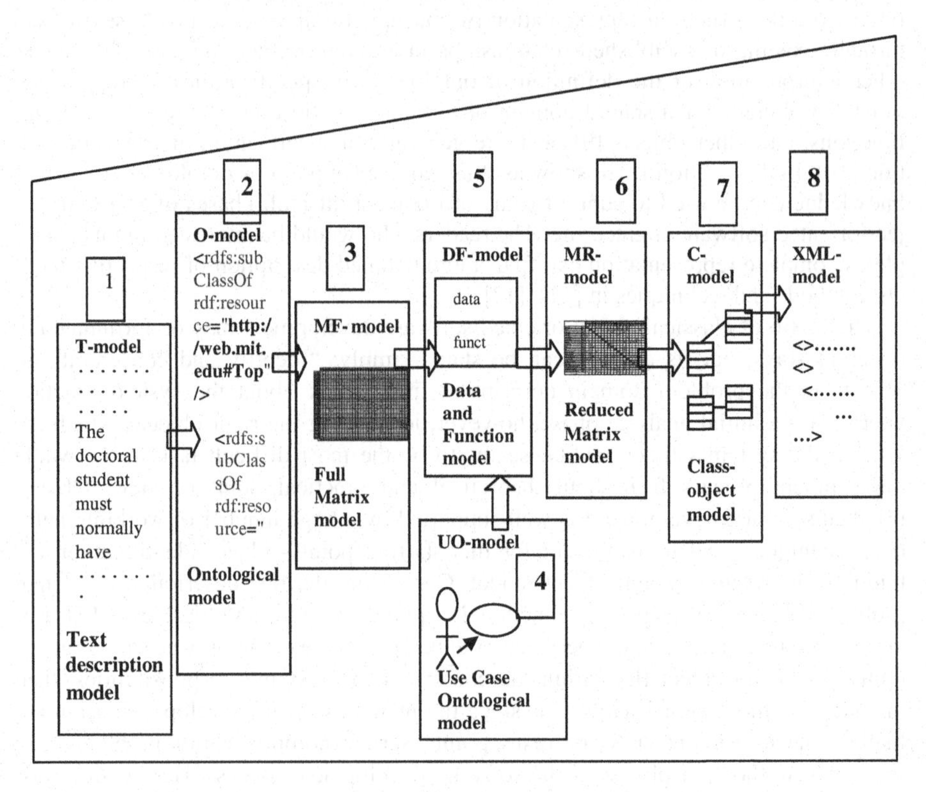

Fig. 3. Models for converting a text description into an object model

Models are inseparable and one of the most significant parts of any methodology. They help developers to better understand complex tasks and represent in a simpler way the work they should do to solve those tasks. Fig. 3 shows the models, which we use to transform requirements specification to object model of the system. The starting point of transformation is the text model (T-model), which represents a concise description of the problem domain, where the software system under development will work, written in a natural language, usually English. If not available, the T-model should be created by the developer describing the general user requirements for the system functionality. The presumption is that this problem domain description contains the main objects, which will participate in ensuring the system's functionality. Of course, at this level the objects are represented by their natural names only and as such are very far from the form we need to reach - represented as ADTs.

To help this process we refer to a tool of conceptualization - an ontological engine, which applied to the T-model, generates an ontology model (O-model) of the problem domain. The O-model is a straightforward and practically useful source of information for identifying the participating objects. We use this information to build a so-called Full Matrix model (MF-model), which represents in a simple form objects along with the linkages (relationships) between them. However, we should say that the processing of the MF-model is semi-formal in nature. This means that at this phase the developer should take important decisions about which objects could be considered as basic ADTs and which, and where, could play a role of attributes of other ADTs. The idea is simple but not very easy for implementation - to reduce the full object matrix to a reduced matrix (we call this model MR-model), which contains only the basic objects represented later as ADTs containing other ADTs as attributes. The implementation is not very easy because we need more information here, which relates to expected functionality of participating objects. This information, however, is available or can be extracted from the Use Case model of the system under development. Note that at this phase we can also use the problem domain ontology. Along with showing the concepts hierarchy (possible objects in the system) the ontologies also analyze the verbs linking those concepts, which can be considered as functions (operations) belonging to respective objects. We actually use the text descriptions of different Use Cases to extract different functionality of the system by the ontological engine and as a result we get the so-called Use Case Ontological model (UO-model). The functionality, expressed by the UO-model, can be used at this phase along with the ontological information about the objects in the MF-model to create the Data and Function model (DF-model). As a matter of principle DF-model can be used for each of the objects in the DF-model but this would lead to a high degree of redundancy and quite complicated matrix presentation even for relatively simple T-models. To avoid this we propose to use so called business object patterns. The representation of the C-model is significantly different from MR-model however, as far as the former shows not only the object hierarchy but the objects' structure as well. In other words, the C-model is a model representing ADTs. The last model, the XML-model is optional but can be very important in practice because it allows the C-model to be published on the Web in a unified (XML-based) format supporting in this way the collaborative work, which is a commonly accepted technology nowadays.

The shown models and the process of their transformation can help developers of complex object-oriented software systems to: (a) transform user requirements (represented as text description) into an object model of the software system based on the use of ontologies; (b) improve the existing methods and techniques for creating a specific ontology from a text description of the system problem domain; (c) work out implementation techniques and tools for semi-automated or automated generating and editing of ADTs for object-oriented application software development, and (d) improve the effectiveness and efficiency of the existing methodology for high-level system analysis in object-oriented software engineering.

4 Use of Tacit Knowledge

For all phases of SDLC without any exceptions the developers are forced to make decisions, which are vital for the quality of the final product. Give one and the same user requirements and specifications to, let us say ten different teams, and you will get certainly ten absolutely different systems as result. Not different in required functionality but different as user acceptance, performance, cost, reliability, etc., generally speaking different in quality. This is not quite normal for other conventional products but for software the opposite would be not normal. This is the uniqueness of software product. And this is because the people who make decisions at some points of SDLC are different – as background, qualification, experience or as knowledge and skills in the field of software engineering. Practically most of this knowledge is tacit in nature – knowledge, which is hidden, difficult to express, explain, share with others and formalize. Tacit knowledge exists usually in two forms [14]: (1) knowledge embodied in people and social networks, and (2) knowledge embedded in the processes and products that people create. We are interested here in the second form. Because the people develop and use tacit knowledge before they are able to formalize or codify it the problem is how to extract this knowledge from those who possess it and after that to represent and process it in a computerized environment in order to implicitly embed it in the software product. The artificial intelligence offers different representation schemes and respective methods and techniques for manipulating such type of knowledge but the most popular and relatively easy for implementation way remains the use of rule-based expert systems. Unfortunately, especially in the field of software engineering there are only a few examples of such systems, which are attempts to support some of the phases of SDLC. We used one of these examples - CASSANDRA to support the phase of requirements analysis and specification. This phase is one of the most difficult and ambiguous and at the same time of vital importance for the success of the project. As it is stated in [13] "*Research indicates that nearly 50% of all software project defects originate in the requirements gathering process and that 60% to 80% of project failures can be attributed directly to poor requirements gathering.*" Obviously any guidance during this phase would be extremely useful.

CASSANDRA [15] is an ambitious project for developing an automated software engineering coach and the name stands for **C**assandra – an **A**ssistant for **S**ystem **S**pecification **AND** **R**equirements **A**nalysis. The idea is the support to be in a style that resembles a human coach – proactive, asks the user questions, gives advices and recommendations and explains them in the case of users' request. Everything in CASSANDRA is implemented in Prolog – the user interface, knowledge base, persistency and CASE tool access. Having the framework however, the contents of the inserted facts and questions asked can be easily changed, which allows for adjusting the expert system to the needs of different categories of users. We, for example, developed an expert tool for novice and junior developers with the idea to apply it to the educational process. In addition we changed completely the user interface implementing it in C# in .NET environment, which has connectivity to Prolog, giving in this way an opportunity for more attractive and convenient interaction between the user and system.

The architecture of CASSANDRA is quite complicated and it is not the aim of this paper to consider it in more details. The generation of recommendations however follows the classical mechanism of rule-based expert systems. All questions, facts and recommendations, which are the basic elements of knowledge base, are organized in functionally well-defined groups, such as goals of the system, system users, system constraints, system architecture, functionality definitions, design, performance, maintenance and support. Below are two examples of related fact and recommendation as illustration:

> ***fact (es1,no):-*** *fact ('present software', no), recommend ('The fact that the company is not using any present software at the moment has both its advantages and its disadvantages. That is, you are going to be the first one who will try to find out all the needs of the company therefore more work should be done. On the other hand, it means that you have no competitor to compete with on the functionality of the software.').*
>
> ***recommendation (existing_software):-*** *fact(es1,no), recommend ('You need to develop very careful strategies from which you will retrieve information about how the company is working and for what reason it needs the software for in order to be able to provide the best possible software solution for it. Some of those methods and techniques you can find out in the book of .... on page ......').*

The inference engine first calls the *recommendation* rule which on its turn calls the *fact* rule in order to be satisfied. The *fact* rule then starts the process to satisfy itself. That is, it starts calling all the facts that are included in its rule.

There are a good number of environments for development of rule-based expert systems. It seems to me however that the problem with having so small number of working examples of expert systems for support of software engineering is to find experts for different phases of SDLC ready for sharing their tacit knowledge. On the other hand the process of creating a solid knowledge base is costly and very time consuming. The decision maybe is the leading organizations in computing as IEEE and ACM to start coordinating this hard work as well as collaboration between universities and respective funding at national level.

5 Conclusion and Recommendations

The long time existing ambiguity of the notion of software engineering is mainly due to the fact that it is based on and depends on knowledge. A new definition of the term is proposed in this paper, which states: *"Software engineering is a systematic approach to the management and development of software systems based on use of all kinds of knowledge, which is embedded in the final software product"*. More specifically, three types of knowledge are identified as basic for support of developers' activities in software engineering: explicit declarative knowledge, which are unstructured in nature, ontologies as representatives of declarative structured knowledge and tacit knowledge. The main body of the paper is dedicated to illustrative examples of and respective comments on the use of each one of those kinds of knowledge in the education and practice of software engineering.

Using explicit declarative knowledge, accumulated and disseminated in form of collections of models, methods, techniques and tools, related books, papers, reports on good and bad practices, curricula, training and certification programs with respective teaching and learning materials, manuals, standards and so on, is still dominant. Nowadays Web technologies and more specifically Web services and currently emerging cloud computing can open new opportunities for more efficient and effective use of world repository in the field of software engineering. This can be considered as a challenging research and implementation topics for knowledge-based support of software engineering.

Most interesting, challenging and promising approach to knowledge-based support of practically all phases of SDLC is using ontologies. They either exist or can be created for different specific problem domains and offer good opportunities for merging with software engineering. The example shown and briefly discussed in the paper is dedicated to creating the object model of the software system from a given textual description of the problem domain and system functionality. It is expected more research to be carried out in this promising area.

Finally, an example of using tacit knowledge in an expert system serving as an automated coach in software engineering, and particularly in the requirement analysis phase, and built on the basis of CASSANDRA environment, is shown and discussed. Such type of expert systems could be very helpful for both education and practice in software engineering but their development requires collaborative work and availability of solid resources.

Acknowledgements. My thanks go to the hundreds of my students, who during the years have always inspired me to teach and supervise them better in the fields of software engineering and knowledge-based systems.

References

1. Batanov, D., Eloranta, E.: Advanced Web technologies for industrial applications, Guest Editorial. Computers in Industry 50(2), 123–125 (2003) (special issue)
2. Computing Curriculum – Software Engineering, Public Draft 1, The Joint Task Force on Computing Curricula. IEEE Computer Society, ACM (2003)
3. Shute, G.: The Nature of Software Engineering (2007), `http://www.d.umn.edu/~gshute/softeng/nature.html`
4. Fairley, R.E.(D.), Tripp, L.L.: Software Engineering: from Craft to Profession (2002), `http://cs.wm.edu/~coppit/.../papers/CraftToProfession.pdf`
5. Kruchten, P.: The Nature of Software: What's So Special About Software Engineering (2004), `http://www.ibm.com/developerworks/rational/library/4700.html`
6. Jacobson, I., Spence, I.: Why We Need A Theory for Software Engineering, Dr. Dobb's Digest (October 2009)
7. Connel, C.: Software Engineering? Computer Science, Dr. Dobb's Digest (June 2009); Jacobson, I., Meyer, B.: Dr. Dobb's Digest (August 2009)
8. Gruber, T.R.: A translation approach to portable ontology specifications. Knowledge Acquisition 5, 199–220 (1993)
9. Gruber, T.R.: Towards Principles for the Design of Ontologies Use for Knowledge Sharing. In: Proceedings of IJHCS 1994, vol. 5(6), pp. 907–928 (1994)

10. Batanov, D.N.: Merging ontologies and object-oriented technologies for software development. In: Proceedings of the 20th International Conference SAER 2006, Plenary paper, Varna, Bulgaria, September 23-24 (2006)
11. Batanov, D.N., Vongdoiwang, W.: Using Ontologies to Create Object Model for Object-Oriented Software Engineering. In: Charman, R., Kishore, R., Ramesh, R. (eds.) Ontologies. A Handbook of Principles, Concepts and Applications in Information Systems, ch. 16, Part 3, pp. 461–487. Springer, US (2007), ISBN: 978-0-387-37019-4 , 978-0-387-37022-4
12. A Practical Guide to Effective Requirements Development, SearchSoftwareQuality.com, E-guide (2007)
13. Horvath, J.A. (2000), `http://providersedge.com/docs/km_articles/`
14. Schaher, M.: CASSANDRA: An Automated Software Engineering Coach. KnowGravity Inc. (2001)

A Hybrid Searching Method for the Unrelated Parallel Machine Scheduling Problem

Christoforos Charalambous[1], Krzysztof Fleszar[2], and Khalil S. Hindi[2]

[1] Department of Computer Science and Engineering, Frederick University, Cyprus
c.charalambous@frederick.ac.cy
[2] Olayan School of Business, American University of Beirut (AUB), Lebanon
kf09@aub.edu.lb, khalil.hindi@aub.edu.lb

Abstract. The work addresses the NP-hard problem of scheduling a set of jobs to unrelated parallel machines with the overall objective of minimizing makespan. The solution presented proposes a greedy constructive algorithm followed by an application of a Variable Neighborhood Decent strategy that continually improves the incumbent solution until a local optimum is reached. The strength of the approach lies in the adoption of different objectives at various stages of the search to avoid early local optimum entrapment and, mainly, in the hybridization of heuristic methods and mathematical programming for the definition and exploration of neighborhood structures. Experimental results on a large set of benchmark problems attest to the efficacy of the proposed approach.

Keywords: parallel machine scheduling, hybrid optimization, variable neighborhood search, mixed-integer programming.

1 Introduction

The unrelated parallel machines scheduling problem is that of scheduling without preemption n jobs available at time zero on m unrelated machines to minimize the makespan (maximal machine completion time), $C_{\max}$. Henceforth, we will also use the term 'span' to signify the completion time of a machine.

The machines are unrelated in that the processing time of a job depends on the machine to which it is assigned. The practical importance of the problem stems from the fact that it is common in many industrial applications to have parallel resources with different capabilities, perhaps procured at different times. These resources would be capable of carrying out the same tasks, but the time taken to perform a task would depend on the resource on which it is performed. Such applications can be found in industries like painting, plastic, textile, glass, semiconductor, chemical, and paper manufacturing, as well as in some service industries [1].

Literature on parallel machine scheduling problems is extensive; for general references and surveys, see [2,3,5,7,10]. However, most of the literature addresses identical machines, where the processing time of a job is the same regardless of the machine to which it is assigned, or uniform machines, where the processing

H. Papadopoulos, A.S. Andreou, and M. Bramer (Eds.): AIAI 2010, IFIP AICT 339, pp. 230–237, 2010.

time of a job is proportional to the speed of the machine. Of the relatively small number of works that address unrelated machines, most deal with the case without setup times. For minimization of the makespan, Martello et al offered exact and approximation algorithms [6]; Mokotoff and Chretienne a cutting plane algorithm [8]; Mokotoff and Jimeno [9] heuristics based on partial enumeration; Ghirardi and Potts [4] a recovering beam search method; and Shchepin and Vakhania [11,12] approximation algorithms.

Throughout, the following notation is used:

$N = \{1, \ldots, n\}$	set of jobs, n being the total number of jobs.
$M = \{1, \ldots, m\}$	set of machines, m being the total number of machines.
p_{jk}	processing time of job j on machine k
$M(j)$	the machine to which job j is assigned to.
J_k	set of jobs assigned to machine k.
C_k	span (completion time) of machine k.
$C_{\max} = \max_{k \in M} C_k$	makespan (maximum machine completion time).

2 Definition of Objective Criteria

The problem's main objective is to minimize the solution makespan (C_{max}). However, focusing solely on the main objective may inhibit the development of the search process as other desirable characteristics are likely to be ignored. In this work, in addition to the main objective, auxiliary criteria are employed to assist the development of high quality solutions. The criteria used are:

1. **Sum of spans:** To be able to minimize the makespan it is helpful to associate jobs to machines that process them efficiently. The lower the sum of spans (an average measurement on how efficiently the jobs have been assigned), the more slack can be expected on non-makespan inducing machines, leading to an increased capacity for solution improvement through job reassignments.
2. **Number of makespan-inducing machines:** Neighborhood search structures operate by altering incumbent solutions through a small number of job reassignments aiming at an overall improvement. The fewer the number of machines that have a span equal to the makespan, the more probable the existence of a job reallocation pattern that would reduce the makespan.

The mode in which objective criteria are used depends on the neighborhood structure, as described in Section 4.

3 Initial Solution

Numerous constructive algorithms have been proposed in the literature including the efficient APPROX method of Martello [6]. These algorithms focus on finding techniques for developing solutions that minimize the objective function. In this work the primal goal of the solution construction is not to minimize makespan but rather to seed the searching phase of the algorithm with a solution that resides in a promising region. The rational against opting for makespan minimization is that fast constructive algorithms invariably adopt a greedy approach in the solution construction that often opts for short-sighted moves which worsen the objective function least. Such an approach leads to relatively good solutions but, due to the naive nature of choices, there is little room left for further improvements.

The initial solution is calculated based on the linear relaxation of the UPMSP

$$\min C_{max} \tag{1}$$

subject to

$$\sum_{k \in M} x_{jk} = 1 \;\; \forall j \in N \tag{2}$$

$$C_{max} \geq \sum_{j \in N} p_{jk} x_{jk} \;\; \forall k \in M \tag{3}$$

$$x_{jk} \in \{0,1\} \;\; \forall j \in N, k \in M \tag{4}$$

where x_{jk} is a binary variable signifying whether job j is assigned to machine k.

When the integrality constraint in 4 is relaxed, the solution generated by the solver will consist of a large set of x_{jk} variables set to zero, and the rest being either one or set to a fractional value. The initialization heuristic sets the upper bound of all x variables that have value zero to zero (i.e. forbidding the corresponding assignment). Subsequently, all fractional variables are sorted based on *regret*. The regret of a variable is defined as the added processing time that would incur in the solution if the corresponding variable was set to one and is calculated by $p_{jk}(1-x_{jk})$. The variable with the maximum regret has its upper bound set to zero and the model is resolved until no variable has a fractional value. Note that if the model has no fractional variables, the corresponding solution is feasible.

4 Neighborhood Moves and Variable Neighborhood Decent

Improving an incumbent solution can often be achieved through the reassignment of jobs to machines. The means of these reassignments, henceforth called moves, is central to the definition of the neighborhood structures used in the searching approach. All moves are based on a series of additions (allocating a job to a machine) and removals (deallocating a job from a machine) and can be categorized as follows:

Transfer removes job j from its machine, $M(j)$, and adds it to machine $k \neq M(j)$ (one removal and one addition).

Swap removes jobs i and j ($i \neq j$) from their respective machines ($M(i) \neq M(j)$), and adds i to $M(j)$ and j to $M(i)$ (two removals and two additions).

Closed ejection chain involves a closed chain of $h \geq 3$ machines. A job is removed from each machine of the chain and added to the next machine in the chain (h removals and h additions).

Open ejection chain involves an open chain of $h \geq 3$ machines. A job is removed from each but the last machine of the chain and added to the next machine in the chain ($h-1$ removals and $h-1$ additions).

Compound move is any collection of transfers, swaps, closed, and open ejection chains such that on each machine at most one addition (restriction 1) and at most one removal (restriction 2) is performed (h removals and h additions, $1 \leq h \leq m$). Furthermore, an expanded compound move can be effected if restriction 1 or restriction 2 is removed, but not both.

The valid moves of a neighborhood define its structure and size. The simpler the moves that can be employed the faster the exploration of the neighborhood. On the other hand, simpler structures are less likely to be capable of identifying improving moves than more involved ones.

The proposed algorithm defines three neighborhood structures of increasing complexity named small (SNS), medium (MNS) and large (LNS). SNS is applied first and upon local optimum entrapment the search progresses to the MNS, altering also the objective criterion. When MNS also reaches a local optimum, LNS is employed. If one iteration of LNS leads to an improvement, the cycle is repeated otherwise the incumbent solution is the algorithm's proposed solution.

4.1 Small Neighbourhood Search

The initial solution is often quite poor but with high potential, given the high slacks. Improvements are obtained through a small neighborhood structure that is searched exhaustively, adopting the best improving move each time. In SNS, the neighborhood structure is defined as any transfer or swap that involves the machine inducing the makespan (in case more that one machines have spans equal to the makespan, one is randomly designated as the makespan machine).

A move is considered improving if, for all the machines affected by it, the new span is smaller than the incumbent makespan. Given the set of improving moves of a small neighborhood, the best is considered the one that minimizes the sum of spans (see Section 2). To further assist the guidance of the search to promising regions, in cases where no improving move can be found through transfers or swaps, or where the move increases the total span to more than 10% of the lower bound, the neighborhood structure is augmented to include ejection chains (open or close) of size 3. SNS terminates when no improving move can be found.

The adoption of a dynamic neighborhood structure proved necessary to balance the effectiveness and efficiency of the search. Exhaustively examining transfer and swap moves can be done very fast, given the simplicity of the moves. If a satisfactory move is identified, something likely in the early stages, it is adopted, thus allowing a speedy convergence without sacrificing the solution characteristics. Only upon non-satisfactory moves is the, more involved, ejection chain move examined.

4.2 Medium Neighbourhood Search

Focusing only on moves that involve the makespan machine inhibits the search from identifying more promising solutions. Clearly, moves in which the makespan machine is not involved cannot improve the overall objective of makespan minimization. However, moves involving the remaining machines can lead to improvements by increasing slack space (reducing the sum of spans) or reducing the number of makespan machines, both of which are desirable as explained in Section 2. In MNS, valid moves include any transfer or swap between any pair of machines. A move is considered undesirable if it worsens any of the two auxiliary objectives. MNS adopts a first-improving approach, committing the first move identified that reduces the sum of spans without increasing the number of makespan machines. Using a first-improving instead of a best-improving approach and limiting the moves to transfers and swaps was preferred to economize on computational effort, given the increase in the set of machines that could be involved in MNS moves.

4.3 Large Neighbourhood Search

As explained, neighborhood structures that employ limited scope moves have limited capabilities of yielding improvements. This is particularly true when the incumbent solution is relatively tight, as solutions obtained after SNS and MNS are likely to be. To escape from such solutions, a large neighborhood structure is defined that explores all feasible compound moves.

Such structures can identify complicated moves that may yield improvements but the difficulty lies in the development of efficient mechanisms for facilitating the search. To achieve this a MILP model is developed as described below.

$$\min C_{max} + \sum_{k \in M} \frac{C_k}{1000|M|} \tag{5}$$

subject to

$$\sum_{j \in N} y_{jk} \leq 1 \ \ \forall k \in M \tag{6}$$

$$\sum_{j \in J_k} \sum_{l \in M} y_{jl} \leq 1 \ \ \forall k \in M \tag{7}$$

$$C_k = C'_k + \sum_{j \in J} p_{jk} y_{jk} - \sum_{j \in J_k} \sum_{l \in M} p_{jk} y_{jl} \ \ \forall k \in M \tag{8}$$

$$C_{max} \geq C_k \ \ \forall k \in M \tag{9}$$

$$C_{max} \leq C'_{max} - 1 \tag{10}$$

where

y_{jk} — boolean variable set to 1 if job j is moved to machine k
C'_k — incumbent span of machine k
C'_{max} — incumbent makespan

The objective function (equation 5) minimizes the makespan using also the auxiliary objective of sum of spans minimization as a secondary objective. Constraint 6 ensures that at most one job can be added to any machine whereas constraint 7 limits the number of jobs that can be removed from a machine to one. Constraint 8 calculates the new span of each machine by adding to the incumbent span the processing times of all jobs moved to that machine and subtracting the processing times of all jobs that were allocated to the machine and have been moved to another machine. Constraint 9 links the makespan to the new spans and constraint 10 ensures that the makespan is improved.

LNS commences by restricting the compound moves to those that allow at most one insertion and one removal on each machine. If no such move is found, the constraint 6 is relaxed, allowing for more than one job to be added to a machine. If again no improving compound move is identified, constraint 6 is reinstated and constraint 7 is relaxed, allowing for more than one jobs to be removed from a machine but limiting the jobs to be added to a machine to one.

Given the large number of binary variables defined by the model, it is crucial to reduce as many as possible so that the model can be solved by a MP solver in realistic execution times. The following reductions are defined:

- $y_{jk} = 0 \quad \forall j \in J_k, k \in M$
- Assuming p_k^{max} represents the largest processing time on machine k of the jobs in J_k, then $y_{jk} = 0 \ \forall j \in N$ if $C'_k + p_{jk} - p_k^{max} > C'_{max} - 1$
- Given two jobs $i, j \in J_k$, $y_{j,l} = 0 \ \forall l \in M \backslash \{k\}$ if $p_{ik} \geq p_{jk} \wedge p_{il} \leq p_{jl}$

The first reduction ensures that a job cannot be moved to its current allocation. The second reduction reduces a move if even assuming the removal of the job with the largest processing time from the destination machine, the move would still be bigger than or equal to the incumbent makespan. Finally, the third reduction states that if two jobs are currently assigned to the same machine, then if the first has bigger or equal processing time on the incumbent machine and for a given destination machine it has less or equal processing time, such a move would always be preferable therefore the variable corresponding to moving the second is reduced. Note that if constraint 7 is relaxed, the second and third reductions do not apply.

Since the incumbent solution is likely to be tight given the prior application of SNS and MNS, the amount of variables reduced is high and experimentation has shown that relevant models can be addressed within acceptable execution times. As a safety net, the time assigned to the MP solver for a model instance has been capped to 1 second.

5 Computational Results

The proposed algorithm was implemented in C#.NET. The MIP problems were solved using ILOG CPLEX version 12.1, with the number of threads limited to 1 and the relative and absolute gaps set to zero. All tests were executed on a x5570 2.93GHz Intel Xeon processor using a single thread on a single core.

Three classes of random problem instances were generated, adopting the same pattern as that used in the literature. Specifically, for the first two classes processing times were taken randomly from uniform distributions $U(10, 100)$ and $U(10, 1000)$, respectively. For the third, *speeds* are assigned to machines and *lengths* to jobs and a weighted randomized processing time is calculated for each job-machine pair (see. [4] for more details).

For each problem class, 20 different instances were generated with the machine number obtained from $\{10, 20, 30, 40, 50\}$ and the job number from $\{100, 200, 500, 1000\}$. Computational results aggregated over classes, machines, and jobs are provided in Table 1. All presented results refer to the average over the 20 instances. The proposed algorithm was compared against the APPROX algorithm of Martello et al [6] and the Recovering Beam Search method with beam width 3 of Ghirardi and Potts [4] (denoted by RBS). For each algorithm the deviation from the lower bound (obtained through applying Langrangean relaxation on the UPMSP model) and the average and maximum execution time are provided. Finally, the last table column (denoted as 'Imp') shows the percentage of improvement (or deterioration) of the proposed solution against the best known deviation. For APPROX and RBS the deviation and times presented in [4] were used (implemented in C++ and executed on a PC-Pentium III/866).

Table 1. Comparative results on set of problem instances

Aggr. by		APPROX			RBS			VND			
		Dev(%)	avT(s)	maxT(s)	Dev(%)	avT(s)	maxT(s)	Dev(%)	avT(s)	maxT(s)	Imp(%)
cl.	1	6.1	17.7	83.0	5.1	79.6	392.2	4.0	1.3	3.8	31.7
	2	12.0	13.5	61.9	10.7	74.0	351.5	10.1	1.7	5.4	25.2
	3	3.1	54.6	212.4	2.1	92.7	427.8	1.4	2.5	7.8	51.2
m	10	1.9	22.8	174.4	1.2	63.8	262.1	0.9	0.8	2.1	41.5
	20	4.0	24.4	176.1	3.0	65.3	260.4	2.3	1.4	3.2	40.5
	30	6.5	28.0	181.8	5.2	75.1	291.2	4.7	2.0	5.2	33.6
	40	9.7	33.0	201.7	8.3	98.9	391.7	7.0	2.3	6.6	35.5
	50	13.2	34.8	212.4	12.1	107.4	427.8	11.0	2.6	7.8	29.0
n	100	17.2	< 1	< 1	15.7	< 1	< 1	15.3	0.4	1.1	7.8
	200	6.6	1.4	3.1	5.2	1.8	3.7	4.6	1.0	2.4	22.2
	500	2.7	15.5	31.3	1.9	29.3	41.9	1.0	2.1	4.1	52.8
	1000	1.7	97.0	212.4	1.1	285.7	392.2	0.4	3.8	7.8	61.2
overall		**7.1**	**28.6**	**212.4**	**6.0**	**82.1**	**427.8**	**5.2**	**1.8**	**7.8**	**36.0**

As shown, the proposed VND greatly outperforms the other algorithms when compared by either class, machines or jobs. VND performs best in the third class which also has the highest time demands. Comparing by number of machines, it seems that VND consistently outperforms the other algorithms independent of machine size. On the contrary, when comparing by job number, it is evident that the performance improvement is linked with the problem size. Also, taking into account processor performance benchmarks, VND achieves comparable execution times with RBS.

6 Conclusion

A variable-neighborhood descent, hybridized with mathematical programming elements, that addresses the NP-hard problem of scheduling unrelated parallel machines with the objective of minimizing the makespan, has been presented. Neighborhood search structures, of increasing complexity, are investigated to guide the search to promising regions. Local optimum entrapment is avoided through the use of large neighborhoods which are explored with the use of mathematical programming. The performance of the proposed scheme has been evaluated by comparing its solutions on a set of benchmark problem instances to those generated by the best available algorithms. The results show that the proposed scheme provides considerably better solutions.

References

1. Arnaout, J.P., Rabadi, G., Musa, R.: A two-stage ant colony optimisation algorithm to monomize the makespan of unrelated parallel machines with sequence-dependent setup times. Journal of Intelligent Manufacturing (2009) (forthcoming)
2. Blazewicz, J., Ecker, K.H., Pesch, E., Schmidt, G., Weglarz, J.: Scheduling computer and manufacturing processes. Springer, Berlin (1996)
3. Brucker, P.: Scheduling algorithms. Springer, Berlin (2004)
4. Ghirardi, M., Potts, C.N.: Makespan minimization for scheduling unrelated parallel machines: A recovering beam search approach. European Journal of Operational Research 165(2), 457–467 (2005)
5. Lee, C.-Y., Pinedo, M.: Optimization and heuristics of scheduling. In: Pardalos, P.M., Resende, M.G.C. (eds.) Handbook of applied optimization. Oxford University Press, New York (2002)
6. Martello, S., Soumis, F., Toth, P.: Exact and approximation algorithms for makespan minimization on unrelated parallel machines. Discrete Applied Mathematics 75, 169–188 (1997)
7. Mokotoff, E.: Parallel machine scheduling problems: A survey. Asia-Pacific Journal of Operational Research 18(2), 193–242 (2001)
8. Mokotoff, E., Chretienne, P.: A cutting plane algorithm for the unrelated parallel machine scheduling problem. European Journal of Operational Research 141(3), 515–525 (2002)
9. Mokotoff, E., Jimeno, J.L.: Heuristics based on partial enumeration for the unrelated parallel processor scheduling problem. Annals of Operations Research 117(1-4), 133–150 (2002)
10. Pinedo, M.: Scheduling: Theory, Algorithms, and Systems, 2nd edn. Prentice Hall, New Jersey (2002)
11. Shchepin, E.V., Vakhania, N.: An absolute approximation algorithm for scheduling unrelated machines. Naval Research Logistics 53(6), 502–507 (2006)
12. Shchepin, E.V., Vakhania, N.: An optimal rounding gives a better approximation for scheduling unrelated machines. Operations Research Letters 33(2), 127–133 (2005)

Aiding Interactive Configuration and Planning: A Constraint and Evolutionary Approach

Paul Pitiot[1], Michel Aldanondo[1], Elise Vareilles[1], Paul Gaborit[1], Meriem Djefel[1,2], and Claude Baron[2]

[1] Toulouse University Mines Albi, [2] Toulouse University INSA, France
{paul.pitiot,michel.aldanondo,elise.vareilles}@mines-albi.fr,
{paul.gaborit,meriem.djefel}@mines-albi.fr,
{meriem.djefel,claude.baron}@insa-toulouse.fr

Abstract. This communication aims to propose a two step interactive aiding system dealing with product configuration and production planning. The first step assists interactively and simultaneously the configuration of a product and the planning of its production process. Then a second step complete the two previous tasks thanks to a constrained multi-criteria optimisation that proposes to the user a set of solutions belonging to a Pareto front minimizing cost and cycle time. The first section of the paper introduces the problem. The second one proposes a solution for the first step relying on constraint filtering for both configuration and planning. The following ones propose an evolutionary optimisation process and first computation results.

Keywords: product configuration, process planning, constraint satisfaction problem, evolutionary algorithm.

1 Introduction

This paper presents an integrated support tool which at first allows interactive configuration of a product and interactive planning and scheduling of its production process, and then minimizes conflicting criteria cost and cycle time. The configuration of a private aircraft will be used as an example to illustrate our research work.

In literature, most of the research into product configuration and production planning treats them independently. However, the decisions of product configuration obviously have strong consequences on the planning of its production process (for example, a luxury finish requires at least two additional months. On the other hand, planning decisions can provide hard constraints to product configuration (for example, such assembly duration forbids the use of such a kind of engine). Therefore, we propose to associate these two problems so that (i) the consequences of each decision of product configuration can be propagated toward the planning of its production process and (ii) the consequences of each process planning or scheduling decision can be propagated towards the product configuration. As we target interactive assistance in order to allow some kind of "what if" operating mode, we need to be able to

H. Papadopoulos, A.S. Andreou, and M. Bramer (Eds.): AIAI 2010, IFIP AICT 339, pp. 238–245, 2010.

show the consequences of each user's elementary requirement. A user's elementary requirement can be defined as a restriction of the domain of a variable involved in configuration (for example, number of seats belongs to [6, 12]) or in planning (for example, due date is prior to 31/10/2010). We consequently do not intend to process all the requirements simultaneously in a single shot to get a solution for both problems but rather to progressively lead the user to a solution for both product configuration and planning of its production.

In the field of configuration, many authors, among whom [1] or [2], showed that product configuration can be efficiently modeled and aided when it is considered as a Constraints Satisfaction Problem (CSP). In a same way, authors interested in planning and scheduling as [3] or [4] have shown that theses problems could be also modeled and aided when considered as a CSP. We therefore propose to consider configuration and planning problems as two constraint satisfaction problems. We assume that a constraint based model of a generic product and the same kind of model for a generic production plan can be established and we restrict configuration and planning tasks to the instantiation of these two models. We also limit the scope of this paper to infinite capacity planning. To support interactive assistance, we only use the filtering or constraint propagation capabilities of the CSP framework. We finally link the two problems (configuration and planning) and the coupling constraints proposed in [5] together to propagate the consequences in both directions.

In the previous system, a product can be entirely configured and its production process entirely planned. "Entirely" means to restrict the solution space to a single solution, each problem variable having a single value. But we are not interested in this operating mode. We assume that it is possible to decompose the set of user's requirements in two sub-sets: non-negotiable requirements and negotiable ones. Our idea is to process interactive configuration and planning with the first sub-set or requirements only (non negotiable) and achieve a first reduction of the solution space. Remaining variable affectations (remaining solution space) are kept for multi-criteria considerations in the second step of our proposition.

In most industrial cases, the resulting products configured are characterized by criteria such as performance and product cost while relevant production plans are associated with cycle time and production cost. A solution is always a compromise of somehow contradictory criteria. In this presentation we only consider two criteria: cost (product cost and production cost) and cycle time for production planning. Hence, the next step is to find solutions that belong to the Pareto front (time/cost) among the solution space restricted during the first step. Multi-criteria optimization techniques and more accurately Evolutionary Algorithm (EA) (see [6] and [7]) have the advantage to avoid the aggregation of criteria and can provide solutions on a Pareto front in a rather simple way. Thanks to an evolutionary approach, the second step will perform a second reduction of the solution space for both product and plan and provide solutions on the Pareto front. Finally, the user can finish the process by selecting the solution that fits his specific time/cost compromise. The next section describes our proposition for the first step with an example, then our proposition for the second step with an evolutionary approach is detailed.

2 Configuration and Planning Models and Constraint Processing

The configuration model (left part of figure 1) gathers product descriptive variables (for example: aircraft range, number of engines, type of finish…) and product cost variables (finish cost, engine cost…) that are either symbols or discrete numbers. Configuration constraints (for example black solid lines that can link aircraft range and engine type together) and cost definition constraints (for example grey solid lines that can link engine type and number of engines with engine cost) correspond most of the time with discrete tables showing allowed combinations of allowed values. In this discrete problem, the associated CSP is discrete and the filtering provided by arc consistency technique [8] allows interactive configuration and cost estimation.

The planning model (right part of figure 1) gathers a set of planning operations (like manufacturing, assembling…) linked with ordering constraints. Each operation is defined with three operation temporal variables (starting date, ending date, possible duration) and eventually resource variables (required resource, quantity of required resource). We assume that the three temporal variables are real variables defined with intervals while resource type is symbolic and resource quantity a real variable. The cost of an operation is a real variable that depends on the resource type or quantity and the operation duration. As we consider planning with infinite capacity of resource, the constraints are as follows. Ordering constraints between operations (if task Y is after task X then starting date of Y is greater than or equal to ending date of X) and operation duration constraints (ending date equals starting date plus possible duration) are numerical constraints (black solid lines). The constraints that link possible duration with required resource and/or quantity of required resource and/or cost are mixed constraints (black and grey solid lines). Our numerical constraints are simple calculations (+,-,*, /, =, >, <), therefore they respect the hypothesis of bound Consistency proposed by Lhomme in [9]. Based on interval arithmetic, bound consistency is gathered with arc consistency and allows interactive planning and scheduling.

A mixed constraint between product description variables and planning operation variables (black doted lines) allows to propagate decision consequences from configuration to planning and from planning to configuration. [5] can be referred for more details. The first global criterion, i.e. total cost, is calculated thanks to a numerical constraint as the sum of product cost and operation cost (grey doted lines of figure 1). The second one, the production cycle time corresponds with the ending date of the last operation.

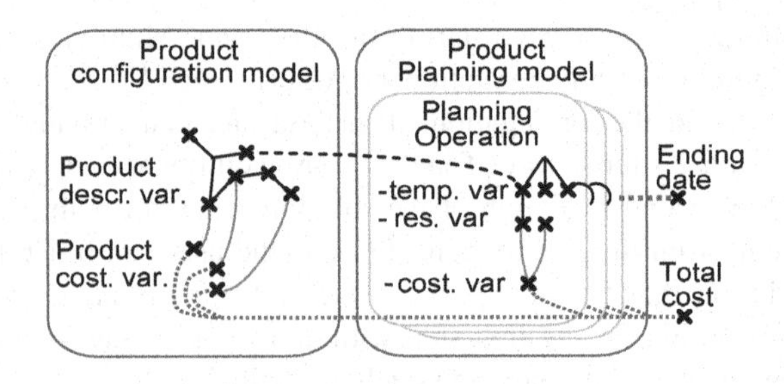

Fig. 1. Configuration and planning model

3 Optimization Problem to Assign

3.1 Definition of the Optimization Problem

The first interactive filtering step leads to the restriction of the initial feasible space (noted (a) on figure 2) to a restrained area (noted (b) on figure 2). This corresponds to the filtering of a customer's non-negotiable requirements. The filtering system provides domain bounds for every criteria variable (minimal and maximal values for cost and cycle time). The restrained area contains solutions corresponding to different remaining decisions to fulfil according to the remaining requirement. Notice that this area also contains unfeasible space where there is no solution due to the constraints of the problem to solve. The aim of the optimization process is to find a selection of the closest solutions to the Optimal Pareto front. Various metaheuristics may be used to solve this problem. In this research work, we focus on evolutionary algorithms for their ability to propose multiple solutions while solving a multiobjective problem. But classical evolutionary algorithms have to be adapted to take into account the constraints of the problem.

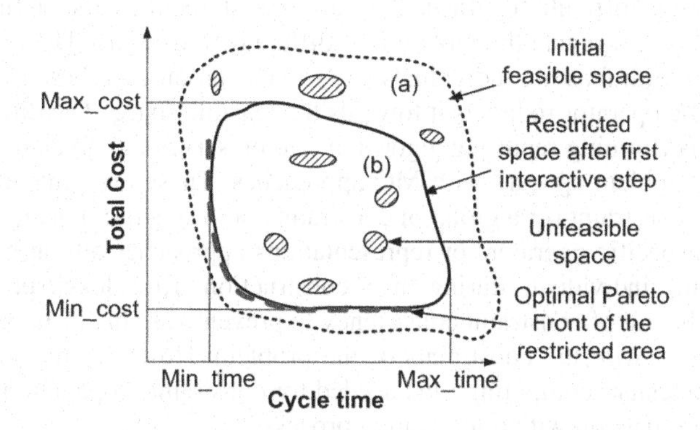

Fig. 2. Illustration of the reduced search space to optimize (b) from the initial feasible space (a) after the filtering process on non-negotiable requirements. Optimization process aims finding the Pareto-Optimal solutions in this restricted area.

3.2 Formal Definition of Optimization Problem

The constrained optimization problem (O-CSP) is defined by the quadruplet <V, D, C, f >, where V is the set of decision variables, D the set of domains linked to each variable of V, C the set of constraints on variable of V and f the multi-valuated fitness function. Here the aim is to minimize both cost and cycle time. The set V gathers: the product descriptive variables and the resources variables (required resource and choice or quantity of required resource). Constraints, as seen in section 2, link the variables of V with all the variables of the two problems. The filtering system allows dynamically computing modification on the domain of variables. This OCSP is a difficult problem to solve. The existing methods to handle constraints in EA are often computationally expensive. The next section briefly presents existing.

3.3 Overview of Existing Constrained Optimisation Approach

Initially, EA deal with large combinative unconstrained problem. They are used to solve multiple problems like project planning or project configuration. But real world problems are generally constrained. Many research studies try to integrate constraints in EA. C. Coello Coello comprehensively surveys the state of the art of these methods [10] (more than eight hundred references). It seems that classical EA performance decreases while the number of constraints increases. Constraints imply feasible and un-feasible areas in search space. Then the ratio between the size of feasible and un-feasible space guides the choice of specific methods to handle constraints in EA [11]. Four kinds of methods deal with this problem: penalty functions [12], repair methods [13], approaches that separate objectives and constraints (Multiobjective Optimization (MO) techniques) [14] and specific representations or operators [15], [16].

The first one, penalty functions, is the most common way to integrate violation of constraints in the objective function. For each individual, the level of violation of constraint is added to the fitness function during the evaluation phase. The aim is to increase the probability of selection for feasible or near feasible solutions. The main drawback of such an approach is that the boundary between feasible and unfeasible regions is usually difficult to grasp. Furthermore, it requires the definition of the weights needed to aggregate the violation of different constraints. The repair methods only try to deal with feasible individuals. As soon as an unfeasible individual is generated, a specific operator redirects it towards the feasible space. The difficulty is thus to elaborate a performing repairing algorithm that preserves the diversity of individuals. The same problem appears with MO approaches. These ones integrate the satisfaction of each constraint (or a group of constraints) as a specific objective.

Finally, the specific operators or representations approaches aim at preserving the feasibility of the individuals during their construction. Kowalczyk previously proposed in [15] the use of constraint consistency to prevent variable instantiations which are not consistent with the constraints of the problem. However, his study suggests that the supplementary computing time needed for constraint propagation may be very expensive in comparison with optimization process.

In this paper, we will focus on specific evolutionary operators that prune search space using constraints filtering (with our own filtering system named Cofiade while Kowalczyk used Ilog-solver, a commercial software, see [17] for a software comparison). The arc-consistence is less time consuming than other CSP solving system but it doesn't guaranty global coherence of the model.

3.4 Proposed Approach

Overview of modified EA used. EA used is adapted from the SPEA2 method [18] with classical evolutionary steps (initialisation, evaluation, selection, stopping criterion and perturbation operators). It is one of the most useful Pareto-based methods founded on preservation of a selection of best solutions in a separate archive. It includes performing evaluation strategy witch bring a well-balanced population density on each area of search space, and uses an archive truncation process that preserve boundary solution. It ensures both a good convergence speed and the preservation of diversity of solutions. Zitzler shows that the maximal complexity of the overall algorithm is $O(N_{pop}^3)$ with N_{pop} the number of solutions in current population. We completed this method with specific evolutionary operators (initialisation, uniform mutation and uniform crossover) presented in next sections. Finally, the stopping criterion is a fixed number of generations.

Initialisation operator. Thanks to this operator, a well-diversified set of initial individuals is obtained. For each individual to create, every gene (decision variable) is randomly instantiated into its current domain. In order to avoid the generation of unfeasible individuals, the domain of every gene is dynamically updated by filtering after each instantiation. If an individual is incoherent, a limited backtrack process cancels one of the previous choice, then the individual is filtered again until the values of the remaining variable are consistent with the constraints. If the backtrack limit is reached, the individual is abandoned to bound the number of backward step and thus the computational time spent by the filtering. This process (random instantiation then filtering) is repeated until all the genes of every individual are instantiated.

Uniform mutation operator. This operator introduces a random perturbation on the evolutionary process that allows escaping from sub-optimal areas and thus the exploration of search space. It modifies the instantiation of some genes on individuals selected according to the mutation probability. Every individual and their genes selected for the mutation are beforehand chosen randomly. The individuals are selected then selected genes are un-instantiated. The filtering system updates the domain of these variables according to the instantiation of others genes. Finally, the mutation of the selected genes is achieved in the same way as during the initialisation (instantiation, filtering and backtrack limit). When every gene is instantiated, the mutation process end and the individuals are added to the next generation.

Uniform crossover operator. This operator allows to shuffle randomly and uniformly the genes of two individuals (named parents) selected according to the crossover probability. As for the mutation operator, couple of parents and crossover of their genes are randomly selected before the crossover operation. To achieve this task, a crossover table (table of crossover flags for every gene) is filled. It allows to select beforehand which genes will be exchange between parents. Indeed, the crossover corresponds to a selected way on a binary tree where each branch is linked to the crossover of a particular gene. An instantiation of the crossover table is equivalent to the selection of a path on the crossover tree. Notice that position of a specific gene is chosen randomly to avoid dominance of genes by their position in chromosome. During crossover operation, the system tries to achieve random crossover. At every gene instantiation, the filtering system updates the domain of remaining genes. The crossover table is initially filled identically for both child, but if an individual become unfeasible, a specific backtrack is done by changing some crossover flag in the table. A supplementary flag is added to the crossover table that memorise unfeasible way on the tree in case of backtrack. Then a backtrack counter limits the number of backward steps. If the backtrack limit is reach, the corresponding child is abandoned. Finally, every feasible child is added to the next generation.

4 Preliminary Results

We perform various tests on the aircraft problem[1] (42 variables, ten of whom are taken into account in EA, 42 constraints, about two millions feasible solutions) to evaluate the proposed approach. Various evolutionary settings were investigated. In most runs, the modified EA was able to provide very good solutions, near Pareto-optimal solutions. Figure 3 presents the Pareto-optimal front and the archive founded after one run of the proposed EA (10 generations, population size: 20, archive size: 21, crossover rate: 0.7, mutation rate: 0.3). During this run, 71.4% of the optimal Pareto front is reach for a computation time of 1730 seconds (around 28 minutes). This result is very interesting, because it clearly shows that in a reasonable amount of time, it is possible to propose a set of solutions that permit the user to decide about his own compromise cost/cycle time. This is much more accurate than solutions frequently proposed in industrial configuration software that are based on the selection of default values that try to minimize either cost or cycle time.

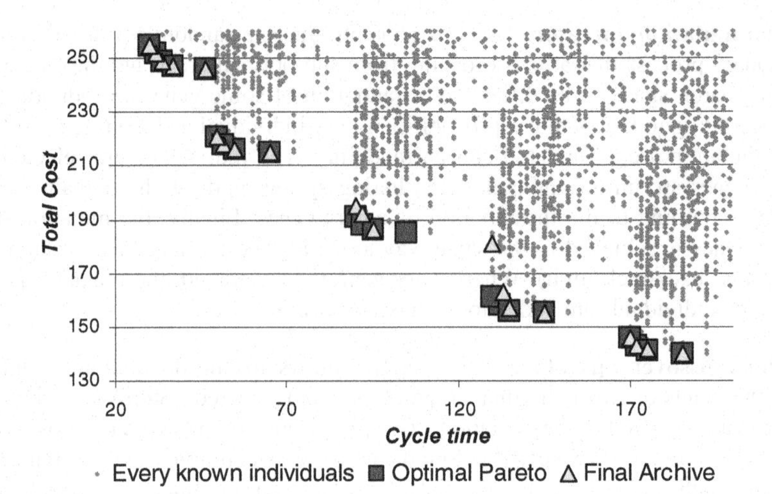

Fig. 3. Detail of archive resulting from one run of modified EA (triangles) and best known Pareto Front (squares) and every individual founded for this problem (points).

5 Conclusions

In this paper, we present an efficient aiding system for coupled product configuration and project planning, using an interactive constraint filtering system then a modified evolutionary optimization system. The modified EA is based on interaction with the filtering system that prunes the search space and thus reduces the search effort by limiting it to the feasible individuals. Standard evolutionary operators are adapted to take advantage of filtering. First experiments indicate that this method is well adapted

[1] Aircraft model could be investigated on-line, select model Aircraft-CSP-EA at: http://cofiade.enstimac.fr/cgi-bin/cofiade.pl

for an interactive aiding system. It generates near-optimal Pareto-solutions in reasonable computing time. It allows the user to decide efficiently about his cost/cycle-time compromise when dealing simultaneously with configuration and planning.

References

1. Soininen, T., Tiihonen, T., Männistö, T., Sulonen, R.: Towards a General Ontology of Configuration. AIEDAM 12(4), 357–372 (1998)
2. Junker, U.: Handbook of Constraint Programming, ch. 26. Elsevier, Amsterdam (2006)
3. Dechter, R., Meiri, I., Pearl, J.: Temporal Constraint Satisfaction Problems. Artificial Intelligence (49), 61–95 (1991)
4. Barták, R., Salido, M., Rossi, F.: Constraint satisfaction techniques in planning and scheduling. J. Intell. Manuf. 21, 5–15 (2010)
5. Vareilles, E., Aldanondo, M., Djefel, M., Gaborit, P.: Coupling interactively Product and Project Configuration: a Proposal using Constraints Programming. In: IMCM PETO 2008, Copenhagen, Denmark (2008) ISBN: 978-87-90855-12-3
6. Li, B., Chen, L., Huang, Z., Zhong, Y.: Product configuration optimization using a multiobjective GA. I.J. of Adv. Manufacturing Technology 30, 20–29 (2006)
7. Chelouah, R., Baron, C., Zholghadri, M., Gutierrez, C.: Meta-heuristics for System Design Engineering. Studies in Computational Intelligence, vol. 203, pp. 387–423. Springer, Heidelberg (2009)
8. Bessiere, C.: Handbook of Constraint Programming, ch. 3. Elsevier, Amsterdam (2006); Richardson, J.T., Palmer, M.R., Liepins, G., Hilliard, M.: Some guidelines for GA with penality functions. In: Schaffer, J.D. (ed.) Proc. of 3rd int. conf. on G.A., pp. 191–197 (1989)
9. Lhomme, O.: Consistency techniques for numerical CSPs. In: IJCAI 1993, Chambéry France, pp. 232–238 (1993)
10. Computer science department: CINVESTAV, `http://www.cs.cinvestav.mx/~constraint/`
11. Coello Coello, C.: Theoretical and numerical constraint-handling techniques used with evolutionary algorithms: A survey of the state of art. Computer Methods in Applied Mechanics and Engineering 191(11-12), 1245–1287 (2002)
12. Richardson, J.T., Palmer, M.R., Liepins, G., Hilliard, M.: Some guidelines for GA with penality functions. In: Schaffer, J.D. (ed.) Proc. of 3rd int. conf. on G.A., pp. 191–197 (1989)
13. Salcedo-Sanz, S.: A survey of repair methods used as constraint handling techniques in evolutionary algorithms. Computer science review, 175–192 (2009)
14. Clevenger, L., Ferguson, L., Hart, W.E.: Filter-based Evolutionary Algorithm for Constrained Optimization. Evolutionary Computation 13(3), 329–352 (2005)
15. Kowalczyk, R.: Constraint Consistent Genetic Algorithms. In: Proc. of IEEE conf. on evolutionary computation, pp. 343–348 (1997)
16. Michalewicz, Z., Nazhiyath, G.: Genocop III: A co-evolutionary algorithm for numerical optimization with non linear constraints. In: Fogel, D.B. (ed.) Proc. of the second IEEE conf on evolutionary computation, pp. 647–651 (1995)
17. Vareilles, E., Carbonnel, S., Djefel, M., Aldanondo, M., Rochet, S., Auriol, G., Baron, C.: Coupling Product and Project Configuration with Constraints: a CSP Software Comparison. In: International Conference SKIMA 2009 (2009)
18. Zitzler, E., Laumanns, M., Thiele, L.: SPEA2: Improving the Strength Pareto Evolutionary Algorithm. Technical Report 103, Computer Engineering and Communication Networks Lab (TIK), Swiss Federal Institute of Technology (ETH), Zurich (2001)

Decentralized Services Orchestration Using Intelligent Mobile Agents with Deadline Restrictions

Alex Magalhães, Lau Cheuk Lung, and Luciana Rech

Computer Science and Statistics Department- INE - CTC,
Federal University of Santa Catarina – UFSC
P.O.B 476, post code 88040-900 – SC, Brasil
{alex,lau.lung,luciana.rech}@inf.ufsc.br

Abstract. The necessity for better performance drives service orchestration towards decentralization. There is a recent approach where the integrator - that traditionally centralizes all corporative services and business logics - remains as a repository of interface services, but now lacks to know all business logics and business workflows. There are several techniques using this recent approach, including hybrid solutions, peer-to-peer solutions and trigger-based mechanisms. A more flexible approach regarding environment configuration and not fully explored in services orchestration technology is the use of intelligent mobile agents to execute it. In this paper, we present new adaptive heuristics for mobile agents to execute the decentralization of orchestration through missions (services) that correspond to the stages of business flow, with the ability to trade-off the quality of the result with the deadline of the mission. Some test case scenarios are presented and collected data are analyzed pointing the advantages and disadvantages of each heuristic.

Keywords: Intelligent Mobile Agents, Real-Time, Distributed Systems.

1 Introduction

The evolution of the WWW brought the need for more complex and scalable systems. To address this need, software architects began to use a composition of services known as service orchestration. Service orchestration provides a unique interface for each composite service and coordinates several simpler services using additional business logic to execute this task of coordination. Strongly based on XML and Web Services, service orchestration quickly became the reference design in Service Oriented Architecture (SOA [1]), centralizing all workflow at an integrator that uses languages as BPEL (Business Process Execution Language) for service coordination.

As critical systems began to use SOA, performance became an issue and there was room for improvement. To achieve better performance, decentralization and orchestration were combined and this new approach was considered very promising by authors [2] [3] [4]. In decentralized orchestration, the integrator still holds all composite services interfaces, but it no longer has control over the composite services workflows. In this approach, all business logic and workflow logic is distributed among the services that forms the composite services.

H. Papadopoulos, A.S. Andreou, and M. Bramer (Eds.): AIAI 2010, IFIP AICT 339, pp. 246–253, 2010.

The main motivation for decentralized orchestration is the achievement of better performance by critical mission and real time systems with a deadline. These systems need the benefits of a centralized repository of interfaces for composite services, but also need to address the deadline restrictions of such systems. Bearing this in mind, there has been proposed hybrid solutions based on popular technologies [4] [5].

In this paper, we propose a hybrid approach using intelligent mobile agents (MA). The integrator deploys a MA in the network that has a mission, for instance, to calculate the final price. The MA will have options such as obtaining the currency exchange rate directly going to the international bank web service or go to a local server that does not have this information in real-time. In this scenario, the sequence of nodes visited by the MA to complete its mission is called itinerary, and the itinerary will be defined by the trade-off between deadline of the mission and the desired QoS level. A MA may have previous knowledge of all the nodes it will need to visit to complete its mission or it does not have any previous knowledge of the network and will discover its itinerary from node to node, in such case, the MA is called myope.

The imprecise computing for MA with a deadline was originally presented in [6], along with some simple heuristics to define the MA's itinerary. Differently of the previous mentioned works, in this paper we will present new adaptive heuristics for MA, bringing as contribution the combination of code mobility and real-time constraints in decentralized services orchestration. With this objective, this paper presents a MA hybrid solution for decentralized orchestration.

The hybrid solution for orchestration using MA fulfills the requirements of high performance and real-time deadlines of a decentralized orchestration. However, this paper does not have as objective the specification of the architecture of a decentralized orchestration using MA. The architecture used in this paper is similar to the solutions presented in [2] [3] [4], with the advantage that MA architecture is already control-centric (through the MA container), resolving one of the major problems of decentralization: the environment configuration.

The main objective of this paper is to present new heuristics that define the MA itinerary during the flow of a composite service. These composite services workflows will be represented here by the MA missions. Each mission is composed of several tasks, which correspond to the internal services of each composite service. In this context, the heuristics for itinerary definition were designed to guarantee that the MA missions are completed in time so that its results are useful for the application.

In this paper, we present three heuristics of low computational cost and the performance of these heuristics will also be compared in different missions and with different deadline constraints. A real MA platform based in Java has been chosen, the JADE framework [7], in order to evaluate the performance of the system.

2 Related Works

Several works consider decentralized orchestration, each one is singular, presenting its advantages and disadvantages, but all have the merit to address the performance issue that traditional (centralized) orchestration is associated with.

In [4] is presented a hybrid solution using the peer-to-peer model. It is based on the Montage workflow, a system for analysis of astronomical images developed at Caltech. Although it has better performance than centralized orchestration, the complexity of this environment maintenance and the fact that each peer of the peer-to-peer network is an integrator itself make this design more complex than attractive.

In [3] is presented another hybrid solution using the concept of Service Invocation Triggers. In this solution, a lightweight trigger technology is proposed to execute the data flow, using a decentralized service control. Although it has a consistent architecture, the fact that the new trigger technology is not based on any mature technology presents a problem, making the biggest contribution of [3] the solution design architecture and not its new technology. The authors mentioned the MA technology, but they argue about MA security instead of evaluating it as potential solution for the problem. In the present paper we do not address the security issue.

In [5] is presented a hybrid solution for decentralization based on the concept of federated systems. They use Proxies in order to communicate the integrator and each task of the service, and each Proxy groups a few tasks that compose its federation. This solution is very interesting from the organizational point of view, since each federation can hold an entire area of knowledge of the corporation, but it lacks the commitment to improve the performance of the services.

In [8] is proposed a system based on agents and on the ANAISOFT framework, which is a repository for business models. The work consists of using MA to integrate several repositories located in distinct servers and to allow the interchangeability of business models among them. Despite the fact that it does not explicitly mention the orchestration aspect of the work, the article presents an effort to decentralize the ANAISOFT repository, with the advantage of using MA to do this.

3 Model Description

This section presents a new approach to assist the MA in the itinerary definition for each new mission. In orchestration, each request to the integrator corresponds to a new MA mission. The three new adaptive heuristics had been developed to define the MA's itinerary and also achieve the performance and deadline requirements of the decentralized orchestration. The main objective of these heuristics is to achieve a good performance on every deadline band, i.e., to respond to soft, medium and hard deadlines while attending to the QoS defined by the client.

In [6], after the performance's evaluation, we were able to correlate each heuristics to a type of deadline. The loose deadline was defined as the one where the Greedy Heuristic had the best performance; for tight deadlines, where the Lazy Heuristic had optimal performance. The adaptive heuristics presented in this paper are based on the simple heuristics described in [6]. We will briefly review these simple heuristics that were the inspiration for the heuristics proposed in the present paper, in order to allow a better understanding of the adaptive heuristics with variation at the departure that will be presented later on. The Lazy Heuristic executes the service as fast as possible, disregarding the level of QoS. The Greedy Heuristic – whenever there is an alternative route – will choose the one with the highest level of QoS, despite the execution time, but respecting the rules of the mission.

It is important to point out that differently from [6], in this paper we are considering services with variable QoS, i.e., all the services that compose the mission have a variable QoS proportional to execution time. This concept of variable QoS, called anytime algorithms, was presented in [9].

3.1 Greedy Heuristic with Variable Decrease at Start (Greedy-VD)

The Greedy-VD heuristic is a variation of the Greedy heuristic [6]. As this new version is based on the concept of anytime algorithms [9], the level of QoS is variable depending on the execution time of each service. The GREEDY-VD strategy is to start with the classical behavior of Greedy heuristic [6] and if this behavior is not efficient enough for the mission's objective, the heuristic will limit the execution time for each service at departure until it reaches an admissible level of QoS in agreement with the mission deadline. This gradual reduction of the execution time is directly associated with an adjustment factor - in this article the adjustment factor was of 10% for each new trip of the intelligent agent. I.e., the heuristics starts with the objective of reaching the maximum level of QoS (100%) and gradually adjusts it until it is capable of meet the deadline and finishing the mission successfully.

As the MA in the next mission uses the same behavior (decision making) and considering that it is a myopic MA [10], this heuristics goal is to find the "optimal percentage", that will allow the MA to reach the maximum level of QoS, respecting deadline of the current mission and spending the minimal number of trips (consecutive missions) possible. For example, using an adjustment factor of 10%, the MA in its third trip, if it has not found the optimal point, it can perform only 80% of the resources of the mission. This is possible because we are using variable resources and the concept of anytime algorithms.

3.2 Lazy Heuristic with Variable Increase at Start (Lazy-VI)

The Lazy-VI is a variation of the Lazy heuristic previously adapted for MA. This new version is also based on the concept of anytime algorithms, so the execution time of the services are variable, allowing the heuristic not to reach the maximum level of QoS for every service of the mission. The Lazy heuristic has as main objective to finish the mission in the best time possible, despite the level of QoS. The main problem with this strategy is that whenever it has room to improve the level of QoS of a mission, it will ignore it and will keep going for the fastest execution time possible. The need to repeat the mission allowed us to modify the classical Lazy heuristic in order to maximize the level of QoS, since the acquired QoS after each MA's visit to a node is proportional to the execution time of each service.

Lazy-VI heuristic has the inverse behavior of the Greedy-VD heuristic. Initially, it uses the classical Lazy behavior (executing the minimal number of blocks allowed for each service of the mission) and in the next trips of the same mission it gradually increases the percentage of execution for each service, considering the predefined adjustment factor. The strategy of this heuristics is to start as a traditional Lazy algorithm, trying to improve the level of QoS for each mission, trip by trip. After the first trip of the mission, the heuristic increases the percentage of the level of QoS until the mission does not meet its deadline or it reaches the maximum level of QoS for each service. In case the mission does not meet its deadline, the heuristic assumes the last percentage state as the optimal percentage of execution of the level of QoS.

3.3 Greedy Heuristic with Utility Function Based on Bipartition (Greedy-FB)

Greedy-FB presents a new approach for this kind of problem. This new version is based on the Greedy heuristic, on the concept of anytime algorithms and also on the concept of bipartition of the graphs theory. This heuristic is capable of executing the services partially and uses a strategy (utility function) of bipartition of the level of QoS, in order to meet the deadline and to get the maximum level of QoS for each type of deadline. The strategy of this heuristic is to initiate as a traditional Greedy heuristic in search of the optimal point of the level of QoS for each service. To accomplish this, the Greedy-FB heuristic uses a utility function based on bipartition, in order to speed up the determination of this optimal point.

When executing its first trip, in case the mission does not meet its deadline, the heuristic bipartitions the percentage interval of blocks of the service to be executed. In the second trip, this interval corresponds to [0, 100], then the heuristic will try the execution of 50% of the blocks for each service. In the third trip, in case the deadline is not met yet, the heuristic partitions one more time the interval in order to meet the deadline, and now the interval corresponds to [0, 50] and the new percentage of blocks for execution is of 25%. However, in case the mission's deadline is met in the third trip, the heuristic try to improve the accumulated level of QoS for the mission and it bipartitions the superior interval (the interval corresponding to [50, 100] and thus the new percentage of execution is 75%). The heuristic follows this strategy of bipartitions until it meets the mission deadline with a previously established range.

4 Performance Analysis

The purpose of the experiments is to compare the heuristics behavior and to evaluate its efficiency in different scenarios considering its deadlines.

4.1 Experiments Conditions

For these experiments we used 18 computers of 64-bit with clock speed between 2.0GHz and 3.2GHz, and 2.0 Gb of RAM. The adopted network was a 802.11g wireless network of 54 Mb/s. Each one of the computers was equipped with a Java Virtual Machine JDK 6.15 to support the JADE platform, the virtual environment supported by FIPA/IEEE to execute applications based on Mobile Agents.

4.1.1 Environment Configuration

For this experiment we considered 18 services and 18 nodes, where each node hosts a service in the environment. To the heuristics demonstrate its flexibility and capabilities, it is necessary that they can choose the optimal itinerary for their missions, and for that the 18 services are divided in 7 kinds of services (each kind contains the same kind of information, regardless of the node), i.e., each kind of service is represented in more than a computer of the network. Whenever possible, we placed each node of the same kind of service in computers with different clock speed, in order to better represent a real environment.

The execution time of a redundant service is considered different in every computer where it is replicated, in order to better represent a real scenario where different machines have different versions of the software that corresponds to the service. Each mission of the MA is composed of five kinds of services, whether there is an order of execution or not. Each of the heuristics evaluates the itinerary according to its algorithms, service after service, or evaluates more than a service at a time, whenever the order of execution does not exist.

We randomly generated 20 missions, all composed of five different kinds of services chosen amongst the 7 kinds of available services. We did not allow repetition. The order of execution is also variable, where a mission may have groups of kinds of services with and without precedence order. The combinations of kinds of services for the mission offer thousand of possible arrangements.

For each mission, we compute the total time of the mission and the accumulated level of QoS of the mission. The total time of the mission is composed by the time of each service execution, represented by the queuing time of the processor and the execution time of the service, and also by the MA's travel time from one node to another, respecting the network latency at the moment of the trip. The accumulated level of QoS of the mission is represented by the sum of every level of QoS of every service executed. Each service is divided in blocks of execution and each block executed represents a part of the sum in the individual level of QoS of the executed service. This division in blocks follows the anytime algorithms strategy [9], in order that the individual level of QoS for each service is a value between the maximum and the minimum level of QoS for this service on a specific node.

4.2 Evaluation Results

To show these heuristics performance in the experiment, we considered 3 different missions for 3 different deadline bands. The main objective of this experiment using different itineraries and deadlines is to verify the behavior of the MA heuristics on different missions. As this MA does not carry a history of the trips, it needs a short time to calibrate itself until if it defines the optimal percentage of execution time for each service that composes the mission, which will reflect in the best level of QoS for the heuristic in that kind of mission and deadline. Table 1 presents the performance of the three adaptive heuristics on hard deadlines. To better understand of the results, we have chosen a single deadline of this band (D=15). In Table 1, NS represents the number of steps (each step represents a full MA trip), columns T and Q represents, respectively, the traveling time and the level of QoS of each mission. The shadowed tuples show the best indices reached for each heuristic. However, they bold values indicate the best performances of each mission for the three heuristics. For example, for the first mission, the heuristic that best performance considering a balance of level of QoS and meeting the deadline was the Greedy-FB. It has taken seven trips (executions using different percentages of execution) to reach its optimal stage. Analyzing the number of steps (number of complete trips of the intelligent agent), the Lazy-VI heuristic was fastest for the first mission, needing fewer steps to reach the optimal percentage of execution for each service.

For second and third missions, the heuristic that presented the best performance was the Greedy-FB heuristic, reaching the optimal level of QoS in the fewest number of steps. Using a Bipartition Utility Function, the Greedy-FB heuristic was expected to reach the optimal percentage quicker than the other heuristics when the band of deadline was not too hard. The results of the experiment proved this expectation. And also as expected, the Greedy-VD heuristic, did not reach the best indices for the hard deadline, needing a higher number of steps to adjust the percentage of the level of QoS and reach its optimal point. It is important to point out, to facilitate the understanding of the results, on Table 1 we only showed the results for deadline 15. Other hard deadline values presented similar results.

Table 1. Performance of the Adaptive Heuristics for missions with hard deadlines

NS	M1						M2						M3					
	Greedy-VD		Lazy-VI		Greedy-FB		Greedy-VD		Lazy-VI		Greedy-FB		Greedy-VD		Lazy-VI		Greedy-FB	
	T	Q	T	Q	T	Q	T	Q	T	Q	T	Q	T	Q	T	Q	T	Q
1	58,4	55	58,4	55	3,8	5	54,4	53	54,4	53	3,7	5	52,4	55	52,4	55	3,8	5
2	54,4	51	29,1	28	7,6	10	47,9	50	27,9	27	7,4	10	48,9	51	26,9	28	7,6	10
3	49	46	16,2	15	9,6	13	43	44	15,6	15	9,3	13	44	46	15,5	16	9,6	13
4	40,7	39	10,8	10	13,4	18	38,1	39	10,4	10	13	18	38,3	40	9,8	10	13,4	18
5	29,9	28	13,7	13	14,8	20	33,2	34	12,3	12	14,9	21	33,4	35	11,4	12	13,4	18
6	24,1	23	14,8	14	18,6	25	25,7	27	14,2	14	14,9	21	26,9	28	14,1	15	13,4	18
7	19,1	18	14,8	14	14,8	20	21,5	22	14,2	14	14,9	21	22	23	14,1	15	13,4	18
8	13,7	13	14,8	14	14,8	20	15,6	17	14,2	14	14,9	21	17,1	18	14,1	15	13,4	18
9	13,7	13	14,8	14	14,8	20	11,7	12	14,2	14	14,9	21	11,4	12	14,1	15	13,4	18

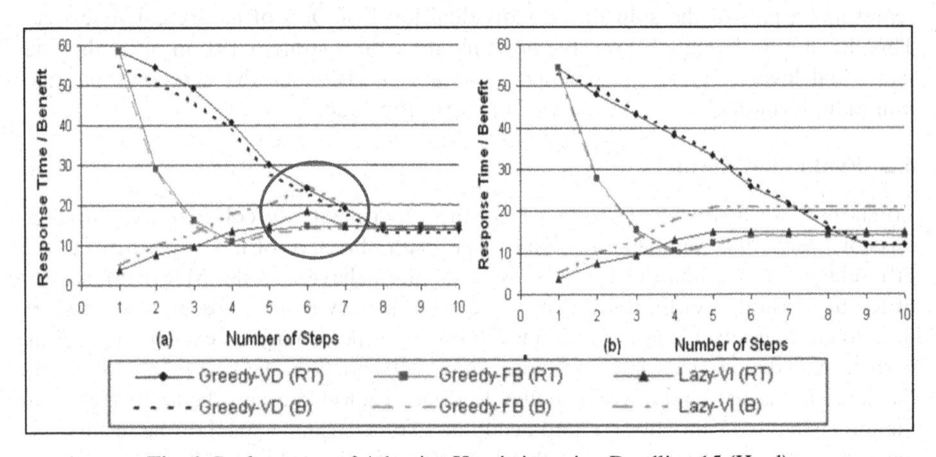

Fig. 1. Performance of Adaptive Heuristics using Deadline 15 (Hard)

Figure 1 present graphics that will contribute to the understanding of the behaviors of each adaptive heuristic. Observing the graphics we can notice that, even for different missions, the heuristics behavior is similar. The graphic 1(a) illustrates the behavior for the first mission. We can notice that the Greedy-FB heuristic is the first one to reach a travel time close enough to the deadline to the mission (step seven). The ellipse evidences in the graphic the "optimal point" reached by each of the heuristics. We can also point out that after the "optimal point" the level of QoS of the missions remains the same. Graphic 1(b) represents the indices reached by the adaptive heuristics on mission two.

5 Conclusion

In a corporative environment, critical mission and real time systems generally suffer with centralized architectures as is service orchestration. In this article the use of MA was proposed to decentralize the orchestration. Using a similar architecture for decentralized orchestration already presented in previous works, we present three new heuristics for the MA's itinerary definition that will assist on orchestration. These heuristics use an adjustment factor to change the number of blocks executed in each service and therefore the level of QoS of the mission, respecting the mission's deadline and attending to the client's specifications.

All these heuristics are of lightweight computational cost, a necessary condition to prove MA's viability as an architecture component for decentralized orchestration, reminding that is not the objective of this article to present this new architecture, so we need to assume that some of the nodes that will be visited by the MA are of low computational power. All the heuristics used the premise of the myopia of the MA, i.e., the agent doesn't know the complete workflow of the composite service and depends on this hybrid architecture of centralized orchestration, to create composite services, and decentralized orchestration, so the services have knowledge of the data flow. The experiment was able to prove that MA is a flexible approach for decentralized orchestration, where the agent is capable of adapting its behavior to reach different levels of QoS and successfully meets the desired deadlines.

References

1. OASIS. Reference Model for Service Oriented Architecture 1.0. Committee Specification (2006)
2. Barker, A., Weissman, J., Hemert, J.: The Circulate architecture: avoiding workflow bottlenecks caused by centralized orchestration. Cluster Computing 12(2), 221–235 (2009)
3. Binder, W., Constantinescu, I., Faltings, B.: Decentralized Orchestration of Composite Web Services. In: Proceedings of the IEEE International Conference on Web Services, pp. 869–876 (2006)
4. Barker, A., Weissman, J., Hemert, J.: Eliminating The Middleman: Peer-to-Peer Dataflow. In: Proc. of the 17th international symposium on High performance distributed computing, pp. 55–64 (2008)
5. Kyprianou, N.: Hybrid Web Service Orchestration, MSc Thesis, The University of Edinburgh (2008)
6. Rech, L., Oliveira, R., Montez, C.: Dynamic Determination of the Itinerary of MA with Timing Constraints. In: IAT 2005 IEEE/WIC/ACM Int. Conf. on Intelligent Agent Technology, pp. 45–50 (2005)
7. Bellifemine, F., Poggi, A., Rimassa, G.: Developing Multi-agent Systems with JADE. In: Intelligent Agents VII Agent Theories Architectures and Languages. Springer, Heidelberg (2001)
8. Schacke, A., Dittrich, K., Schönhoff, M.: Realization of an Agent-based federated system for trading workflows. The University of Zurich (2001)
9. Garvey, A., Lesser, V.: A survey of research in deliberative real-time artificial intelligence. The Journal of Real-Time Systems (1994)
10. Rech, L., de Oliveira, R., et al.: Determination of the Itinerary of Imprecise Mobile Agents using an Adaptive Approach. In: 13th IEEE Int. Conf. on Emerging Technologies and Factory Automation (2008)

Mobile Robot-Assisted Cellular Environment Coverage

Georgios Siamantas, Konstantinos Gatsis, and Antony Tzes

University of Patras, Electrical and Computer Engineering Dept.,
Rion, Patras, Greece
gsiama@upatras.gr, tzes@ece.upatras.gr

Abstract. The robotic coverage problem of a known rectangular cellular environment with obstacles is considered in this article. The robot can move only at directions parallel and perpendicular to the sides of the rectangle and can cover one cell at each time unit. A suboptimal minimum-time complete area coverage path-planning algorithm is presented. This algorithm minimizes a distance cost metric related to its current position and the quadtree-based decomposed blocks of the unexplored space. The efficiency of the suggested algorithm is shown through simulation studies and the implementation of a non-linear controller designed for a two-wheeled robot to achieve tracking of reference trajectories. For the implementation of the controller, the localization of the robotic vehicle was necessary and it was achieved via image processing.

Keywords: Robotics, Control, Planning.

1 Introduction

Robotic path-planning algorithms tackle the problem of finding a path in an environment that enables a robot to visit or sense with its sensors as much as or all if possible of this environment. Many applications of such algorithms appear in the literature [6]. In the present work we consider a robot placed inside a cellular environment with obstacles. We seek to find a path that minimizes the time that takes the robot to visit each cell of this unexplored environment. The entry and exit points of this path can be different cells. The optimal solution for the suggested l_1-distance cost metric rectilinear environment is NP-hard [5].

Similar works were presented by Zelinsky [3] and Gabriely and Rimon [8]. Zelinsky used the conventional wavefront algorithm to determine a coverage path (distance transform coverage-DTC). In DTC the starting and ending (goal) robot positions are specified in order to create an expanding field of increasing distance values from the end position extending to the whole grid. The robot then finds a cover path from start to goal performing a "pseudo-gradient ascent", on the distance field. Gabriely and Rimon describe a Spanning Tree Coverage (STC) algorithm. The STC works not on the original cellular environment but on a coarsen grid which is a tool-based approximation of the work-area and achieves a covering path in linear time $O(N)$, where N is the number of cells comprising

H. Papadopoulos, A.S. Andreou, and M. Bramer (Eds.): AIAI 2010, IFIP AICT 339, pp. 254–261, 2010.

the approximate area. The DTC method is very well suited for applications where we want to create a covering path in $O(n)$ time (n the grid area) and finish in a specific cell in the grid. But it does not guarantee that this path will be as close to the o as possible. In this paper the proposed coverage method produces close to optimal paths most of the time, overcomes the need of specifying a goal position as required in the DTC algorithm and works directly in the original cellular environment grid in contrast to the STC algorithm.

The paper is organized as follows. The model of the environment, the robot motion and associated metrics and methods are described in Section 2. The proposed algorithm is provided in Section 3 followed by simulation results and robot experiments in Section 4, while conclusions are drawn in the last Section 5.

2 Problem Statement

2.1 Preliminaries, Concepts

We consider a cellular environment to be a rectangular grid of n x n dimensions where $n = 2^k$, $k > 0$. The grid cells can be empty (free) and thus possible for the robot to visit them or occupied by an obstacle and thus inaccessible to the robot. The topology of the grid is dictated by the number and positions of the obstacles inside the grid. Each grid cell has coordinates of the form: (i, j), where $1 \leq i \leq n$ and $1 \leq j \leq n$. The robot can move under the Manhattan or l_1-metric motion model i.e. parallel or perpendicular to the sides of the rectangle and only one grid cell at each time unit, thus the set of actions (moves) that the robot can make is $U = \{(0,1),(1,0),(0,-1),(-1,0)\}$ where the robot's position is changed at the kth time instant to $X^{k+1}(i,j) = X^k(i,j) + U$. Each empty cell has from 1 to 4 neighbor free cells depending on adjacent obstacles and on whether is located at a corner (1-2 neighbors), an edge (1-3 neighbors) or inside the grid (1-4 neighbors). Each empty cell is accessible from any other empty cell in the grid, i.e. obstacles do not form closed regions of empty space inside the grid. The following notation is used in the remaining of this article: C the set of grid cells, C_{free} the set of free grid cells, C_{obs} the set of cells occupied by obstacles, C_{feas} the set of cells that the robot can visit from the current position, C_{vis} the set of already visited cells, C_{unvis} the set of remaining unvisited cells, $X_{i,j}$ the grid position at coordinates (i,j), X_{cur} the robot's current position coordinates, X_{next} the robot's next move position coordinates, X_I the robot's initial position coordinates.

Under the above environment formulation the minimum-time full area coverage path must intersect itself as less times as possible if not at all. One known way [7] to achieve this (minimizing intersections) is to execute a serpentine like motion consisting of straight moves followed by 180 degree U-turns, the so called boustrophedon (ox turning-like) motion and try to avoid visiting already visited space. In order to achieve as much as possible boustrophedon motion in a known topology environment is to ensure that at each path step the sum of the distances from the next robot position to as many as possible landmarks in the environment is maximized. An equivalent to this maximization criterion is to minimize the sum of the distances from all the other possible move positions except from the current pick position, e.g. if the robot can move to four adjacent cells then pick one possible move at a time and measure the distance of the other three possible move cells to the landmarks and then move to the cell with the minimum sum of these distances. We used the minimization criterion instead of the maximization because in the l_1 metric cellular environment, the simulation studies indicated a reduced sensitivity when dealing with moves that are equidistant from the landmarks. In our environment formulation all the possible robot positions are countable, so it is possible to sum the distances from the next position to all free cell positions in the grid. Although this is possible it is time consuming for large grid sizes. An equivalent criterion is to decompose free cell space to blocks and then sum the distances from the next position to some cell inside these blocks. The distances should be weighted according to the size of each block to ensure that the mean block distance is approximately the same with that of summing the distances of all the cells inside the block. The concept is shown in Figure 1(a). If the current robot position is in cell C and we choose to move to cell N the two other possible moves are to cells R. We compute the sum of all distances q_1, q_2 etc. from cells R to the decomposed free space blocks and repeat the same computation for each of the 3 possible moves. The criterion to move to the next cell is decided by the one that makes the sum of these distances minimum.

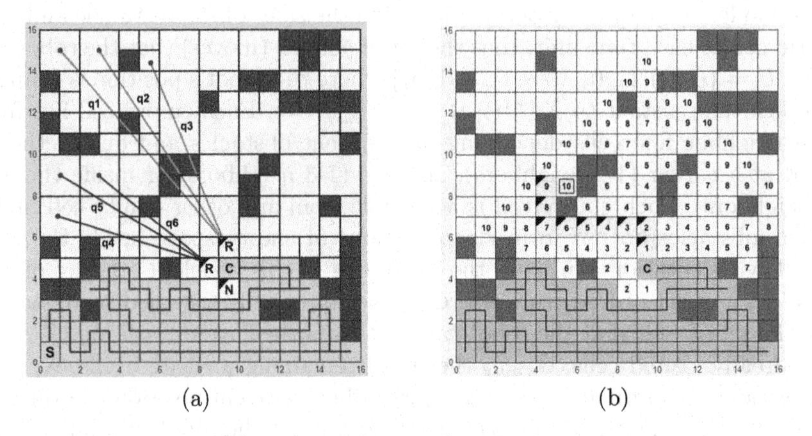

Fig. 1. (a) Sum of distances computation (b) The propagation and retrace phases

2.2 Quad-Tree Decomposition

An n x n grid matrix G is defined such as:

$$G(i,j) = \begin{cases} 1 \text{ if } X_{i,j} \in C_{obs} \text{ or } X_{i,j} \in C_{vis} \\ 0 \text{ if } X_{i,j} \in C_{unvis} \end{cases} \quad (1)$$

where the grid cells that are considered obstacles and the already visited cells (soft obstacles) take value 1 and all other unvisited cells take value 0.

In this paper the Quad Tree Decomposition (QTD) algorithm is used for decomposing the environment grid matrix into rectangular blocks of similar value cells. The QTD is a well established method [4] of decomposing an image into blocks that meet some criterion of homogeneity and has been used extensively in robotics [9]. The motivation for decomposing the grid matrix into blocks is to reduce the number of cell distance calculations by some constant factor. The quad tree method decomposes the grid matrix into two parts. The first part is a tree structure that shows the location and size of each homogeneous region. The second part specifies the cell value at each leaf node of the tree, representing if it is an obstacle, visited space or unvisited space.

Quad tree decomposition is based on the successive subdivision of the grid matrix into four equal-sized quadrants. If the matrix does not consist entirely of 1s or entirely of 0s, then it is subdivided into quadrants, sub-quadrants, and so on, until blocks are obtained that consist entirely of 1s or entirely of 0s; i.e. each block entirely contains cells that belong to C_{obs} and C_{vis} or entirely contains cells that belong to C_{unvis}. The grid quad tree can be characterized as a variable resolution data structure. QTD is an $O(n)$ complexity process, where n is the grid area [9]. The worst case for the QTD in terms of distance calculations occurs when the topology of the grid corresponds to a checkerboard pattern. The amount of distance calculations required is a function of the resolution (i.e., the number of levels in the quad tree), the number of obstacles and visited cells, and their relative positioning inside the grid. Thus, when calculating the sum of distances instead of dealing with all the free grid cells we have to deal only with the nodes of the QTD, a substantial reduction.

2.3 Shortest Path Computation

In order to measure distances in the l_1-metric grid we used an algorithm proposed by [2] that it is widely used in routing two terminal nets in circuit board design problems. This algorithm has $O(n)$ time and space complexity, where n is the area of the grid matrix and guarantees of finding an optimal solution if one exist. The used algorithm consists of two phases [1]. The propagation phase is based on breadth-first search. The search resembles a wave propagating from the starting cell. The start cell is labeled '0' and the wave front propagates to all the unblocked vertices adjacent to it. Every unblocked cell adjacent to the start cell is marked with a label '1' corresponding to its l_1 distance from start. Then, every unblocked cell adjacent to the cells with distance 1 is marked with a label '2' because its distance from the start cell is 2, and so on. This process continues until the target cell is reached or no further expansion of the wave can be carried out. An example of the algorithm is shown in Figure 1(b). Due to the breadth-first nature of the search, this algorithm is guaranteed to find a path between a source and a target cell, if one exists. In addition, it is guaranteed to be the shortest path between these cells [1]. The second phase is the retrace phase where from the target cell we follow back the cell labels in descending distance order till we reach the starting cell. In our formulation for the sum of distances computation the propagation phase is computed once in each robot's move. The retrace phase is not needed because we only need to find the minimum distance path between the starting and ending cells if such a path exists.

3 Environment Coverage

Let B the set of all QTD blocks, B_{free} the set of QTD blocks of cells with value 0, B_{occ} the set of QTD blocks of cells with value 1, $B_{i,j}$ the position of a block as the upper left cell position $X_{i,j}$ of this block, $A^{B}_{i,j}$ the area of a block $B_{i,j}$ given from the QTD. $d^{m,n}_{k,l}$ is the l_1-metric distance from $X_{k,l}$ to $X_{m,n}$ and is returned by the aforementioned in Subsection 2.3 algorithm. If there is no path connecting $X_{k,l}$ to $X_{m,n}$ because of the presence of obstacles and/or cells that are already visited then the implemented algorithm returns a very large value to bias the next move properly.

In order to choose which move to select the following quantity is defined:

$$D^{m,n}_{i,j} = \sum_{\forall X_{k,l} \in \{C_{feas} \backslash \{X_{i,j}\}\}} d^{m,n}_{k,l}. \tag{2}$$

This quantity reflects the definition of the sum of the distances from some cell $X_{m,n}$ to all the other possible next moves except the one for which we compute the sum.

Similarly, let:

$$\Delta_{i,j} = \sum_{\forall B_{m,n} \in B_{free}} A^{B}_{m,n} D^{m,n}_{i,j}, \tag{3}$$

where $\Delta_{i,j}$ is the sum of the previously defined D distances to all the free block positions scaled by each block's area as explained in section II.

If the set M is defined as:

$$\{ \Delta_{i,j} \mid X_{i,j} \in C_{feas} \}, \tag{4}$$

then the position coordinates of the robot's next move in order to fulfill the minimization criterion of the proposed algorithm can be set as:

$$X_{next} = \arg \inf_{X_{i,j}} M. \tag{5}$$

In the algorithm *Grid_Cover* shown below function QuadTreeDecomp() performs quad tree decomposition on the given grid. Function FindUnvisitedCellBlocks() finds in the quad tree structure the QTD blocks of cells with value 0 i.e. that correspond to unvisited cells. Function FindAdjacentUnvisitedCells() finds the adjacent to X_{cur} unvisited cells that are not obstacles. Function GridIsCovered() checks if the grid is covered by the robot. Function FindNearestUnvisitedCellTo() finds the nearest to X_{cur} unvisited cell (even if it is necessary to pass over already visited cells) thus implementing a backtracking mechanism that guarantees the grid's full coverage by the robot. Function FindMinDistancePathTo() finds a minimum distance path to that nearest cell using the suggested shortest path algorithm.

Algorithm *Grid_Cover*(G, X_I)
Input: A grid matrix G; the starting cell position X_I.
Output: An array *Path* with the covering path cell positions.
1. $X_{cur} \leftarrow X_I$
2. $S \leftarrow$ QuadTreeDecomp(G)
3. $B_{free} \leftarrow$ FindUnvisitedCellBlocks(S)
4. $C_{feas} \leftarrow$ FindAdjacentUnvisitedCells(X_{cur})
5. **if** sizeof(C_{feas}) = 1 **then**
   $X_{Next} \leftarrow C_{feas}$; **goto** 8.
6. **if** sizeof(C_{feas}) > 1 **then**
   With C_{feas} and B_{free} compute equations (2)-(4)
   Set X_{Next} according to equation (5); **goto** 8.
7. **if sizeof**(C_{feas}) = 0 **then**
   if GridIsCovered() = True **then**
     return *Path*
   else
     $X_{near} \leftarrow$ FindNearestUnvisitedCellTo(X_{cur})
     Path $\leftarrow$ *Path* + FindMinDistancePathTo(X_{near})
     $X_{Next} \leftarrow X_{near}$
8. Update *Path*, G, X_{cur} with X_{Next}; **goto** 2.

4 Simulations and Experiments

In order to evaluate the performance of the algorithm we define as MNP the minimum number of path steps, as ANP the algorithm number of path steps and as a performance index $PI_{opt} = MNP/ANP$. The numerator refers to the number of robot moves needed to cover optimally the whole grid in minimum time and the denominator refers to the number of moves proposed by the algorithm for full area coverage. So the grid will be covered optimally if $PI_{opt}=1$. Several tests were performed with different grid areas, obstacle topologies and robot start positions and the algorithm performed acceptably with $PI_{opt} > 0.87$ most of the time. The proposed algorithm shows $O(n^2)$ time complexity where n corresponds to the grid area. This is because wavefront propagation algorithms have an $O(n)$ time complexity and that, without taking QTDs into account, it takes at most $(n^2 + n)/2$ distance propagation related operations to cover the whole area. DTC computed for all possible end positions in the grid (full DTC) has also $O(n^2)$ time complexity.

In Figure 2 we see a comparison of the performance of the proposed algorithm with full DTC. In an 8x8 grid 100 simulation experiments were performed in MATLAB with 15 uniformly randomly distributed obstacles each time. In Figure 2(a) we see that the proposed algorithm (solid line) creates cover paths that have approximately the same number of steps as the minimum-step paths and considerably faster paths than the maximum-step ones created by the full DTC algorithm. Also in Figure 2(b) we see the execution times in the same experiment. The proposed algorithm (dashed line) was almost 7 times faster than full DTC.

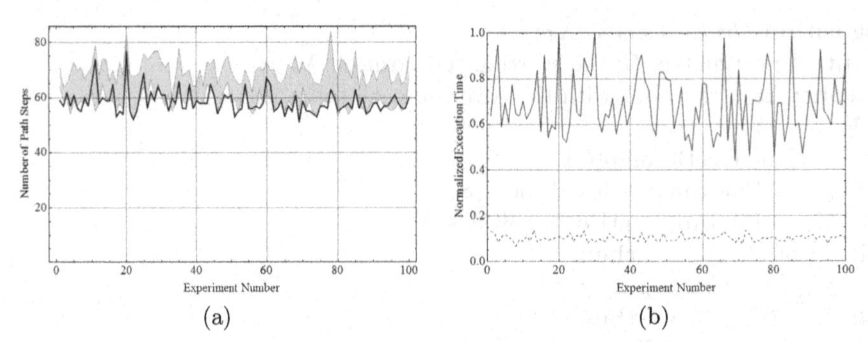

Fig. 2. Simulation results. The gray region represents the minimum and maximum path steps generated by full DTC.

In Figure 3 we see the Amigo differential drive two-wheeled robot that was used to test the proposed algorithm in real time. In order to do this a non-linear controller was implemented, based on [10], in order to achieve tracking of the required by the algorithm reference trajectories. For the implementation of this controller, the localization of the robotic vehicle was necessary and it was achieved via image processing during the experiments. The robot was recognized in images taken by a surveillance camera Figure 3(b), and after appropriate geometric analysis, it was possible to estimate the position and the orientation of the robot. In order to make it easier to recognize the robot's position and orientation inside the image, the robot's external surface was covered by white material and three small black spheres were placed on it, as shown in Figure 3(a). As shown in Figure 3(c) the robot followed the reference path computed by the algorithm with acceptable accuracy.

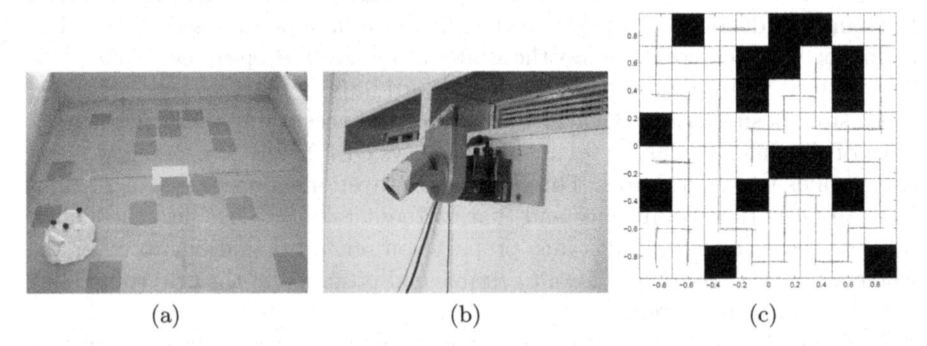

Fig. 3. (a) Robot in test environment (b) Camera (c) Reference and real robot paths

5 Conclusions

In this work a suboptimum complete area coverage algorithm is presented for rectangular l_1-metric grid environments. The start and finish positions may not be the same and the grid may have an arbitrary number of obstacles in arbitrary positions. The algorithm showed good performance in various grid topologies with acceptable time complexity taking into account its simplicity and hardness of the problem. The algorithm works on the original cellular environment and not on a coarsen approximation grid like the STC algorithm. It also guarantees the grid's full coverage by the robot because of its implemented backtracking mechanism. For comparison we performed various experiments with random topologies and saw that the proposed algorithm produced equally fast cover paths with the minimum-step paths produced by the full DTC algorithm but with considerably less computation time. The full DTC algorithm is an extension of the original DTC algorithm that we derived in order to compare it with the proposed algorithm. In the original DTC the goal position must be specified and produces cover paths with varying step sizes depended on this position. Finally the performance of the algorithm was validated in real time experiments with a differential drive two-wheeled robot moving according to the algorithm output in a laboratory test environment.

References

1. Sherwani, N.: Algorithms for VLSI Physical Design Automation, ch. 8. Kluwer, Dordrecht (2002)
2. Lee, C.Y.: An Algorithm for path connections and its applications. IRE Transactions on Electronic Computers (1961)
3. Zelinsky, A., Jarvis, R.A., Byrne, J.C., Yuta, S.: Planning Paths of Complete Coverage of an Unstructured Environment by a Mobile Robot. In: Proc. of Intern. Conf. on Advanced Robotics, Tokyo Japan (November 1993)
4. Samet, H.: Region Representation: Quadtrees from boundary codes. Comm. ACM 23, 163–170 (1980)
5. Deng, X., Kameda, T., Papadimitriou, C.: How to learn an unknown environment I: the rectilinear case. J. ACM 45(2), 215–245 (1998)
6. Choset, H.: Coverage for robotics-A survey of recent results. Annals of Mathematics and Artificial Intelligence 31, 113–126 (2001)
7. Choset, H.: Coverage of known spaces: The boustrophedon cellular decomposition. In: Autonomous Robots, vol. 9, pp. 247–253. Kluwer Academic Press, Dordrecht (2000)
8. Gabriely, Y., Rimon, E.: Spanning-tree based coverage of continuous areas by a mobile robot. Annals of Mathematics and Artificial Intelligence 31, 77–98 (2001)
9. Kambhampati, S., Davis, L.: Multiresolution Path Planning for Mobile Robots. IEEE Journal of Robotics and Automation RA-2(3) (September 1986)
10. Klancar, G., Matko, D., Blazic, S.: Mobile Robot Control on a Reference Path. In: Proc. 13th Mediterranean Conf. on Control and Automation, Cyprus, June 27-29 (2005)

A Novel Feature Selection Method for Fault Diagnosis

Zacharias Voulgaris and Chris Sconyers

ICSL lab, Georgia Institute of Technology, 813 Ferst Drive NW,
Atlanta, GA 30332, USA
{zvoulgaris,csconyers}@gatech.edu

Abstract. A new method for automated feature selection is introduced. The application domain of this technique is fault diagnosis, where robust features are needed for modeling the wear level and therefore diagnosing it accurately. A robust feature in this field is one that exhibits a strong correlation with the wear level. The proposed method aims at selecting such robust features, while at the same time ascertain that they are as weakly correlated to each other as possible. The results of this technique on the extracted features for a real-world problem appear to be promising. It is possible to make use of the proposed technique for other feature selection applications, with minor adjustments to the original algorithm.

Keywords: feature selection, dimensionality reduction, fault diagnosis, classification.

1 Introduction

The extraction of features is a relatively manageable and straightforward process, especially in cases where there is abundance of data. Also, as there have been developed methods for automating this process, the feature set size increases dramatically (leading to what is commonly refer to as the *curse of dimensionality*). However, the often large number of features does not necessarily mean better results in one's analysis of the problem at hand. Many of the extracted features are usually redundant and even relatively weak. Since the fault diagnosis process, in order to be carried out effectively requires one or more robust features; it is essential that an efficient feature selection step is required and demands the researcher's attention. Fault diagnosis based on the selected feature is usually carried out after the stage of feature fusion, which involves taking the selected features and creating a super-feature by combining them. However, in this paper the selected features are used as they are, so that they can be evaluated through their diagnosis classification performance.

Feature selection has been tackled mainly as an optimization problem. The relevant research includes supervised, unsupervised and semi-supervised ways to select the optimum subset of features [1]. Yet, the general approaches to feature selection are two-fold: either rank the features according to some evaluation metric and select the top k ones, or select a minimum subset of features that allows for good performance (i.e. the quality of the dataset, in terms of relevant information, does not deteriorate). [1]. Furthermore, the feature selection techniques, particularly in the fault diagnosis/prognosis

H. Papadopoulos, A.S. Andreou, and M. Bramer (Eds.): AIAI 2010, IFIP AICT 339, pp. 262–269, 2010.

domain, can be categorized in two broad groups: optimal and sub-optimal methods, the former including exhausting search and the Brand & Bound algorithm, while the latter using methods like sequential search, floating search, plus-L-minus-R search [2].

The rest of the paper is structured as follows. In section 2 the related work to feature selection methods is presented and briefly analyzed, after the evaluation criteria for the features at hand are defined. Based on the analysis of the feature selection methods, the necessity to adopt an alternative approach is pinpointed for the application at hand. This approach is described, on a theoretical level, in section 3. In the section that follows (4), the application of the proposed method is exhibited, using some real world data from a project the authors are involved in. The results of the relevant simulations are also presented in this section. Afterwards, a discussion of this research is carried out in section 5. Conclusions are drawn and avenues of future research based on this work are mentioned in the final section of this paper (6).

2 Evaluation Criteria and Related Work

There have been numerous attempts to solve the feature selection problem using various strategies for obtaining a robust reduced feature set. Although most of these approach are classification-oriented (including data mining) [3-5], some are designed having a fault diagnosis/prognosis problem in mind [2, 7-9].

Although the basic principles are the same in these disciplines, as regards the feature selection problem, in the later case the evaluation of a feature is more complicated. Namely, a robust feature in fault diagnosis (and prognosis) is characterized not merely by good distinguishability (discernibility, class separability, etc.) but by a sense of monotonicity as well (the more monotonic in relation to wear level the better). As monotonicity can be modeled by taking the relationship between consecutive wear levels (classes) so that the feature values form a monotonic pattern, it can be thought of as *ordered* discernibility. So far, the most widely-used and academically acceptable measure of monotonicity is the correlation coefficient.

There are several methods for dimensionality reduction, which are not related to the problem at hand (e.g. PCA yielding a small number of features to express almost the same information through different combinations of the original features). The term feature selection which is used in this research denotes that a subset of the *original* features is used, thereby maintain many of the original dataset characteristics and yielding more meaningful features in the new feature set.

Classification-oriented methods for feature selection include, but are not limited to, vertical compactness of data, the relevance-based algorithm approach, statistics-based techniques, and Genetic Algorithms [3]. Alternative approaches in the same category of methods for feature selection include the Receiver Operating Characteristics curve approach and class separability measures, through certain searching techniques (which are usually suboptimal) [4]. As regards class separability measures, one of the authors of this paper has conducted a thorough research on one of them (which he coined with the term Discernibility) with applications in feature selection, among a number of other research topics in the field of classification [5].

Feature selection for fault diagnosis/prognosis [6], though exhibiting similarity in the methods it employs, often makes use of alternative ways to tackle this problem.

Namely, in [2] an enhanced version of the Brand & Bound algorithm is proposed and implemented. Also, the use of consistency as a measure of feature evaluation for feature selection has been successfully employed, as it exhibits monotonicity and speed [2]. A Genetic Programming based approach has also been applied for feature selection, as it yields good results without the need for large samples, nor assumptions used in e.g. statistical approaches to the problem at hand [7]. Also, this approach is preferred by some researches due to its efficiency since it bypasses the analytical way of creating all possible feature subsets, which is an NP-hard problem [7]. Finally, a rough sets approach has also been employed for feature selection in this domain, with promising results [8].

It should be noted that feature selection is application dependent, especially in fault diagnosis/prognosis problems [9]. Yet, there are some recurring patterns that are found across the different applications in the field. These are the basis for feature analysis and evaluation, and comprise of the following criteria [9]: Distinguishability (often referred to as class separability or Discernibility), Detectability/isolatability, Identifiability, Degree of Certainty. The use of at least one of these criteria is essential for conducting a proper feature selection for the field of fault diagnosis/prognosis.

Note that most of the aforementioned methods for feature selection assume quantitative feature values. This is the most commonly encountered feature type in applications related to fault diagnosis. However, there exist several other feature types, which may require a different strategy for feature selection [10].

Though all of the above methods show promising results in the applications they were tested, a generic and therefore versatile method of feature selection is yet to be found. The fact that feature selection often requires some type of threshold to filter out the redundant features is part of the problem. Another aspect of the problem is that the feature selection methods under consideration are based on one or, less often, more evaluation metrics that model the four criteria mentioned earlier. If the user of these methods wishes to experiment with alternative measures for ranking the features, this would be a time-consuming task which in some cases it is even infeasible, due to the nature of the feature selection method. An attempt to address these issues is made using an alternative, more generic approach to feature selection, which is presented in this paper.

3 Methodology

The proposed method performs feature selection in an efficient and automated way (which is why it is called Automated Feature Selector or AFS for short). AFS takes as inputs the feature matrix F (input values in classification terminology), the wear level vector W (target values) and, optionally, a correlation threshold th. The later is there in case the user wishes to obtain only features that meet a certain quality standard (e.g. strong correlation with the wear) and can be particularly useful in cases of very large feature sets.

The AFS method functions as follows. Initially, a dimensionality check is performed, to ensure that both F and W contain the same number of points (N). Once this is attained, the features are individually evaluated to determine how useful they are for the objective at hand. This is done by first calculating the correlation between each feature i (F_i) and W and storing it in a vector C. Then, the standardized Fischer Discriminant Ratio of Fi is calculated as well and stored in vector DR. By the term standardized we mean a variation of FDR such that it yields values in the interval (0, 1], for reasons that will become apparent later on.

Following this stage, a series of similarity matrices for all feature combinations is calculated, in order to determine how related these features are in terms of the information they yield. This step comprises of the computation of three distinct metrics (in the form of three matrices). The metrics used are absolute correlation, similarity (with $\alpha = 1$) [6], and 1 – cosine of angle they form [10]. These metrics were selected so that a significant amount of information about the features can be extracted. Yet, even though this information is more than the information yielded by a single metric, it can be increased if more metrics are added to the list, as long as they all yield values in the interval [0, 1].

After this, a likeliness metric L is calculated by taking the point-to-point product of these matrices and adding I_n to it, where n is the number of features in the original feature set. A (feature) potency metric G is also computed, by taking the point-to-point product of vectors C and DR. By taking the ratio of the aforementioned two composite metrics, L and G, yields a distance-like metric D, which constitutes the basis of the feature selection process that ensues. This metric is in essence the evaluation of all the individual features of the original feature set and its presence reduces the original problem into a relatively simple mathematical problem, which is solved in the steps that follow.

Afterwards, the most correlated feature f is selected and its index is stored in vector *in* (which is the index vector, one of the outputs of the AFS method). This is done because in fault prediction correlation with the wear level plays a vital role for assessing the quality of the condition indicator at hand, so as a starting point one would prefer to have something quite robust. Alternatively, the feature with the highest potency could be selected instead. The inverse of the correlation value of this feature with the wear is stored as the first element of vector X and all "distances" from f to the other features are set to infinite. This is done so that feature f is not selected again. Following that, the "closest" feature to f (let us call it g) is found and its index is stored in *in*. The "distance" between g and f, in terms of the distance-like metric D, is stored in variable m. The average of X_1 and m is stored in X_2 and all "distances" to g are set to infinite (so that feature g is not selected again either). By repeating the above process for all n features a distance-based feature structure is formed in matrix X and vector *in* contains the corresponding indexes.

If a correlation threshold *th* is given, the features from *in* that yield a C value greater or equal to *th* are selected while the rest of the features are discarded. Otherwise, the smallest value of X is found (as well as the corresponding index *ind*) and all features having an *in* value between 1 and *ind* are selected. In essence this means that the feature set is partitioned in two clusters, based on D instead of their spatial distances. One of these clusters contains the "good" features, which are taken as an output of the method, while the other contains the "bad" ones, which are discarded. Alternatively, other methods of clustering could be used to achieve this binary partitioning, but this one was selected for its simplicity and effectiveness in practice.

4 Experiments, Results and Discussion

To test the effectiveness of this method, a set of features extracted from a real-world dataset was used. The features were based on vibration data of axial and radial orientation of the sensors in relation with the bearing examined. The wear levels, four in total including the baseline, corresponded to different corrosion levels of the bearing. Also, the data included two distinct working conditions and were collected over a number of experiment rounds. The feature sets used in this research comprised of 16 features for radial and 16 features for axial data. In both cases they comprised of 240 data points (larger dataset could have been tested by increasing the sampling rate of the features during the extraction process, but this would not change anything in the whole feature selection process, due to the nature of the features in this field). From these feature sets two feature selections were made, one for each set. In the first selection (SF_1) which corresponds to the axial features, six features were selected by the AFS method, while for the radial features, the new feature set (SF_2) comprised of four features. An additional feature selection was carried out afterwards, this time making use of the union of the original two feature sets. This yielded a subset (SF_3) made up of four features (two from each one of the original feature sets).

The experimental setup for the feature selection part is as follows. Using the 240 data points of the dataset, we split them into 10 equal subgroups, using k-fold cross validation, allowing for 216 training points and 24 test points in 10 cross validation groups. These groups are all separately classified across all features within the feature set through a number of classifiers to compare classification performance.

Four classifiers were chosen based on relative speed and accuracy of classification—three of them different flavors of the *k*-Nearest Neighbor algorithm, thoroughly described in [5]. For the variants that made use of a k parameter (number of neighbors), the value of 5 was chosen, since it is one commonly used. The classifiers used are the following:

- C4.5 Decision Tree Classifier — (C45)
- Variable *k*-Nearest Neighbor — (VKNN)
- Weighted *k* -Nearest Neighbor — (WKNN)
- Discernibility *k* -Nearest Neighbor — (DKNN)

All six feature sets are classified separately: reduced feature sets for radial, axial, and both features, and total feature sets for the same. For each classification is computed the classification accuracy, measured as percent correct test point classifications; and the CPU time, measured as the number of seconds per classification round. These will measure the ability of the reduced feature set and classifier to properly diagnose the wear level, as well as how long a classifier requires to train itself for diagnosis. These are computed and averaged across 30 iterations of the above procedure, resulting in a total of 300 classification experiments for each combination of feature set and classifier.

A brief analysis of the features selected from the above subgroups of the original feature set is shown in Table 1.

Table 1. Feature Characteristics. Discernibility refers to the Spherical Index of Discernibility [5], a measure of class distinguishability.

Feature	Correlation with Wear	Fischer Discriminant Ratio	Discernibility (SID)	Comments
# 3, Axial	0.8542	0.9880	0.7292	SF_1, SF_3
# 4, Axial	0.6242	0.8286	0.2833	SF_1, SF_3
# 9, Axial	0.6859	0.8198	0.3667	SF_1
# 12, Axial	0.6468	0.7211	0.2375	SF_1
# 13, Axial	0.8194	0.9733	0.5833	SF_1
# 16, Axial	0.6759	0.7787	0.4792	SF_1
Mean, Axial	*0.6853*	*0.7996*	*0.3352*	*All features*
# 3, Radial	0.7460	0.9662	0.2583	SF_2
# 5, Radial	0.6400	0.9992	0.5375	SF_2
# 8, Radial	0.7082	0.7930	0.2583	SF_2
# 14, Radial	0.7229	0.9016	0.1833	SF_2, SF_3
# 15, Radial	0.6312	0.9993	0.5458	SF_3
Mean, Radial	*0.6871*	*0.8514*	*0.3003*	*All features*

Further analysis in the feature subsets yielded by AFS revealed some interesting information that provides insight to the collaborative potential of the selected features. Specifically, the Spherical Index of Discernibility (SID) [5] was employed, first on the whole feature sets and then on the subset of selected features. As it is expectable, there was a drop in most cases in the SID value when the feature set was reduced, yet this drop was not dramatic, considering the dimensionality reduction ratio (Table 2). Yet, in the case of the combined feature set of 32 features, the reduced feature set yielded a surprising increase in the SID, despite the significant dimensionality reduction ratio exhibited (Table 2).

Table 2. Results of Classification Experiments and Feature Set Evaluation. The Index of Discernibility refers to the Spherical one, introduced and described in [5]. The numbers in bold denote an improvement.

Feature Set	Mean Accuracy	CPU time	SID	Dimensionality Reduction Ratio
Axial	94.3%	0.530s	0.6625	-
Radial	72.0%	0.534s	0.2792	-
Both	85.6%	0.984s	0.2958	-
SF_1 (Axial)	85.5%	**0.222s**	0.5000	62.5%
SF_2 (Radial)	69.9%	**0.152s**	0.2583	75.0%
SF_3 (Both)	**86.7%**	**0.160s**	**0.6208**	87.5%

Figure 1 shows the accuracy rate as a percentage of the number of test cases classified correctly per reduced feature set and for each classifier. The feature sets for radial features and for both features, when classified with a KNN-based classifier, show around 90% accuracy in diagnosing the correct wear level.

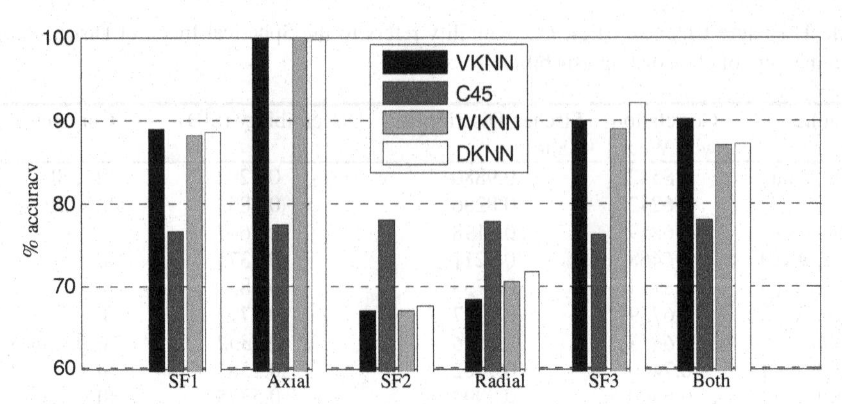

Fig. 1. Accuracy of classifiers on reduced and complete feature sets

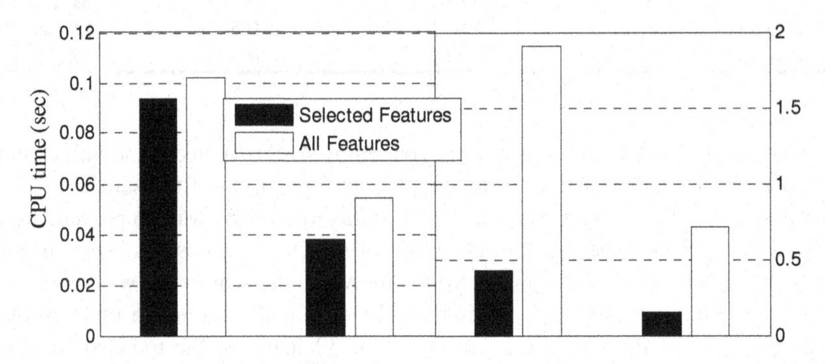

Fig. 2. Average classification CPU time. Classifiers are, from left to right: Variable kNN, Discernibility kNN, C45, and Weighted kNN.

The time it takes each classifier to train on each feature, as a measure of processing time in seconds, is computed and averaged for all reduced feature sets, and all complete feature sets. The CPU time (Figure 2) shows definite improvement in training time for all four classifiers. Particularly, The C45 and Weighted kNN classifiers, which exhibit an almost 4 times speed up in training time.

5 Conclusions and Future Work

Based on the previous analysis, it can be concluded that the proposed method yields robust feature subsets, something that reflects in good accuracy rates for fault diagnosis, using a number of classifiers. As one would expect, the accuracy rates in the reduced feature sets are not as high as in the original feature set, since there is some information loss in the feature selection process. However, in the case of the complete feature set of 32 features not only there is no decrease in the accuracy but the opposite is observed. It is also noteworthy that the reduced feature sets are much smaller than the original ones, something that yields significantly better classification speed (based on CPU time).

Moreover, the Discernibility of the reduced feature sets appears to not drop significantly, while in one case even increase, after the feature selection takes place.

Future work will include alternative measures of (dis)similarity to incorporate in the AFS method. Also, further testing of this method, in different fault feature sets will be conducted to test its generality. Furthermore, it is planned to make use of AFS as a module of a larger system, which will be implemented as an autonomous, fully automated feature generation system taking as inputs preprocessed data and yielding a single fused feature, along with its characteristics, as outputs.

Acknowledgement

The authors of this paper would like to acknowledge the sponsor of the AVDPIP project, from which the feature data used in this research stems. Also, they would like to thank the rest of the research team involved in AVDPIP, in both the Georgia Institute of Technology and in Impact Technologies.

References

1. Liu, H., Motoda, H.: Computational Methods of Feature Selection. Chapman & Hall/CRC, Boca Raton (2008)
2. Liu, X.: Machinery Fault Diagnostics Based on Fuzzy Measure and Fuzzy Integral Data Fusion Techniques. PhD thesis, Queensland University of Technology (2007)
3. Liu, H., Motoda, H. (eds.): Feature Extraction, Construction and Selection – A Data Mining Perspective. Kluwer Academic Publishers, USA (1998)
4. Theodoridis, S., Koutroumbas, K.: Pattern Recognition. Academic Press, USA (1990)
5. Voulgaris, Z.N.: Discernibility Concept in Classification Problems. PhD thesis, University of London (2009)
6. Vachtsevanos, G., Lewis, F.L., Roemer, M., Hess, A., Wu, B.: Intelligent Fault Diagnosis and Prognosis for Engineering Systems. John Wiley & Sons, Inc., Hoboken (2006)
7. Sun, R., Tsung, F., Qu, L.: Combining Bootstrap and Genetic Programming for Feature Discovery in Diesel Engine Diagnosis. International Journal of Industrial Engineering 11(3), 273–281 (2004)
8. Lee, S.: An Architecture for a Diagnostic/Prognostic System with Rough Set Feature Selection and Diagnostic Decision Fusion Capabilities. PhD thesis, Georgia Institute of Technology (2002)
9. Saxena, A., Vachtsevanos, G.: Optimum Feature Selection and Extraction for Fault Diagnosis and Prognosis. In: Proceedings of the 2007 AAAI Fall Symposium on Artificial Intelligence for Prognostics, Arlington, VA (2007)
10. Pekalska, E., Duin, R.P.W.: The Dissimilarity Representation for Pattern Recognition Foundations and Applications. World Scientific, Singapore (2005)

Dimensionality Reduction for Distance Based Video Clustering

Jayaraman J. Thiagarajan, Karthikeyan N. Ramamurthy, and Andreas Spanias

SenSIP Center, School of ECEE, Arizona State University,
Tempe, AZ 85287-5706, USA
{jjayaram,knatesan,spanias}@asu.edu

Abstract. Clustering of video sequences is essential in order to perform video summarization. Because of the high spatial and temporal dimensions of the video data, dimensionality reduction becomes imperative before performing Euclidean distance based clustering. In this paper, we present non-adaptive dimensionality reduction approaches using random projections on the video data. Assuming the data to be a realization from a mixture of Gaussian distributions allows for further reduction in dimensionality using random projections. The performance and computational complexity of the K-means and the K-hyperline clustering algorithms are evaluated with the reduced dimensional data. Results show that random projections with an assumption of Gaussian mixtures provides the smallest number of dimensions, which leads to very low computational complexity in clustering.

Keywords: Clustering, Random projections, Gaussian mixtures, Video summarization.

1 Introduction

Classification of data is an important problem in machine learning, where data sets are separated into several disjoint classes based on predefined criteria. The predefined criteria, referred as the hypothesis, can be supplied by the user or learned by the machine itself from classes of labeled training samples. In supervised learning, hypotheses of multiple classes are learned from a set of labeled training data for each class [1]. Clustering is a more general problem in machine learning where the observed unlabeled data need to be grouped into different clusters. A cluster is a group of similar data where the similarity is quantified based on a well-defined measure. A useful similarity measure for clustering is the Euclidean distance measure and clustering based on Euclidean distance is an NP-hard problem [2].

Clustering of video frames is more than just a generalization of clustering of images. This is because the video frames that convey meaning as a group are both statistically and semantically related. One of the popular approaches to video clustering involves extracting the keyframes by shot boundary detection and clustering the keyframes together to derive a semantic interpretation [3]. However, it is

H. Papadopoulos, A.S. Andreou, and M. Bramer (Eds.): AIAI 2010, IFIP AICT 339, pp. 270–277, 2010.

important to understand that extraction of the keyframes by detecting the shot boundaries itself is a fundamental clustering problem which we address in this paper. Video frames of a single shot have similar background structure and they can be clustered together using color histograms or distance measures.

In this paper, we address the problem of clustering high dimensional long video sequences. In general, this kind of video clustering involves grouping the frames with similar background together for the purpose of extracting keyframes. Using the fact that the video frames that have similar backgrounds are close together in terms of the Euclidean distance measure (l_2 norm), we perform distance based clustering. It is important to clarify the notion of background in this problem. Background is the region in a frame that remains relatively motionless. Even if some objects in the foreground are relatively motionless they can be treated as background. The very high spatial and temporal dimensionality of the video data makes l_2 norm based clustering intractable in the absence of tremendous computational power. Therefore, it becomes essential to reduce the dimensionality of the video data in order to perform clustering with low complexity. This problem is highly significant in scenarios where fast summarization of video needs to be performed at a reduced computational cost. In this paper, we propose a framework for dimensionality reduction to cluster video frames having similar background structure. The framework is based on non-adaptive dimensionality reduction using the theory of random projections [4] and assumption of Gaussian mixture (GM) models for data.

2 K-Means and K-Hyperline Clustering

The K-means clustering problem seeks to cluster the T data samples into K clusters by minimizing the sum of intracluster distances across all clusters. It converges to a locally optimal solution closest to the initial values. This is a 2-step alternating minimization problem where the samples are associated to the cluster centroids in the first step and the centroids are recalculated in the second step using the associated member samples. The member sample is associated to a cluster centroid that is closest in terms of the Euclidean distance measure. The centroid that minimizes the sum of distances to all its member samples is computed by solving,

$$\bar{\mathbf{x}}_j = \min_{\mathbf{r}} \sum_{i \in \Lambda_j} \|\mathbf{x}_i - \mathbf{r}\|^2, \tag{1}$$

where $\bar{\mathbf{x}}_j$ is the cluster centroid and Λ_j is the index set containing the memberships of the jth cluster. The solution obtained for $\bar{\mathbf{x}}_j$ is the mean of member samples.

K-hyperline clustering seeks to compute a rank-1 subspace using Singular Value Decomposition (SVD) for each cluster that minimizes the sum of distance of the member data to the subspace [5]. K-hyperline clustering is more accurate and general than K-means clustering in that the minimization problem yields lesser sum of distances than the K-means clustering and it can easily generalize to higher rank subspaces. The association rule for the member sample for the nearest rank-1 subspace is based on a maximum correlation measure. In K-hyperline clustering the rank-1 subspace is of unit norm, whereas no such constraint is imposed in K-means clustering on the cluster centroid.

3 Random Projections

Consider a high dimensional data matrix, $\mathbf{X}$, with dimensions $M \times T$, where each column represents a single data observation. In order to project this onto a random low dimensional space, we define a matrix $\mathbf{R}$ of dimensions $M \times N$ with $N < M$, whose entries are chosen independently from the standard normal distribution $\mathcal{N}(0, 1)$. The Random Projection (RP) of the data vectors is,

$$\mathbf{Y} = \frac{1}{\sqrt{N}} \mathbf{R}^T \mathbf{X}. \tag{2}$$

RP reduces the dimensionality of the data from M to N while approximately preserving pair-wise distances with high probability [4]. This is formalized by the *Johnson-Lindenstrauss* (JL) lemma, which states that for a large enough N ($N \geq C \frac{\ln T}{\epsilon^2}$) (3) holds with high probability.

$$(1 - \epsilon) \|\mathbf{y}_i\|^2 \leq \frac{1}{N} \|\mathbf{R}^T \mathbf{x}_i\|^2 \leq (1 + \epsilon) \|\mathbf{y}_i\|^2, \tag{3}$$

where $\mathbf{y}_i$ and $\mathbf{x}_i$ represent the columns of the matrices $\mathbf{Y}$ and $\mathbf{X}$ respectively and $0 < \epsilon < 1$. It is important to note that the JL lemma does not depend on the actual dimensionality of the data and depends only on the number of data vectors. The K-means clustering defines the cluster centroid as the mean of data vectors in a cluster. It is easy to observe from JL lemma that the distances between the cluster centroid and the data vectors will be approximately preserved even after random projections. Hence we can use the JL lemma to reduce the dimensionality of the data matrix for use in K-means clustering.

3.1 Computation of SVD Using Random Projections

The computation of SVD can also be performed using the reduced dimensional matrix from random projections [4]. It can be shown that for the same low rank approximations of $\mathbf{Y}$ and $\mathbf{X}$, the Frobenius norms will be approximately preserved with high probability. This can be mathematically expressed as,

$$\sum_{i=1}^{s} \lambda_i^2 \geq (1 - \epsilon) \sum_{i=1}^{s} \sigma_i^2, \tag{4}$$

where λ_i and σ_i are the singular values of $\mathbf{Y}$ and $\mathbf{X}$ respectively and s is the desired rank of the approximation [4]. In particular, considering a rank-1 approximation, it can be shown that, with high probability [4,6]

$$(1 - \epsilon)\sigma_1^2 \leq \lambda_1^2 \leq \sigma_1^2. \tag{5}$$

The existence of such upper and lower bounds motivates the use of K-hyperline clustering on $\mathbf{Y}$ instead of $\mathbf{X}$.

4 Gaussian Mixture Models for Clustering

Statistical clustering algorithms assume that the data is a realization from a mixture of probability distributions. In the case of mixture of K arbitrary distributions, the overall probability density is given by,

$$f = \sum_{i=1}^{K} w_i f_i \quad \text{s.t.} \quad w_i \geq 0 \quad \text{and} \quad \sum_{i=1}^{K} w_i = 1, \tag{6}$$

where f_i is the probability distribution function (pdf) and w_i is the non-negative weight of the ith distribution.

The best SVD subspace for a spherical Gaussian distribution is any subspace through its mean [2]. More importantly, the best K-dimensional SVD subspace for a mixture of K Gaussians whose covariance is a scalar multiple of identity, contains the span of the means of the component distributions. This can also be extended to a mixture of arbitrary distributions. In general, we do not have the exact statistics of a GM and we have only the samples of realizations. The covariance matrix is also not a scalar multiple of identity. Even under these conditions, it has been proved that the SVD subspace of the sample matrix is not far from the subspace spanned by the actual component means [2].

4.1 Separation between Spherical Gaussians

Two spherical Gaussians $\mathcal{N}(\mu_1, \sigma_1^2\mathbf{I})$ and $\mathcal{N}(\mu_2, \sigma_2^2\mathbf{I})$ are considered to be c-separated if $\|\mu_1 - \mu_2\|_2^2 \geq c^2 M \max(\sigma_1^2, \sigma_2^2)$, where M is the dimension of the Gaussian [6]. A 2-separated mixture corresponds to almost completely separated Gaussians, whereas a 1- or 1/2-separated mixture contains Gaussians which overlap significantly. By projecting the Gaussian mixtures on to a K-dimensional subspace spanned by the means of K Gaussians, we are equivalently projecting the Gaussian mixtures onto their best rank-K SVD subspace. This preserves the distance between the means (intercluster distance), whereas the intracluster distance reduces drastically [2]. Therefore, the separation between the Gaussians in the mixture increases and the clustering performance improves. Similar results can also be shown for mixtures of Gaussians with arbitrary covariances [2].

5 Proposed Clustering Framework

In this paper, we use both the K-means and the K-hyperline clustering algorithms for clustering the video data. We consider four different approaches: a) basic K-means/K-hyperline clustering on the high dimensional data, b) reducing the dimensions of the data using random projections prior to clustering, c) reducing the dimensions of the data assuming it as a mixture of Gaussians prior to clustering and d) reducing the dimensions, first using RP and then under the mixture of Gaussians assumption, prior to clustering. Both centroid and left singular vector of a group of video frames retrieve the background information effectively. This motivates the use of both K-means/K-hyperline clustering in our approaches. We will assume that we have K clusters of the T video frames and we vectorize each video frame into a M dimensional vector thereby generating the $M \times T$ matrix $\mathbf{X}$. The K index sets of the clusters are given by $\{\Lambda_i\}_{i=1}^{K}$ and $T_i = |\Lambda_i|$. In the remaining part of this section, we describe the different approaches for clustering.

5.1 Random Projection Based Clustering

The RP method can be used to reduce the dimensionality of the data matrix $\mathbf{X}$ according to (2), preserving the length of the data vectors, pairwise distances and angles with high probability. We have also seen in Section 3 that the centroid and SVD of a set of data vectors are approximately preserved with a high probability, given a sufficiently large number of measurements N.

Assuming that $K = 1$, the centroid and the first singular vector of $\mathbf{Y}$ are approximately equal to that of $\mathbf{X}$ for a sufficiently large N. If $K > 1$, the centroid and first singular vector of each cluster will still be approximately preserved because $T_i < T$. Therefore, the RP method will be useful regardless of the number of clusters, provided we choose N based on the assumption of single cluster. The linear increase in the number of data vectors T will not change N significantly because $N \propto \ln T$. Therefore, in order to perform RP based clustering, we use either the K-means or the K-hyperline clustering algorithm on the reduced dimensional data.

5.2 Gaussian Mixture Based Clustering

We know from Section 4 that the rank-K SVD subspace of the sample matrix is not far from the space spanned by the K component means even when the Gaussians are not spherical. In this approach for clustering, we assume that $\mathbf{X}$ contains realizations from a mixture of K Gaussians, not necessarily spherical, where each Gaussian represents a cluster. We compute the best rank-K subspace of $\mathbf{X}$ using SVD and denote the basis vectors of the rank-K subspace (first K left singular vectors of $\mathbf{X}$) by $\mathbf{U}_K$. The projection to the rank-K subspace is given by,

$$\mathbf{W} = \mathbf{U}_K^T \mathbf{X}, \tag{7}$$

where $\mathbf{W}$ contains the K dimensional data vectors after projection. Because of the reasoning provided in Section 4.1, the clusters in the K dimensional space are more separated than the clusters in the M dimensional space. Therefore, we perform K-means or K-hyperline clustering on $\mathbf{W}$ and identify the index sets of the clusters.

5.3 Random Projection and Gaussian Mixture Based Clustering

Similar to the previous case, in this approach we assume $\mathbf{X}$ to be a set of T realizations of K Gaussians. Furthermore, as RP approximately preserves the pairwise distances of the samples, realizations from a mixture of K Gaussians in M dimensions preserve their structure in N dimensions. This fact is used to further reduce the computational complexity of the framework.

In this approach, we first project $\mathbf{X}$ to obtain $\mathbf{Y}$ according to (2). The elements of $\mathbf{Y}$ are treated as realizations of a mixture of K Gaussians in N dimensions. From the arguments in Section 5.2, we project $\mathbf{Y}$ onto a K dimensional SVD subspace to obtain $\mathbf{Z}$. K-means or K-hyperline clustering can be performed on $\mathbf{Z}$ to identify the index sets of the clusters. Note that in this case, we first perform an initial level of dimensionality reduction using RP, which aids in a faster computation of the SVD subspace. In the second stage, we reduce the dimensionality further using the GM assumption. Hence, this approach combines the advantages of both the previous approaches in terms of a much reduced computational complexity due to RP and improved clustering performance along with further reduction in computational complexity due to the assumption of Gaussian mixtures. The outline of the algorithm to perform video clustering and identify the keyframes is shown in Table 1.

To improve the clustering performance, we adopt a two stage approach to clustering. This reduces the possibility of the clustering algorithm being stuck in a local minima. Initially, we solve the clustering problem for a number of clusters J that is larger than the actual number K. Then, we greedily combine the columns of the cluster centroid matrix $\mathbf{B}$ in order to obtain the matrix

Table 1. Algorithm to cluster video data and identify keyframes

Goal: To perform clustering of high-dimensional long video sequences using the approach given in Section 5.3

Variables

High-dimensional data matrix, $\mathbf{X}$ of size $M \times T$.
Intermediate data matrix after RP, $\mathbf{Y}$ of size $N \times T$.
Final data matrix used for clustering, $\mathbf{Z}$ of size $K \times T$.
Initial number of clusters, J.
Actual number of clusters, K.
Cluster centroid matrix (J Clusters), $\mathbf{B}$ of size $K \times J$.
Cluster centroid matrix (K Clusters), $\mathbf{A}$ of size $K \times K$.
Index set for the ith cluster, $\mathbf{\Lambda}_i$.
Gaussian i.i.d. random matrix, $\mathbf{R}$ of size $M \times N$.

Algorithm

1. Compute the RP, $\mathbf{Y} = (1/\sqrt{N})\mathbf{R}^T\mathbf{X}$.
2. Compute rank-K SVD, $[\mathbf{U}_K, \mathbf{S}_K, \mathbf{V}_K] = \text{SVD}(\mathbf{Y}, K)$.
3. Project on to the K-dimensional SVD space, $\mathbf{Z} = \mathbf{U}_K^T\mathbf{Y}$.
4. Initialize $\mathbf{B}$ with J randomly chosen columns of $\mathbf{Z}$.
5. Perform K-means/K-hyperline clustering for J clusters.
6. Using greedy combinations of columns of $\mathbf{B}$, create $\mathbf{A}$.
7. Perform K-means/K-hyperline clustering for K clusters using $\mathbf{A}$ as initial centroids.
8. Using the index sets $\mathbf{\Lambda}_i$ obtained from clustering, identify the keyframes.

A (step 6 of Table 1). In this greedy combination method we first choose the two most similar vectors of **B** and combine them. We repeat this procedure for two columns at a time until we are left with only K columns. In the K-means method two most similar vectors are the ones that have the minimum pairwise Euclidean distance and the combined vector is the mean of the two. In K-hyperline clustering two most similar vectors are the ones that have the maximum correlation and the combined vector is principal left singular vector of the matrix of the two vectors.

6 Experimental Results

The video sequences in QCIF format, used for evaluating the performance of the algorithms, were obtained from [7] and the spatial resolution was changed to 128×128. The first test data set was generated by stitching 10 different video sequences and it contains 1900 frames in total. The second test data set has a total of 550 frames obtained from 3 different video sequences. The initial number of clusters J is set to 3 times the actual number of clusters K. For the first data set $K = 10$ and for the second data set $K = 3$. The keyframes are identified using all the four approaches for both K-means and K-hyperline clustering. The keyframes obtained for the first data set are shown in Figure 1. The keyframes identified are similar with the all the four approaches and we also obtain 100% clustering performance.

The running times for the different approaches in MATLAB (version R2007b) to cluster the test data are listed in Table 2. It can be seen that the approach based on RP and GM is of least computational complexity. The running time for the K-means and the K-hyperline clustering algorithm are close to each other except for the case of the basic approach, which however is not the choice when clustering high dimensional data.

Fig. 1. Keyframes obtained by clustering the test data using the algorithm in Table 1

Table 2. Comparison of running time for the different clustering approaches in MATLAB. Wherever applicable, the running times include dimensionality reduction phase also. The results for the first and the second data sets are separated by a slash (/). The third approach could not run for the first data set owing to memory issues because of high dimensionality.

Approach	Running time(s)		Number of Dimensions
	K-means	Hyperline	
Basic	696.51/37.69	774.23/102.22	16384/16384
RP	33.87/7.59	32.17/10.15	400/400
GM	-/10.18	-/10.30	10/3
RP and GM	29.41/7.36	26.84/7.45	10/3

7 Conclusions

In this paper, we proposed different approaches for dimensionality reduction based on random projections in order to cluster video data. These approaches provide a practical solution to clustering video frames with similar background for fast video summarization. Incorporation of outlier rejection and compensating for global motion between the video frames are possible extensions to the proposed dimensionality reduction approaches for robust clustering.

References

1. Alpaydin, E.: Introduction to Machine Learning. In: Adaptive Computation and Machine Learning. The MIT Press, Cambridge (2004)
2. Kannan, R., Vempala, S.S.: Spectral Algorithms. In: Foundations and Trends in Theoretical Computer Science. Now Publishers Inc. (2009)
3. Vailaya, A., Jain, A.K., Zhang, H.: Video Clustering. Technical report, Michigan State University (1996)
4. Vempala, S.S.: The Random Projection Method. Discrete Mathematics and Theoretical Computer Science. American Mathematical Society, Providence (2004)
5. He, Z., Cichoki, A., Li, Y., Xie, S., Sanei, S.: K-hyperline clustering learning for sparse component analysis. Signal Processing 89, 1011–1022 (2009)
6. Dasgupta, S.: Learning Mixtures of Gaussians. In: Proceedings of the 40th Annual Symposium on Foundations of Computer Science, Washington, DC, USA, pp. 634–644 (1999)
7. YUV Video Sequences, `http://trace.eas.asu.edu/yuv/index.html`

Towards Stock Market Data Mining Using Enriched Random Forests from Textual Resources and Technical Indicators

Manolis Maragoudakis[1,2], and Dimitrios Serpanos[2,3]

[1] Department of Information and Communication Systems Engineering, University of Aegean, Samos, 82000, Greece
[2] I.S.I. - Industrial Systems Institute Patras Science Park building Platani, PATRAS, Greece, 26504
[3] Department of Electrical and Computer Engineering, University of Patras, Rion, 26500, Greece
mmarag@aegean.gr, serpanos@ece.upatras.gr

Abstract. The present paper deals with a special Random Forest Data Mining technique, designed to alleviate the significant issue of high dimensionality in volatile and complex domains, such as stock market prediction. Since it has been widely acceptable that media affect the behavior of investors, information from both technical analysis as well as textual data from various on-line financial news resources are considered. Different experiments are carried out to evaluate different aspects of the problem, returning satisfactory results. The results show that the trading strategies guided by the proposed data mining approach generate higher profits than the buy-and-hold strategy, as well as those guided by the level-estimation based forecasts of standard linear regression models and other machine learning classifiers such as Support Vector Machines, ordinary Random Forests and Neural Networks.

Keywords: Stock return forecasting; Data mining; Expert systems; Random forests; Markov blanket; Trading strategies.

1 Introduction

Stock market prediction has always gained certain attention from researchers. There is a controversy as regards to whether there is a method for accurate prediction of stock market movement, mainly due to the fact that modeling market dynamics is a complex and volatile domain. Stock market research encapsulates two main philosophical attitudes, i.e. fundamental and technical approaches [1]. The former states that stock market movement of prices derives from a security's relative data. Fundamentalists are of the belief that numeric information such as earnings, ratios, and management effectiveness could determine future forecasts. In technical analysis, it is believed that market timing is the key. Technicians utilize charts and modeling techniques to identify trends in price and volume. These latter individuals rely on historical data in order to predict future outcomes. However, according to several researchers, the goal is not to question the predictability of financial time series but to discover a good model that is capable of describing the dynamics of stock market.

H. Papadopoulos, A.S. Andreou, and M. Bramer (Eds.): AIAI 2010, IFIP AICT 339, pp. 278–286, 2010.

There is a plethora of proposed methods in stock market prediction. The majority of them are strongly related to structured, numerical databases and domain expertise rules. In the field of trading, most of decision support tools focus on statistical analysis of past price records. Nevertheless, throughout recent studies, prediction is also based on textual data, based on the rational assumption that the course of a stock price can be influenced by news articles, ranging from companies releases and local politics to news of superpower economy [2].

However, unrestricted access to news information was not possible until the early 1990's. Nowadays, news are easily accessible, access to important data such as inside company information is relatively cheap and estimations emerge from a vast pool of economists, statisticians, journalists, etc., through the World Wide Web. Despite the large amount of data, advances in Natural Language Processing and Knowledge Discovery from Data (also known as Data Mining) allow for effective computerized representation of unstructured document collections, analysis for pattern extraction and discovery of relationships between document terms and time-stamped data streams of stock market quotes.

Nevertheless, when data tend to grow both in number of records and features, numerous mining algorithms face significant complications, resulting in poor prediction ability. The aim of this study is to propose a potential solution to the problem, by considering the well-known algorithm of Random Forests [3] and altering their construction phase by utilizing a Markov Blanket approach which discards irrelevant features, thus improving classification results. The importance of this study lies to the fact that technical analysis contains the event and not the cause of the change, while textual data may interpret that cause. Certainly, as it is tedious for a human investor to read all daily news concerning a company and other financial information, a prediction system that could analyze such textual resources and find relationships with price movement at future time windows is beneficial.

The paper is structured as follows: section 2 provides an overview of literature concerning Stock Market prediction using Data Mining techniques. Section 3 describes the proposed Markov Blanket Random Forest utilization. Section 4 provides an overview of our experimental design and discusses the evaluation outcome.

2 Previous Work

Due to numerous studies in traditional technical analysis, we shall emphasize on researches that study the influence of news articles on stock markets. Chang et al., [4] were among the first to confirm the reaction of the market to news article. They had shown that economic news always has a positive or negative effect in the number of traded stock. They used salient political and economic news as proxy for public information. Klibannof et al., [5] deal with closed-end country fund's prices and country specific salient news. They stated that there is a positive relationship between trading volume and news. Similar to the aforementioned approach, Chan and Wei [6] founded that news that is placed in the front page of the *South China Post* increase the return volatility in the Hong Kong stock market. Mitchell and Mulherin [7] used the daily number of headlines of Dow Jones as a measure of public information. They mentioned the positive impact of news on absolute price changes. Mittermayer [8] proposed a prediction system called NEWSCATS, which provides an estimate of the price after the publication of press releases. Schumaker and Chen [9] examined three different textual representation formalisms and studied their abilities to predict discrete stock prices 20 minutes after an article release.

3 Markov Blanket Random Forests

A problem arises when the number of possible features is vast and the percentage of actually informative features is small, i.e. the performance of the base classifiers degrades. This phenomenon is particularly present in financial data sets, where most attributes represent technical indicators with little or unknown certainty about their correlation to the true course of a stock. Technically, in the case of a Random Forest classifier, this problem arises due to the fact that, if simple random sampling is used for selecting the subset of m eligible features at each node, almost all these subsets are likely to contain a predominance of non-informative features.

The solution proposed in this paper is based on the notion of a feature selection and reasoning algorithm, i.e. the Markov Blanket of the class attribute. The identification of relevant variables is an essential component of construction of decision support models, and computer-assisted discovery. In financial decision systems for example, such as the task at hand, elimination of redundant features could increase the computational performance significantly. The problem of variable selection in financial domains is more pressing than ever, due to the recent emergence of many news portals, on-line financial services, etc. Similar cases are also common in biomedical engineering, computational biology, text categorization, information retrieval, mining of electronic medical records, consumer profile analysis, temporal modelling, and other domains [10]. Several researchers [11] have suggested, intuitively, that the Markov Blanket (MB) of the target variable t, denoted as $MB(t)$, is a key concept for solving the variable selection problem. $MB(t)$ is defined as the set of variables conditioned on which all other variables are probabilistically independent of t. Thus, knowledge of the values of the Markov Blanket variables should render all other variables superfluous for classifying t.

3.1 Bayesian Networks and Markov Blanket

In order to better capture the significant properties of a Markov Blanket, a brief introductory section of Bayesian networks is included. Bayesian networks graphically represent the joint probability distribution of a set of random variables. A Bayesian network is composed of a qualitative portion (its structure) and a quantitative portion (its conditional probabilities). The structure BS is a directed acyclic graph where the nodes correspond to domain variables $x_1,\ldots, x_n$ and the arcs between nodes represent direct dependencies between the variables. Likewise, the absence of an arc between two nodes x_i and x_j represents that x_j is independent of x_i given its parents in BS. Following the notation of Cooper and Herskovits [12], the set of parents of a node x_i in BS is denoted π_i. The structure is annotated with a set of conditional probabilities (BP), containing a term $P(x_i=X_i|\pi_i=\Pi_i)$ for each possible value X_i of feature x_i and each possible instantiation Π_i of π_i. A Markov Blanket of a node x_i, denoted as $MB(x_i)$, is a minimal attribute set, such that every other attribute is independent of x_i given its Markov Blanket. Mathematically, the above statement is translated into:

$$\forall x_i \in \{x_1,...,x_n\} \setminus MB(x_i) \cup \{x_i\}, x_i \vdots x_k \mid MB(x_i), \quad (1)$$

where $\vdots$ denotes the conditional independence of x_i with x_k given $MB(x_i)$.

Suppose B_i and B_j are two Bayesian networks that have the same probability distribution, then $MB_{Bi}(x_k) = MB_{Bj}(x_k)$ for any variable x_k. Certainly, MBs are not exclusive and may vary in size, but any given BN has a unique $MB(x_i)$ for any x_i, which is the set of parents, children and parents of children of x_i. In Fig. 1, a Bayesian network is depicted along with the Markov Blanket of a target node x, colored in blue. As regards to the dataset interpretation, feature x is independent of all other features given its $MB(x)=\{U_i, U_j, Y_k, Y_l, Z_{km}, Z_{ln}\}$.

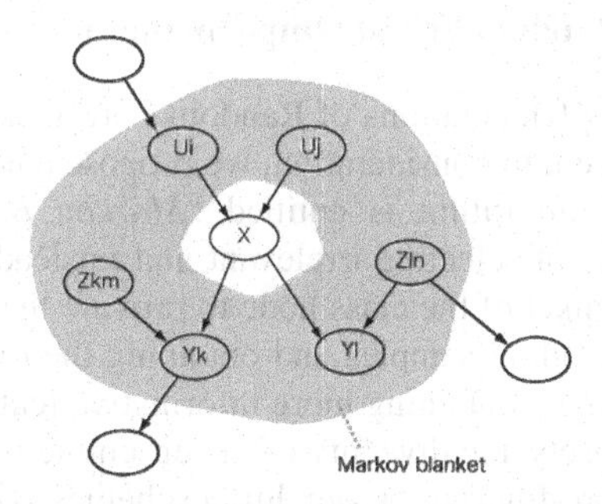

Fig. 1. An example of a Bayesian Network with the Markov Blanket of node x

3.2 Random Forests

Random Forests, in general, are a combination of decision tree classifiers such that each tree depends on the values of a random vector sampled independently and with the same distribution for all trees in the forest. Given a training set X comprised of N instances, which belong to two classes, and F features, a Random Forest multi-way classifier $\Theta(x)$ consists of a number of decision trees, with each tree grown using some form of randomization, where x is an input instance. The leaf nodes of each tree are labelled by estimates of the posterior distribution over the data class labels [13]. Each internal node contains a test that best splits the space of data to be classified. A new, unseen instance is classified by sending it down every tree and aggregating the reached leaf distributions. The process is described in Fig. 2. Each tree is grown as follows:

- If the number of cases in the training set is N, sample N cases at random but with replacement, from the original data. This sample will be the training set for growing the tree.
- If there are F input features, a number $m \ll F$ is specified such that at each node, m variables are selected at random out of the F and the best split on these m is used to split the node. The value of m is held constant during the forest growing.
- Each tree is grown to the largest extent possible. Therefore, no pruning procedures are applied.

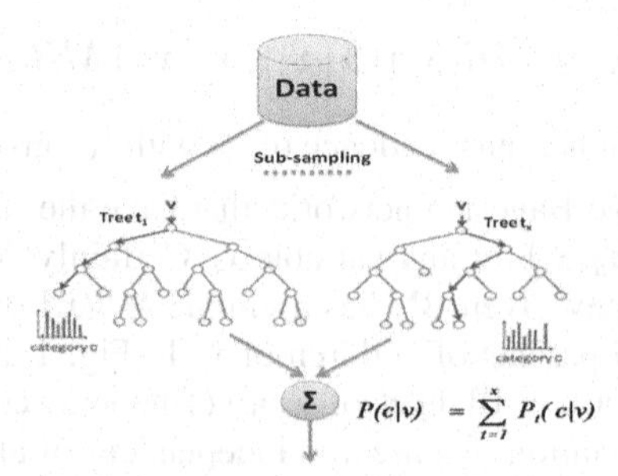

Fig. 2. Hierarchical decomposition of a Random Forests classifier on a data set

3.3 Markov Blanket Random Forests Implementation

Based on the existing implementations of Random Forests and taking our initial concerns on feature relevance into consideration, we propose a novel algorithm for classification using RF. The algorithm is entitled "*Markov Blanket Random Forests-MBRF*", since the danger of selecting irrelevant and misleading features is remedied by using the Markov Blanket of the class node to provide the best splitting criteria for each tree. By selecting random samples and obtaining the extracted *MB* of the target node, the probability of tree containing more informative features is increased. In case of high-dimensional datasets, the diversity of the ensemble is not compromised and is more robust that other, pre-filtering or weighting schemes. The algorithm is consisted of two distinct phases; the former regards the construction of the Markov Blanket and the latter deals with constructing the trees. Its basic procedure can be sketched in the following phases:

MBRF (Data D, Features F, n_{tree} trees, Target C)

1. Draw n_{tree} bootstrap samples from the original data D.
2. Build an unconstrained Bayesian network without learning the conditional probability table.
3. Obtain the MB of the class node C.
4. For each of the bootstrap samples, grow an un-pruned classification or regression tree, with the following modification: at each node, rather than choosing the best split among all predictors, use m_{try} of the Markov Blanket and choose the best split from among those variables.
5. Predict new data by aggregating the predictions of the n_{tree} trees (i.e., majority votes for classification, average for regression).

4 Experimental Design and Evaluation

As mentioned earlier, articles containing financial news were combined with a plethora of technical indices in order to search for direct influence patterns of the former to the latter. More specifically, we focused on three heterogeneous stock securities from the Greek stock market (*Athens Stock Exchange, .ATG*), a major Greek bank (*Piraeus Bank, .TPEIR*), the main telecommunication provider of Greece (*OTE, .OTE*) and one of the biggest Greek airline companies (*Aegean, .AEGN*). We incorporated past data from the major European, Asian and American stock markets, as well as data from energy and metal commodities. Finally, for each of the aforementioned three stock securities, a variety of major technical indices was utilized. News was automatically extracted from the electronic versions of the leading Greek financial newspapers, i.e. "*Naftemporiki*" (www.naftemporiki.gr) and "*Capital*" (www.capital.gr). The time period for all collected data was from November 2007 to January 2010. The technical indices were calculated using the *AnalyzerXL* tool. Table 1 tabulates data regarding the three benchmark stocks and their corresponding articles that were collected, while Table 2 contains data about historical data of other, main markets and commodities. Finally, Fig. 3 depicts a categorized list of the technical indices that were also taken into consideration.

Stock quotes are gathered on a per day basis and articles are aligned accoriding to their release date. In case an article was pubished on a Friday evening (after the closing of the Athens stock market) or during the weekend, it was considered as published on a Monday. The textual analysis phase consisted of three activities:

Table 1. The benchmark tickers

Name	Category	#Articles	#Days	Symbol
O.T.E.	Telephony	1779	519	.OTE
Bank of Piraeus	Bank	1607	514	.TPEIR
Aegean Airlines	Airline	308	547	.AEGN

Table 2. Market and commodities data

Category	Description	#Days	Symbol
European Markets	^FCHI-CAC 40 Index	542	.FCHI
	^FTSE-FTSE 100 Index	569	.FTSE
	^GDAXI-Xetra Dax Index	537	.GDAXI
	^ATG-Athens Stock Exchange	511	.ATG
Asia/Pasific Markets	^HIS-Hang Seng Index	556	.HSI
	^AORD-All Ordinaries Index	569	.AORD
	^N225-Nikkei 225 Average Index	556	.N225
United States Markets	^GSPC-S&P 500 Index	558	.GSPC
	^IXIC-Nasdaq Composite Index	531	.IXIC
	^DJI-Dow Jones Industrial Average Index	539	.DJI
Energy	Brent DTD	566	BRT-
	WTI CUSHING	521	WTC-
Metals	Silver	537	XAG-HH
	Gold Bullion	537	XAU-B-HH

Group Name	Function or Indicator Name
Basic Functions	Median Price (AKA Typical Price Indicator)
Statistical Functions	Standard Deviation
Trend Indicators	MACD Indicator
	Simple Moving Average
	Exponential Moving Average
	Line Weighted Moving Average
Volatility Indicators	Average True Range
	Bollinger Band Width
Momentum Indicators	Williams %R
	TRIX Indicator
	Wilder RSI Indicator
	Chande Momentum Oscillator
	Price Rate-Of-Charge Indicator
	Cutler's Relative Strength Index
	DX (Directional Movement Indicator)
	Stochastic Oscillator
	Price Oscillator Percentage Difference
Market Strength Indicators	Chaikin A/D Oscillator
	Average of Volume ROC
	Market Facilitation Index (MFI)
Support and Resistance Indicators	Envelope

Fig. 3. The technical indices considered

(a) removal of stop words (i.e. articles, special characters, etc), (b) lemmatization of words using a Levenshtein distance based Greek lemmatizer [14], (c) removal of terms appearing less than 30 times within the complete article corpus and taking the 150 most frequent of them. Upon completion of the aforementioned phases, we kindly asked a domain expert (financial journalist) to annotate terms according to their genre. More specifically, she annotated each word with a signed integer according to whether it encompasses a very positive (+2), positive(1), neutral(0), negative(-1) of very negative(-2) sense. Examples of such terms respectively are: *κερδοφορία* (*profitability*, +2), *ισχυρή* (*powerful*, +1), *πορεία* (*course*, 0), *υποχώρηση* (*downgrading*, -1) and *κρίση* (*crisis*, -2). The predicted class attribute contained three discrete values, namely *UP*, *STEADY* and *DOWN*, if the stock quote closed at a price more than 1%, between 1% and -1% and less than -1% in the following day respectively. A window of 5 days was used in order to predict the class, resulting in a high-dimensional dataset of more than 620 features. Article as well as stock quotes data was processed by our proposed methodology (MBRF), regular Random Forests (RF), Radial Basis Functions neural networks (RBF) and a derivative of Support Vector Machines, namely Sequencial Minimal Optimization (SMO) which can handle discrete values and acts similar to regression. Since the latter Machine Learning algorithms do not reduce features by default, in order to compare the MBRF technique against them, a PCA analysis approach was followed using the Nmath library for .NET platforms (http://www.centerspace.net/products/nmath).

Regarding the experimental design, two different approaches were followed. The former dealt with standard, 10 fold cross validation, classification in terms of stock quotes closeness, using datasets with articles and without articles, in order to evaluate the impact of articles on the predictability of a stock quote. We used the *F-measure* metric for evaluation, which acts as the harmonic mean of precision and recall. Table 3 tabularizes the *F-measure* score of all machine learining algorithms against linear regression (LR). From these outcomes, we could initially observe that combining information from both time series and textual data leads to improvement of the performance for all methodologies. Furthermore, by using only technical analysis data, SMO perform similar to MBRF and significantly outperform all other approaches, while when incorporating textual information, MBRF is noticeably the best classification approach, a fact that could be attributed to the dimensionality reduction when applying the Markov Blanket preprocessing step. According to Table 3, the performance of MBRF is one of the highest ever reported, with the drawback of a very time and recource consuming training phase.

The latter experimental design developed was a simulated trading strategy, in an effort to further examine if the MBRF model could practically be applied to generate higher profits than those earned by employing the traditional regression model of by simply following a buy-and-hold (passive) investment strategy. The operational details of the trading simulation are explained as follows: The trading simulation assumes that the investor has 100,000€ to create a portfolio by selecting a balanced

Table 3. Classification performance in terms of F-measure

Dataset	MBRF	RF	SMO	RBF	LR
No articles	64.23	55.25	63.56	53.21	47.54
Including articles	73.44	60.66	67.76	56.80	47.76

percentage of each of the three Greek stock quotes mentioned earlier. Each day, the investor could buy, sell or wait, according to the class prediction of the MBRF model. We assume that transactional costs apply when buying or selling (0,335% and 0,35% respectively) and a random choice between 5% and 10% of the current portfolio can be traded each day. The time period was set to the last 35 weekdays of the aforementioned dataset. As Fig. 4 depicts, the dashed line, which represents the portfolio budget for the MBRF investing strategy is clearly outperforming the solid line of the buy-and-hold investment strategy by a mean factor of 12.5% to 26% for the first 2 weeks and from 16% to 48% for the remaining ones.

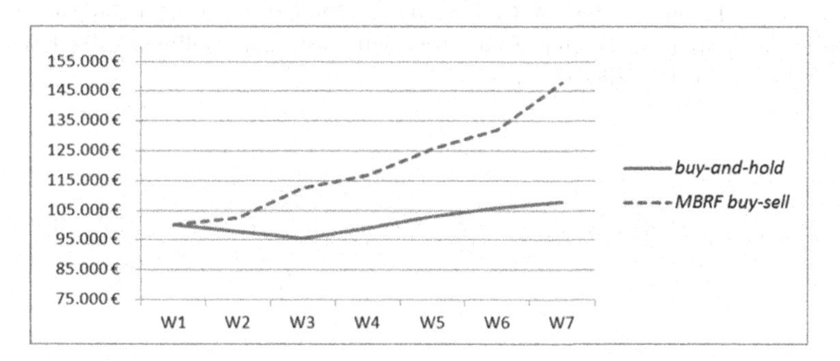

Fig. 4. Plot of portfolio outcomes using the two different trading strategies

References

1. Technical-Analysis. The Trader's Glossary of Technical Terms and Topics (2005), `http://www.traders.com`
2. Ng, A., Fu, A.W.: Mining Frequent Episodes for Relating Financial Events and Stock Trends. In: Whang, K.-Y., Jeon, J., Shim, K., Srivastava, J. (eds.) PAKDD 2003. LNCS (LNAI), vol. 2637, pp. 27–39. Springer, Heidelberg (2003)
3. Breiman, L.: Random forests. Machine Learning Journal 45, 532 (2001)
4. Chung, F., Fu, T., Luk, R., Ng, V.: Evolutionary Time Series Segmentation for Stock Data Mining. In: Proceedings of IEEE International Conference on Data Mining, pp. 83–91 (2002)
5. Klibanoff, P., Laymont, O., Wizman, T.A.: Investor reaction to Salient News in Closed-end Country Funds. Journal of Finance 53(2), 673–699 (1998)
6. Chan, Y., John-Wei, K.C.: Political Risk and Stock Price Volatility: The Case of Hong-Kong. Pacific-Basin Finance Journal 4(2-3), 259–275 (1996)
7. Mitchell, M.L., Mulherin, J.H.: The Impact of Public Information on the Stock Market. Journal of Finance 49(3), 923–950
8. Mittermayer, M.A.: Forecasting Intraday Stock Price Trends with Text Mining Techniques. In: Proceedings of the 37th Annual Hawaii International Conference on System Sciences (HICS), vol. 3(3), p. 30064.2. IEEE Computer Society, Los Alamitos (2004)
9. Shumaker, R.P., Chen, H.: Textual Analysis of Stock Market Prediction Using Financial News Articles. In: On the 12th American Conference on Information Systems, AMCIS (2006)

10. Díaz-Uriarte, R., de Andrés, S.A.: Gene selection and classification of microarray data using random forest. BMC Bioinformatics 7, 3 (2006)
11. Kohavi, R., John, G.: Wrappers for feature subset selection. Artificial Intelligence journal, special issue on relevance 97(1-2), 273–324 (1997)
12. Cooper, G.F., Herskovits, E.: A Bayesian Method for the Induction of Probabilistic Networks from Data. In: Machine Learning, vol. 9, pp. 309–347. Kluwer Academic Publishers, Boston (1992)
13. Strobl, C., et al.: Bias in random forest variable importance measures: illustrations, sources and a solution. BMC Bioinformatics 8, 25 (2007)
14. Lyras, D.P., Sgarbas, K.N., Fakotakis, N.D.: Using the Levenshtein Edit Distance for Automatic Lemmatization: A Case Study for Modern Greek and English. In: 19th IEEE International Conference on Tools with Artificial Intelligence (ICTAI 2007), vol. 2, pp. 428–435 (2007)

On the Problem of Attribute Selection for Software Cost Estimation: Input Backward Elimination Using Artificial Neural Networks

Efi Papatheocharous[1] and Andreas S. Andreou[2]

[1] University of Cyprus, Department of Computer Science. 75 Kallipoleos Street, P.O. Box 2053, CY1678 Nicosia, Cyprus
efi.papatheocharous@cs.ucy.ac.cy
[2] Cyprus University of Technology, Department of Electrical Engineering and Information Technology, 31 Archbishop Kyprianos Street, 3036 Lemesos, Cyprus
andreas.andreou@cut.ac.cy

Abstract. Many parameters affect the cost evolution of software projects. In the area of software cost estimation and project management the main challenge is to understand and quantify the effect of these parameters, or 'cost drivers', on the effort expended to develop software systems. This paper aims at investigating the effect of cost attributes on software development effort using empirical databases of completed projects and building Artificial Neural Network (ANN) models to predict effort. Prediction performance of various ANN models with different combinations of inputs is assessed in an attempt to reduce the models' input dimensions. The latter is performed by using one of the most popular saliency measures of network weights, namely Garson's Algorithm. The proposed methodology provides an insight on the interpretation of ANN which may be used for capturing nonlinear interactions between variables in complex software engineering environments.

Keywords: Software Cost Estimation, Artificial Neural Networks, Connection Weights, Garson's Algorithm.

1 Introduction

Software effort estimation is the process of predicting the required effort to support the software development process by utilising attributes of cost, often called 'cost drivers'. This process usually involves activities after product specification and until software implementation and delivery, and is usually performed at the initiation of a project. The accurate computation of the development effort in software organisations is critical since it enables project managers to effectively deal with uncertainties and risks associated with resource planning and allocation. Specifically, cost overestimations result, in over allocation of resources and budget increase, which may cause loss of contracts and interruption of negotiations. On the other end, cost underestimations cause loss of money upon project completion, misallocation of project resources, quality compromises or budget and schedule extensions.

H. Papadopoulos, A.S. Andreou, and M. Bramer (Eds.): AIAI 2010, IFIP AICT 339, pp. 287–294, 2010.

Artificial Intelligence (AI) techniques are quite popular in software cost estimation and are used for building models and calculating the effort factor. Especially, Artificial Neural Networks (ANN) that have the ability to provide a non-linear mapping among the inputs and the output have been used extensively. Nevertheless, in previous related works ANN are commonly used only as predictors and very rarely quantitative analysis is conducted regarding the influence of the network inputs on the output. In this work we focus on the ability of ANN to capture interactions between the influencing cost factors and effort and, in addition, the input's degree of influence built within the network is examined using Garson's Algorithm [1]. The overall purpose is to examine the prediction accuracy of development effort through the utilisation of different models and variable contributing factors. The values of the cost drivers are located within two widely-known and public databases, namely the Desharnais and the ISBSG, which are selected for experimentation. The contribution of inputs is assessed through a random sampling approach and using the resulting values of internal weights from the ANN. Gradually the contributing weights of inputs whose values do not significantly affect the output of the ANN are removed from the initial complete set of cost factors. The experiments conducted show that in software cost estimation there are several factors which are not critically significant but are commonly used for predicting effort in related research work. The input analysis conducted using Garson's Algorithm helps in removing factors in a backward manner, starting from the least significant ones and until half of the initial cost factors are left in each dataset whereas during this process ANN's prediction performance is continuously assessed.

The rest of this paper is organised as follows: Section 2 discusses the recent work on ANN utilisation for the problem of software cost estimation and also presents common approaches used for simplification and interpretation of ANN in other problem domains. Section 3 specifies the modeling technique and theory behind Garson's Algorithm. Section 4 presents in detail the methodology proposed, accuracy measures used in the experimental process and discusses the main results obtained. Finally, Section 5 summarises the conclusions and future research steps.

2 Related Work

This section initially presents the latest applications of Artificial Neural Networks (ANN) in the software cost estimation literature and identifies that even though the approach is considered promising, one of the most important steps, the identification and inspection of the dominant cost attributes, is not given proper thought.

Recent work of Tronto et al. [2] investigates the application of ANN and stepwise regression for software effort prediction. The experiments were conducted on the COCOMO dataset employing categorical variables whose impact was identified based on the work of Angelis et al. [3] forming new categorical values. The authors identified a strong relationship between the success of each technique and the size of the learning dataset, the nature of the cost function and dataset characteristics, such as existence of outliers, collinearity and number of attributes.

In software cost estimation the comparison of models is a common research goal. Kaur et al. [4] prove the effectiveness of ANN models for the NASA dataset compared to the Halstead, Walston-Felix, Bailey-Basili and Doty models, all of which are popular legacy models used in software cost estimation. Backpropagation ANN were used and reported as the most generalised networks currently in use that present good estimation capabilities. In addition, Reddy and Taju [5] used the popular COCOMO model in software cost estimation mapped to an ANN with minimal number of layers and nodes to increase the performance of the network. They employed a feedforward backpropagation MLP and obtained improved predictions for effort using the COCOMO dataset compared to the COCOMO model. Rao et al. [6] used a Functional Link Artificial Neural Network (FLANN) which does not contain any hidden layers so that the network architecture becomes simple and training does not involve full backpropagation, thus reducing computational complexity. Their method provides more accurate results compared to other methods for software cost estimation on the NASA dataset.

Many researchers applied ANN on software cost estimation and yielded very accurate results. However, when using ANN one crucial step is to identify the dominant cost factors, or attributes, that affect development effort [7]. A number of measures exist to determine the significance of ANN input attributes [1, 8, 9, 10] but we identified that they have never been applied for software cost drivers. For example, sensitivity analysis, fuzzy curves, MSE change, weight elimination and node pruning, and optimal brain damage (OBD) methods are measures that rank input feature importance. Some of these measures are heuristic (forward and backward selection), sensitivity index-based, are based on pseudo weights, rely on Garson's algorithm and some of its modified and extended versions that appear in the literature [11]. More specifically, in this work the concepts described in the following methodologies have been adopted for software cost drivers: Garson [1] proposed a method for partitioning the ANN connection weights to determine the relative importance of each input variable in the network (for more details see Section 3.2). Glorfeld [9] presented a methodology to simplify ANN using a backward selection process to eliminate input variables that are not contributing to the predictive power of accurate networks. According to the author this enables decision makers to understand the resulting effect of each contributing variable in producing accurate predictions. The application is on two classification examples, a commercial loan and a cheque overdraft problem.

3 Modeling Technique and Methodology for Input Elimination

3.1 Artificial Neural Networks

One of the primary applications of ANN involves models to forecast a dependent variable from a given set of independent variables. These are non-linear, model-free and alternative to traditional statistical methods. ANN consist of basic computational elements called neurons organized in groups forming layers. Certain types of neurons organised in multiple layers form the Multi-Layer Perceptron (MLP) [12] which is one of the most popular networks. The number of neurons in the input (first) layer is equal to the number of attributes used as independent variables. The last layer is the network output. Each subsequent layer uses the weights coming from the previous layers and adjusts them so that the accuracy performance error between the actual and predicted values for the dependent variable represented by the output is diminished.

3.2 Methodology with ANN and Garson's Algorithm

There are many methods for measuring the contribution of independent variables within a neural network, but most of which are very complicated and thus are rarely used in the area of software cost estimation. Garson's algorithm [1] is considered a good trade-off example among complexity and effectiveness. It partitions the hidden layer weights into components associated with each input node. Next, the percentage of all hidden nodes weights associated with a particular input node is used to measure the relative importance of that attribute. The interested reader may refer to [13] for a step-by-step example on the algorithm.

A variety of ANN architectures were implemented, starting with a topology which contains a number of neurons in the hidden layer equal to the number of attributes used as inputs in each experiment and continuing with topologies resulting from increasing the number of hidden neurons by 1 until their number becomes twice the size of the input attributes. In addition, the 'weakest' attribute is removed from the sample until the inputs are reduced to half the initial size. Moreover, the following randomisation process was followed for each sample: The initial weights and biases of the network were randomly set and the dataset used was randomly divided into three holdout subsets, training, validation and testing, with the percentages of 60%, 20% and 20% of the total available samples respectively, where each sample participates in only one subset.

The scaled conjugate gradient training function was used which is based on the derivative functions of weights, net inputs and transfer functions. The training process is repeated ten times so that the optimal network that minimizes the prediction error is identified and the weights of each input-hidden-output path are stored for further use by Garson's algorithm. Evaluation of the networks was performed using the testing data samples based on the well known *MMRE* and *pred(.25)* accuracy measures. For each experiment ten holdout random samples were chosen so that validation on random data is performed.

After training is executed and the network is stabilized, for each input j, $j=1,2,\ldots,i$, the Relative Importance (RI_j) is calculated using equation (1), where N_i and N_h are the number of input and hidden neurons, respectively and w is the connection weight, the superscripts 'i','h' and 'o' refer to input, hidden and output layers, respectively and subscripts 'k', 'm' and 'n' refer to input, hidden and output neurons (in our case n=1 as there is only one output neuron). According to Garson's algorithm, for each input node j the relative contribution of j to the outgoing signal of each hidden neuron is calculated and converted to a percentage, which serves as a measure of importance for each input node representing each variable. According to the proposed methodology, each input that makes the smallest contribution to the final output of the network, as this is reflected through the weight connections, is eliminated. Thus, in each repetition the initial number of variables utilised is lowered gradually by one until the necessary number of variables are left in the dataset.

$$RI_j = \frac{\sum_{m=1}^{Nh}\left[\left[\frac{\left|w_{jm}^{ih}\right|}{\sum_{k=1}^{Ni}\left|w_{km}^{ih}\right|}\right]\times\left|w_{mn}^{ho}\right|\right]}{\sum_{k=1}^{Ni}\left[\sum_{m=1}^{Nh}\left[\left[\frac{\left|w_{km}^{ih}\right|}{\sum_{k=1}^{Ni}\left|w_{km}^{ih}\right|}\right]\times\left|w_{mn}^{ho}\right|\right]\right]} \quad (1)$$

4 Experiments and Results

4.1 Datasets Description

The Desharnais (1989) dataset [14] included 81 observations for systems developed by a Canadian Software Development House. The second dataset ISBSG R9 [15] provided by the International Software Benchmarking Standards Group contains an analysis of the cost and other measurements for a large group of software projects, approximately 3,024. The projects come from a broad cross section of industry and range in size, effort, development platform, language, etc. These projects underwent a series of quality checks and pre-processing to create filtered versions of the datasets that do not contain null values and conform to the standards we set for homogeneity and integrity before feeding them as inputs to the ANN. The filtered datasets contained 77 and 113 in the Desharnais and ISBSG datasets and the attributes selected and used in this work, along with their abbreviations are summarised in Table 1.

Table 1. Summary of the attributes in the datasets used

Desharnais		ISBSG	
Team Experience (years)	TE	Functional Size	FS
Manager Experience (years)	ME	Adjusted Function Points	AFP
Duration (months)	DU	Project Elapsed time	PET
Transactions	TR	Project Inactive time	PIT
Entities	EN	Resource Level (ordinal)	RL
Points Adjusted	PA	Maximum Team Size	MTS
Scope	SC	Input count	INC
Points Non Adjusted	PNA	Output count	OC
		Enquiry count	EC
		File count	FC
		Interface count	IFC
		Added count	AC
		Changed count	CC
		Deleted count	DC

4.2 Results

The results reported in this section include the initial and final lowest performance values of the best obtained network architectures in terms of prediction accuracy

(*MMRE* value). The results of the training and testing phases of the various network architectures created are sensitive to the initialisation of weights, bias values and random division of the data samples used for training and testing. Initially, the number of input attributes for each experiment is reduced gradually by one according to importance of the inputs suggested by Garson's algorithm; then, the network is trained again with the reduced variables and the new performance is traced. This process is repeated ten times (on random holdout samples) and Tables 2 and 3 report the *MMRE* and *Pred(.25)* ANN performance figures for the training and testing phases using the Desharnais and ISBSG datasets respectively. The 'Initial' and 'Final' column results report accuracy having the number of inputs being equal to the initial number of attributes in each dataset *p* and being reduced to *p/2* respectively. The order in which attributes are removed in each experiment repetition (first column) is given in the second column, while the rest columns present the forecasting performance observed after removing the 'less important' attributes.

The experiments indicate that quite accurate and successful predictions were obtained, as suggested by the consistently low *MMRE* values in both the Desharnais and ISBSG cases throughout the random holdout validation sampling process. Moreover, comparing the initial and final values of the accuracy measures we observe some performance degradation, something which indicates that maybe a part of useful information is lost when reducing the number of the participating attributes. One may argue that this is expected as the information contributing to the ANN learning process is truncated and hence prediction accuracy is gradually lowered as we move from the initial to the final network state. More specifically, the accuracy degree of the Desharnais dataset decreases 0-29% in the training phase depending on the experiment repetition, while during testing accuracy increases in some cases by 20% and in others decreases by 16%. This also occurs with the ISBSG dataset, where by removing attributes the performance accuracy increases by 6% and decreases by 6% in the training phase depending on the experiment, while during the testing phase accuracy increases by 16% and decreases by 24%.

Another interesting finding is that the attributes that seem to be the 'weakest' effort contributors in the majority of the experiments are TE, ME, DU, SC for the Desharnais case and RL, FC, CC, INC, OC for the ISBSG. We should also report, though, that there were validation cases in which the general picture of input significance was

Table 2. Random sampling and first four attributes removed from the Desharnais dataset

#	Order of Attributes Removed	ANN Training Phase				ANN Testing Phase			
		Initial *MMRE*	Initial *Pred*	Final *MMRE*	Final *Pred*	Initial *MMRE*	Initial *Pred*	Final *MMRE*	Final *Pred*
1	TE,DU,SC,ME	0.384	0.936	0.559	0.936	0.536	0.867	0.600	0.867
2	ME,DU,TR,TE	0.280	0.979	0.343	1.000	0.409	0.933	0.487	0.933
3	SC,EN,TE,DU	0.557	0.936	0.583	0.936	0.198	1.000	0.335	1.000
4	TE,TR,PA,SC	0.474	0.915	0.485	0.894	0.364	1.000	0.387	1.000
5	ME,PA,TE,DU	0.361	0.957	0.360	0.979	1.264	0.867	1.060	0.867
6	TE,SC,DU,ME	0.512	0.915	0.784	0.915	0.386	0.933	0.346	0.933
7	ME,SC,PNA,PA	0.472	0.957	0.572	0.957	0.569	0.800	0.512	0.800
8	TE,ME,SC,EN	0.509	0.957	0.572	0.957	0.351	0.800	0.512	0.800
9	TE,ME,EN,DU	0.358	0.936	0.356	0.936	0.507	0.933	0.578	0.933
10	TE,SC,EN,DU	0.482	0.894	0.768	0.894	0.312	0.933	0.293	0.933
	Mean	**0.439**	**0.938**	**0.538**	**0.940**	**0.490**	**0.907**	**0.511**	**0.907**

Table 3. Random sampling-first seven attributes removed from the ISBSG dataset

#	Order of Attributes Removed	ANN Training Phase				ANN Testing Phase			
		Initial *MMRE*	Initial *Pred*	Final *MMRE*	Final *Pred*	Initial *MMRE*	Initial *Pred*	Final *MMRE*	Final *Pred*
1	CC,AFP,MTS,PET,OC,PIT,EC	0.223	0.957	0.329	0.928	0.358	1.000	0.578	1.000
2	DC,CC,PET,RL,PIT,INC,EC	0.280	0.986	0.352	0.971	0.418	0.955	0.575	0.955
3	INC,CC,DC,RL,OC,AC,FC	0.337	1.000	0.404	0.986	0.255	0.955	0.265	0.909
4	MTS,CC,RL,INC,FC,FS,AC	0.350	0.957	0.493	0.957	0.199	0.955	0.191	0.955
5	FS,RL,FC,EC,IC,INC,CC	0.367	0.971	0.411	0.986	0.202	0.955	0.298	0.955
6	RL,DC,CC,INC,OC,EC,FS	0.362	1.000	0.303	1.000	0.346	0.909	0.385	0.864
7	CC,OC,RL,FC,INC,AFP,IC	0.210	1.000	0.271	0.986	0.377	0.909	0.257	0.909
8	FC,OC,AC,AFP,RL,IC,FS	0.247	0.957	0.344	0.957	0.188	1.000	0.424	1.000
9	FC,IC,RL,AFP,AC,EC,PIT	0.261	0.957	0.305	0.971	0.662	1.000	0.500	1.000
10	FS,IC,PIT,OC,AFP,RL,FC	0.338	0.986	0.279	0.986	0.249	0.955	0.270	0.955
	Mean	**0.297**	**0.977**	**0.349**	**0.972**	**0.325**	**0.959**	**0.374**	**0.950**

a bit disrupted. This may be considered as a weakness of the method as it relies on ANN models that behave differently when the initial conditions (initialisation) and the training/testing data samples change. Therefore, the results must be interpreted cautiously and only after a satisfactory number of repetitions that will enable a statistically safe conclusion.

5 Conclusions

This work investigates the ability of ANN to capture interactions between the influencing cost factors and effort within empirical software engineering project samples and attempt to provide cost predictive models. The main contribution is the understanding of the explanatory value of the inter-relationships between the input variables and the final output (effort) which is extracted from the internal network weights. Thus, it may provide an insight regarding each variable's contribution to the overall prediction.

We performed a backward elimination strategy to minimise the initial inputs and progressively evaluated the significance of connection weights and input variables. The approach was based on Garson's Algorithm which exploits the various ANN models created, trained and tested over ten random sets of training and testing distinct samples. Moreover, from the various architectures created, trained and tested the results obtained from the best networks in terms of *MMRE* of actual vs. the prediction testing samples, i.e. result of the simulation phase of the process provide quite promising results. The approach enables decision makers to understand the resulting effect of each contributing variable in producing accurate predictions.

An interesting issue for future research is the comparison of the contributions of other saliency measures reported in the literature along with Garson's over the same or newer datasets. In addition, a cost-benefit analysis of accuracy declination and the fewer attributes leading to less cost for collecting data and faster cost estimation modeling should be carried out to prove the validity of this work.

References

1. Garson, G.D.: Interpreting Neural-Network Connection Weights. AI Expert 6, 46–51 (1991)
2. Tronto, I.F.D.B., Silva, J.D.S.D., Sant'Anna, N.: An Investigation of Artificial Neural Networks based Prediction Systems in Software Project Management. Journal of Systems and Software 81, 356–367 (2008)
3. Angelis, L., Stamelos, I., Morisio, M.: Building A Software Cost Estimation Model Based On Categorical Data. In: Proceedings of the 7th International Symposium on Software Metrics, pp. 4–15. IEEE Computer Society, Los Alamitos (2001)
4. Kaur, J., Singh, S., Kahlon, K.S., Bassi, P.: Neural Network – A Novel Technique for Software Effort Estimation. International Journal of Computer Theory and Engineering 2(1), 1793-8201, 17–19 (2010)
5. Reddy, C.S., Raju, K.: A Concise Neural Network Model for Estimating Software Effort. International Journal of Recent Trends in Engineering 1(1), 188–193 (2009)
6. Rao, B.T., Sameet, B., Swathi, G.K., Gupta, K.V., RaviTeja, C., Sumana, S.: A Novel Neural Network Approach for Software Cost Estimation using Functional Link Artificial Neural Network (FLANN). International Journal of Computer Science and Network Security 9(6), 126–131 (2009)
7. Park, H., Baek, S.: An Empirical Validation of a Neural Network Model for Software Effort Estimation. Expert Systems with Applications 35, 929–937 (2008)
8. Belue, L.M., Bauer, K.W.: Determining Input Features for Multilayer Perceptrons. Neurocomputing 7, 111–121 (1995)
9. Glorfeld, L.W.: A Methodology for Simplification and Interpretation of Backpropagation-Based Neural Network Models. Expert Systems with Applications 10, 37–54 (1996)
10. Satizábal, H.M., Pérez-Uribe, A.: Relevance Metrics to Reduce Input Dimensions in Artificial Neural Networks. In: de Sá, J.M., Alexandre, L.A., Duch, W., Mandic, D.P. (eds.) ICANN 2007. LNCS, vol. 4668, pp. 39–48. Springer, Heidelberg (2007)
11. Zhang, G.: Neural Networks for Classification: A Survey. IEEE Transactions on Systems, Man, and Cybernetics, Part C: Applications and Reviews 30, 451–462 (2000)
12. McCulloch, W.S., Pitts, W.: A Logical Calculus of the Ideas Immanent in Nervous Activity. Bulletin of Mathematical Biology 5(4), 115–133 (1943)
13. Olden, J.D., Jackson, D.A.: Illuminating the "Black Box": a Randomization Approach for Understanding Variable Contributions in Artificial Neural Networks. Ecological Modelling 154, 135–150 (2002)
14. Desharnais, J.M.: Analyse Statistique de la Productivite des Projects de Development en Informatique a Partir de la Technique de Points de Fonction, MSc. Thesis, Université du Québec, Montréal (1989)
15. The International Software Benchmarking Standards Group. Repository Data Release 9 (2005), `http://www.isbsg.org/`

A Fast Mobile Face Recognition System for Android OS Based on Eigenfaces Decomposition

Charalampos Doukas[1] and Ilias Maglogiannis[2]

[1] University of the Aegean, Samos, Greece
doukas@aegean.gr
[2]University of Central Greece, Lamia, Greece
imaglo@ucg.gr

Abstract. This paper presents a speed-optimized face recognition system designed for mobile devices. Such applications may be used in the context of pervasive and assistive computing for the support of elderly suffering from dementia in recognizing persons or for the development of cognitive memory games Eigenfaces decomposition and Mahalanobis distance calculation have been utilized whereas the recognition application has been developed for Android OS. The initial implementation and the corresponding results have proven the feasibility and value of the proposed system.

Keywords: Mobile face recognition, Android OS, Face Detection, Eigenfaces decomposition.

1 Introduction

Face recognition refers to automatically identifying or verifying a person from a digital image or a video frame from a video source. The process involves the face detection in images or image sequences, proper feature extraction, the creation of a reference set based on known faces and the detection of new faces based on the latter. Mobile face recognition can be utilized either for user authentication or for cognitive assistance in identifying individuals. The latter can be particularly useful for the elderly whose face recognition ability declines while aging. A plethora of face recognition algorithms have been proposed in literature ([1] – [12]). However, the application of the well known and established techniques to mobile devices, such as mobile phones, raises some challenges related to two major issues: the inefficiency of the algorithms and the limited processing capabilities of the mobile devices.

This paper presents a speed-optimized method for performing face recognition on Android mobile phones. Both training and recognition phases are performed on the device. The Android mobile operating system provides built-in functionality for detecting faces on still images. Eigenface decomposition is used in order to extract important features from the images and a training vector is created based on an initial image training set. The Mahalanobis cosine distance is utilized for measuring the similarity of the given train set images and the examined image in the Eigenface projection space. The latter methods have been selected for implementation on a mobile device due to lower complexity and memory requirements compared to other methods

H. Papadopoulos, A.S. Andreou, and M. Bramer (Eds.): AIAI 2010, IFIP AICT 339, pp. 295–302, 2010.

available in literature. In addition to the face recognition process performed thoroughly on the device, a web service has been developed that allows the feature extraction and classification using more advanced dace recognition techniques. The latter method requires network connectivity from the mobile device but can provide more accurate results.

Initial evaluation of the system has been performed on Google's G1 mobile phone and results are quite promising regarding the system's performance and accuracy. The paper presents the aforementioned, discusses related work and includes technical information and implementation details.

2 Related Work and Background Information

In general, the whole process of face recognition consists of detecting and selecting faces from the current image or video frame utilizing techniques like skin detection, process the selected image region and finally applying a recognition algorithm.

Several classifiers have been proposed in the literature e.g. minimum distance classification in the eigenspace ([11], [12]), Fisher's discriminant analysis [3], and neural networks [2]. Global techniques work well for classifying frontal views of faces, however they are not robust against pose changes since global features are highly sensitive to translation and rotation of the face. To avoid this problem an alignment stage can be added before classifying the face. Aligning an input face image with a reference face image requires computing correspondence between the two face images. The correspondence is usually determined for a small number of prominent points in the face like the center of the eye, the nostrils, or the corners of the mouth. The image region is also converted to grey levels and equalized, then cropped with an elliptic mask and normalized. Finally, the recognition algorithm compares the processed image to a set of prerecorded faces, all belonging to the same subject, to find the best match. This algorithm usually returns also a measure of similarity between the current face and the reference template.

2.1 Face Detection and Preprocessing

One of the most crucial parts in the presented face recognition system is the preprocessing of the considered image region. To align a face horizontally, a common technique consists in locating the eyes and then rotating the selected image so that their inclination is null. Also, the distance between the eyes is used to resize the same image to a given value of width and height.

Once the positions (uR, vR) and (uL, vL), of the right and left eye respectively, have been determined, we simply calculate the angle and the distance between them as follows:

$$a_{RL} = \tan^{-1}\left(\frac{vL-vR}{uL-uR}\right) \tag{1}$$

$$d_{RL} = \sqrt{(u_L - u_R)^2 + (v_L - v_R)^2} \tag{2}$$

The face is then rotated of an angle $-a_{RL}$ in order to align the eyes, and then resized to 24x24 pixels, i.e. the size of the templates used for comparison. Rotation and scaling, centered on the right eye (u_R, v_R), are performed with a simple affine transformation as follows:

$$\begin{bmatrix} u' \\ v' \end{bmatrix} = \begin{bmatrix} a & b & (1-a)u_R - bv_R \\ -b & a & bu_R + (1-a)v_R \end{bmatrix} \begin{bmatrix} u \\ v \\ 1 \end{bmatrix} \quad (3)$$

$$a = s\cos(-a_{RL})$$
$$b = s\sin(-a_{RL})$$

where (u, v) are the pixel coordinates of the source image and (u′, v′) those of the destination. In order to avoid possible outliers in the final image after rotation, the affine transformation is actually applied to a sub-region slightly bigger than the original face bounding box. The final step of the image processing consists in cropping the face's area with an elliptical mask. This reduces the influence of hair and background pixels at the four corners of the rectangular region.

2.2 Eigenfaces Decomposition and Similarity Detection

After the image processing described above, the most popular techniques for face recognition, called Eigenfaces [12] is applied. Eigenfaces decomposition and similarity detection relies on measuring the similarity between a new face image and a reference one, projecting both the images into an eigenspace, previously created by training, and calculating the distance between the projections. The idea behind the eigenface technique is to extract the relevant information contained in a facial image and represent it as efficiently as possible. Rather than manipulating and comparing faces directly, one manipulates and compares their representations.

To create a set of eigenfaces, one must prepare a training set of face images. The pictures constituting the training set should have been taken under the same lighting conditions, and must be normalized to have the eyes and mouths aligned across all images. They must also be all resampled to the same pixel resolution. Each image is treated as one vector, simply by concatenating the rows of pixels in the original image, resulting in a single row with r × c elements. For this implementation, it is assumed that all images of the training set are stored in a single matrix T, where each row of the matrix is an image. The mean is then subtracted. The average image *a* has to be calculated and then subtracted from each original image in *T*. The eigenvectors and eigenvalues of the covariance matrix *S* are thereafter calculated. Each eigenvector has the same dimensionality (number of components) as the original images, and thus can itself be seen as an image. The eigenvectors of this covariance matrix are therefore called eigenfaces (see Fig. 1). They are the directions in which the images differ from the mean image. The *D x D* covariance matrix will result in *D* eigenvectors, each representing a direction in the $r \times c$ dimensional image space. The eigenvectors (eigenfaces) with largest associated eigenvalue are kept. The latter can now be used to represent both existing and new faces: we can project a new (mean-subtracted) image on the eigenfaces and thereby record how that new face differs from the mean face. The eigenvalues associated with each eigenface represent how much the images in the training set vary from the mean image in that direction.

The computation of eigenvectors is performed as follows:

Let T be the matrix of preprocessed training examples, where each row contains one mean-subtracted image. The covariance matrix can then be computed as $S = T^T T$ and the eigenvector decomposition of S is given by:

$$Sv_i = T^T T v_i = \lambda_i v_i \tag{4}$$

However TTT is a large matrix, and if instead we take the eigenvalue decomposition of:

$$TT^T u_i = \lambda_i u_i \tag{5}$$

then we notice that by pre-multiplying both sides of the equation with TT, we obtain:

$$T^T TT^T ui = \lambda_i T^T u_i \tag{6}$$

Meaning that, if $\mathbf{u}_i$ is an eigenvector of $\mathbf{TT}^T$, then $\mathbf{v}_i=\mathbf{T}^T\mathbf{u}_i$ is an eigenvector of S.

Regarding measuring the similarity between different eigenvectors and performing image recognition, several methods have been proposed. The most common one, which is considered to give the best performance, is the Mahalanobis Cosine distance ([13], [14]). As the name suggests, this is given by the cosine of the angle θij between the two face projections fi and fj in the Mahalanobis space:

$$\xi(fi, fj) = -\frac{fi \bullet fj}{\|fi\|\|fj\|} = -\cos\theta ij \tag{7}$$

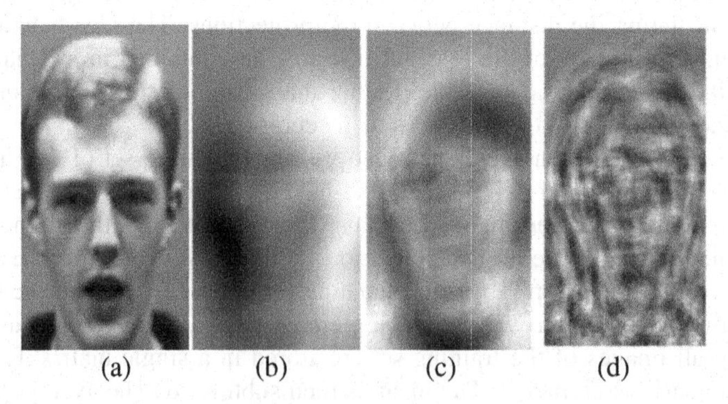

Fig. 1. Example illustrating eigenfaces decomposition: a) original face image, b) first eigenface, c) tenth eigenface, d) hundredth eigenface.

3 Technical Details and System Architecture

The mobile face recognition system has been developed utilizing Google's Android mobile Operating System (OS) [15] and the appropriate software development kit (sdk).

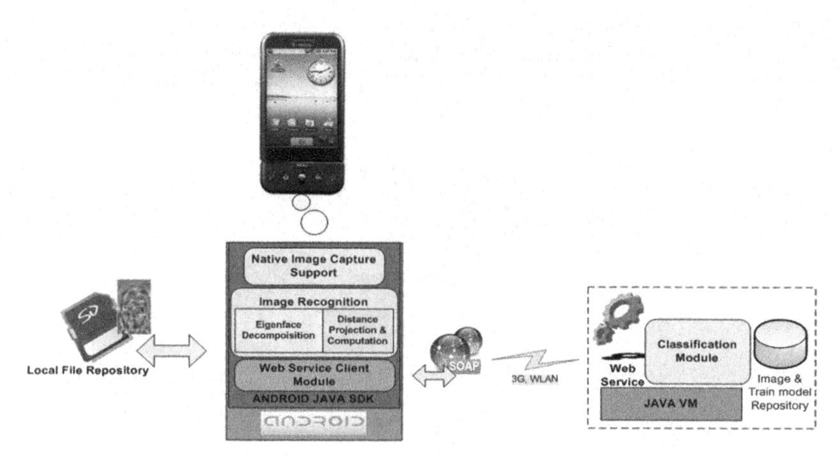

Fig. 2. Architecture of the proposed mobile face recognition system

Android is a mobile operating system running on the Linux kernel. Several mobile device vendors already support it. The platform is adaptable to larger and traditional smart phone layouts and supports a variety of connectivity technologies (CDMA, EV-DO, UMTS, Bluetooth, and Wi-Fi). It supports a great variety of audio, video and still image format, making it suitable for capturing and displaying images. Finally, it supports native multi-touch technology, which increases the application's usability.

Fig. 2 presents an illustration of the proposed system architecture. The mobile application utilizes native Android resources for capturing images using the mobile devices' image recording capabilities. The images can also be allocated on a remote or local repository. Android also provides native face detection functionality. Appropriate java methods can be used directly in any custom application and allow the detection of faces within images. The methods provide the coordinates of the eyes. Appropriate java classes (referred as modules) have been developed for performing the initial image processing (i.e., face cropping, rotation and scaling), the decomposition of the initial image into eigenfaces and the projection of the eigenvactors into the Mahalanobis domain and distance calculation between a given new image instance and the ones used as a training set. The images and the computed eigenvectors of the training set reside within the mobile device's storage repository (e.g., a memory card).

In addition to the face recognition process implemented locally on the mobile device, a web service has bee also developed that enables a more advanced feature extraction and classification of the face images. An appropriate module enables the communication between the Android application and the Web Service utilizing SOAP messages. The Web Service has been developed using Java and Apache Axis [17]. The classification module is based on the same java code developed for the Android for pre-processing the images and extracting the eigenvectors. Classification of the latter is performed using the Weka tool [18] that contains a great variety of tools for pattern recognition. The data exchange between the mobile device and the web service involves the exchange of images and the classification results.

Fig. 3 illustrates the main steps for performing face recognition on an Android-enabled mobile device. When faces are detected in an image, a validation of the faces found and the corresponding eye coordinates and eye distances is performed in order to minimize erroneous cases.

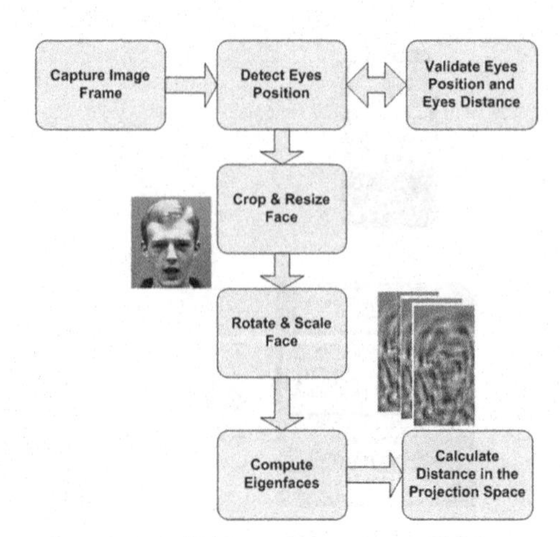

Fig. 3. Data flow diagram illustrating the basic steps for face detection and recognition using eigenface decomposition

4 Initial Evaluation Results

In order to evaluate the proposed system a HTC G1 [16] mobile phone running Android OS version 1.6 has been used in order to host the developed face recognition application. The initial training set consisted of 126 images containing 7 different faces (4 males and 3 males, 18 sample images per face). All faces have been cropped from the background as described in Section 2 and resized to a 100x100 pixels resolution. All images reside into the mobile phone's memory card and the developed face recognition application has been initially used in order to generate the corresponding eigenfaces. The first 50 eigenvalues have been for each image and the corresponding vectors have been saved into the mobile's phone memory card. For the evaluation, 5 different images (i.e. not included in the training set) from each individual have been used.

The creation of the training set has taken an average of 65 seconds for all 126 images on the device, whereas the processing, eigenfaces decomposition and mahalanobis distance calculation and comparison for each evaluation image has taken approximately only 1.2 seconds. The overall system accuracy in detecting faces in images is about 96%, whereas the total accuracy in recognizing faces that do not belong to the training set is 80%. The latter are quite promising considering the size of the evaluation set and the fact that both the training and recognition phases are performed on the mobile device at very reasonable time delays, allowing the usage of the proposed face recognition system in real applications.

Regarding the ability of the system to utilize more advanced classifiers through the Weka tool for more accurate face recognition, a second experiment has been conducted. The same training set has been used, whereas for the evaluation process, the latter set along with new image samples have been used through 10-fold cross validation. Several classifiers like Bayes Networks, Support Vector Machines (SVM), and Neural Networks (NN) have been validated. The best performance has been achieved by SVM approaching a total accuracy of 98%, followed by 96.2% achieved by NN and 95% by the Bayes classifier respectively. The total time to process and classify an

Table 1. Initial evaluation results for the proposed mobile face recognition system. System accuracy in face detection within images and overall face recognition accuracy are presented. Both training and recognition phases have been performed on the mobile device.

	Accuracy in detecting faces in images in training set (%)	Accuracy in Face Recognition (%)
Male 1	94.4	80
Male 2	100.0	80
Male 3	100.0	100
Male 4	88.8	60
Female 1	88.8	80
Female 2	100.0	80
Female 3	100.0	80

image is less than 1 sec at this case, but the overall network transmission of data introduces further delay until the result is displayed at the mobile device.

5 Conclusions

The advances of state of the art smart phones and PDAs have enabled the creation of fast and reasonably robust face detection/recognition applications like the one presented in this work. Such applications may be used in the context of pervasive and assistive computing for the support of elderly suffering from dementia in recognizing persons or for the development of cognitive memory games.

An important issue to be handled in face recognition is the "Face Variation" problem. This is closely related to changes of illumination, pose, and expression. As non uniform illumination affects the acquired face image, it is important to know how the recognition algorithm works under varying illumination. Pose variations include approaches on how to model faces seen from different points of view. Expression variations refer to how we can design algorithms for the recognition of emotions and other facial expressions [10]. Future research will focus on addressing such issues, on improvement of detection precision, the addition of online training module and other novel applications.

References

1. Chellappa, R., Wilson, C.L., Sirohey, C.: Humain and machine recognition of faces: A survey. Proc. IEEE 83(5), 705–740 (1995)
2. Fleming, M., Cottrell, G.: Categorization of faces using unsupervised feature extraction. In: Proc. IEEE IJCNN International Joint Conference on Neural Networks, pp. 65–70 (1990)
3. Belhumeur, P., Hespanha, P., Kriegman, D.: Eigenfaces vs fisherfaces: Recognition using class specific linear projection. IEEE Transactions on Pattern Analysis and Machine Intelligence 19(7), 711–720 (1997)

4. Jonsson, K., Matas, J., Kittler, J., Li, Y.: Learning support vectors for face verification and recognition. In: Proc. IEEE International Conference on Automatic Face and Gesture Recognition, pp. 208–213 (2000)
5. Brunelli, R., Poggio, T.: Face recognition: Features versus templates. IEEE Transactions on Pattern Analysis and Machine Intelligence 15(10), 1042–1052 (1993)
6. Tolba, A.S., El-Baz, A.H., El-Harby, A.A.: Face Recognition: A Literature Review. International Journal of Signal Processing 2(2) (2006)
7. Hong, L., Jain, A.: Integrating faces and fingerprints for personal identification. IEEE Trans. Pattern Analysis and Machine Intelligence 20(12), 1295–1307 (1998)
8. Deniz, O., Castrillon, M., Hernandez, M.: Face recognition using independent component analysis and support vector machines. Pattern Recognition Letters 24, 2153–2157 (2003)
9. Li, Y., Gong, S., Liddell, H.: Support vector regression and classification based multi-view face detection and recognition. In: Proc. IEEE International Conference on Face and Gesture Recognition, Grenoble, France (March 2000)
10. Maglogiannis, I., Vouyioukas, D., Aggelopoulos, C.: Face Detection and Recognition of Human Emotion Using Markov Random Fields. In: Personal and Ubiquitous Computing. Springer, Heidelberg, doi: 10.1007/s00779-007-0165-0
11. Sirovitch, L., Kirby, M.: Low-dimensional procedure for the characterization of human faces. Journal of the Optical Society of America A 2, 519–524 (1987)
12. Turk, M., Pentland, A.: Eigenfaces for recognition. Journal of Cognitive Neuroscience 3(1), 72–86 (1991)
13. Aly, S., Tsuruta, N., Taniguchi, R.-I.: Face recognition under varying illumination using Mahalanobis self-organizing map. Journal of Artificial Life and Robotics 13(1), 298–301 (2008)
14. Beveridge, R., Bolme, D., Teixeira, M., Draper, B.: The CSU Face Identification Evaluation System User's Guide: Version 5.0, Computer Science Department, Colorado State University (May 2003)
15. The Android mobile OS by GoogleTM, `http://www.android.com/`
16. HTC G1 mobile phone,
`http://www.htc.com/www/product/g1/overview.html`
17. The Apache Axis Framework, `http://ws.apache.org/axis/`
18. Hall, M., Frank, E., Holmes, G., Pfahringer, B., Reutemann, P., Witten, I.H.: The WEKA Data Mining Software: An Update. SIGKDD Explorations 11(1) (2009)

Application of Conformal Predictors to Tea Classification Based on Electronic Nose

Ilia Nouretdinov[1], Guang Li[2], Alexander Gammerman[1], and Zhiyuan Luo[1]

[1] Computer Learning Research Centre, Royal Holloway, University of London
Egham Hill, Egham, Surrey TW20 0EX, UK
zhiyuan@cs.rhul.ac.uk

[2] Department of Control Science and Engineering, Zhejiang University,
38 ZheDa Road, HangZhou 310027, Zhejiang, P.R. China

Abstract. In this paper, we present an investigation into the performance of conformal predictors for discriminating the aroma of different types of tea using an electronic nose system based on gas sensors. We propose a new non-conformity measure for the implementation of conformal predictors based on Support Vector Machine for multi-class classification problems. The experimental results have shown the good performance of the implemented conformal predictors.

Keywords: Tea Classification; Electronic Nose; Conformal Predictors; Support Vector Machines; Pattern Recognition.

1 Introduction

Tea is by far the most popular drink consumed in Britain today, with over 165,000,000 cups being drunk in the UK every single day of the year [1]. The major quality attributes of tea are flavour, aroma, colour and strength. Aroma means smell of the tea and is considered as the most important of the attributes. However, characterisation of aroma of tea has been a challenge for tea scientists for long. Tea flavour is traditionally measured through the use of a combination of conventional analytical instrumentation and human or ganoleptic profiling panels. These methods are expensive in terms of time and labour and also inaccurate because of a lack of either sensitivity or quantitative information [5].

Electronic nose has been successfully used for tea classification and grading [2,5,6,12]. In general, an electronic nose is a smart instrument that is designed to detect and discriminate among complex odours using an array of sensors. The array of sensors is exposed to volatile odour vapour through suitable odour handling and delivery system that ensures constant exposure rate to each of the sensors. The response signals of these sensors are recorded and processed for analysis and classification. For tea classification, we want to discriminate the aroma of different types of tea.

Different data analysis and machine learning methods have been tried to analyse and classify the signals from the sensor array of the electronic nose system. For example, these signals were processed using Principal Components Analysis (PCA), Fuzzy C Means algorithm (FCM), Self-Organizing Map (SOM) method along with a Radial Basis Function network (RBF), Probabilistic Neural Network classifier, Back-Propagation network and Olfactory model [3,5,6,8,11,12]. Good performance has been reported in

H. Papadopoulos, A.S. Andreou, and M. Bramer (Eds.): AIAI 2010, IFIP AICT 339, pp. 303–310, 2010.

the literature. However, the learning techniques used can only provide bare predictions, i.e. algorithms predicting labels for new examples without saying how reliable these predictions are. The reliability of a method is often given by measuring general accuracy across an independent test set. Confidence estimation is a well-studied area of both parametric and non-parametric statistics; however, usually only low-dimensional problems are considered.

This paper describes an investigation into the performance of conformal predictors for discriminating the aroma of different types of tea using an electronic nose system. In particular, we study a recently developed theory of algorithmic learning, namely conformal predictors described in [10]. One of the major advantages of these algorithms is that they can be used for solving high-dimensional problems without requiring any parametric statistical assumptions about the source of data (unlike traditional statistical techniques). The only assumption made is the i.i.d. assumption (the examples are generated from the same probability distribution independently of each other). Another advantage of the method is that it also allows to make estimation of confidence in the classification of individual examples. The remainder of this paper is arranged as follows. We briefly describe conformal predictors and present a new non-conformity measure for the implementation of conformal predictors based on Support Vector Machine for multi-class classification problems in the next section. Section 3 discusses the experiments and results of applying the implemented conformal predictors to tea classification. Finally, Section 4 concludes the paper.

2 Conformal Prediction

In machine learning problems, we usually work with examples which are pairs $z_i = (x_i, y_i)$. $x_i \in \mathbf{X}$ is an object, normally represented as a vector and its components (dimensions) are called attributes. $y_i \in \mathbf{Y}$ is its label, which may be real-valued, or taking one of several possible values [4]. For the tea classification problem using electronic nose system, x_i is a vector of response signals from the sensor array and y_i is the finite types of tea.

Let us assume that we are given a training set of l examples $\{(x_1, y_1), \ldots, (x_l, y_l)\}$. Our goal is to predict the classifications for testing examples. We make only one assumption (i.i.d.) about the examples: all the examples have been generated independently by some fixed but unknown stochastic mechanism. Suppose that for a new example x_{l+1} and its corresponding real label y_{l+1} is unknown, we are interesting in predicting its label. The basic idea behind conformal predictors is to try every possible label $y \in \mathbf{Y}$ as candidate for the new example and see how well the resulting sequence $\{(x_1, y_1), \ldots, (x_l, y_l), (x_{l+1}, y)\}$ conforms the i.i.d. assumption. If it does, we will say it is "random". The ideal case is where all $y \in \mathbf{Y}$ but one lead to sequences that are not random and y can be considered as the prediction for x_{l+1}. The general description of the conformal predictors is given in Algorithm 1 [10].

In conformal predictors, checking whether a possible new label violates i.i.d. assumption is done by a test for randomness based on a *non-conformity (or strangeness) measure* $\alpha = A(z_{l+1}, \{z_1, \ldots, z_{l+1}\})$. This is a measure of disagreement or non-conformity between a finite set (subset of $\mathbf{X} \times \mathbf{Y}$) and its element. Specific form of this function depends on a particular machine learning algorithm to be used and can be determined for many well-known machine learning algorithm. In this paper, we investigate and propose a non-conformity function for multi-class Support Vector Machine.

Algorithm 1. Conformal Predictor

Input: example sequence $\{(x_1, y_1), (x_2, y_2), \ldots, (x_l, y_l)\}$
Input: new example x_{l+1}
Input(optional): confidence level γ
Input: non-conformity measure A
for $y \in Y$ **do**
  $z_{l+1} = (x_{l+1}, y)$
  for j in $1, 2, \ldots, l+1$ **do**
    $\alpha_j = A(z_j, \{z_1, \ldots, z_l, z_{l+1}\})$
  end for
  $p(y) = \frac{\#\{j=1,\ldots,l+1 : \alpha_j \geq \alpha_{l+1}\}}{l+1}$
end for
Output(optional): predictive set $R^{\gamma}_{l+1} = \{y : p(y) \geq 1 - \gamma\}$
Output: forced prediction $\hat{y}_{l+1} = \arg\max_y \{p(y)\}$
Output: confidence $conf(\hat{y}_{l+1}) = 1 - \max_{y \neq \hat{y}_{l+1}} \{p(y)\}$
Output: credibility $cred(\hat{y}_{l+1} = \arg\max_y \{p(y)\}$

The conformal predictor calculates $p(y)$ which is a *p-value* associated with a hypothetical completion $y_{l+1} = y$ for the test example. Using these *p-values*, predictions made by conformal predictors can be presented in following two ways:

- Either the user inputs a required degree (level) of certainty and the algorithm outputs a predictive set: a list of classifications that meet this confidence requirement.
- Or the algorithm outputs so called *forced prediction*: an individual predictions together with its confidence, which is the minimal level of certainty at which the output is certain (predictive set consists of only one label) and credibility, which indicates whether the test object is representative of the training set, in order to compare with the other conventional learning methods.

The common way of presenting the prediction results is to choose a significance level ("degree of certainty") $\gamma < 1$ and output the (γ)*-predictive set (region)* containing all labels with p-value equal or greater than $1 - \gamma$:

$$R^{\gamma}_{l+1} = \{y : p(y) \geq 1 - \gamma\} \tag{1}$$

Prediction errors can occur when the prediction set fails to contain the true label. If the size of the predictive set R^{γ}_{l+1} is 1 or 0, then this prediction is *certain*, otherwise it is uncertain and we have multiple predictions. In the context of conformal predictors, uncertain predictions are not errors, but these are indications that the amount of information is not sufficient to make a certain decision at the selected level. Naturally, the higher the confidence level is, the more multiple predictions will appear. If non-conformity measure is specified adequately for the problem at hand, the percentage of uncertain predictions should not be too high. It has been proved that the conformal predictors have validity property [10,7] in online mode where the examples are presented one by one. Each time, the predictors take an object to predict its label and then receive feedback of the true label. The validity property implies that the successive predictions will be right 1- ϵ of the time for a given error probability ϵ, even though they are based on an accumulating dataset rather than on independent dataset.

It is important to note that the prediction region of a conformal predictor in online mode is valid for any used non-conformity measure [10]. However, the quality of prediction depends on the choice of this measure. If it is inappropriate, the confidence in a prediction will be still valid but never high. Thus a non-conformity measure for conformal predictor is usually based on an underlying method of prediction (such as nearest neighbours and Support Vector Machine) which is appropriate enough for the data set.

In the case of finite label space, there is an alternative way to represent the output: single ("forced") prediction and some measure of *confidence* in this prediction. It does not require the confidence level γ as input. The single prediction is selected by largest p-value. The maximal confidence level (degree of certainty) at which the prediction is certain, is called "confidence". It can be calculated as 1 minus the second largest p-value.

2.1 Support Vector Machine

Conformal predictors are a generic framework for making prediction and virtually any classification or regression algorithm can be transformed into a conformal predictor [10]. In this section, we consider the implementation of conformal predictors on the top of Support Vector Machine. Support Vector Machines (SVM) are effective learning systems based on statistical learning theory and have been applied successfully to many real problems [9]. For pattern recognition, SVM finds the hyperplane that separates the data set $(x_1, y_1), ..., (x_l, y_l), x_i \in \mathbf{R}^d, y_i \in \{-1, +1\}$ with maximal margin. If the data are not linearly separable one can introduce slack variables (ζ_i) in the optimisation,

$$\min \frac{1}{2} \parallel \mathbf{w} \parallel + C \sum_{i=1}^{l} \zeta_i \tag{2}$$

under the constraints $y_i[(\mathbf{w} \cdot \mathbf{x}_i) + b] \geq 1 - \zeta_i$ and $\zeta_i \geq 0$, where $\mathbf{w}$ is a normal vector that is perpendicular to the hyperplane, $\frac{\|\mathbf{w}\|}{b}$ determines the offset of the hyperplane from the origin along the normal vector $\mathbf{w}$ and C is a constant chosen a priori. This optimisation problem can then be translated into its corresponding dual problem [9]

$$\max \sum_{i=1}^{l} \alpha_i - \frac{1}{2} \sum_{i,j=1}^{n} \alpha_i \alpha_j y_i y_j K(\mathbf{x}_i, \mathbf{x}_j), \tag{3}$$

under the constraints, $0 \leq \alpha_i \leq C$, $\sum_{i=1}^{l} \alpha_i y_i = 0$, where K is a kernel function defined by $K(\mathbf{x}_i, \mathbf{x}_j) = \mathbf{x}_i \cdot \mathbf{x}_j$. The Lagrangian α_i is assigned to each example in the dual setting of SVM and it is independent on their order. It is positive if the example is a Support Vector (SV), and 0 if it is non-SV. Therefore, these Lagrangian values can be used as non-conformity measure for SVM.

SVMs are inherently two-class classifiers. The common approach to dealing with multi-class SVM is to reduce the multi-class problem into multiple binary classification problems. Then each of the problems can be solved by a binary classifier. In this paper, we propose and use the following non-conformity measure as a tool to link two-class problems together for multi-class classification problems, see Algorithm 2. For example, if we have 4 classes: A, B, C and D, then a binary class non-conformity measure can be applied to each of possible class pairs: A-B, A-C, A-D, B-C, B-D and C-D. When the class pair A-B is considered, non-conformity values are only assigned to those examples from the class A and class B and the examples from other classes are ignored. The same procedure is applied to the other class pairs. Having considered all the possible class pairs, each example now has 3 different non-conformity values. The maximal of them is considered as the final non-conformity measure of this example.

Algorithm 2. Non-conformity measure for multi-class SVM (for $q > 2$ classes)

Input: Data set $\{z_1 = (x_1, y_1), \dots, z_{l+1} = (x_{l+1}, y_{l+1})\}$, where $y_i \in Y = \{1, 2, \dots, q\}$
Input: z_j (one of examples above)
for $r = 1, \dots, q-1$ **do**
  for $s = r+1, \dots, q$ **do**
    Apply dual form SVM for examples from classes r and s only
    Extract Lagrange multipliers $\alpha_j^{SVM(r:s)}$ for j such that $y_j \in \{r, s\}$
    $\alpha_j^{SVM(r:s)} = 0$ for j such that $y_j \notin \{r, s\}$
  end for
end for
$\alpha_j := \max_{r,s} \alpha_j^{SVM(r:s)}$

3 Experiments and Results

Metal Oxide Semiconductor (MOS) sensors are commonly used in electronic nose applications for their convenience in operating and steadiness in features. The MOS gas sensors obtained from Figaro Engineering Inc., Japan (http://www.figaro.co.jp) were used in the experiments. In particular, an array of seven gas sensors: TGS2610, TGS2611, TGS800, TGS813, TGS822, TGS826 and TGS880 were chosen for the tea classification experiments. For illustration purposes, four types of tea which are ready available from the market: Japanese Green Tea (JG), Chinese Green Tea (CG), Indian Black Tea (IB) and Chinese Black Tea (CB) were used and we label them as class 1 to 4, respectively.

Each tea sample is heated before data acquirement. There are 68 data samples in total obtained in the experiments (15 for Indian Black Tea, 19 for Chinese Black Tea, 17 for Chinese Green Tea and 17 for Japanese Green Tea). Each data sample is described by 7 features obtained from those 7 gas sensors of the electronic nose system. A typical sensor response is shown in Figure 1. Originally each of these 7 features was represented as time series and we simply take its maximum value in time for the classification problem.

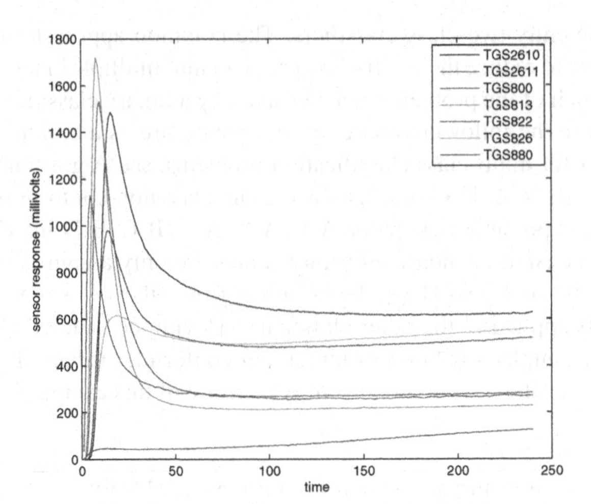

Fig. 1. An example of recorded sensor response

Table 1. Examples of individual predictions for leave-one-out cross validation

No.	T	p(1)	p(2)	p(3)	p(4)	P	Conf	Cred	No.	T	p(1)	p(2)	p(3)	p(4)	P	Conf	Cred
..	..	...	...	...	...	..	...	...	..	..	...	...	...	...	..	...	...
6	2	0.056	0.118	0.063	0.05	2	0.937	0.118	40	1	1	0.056	0.063	0.05	1	0.937	1
7	4	0.222	0.056	0.063	0.053	1	0.937	0.222	41	4	0.056	0.056	0.063	1	4	0.937	1
8	1	1	0.056	0.063	0.05	1	0.937	1	42	4	0.056	0.056	0.063	1	4	0.937	1
9	4	0.167	0.111	0.188	0.316	4	0.812	0.316	43	3	0.056	0.056	1	0.05	3	0.944	1
..	..	...	...	...	...	..	...	...	..	..	...	...	...	...	..	...	...

A 4-class SVM conformal predictor as described above (see Algorithm 2) has been implemented and applied to the tea classification dataset in leave-one-out mode. Table 1 shows example individual predictions with sample data presented in random order, where "T" is the true label, p(1) ... p(4) are calculated *p-values* for different labels, "P" is the forced prediction, "Conf" and "Cred" are the confidence and credibility associated with the forced prediction, respectively. For example, the true label is 4 and the forced prediction is 1 for the 7th sample data. The calculated confidence for the prediction is 0.937 and credibility is 0.222. High (i.e. close to 100%) confidence means that all labels except the predicted one are unlikely. Low credibility means that the whole situation is suspect; perhaps the test example is very different from all examples in the training set. The overall accuracy of forced prediction is 89.7%, as 61 of 68 predictions are correct according to the confusion matrix shown in Table 2.

The performance of forced prediction of the conformal predictor is compared with KIII model (a novel bionic neural network), a conventional artificial neural network (back-propagation network) and volunteers [11]. 30 volunteers were invited to try tea classification. All the volunteers were trained to remember the odour of each kind of tea. Then they made tea classification by smelling without seeing. Table 3 shows the comparison of forced prediction of conformal predictors with other methods on the same

Table 2. Confusion matrix for forced prediction

True Label\Forced Prediction	JG Tea (1)	CG Tea (2)	IB Tea (3)	CB Tea (4)
JG Tea (1)	14	1	1	1
CG Tea (2)	1	16	0	0
IB Tea (3)	1	0	14	0
CB Tea (4)	2	0	0	17

Table 3. Performance comparison (accuracy %)

Method	JG Tea (1)	CG Tea (2)	IB Tea (3)	CB Tea (4)	Average
KIII	93.3%	86.7%	93.3%	80%	88.3%
Back-Propagation	80%	100%	66.7%	93.3%	85%
Volunteers	80%	46.7%	83.3%	50%	65%
Conformal Predictor	82.4%	94.1%	93.3%	89.5%	89.7%

data set. Forced predictions made by conformal predictors have similar performance as those of the KIII model and BP networks. The volunteers performed not so well as the electronic nose, because of some physiological and psychological factors [5].

The predictive sets for a specific confidence level γ can be derived from the list of p-values by Equation (1). For the confidence level of 80%, 85%, 90% and 95%, the performance (in terms of error rate and multiple prediction rate) of the implemented conformal predictor is shown in Table 4. It is clear that the predictions made by the conformal predictor are valid and uncertainty rate becomes larger as the confidence level grows. Validity is true in the sense that error rate is smaller than 1 minus confidence level. The property of validity means that the conformal predictors never overrate the accuracy and reliability of their predictions. From the practical point of view, they allow us to control the error rate by selecting a suitable confidence level. For example, at 80% confidence level, the error rate is just over 16% and less 6% of predictions are uncertain. At 90% confidence level, over 23% of predictions become uncertain.

Table 4. Performance of conformal predictor at different confidence level

Confidence Level	Error Rate	Multiple Prediction Rate
80%	16.2%	5.9%
85%	11.8%	13.2%
90%	5.8%	23.5%
95%	0	100%

4 Conclusion

An investigation into application of conformal predictors to tea classification based on an array of gas sensors has been described. A new non-conformity measure for multi-class tea classification problem using SVM has been discussed. Experiments have been carried out to evaluate the performance of the conformal predictors. Empirical results have showed the good performance of the implemented conformal predictor.

Acknowledgements. This work was supported by a Royal Society International Project, Natural Science Foundation of China (NSFC-RS 60911130129) and ERASysBio grant for SHIPREC project.

References

1. United Kingdom Tea Council, http://www.tea.co.uk/the-home-of-tea-all-you-need-to-know-about-tea (last accessed: June 2010)
2. Bhattacharyya, N., Bandyopadhya, R., Bhuyan, M., Tudu, B., Ghosh, D., Jana, A.: Electronic Nose for Black Tea Classification and Correlation of Measurement with "Tea Taster" Marks. IEEE Transactions on Instrumentation and Measurement 57(7), 1313–1321 (2008)
3. Borah, S., Hines, E.L., Leeson, M.S., Iliescu, D.D., Bhuyan, M., Gardner, J.W.: Neural network based on electronic nose for classification of tea aroma. Sensing and Instrumentation for Food Quality and Safety 2(1), 7–14 (2008)
4. Duda, R.O., Hart, P.E., Stork, D.G.: Pattern Classification. Wiley, Chichester (2001)
5. Dutta, R., Kashwan, K.R., Bhuyan, M., Hines, E.L., Gardner, J.W.: Electronic nose based tea quality standardization. Neural Networks 16(5-6), 847–853 (2003)
6. Gonzalez, E., Li, G., Ruiz, Y., Zhang, J.: A Tea Classification Method Based on an Olfactory System Model. In: Advances in Cognitive Neurodynamics ICCN 2007, pp. 747–751 (2008)
7. Nouretdinov, I., Vovk, V.: Criterion of calibration for Transductive Confidence Machine with limited feedback. Theoretical Computer Science, Algorithmic learning theory 364(1), 3–9 (2006)
8. Tudu, B., Jana, A., Metla, A., Ghosh, D., Bhattacharyya, N., Bandyopadhyay, R.: Electronic nose for black tea quality evaluation by an incremental RBF network. Sensors and Actuators B: Chemical 138, 90–95 (2009)
9. Vapnik, V.N.: Statistical Learning Theory. Wiley, New York (1998)
10. Vovk, V., Gammerman, A., Shafer, G.: Algorithmic Learning in a Random World. Springer, Heidelberg (2005)
11. Yang, X., Fu, J., Lou, Z., Wang, L., Li, G., Freeman, W.J.: Tea classification based on artificial olfaction using bionic olfactory neural network. In: Proceedings of Third International Symposium on Neural Networks, pp. 343–348 (2006)
12. Yu, H., Wang, J.: Discrimination of LongJing green-tea grade by electronic nose. Sensors and Actuators B 122, 134–140 (2007)

Detecting and Confining Sybil Attack in Wireless Sensor Networks Based on Reputation Systems Coupled with Self-organizing Maps

Zorana Banković, David Fraga, José M. Moya, Juan Carlos Vallejo, Álvaro Araujo, Pedro Malagón, Juan-Mariano de Goyeneche, Daniel Villanueva, Elena Romero, and Javier Blesa

ETSI Telecomunicación, Universidad Politécnica de Madrid,
Av. Complutense 30, 28040 Madrid, Spain
{zorana,dfraga,josem,jcvallejo,araujo,malagon}@die.upm.es,
{goyeneche,danielvg,Elena,jblesa}@die.upm.es

Abstract. The Sybil attack is one of the most aggressive and evasive attacks in sensor networks that can affect on many aspects of network functioning. Thus, its efficient detection is of highest importance. In order to resolve this issue, in this work we propose to couple reputation systems with agents based on self-organizing map algorithm trained for detecting outliers in data. The response of the system consists in assigning low reputation values to the compromised node rendering them isolated from the rest of the network. The main improvement of this work consists in the way of calculating reputation, which is more flexible and discriminative in distinguishing attacks from normal behavior. Self-organizing map algorithm deploys feature space based on sequences of sensor outputs. Our solution offers many benefits: scalable solution, fast response to adversarial activities, ability to detect unknown attacks, high adaptability and low consumption. The testing results demonstrate its high ability in detecting and confining Sybil attack.

Keywords: wireless sensor networks, reputation system, self-organizing maps, outlier detection.

1 Introduction

WSNs consist of a large number of sensor nodes (also called motes). These nodes have to be very cheap, so they exhibit very limited power and computational resources, small memory size and low bandwidth usage and usually no tamper-resistant hardware is incorporated with any of them.

The most aggressive and the most evasive of all the attacks on sensor networks is the Sybil attack [1]. In essence, it refers to the scenario when one (or more) node(s) claim to have multiple identities, either fabricated or stolen ones. In this way it is able to affect on various aspects on network functioning, some of them being routing protocols, voting (in trust schemes), fair resource allocation, etc. Thus, it is of highest importance to efficiently detect and confine this attack.

H. Papadopoulos, A.S. Andreou, and M. Bramer (Eds.): AIAI 2010, IFIP AICT 339, pp. 311–318, 2010.

We believe that spatial and temporal characterization of the data coming from the sensors can be of great importance in discovering manipulated data and/or compromised nodes. Any major data inconsistence can be connected to malicious data manipulation.

In this work we propose to detect presence the of the Sybil attack using a self-organizing map (SOM) algorithm for detecting data outliers. The first step in deploying any machine learning technique is to define the model of data. The model consists of certain number of characteristics, i.e. features, that describe all possible aspects of the phenomenon. Furthermore, in our case it is essential to be able to distinguish normal from anomalous behavior.

For that reason, we deploy temporal and spatial models of the sensors using n-grams. The temporal model is defined for each sensor, while spatial model considers groups of close sensors. Each n-gram in the temporal model consists of a predefined number of successive sensor values, while an n-gram in the spatial model consists of outputs of all the sensors that make the group. Therefore, the features are the n-grams and the feature values are the number of occurrences or the frequency of the corresponding n-gram during a certain period of time. Considering that number of n-grams is not constant within consecutive periods of time, SOM deploys methods for measuring distance between sequences presented in [2].

We further propose to couple the system of detection agents based on SOM with a reputation system. In our proposal, the output of an agent affects on the reputation system in the way that it assigns lower reputation to the nodes where it detects adversarial activities and vice versa. We envision a reputation system where every node is being examined by at least one agent that resides on a node in its vicinity and listens to its communication in a promiscuous manner, and executes one of the algorithms for detecting attacks or temporal and spatial inconsistencies. We further advocate avoiding any contact with the nodes that have low reputation (below certain threshold). In this way, the compromised nodes remain isolated from the network and have no role in its further functioning. Comparing to our previous work on the subject [4, 5], in this work we propose improved way of calculating reputation based on the output of the SOM algorithm, which is more flexible and discriminative when it comes to distinguishing attacks from normal behavior. Furthermore, we present more thorough results on the behavior of the SOM algorithm in different scenarios.

The rest of the work is organized as follows. Section 2 present common solutions for treating the problem of the Sybil. Section 3 details the proposed solution, while Section 4 presents obtained results. Finally, conclusions are drawn in Section 5.

2 Previous Work on Coping with the Sybil Attack

The proposed solutions to the Sybil attack include [1]:

1. Radio resource testing which relies on the assumption that each physical device has only one radio;
2. Random key pre-distribution which associates the identity of the node to the keys assigned to it and validate the keys to see if the node is really who it claims to be;
3. Registration of the node identities at a central base station;
4. Position verification which makes the assumption that the sensor network topology is static.

Each of the above solutions has its own drawbacks. For example, we do not know in advance that every physical device is going to have only one radio interface. Moreover, some of the MAC protocols rely on the fact that each node has more than one radio interface. The key pre-distribution is challenging, since attackers can deploy side-channel attacks in order to discover secret keys [3]. Finally, the last solution is applicable only in static networks, which is very uncommon scenario since there is often a number of mobile nodes that change their position.

3 Proposed Solution

3.1 Feature Extraction and Formation of Model

As previously mentioned our idea is to find temporal and/or spatial inconsistence in sensed data in order to detect manipulated data and/or compromised nodes. For this reason, we follow the idea presented in our previous work [4] based on extracted *n*-grams and their frequencies within different time windows. For the purpose of illustration, we will give a short example for a sensor that detects presence. Let the sensor give the following output during the time window of size 20: 1 1 1 1 0 0 0 0 0 0 1 1 1 1 1 0 0 0 0 0. If we fix the *n*-gram size on 3, we extract all the sequences of size 3 each time moving one position forward. In this way we can observe the following sequences and the number of their occurrences within the time window: 111 – occurs 6 times, 110 – 2, 100 – 2, 000 – 6, 001 – 1, 011 – 1. Thus, we can assign them the following sequences: 111 – 0.33, 110 – 0.11, 100 – 0.11, 000 – 0.33, 001 – 0.05, 011 – 0.05. In our model, the sequences are the features and their frequencies are the corresponding feature values. Thus, the sum of the feature values is always equal to 1. In our algorithm this characterization is performed in predefined moments of time and takes the established amount of previous data, e.g. we can perform the characterization after every 40 time periods based on previous 40 values.

In a similar fashion, we form features for spatial characterization. The first step is to establish vicinities of nodes that historically have been giving consistent information. Furthermore, since an agent is supposed to reside on a node, vicinities are established using the nodes which information can reach the agent. In this way, an *n*-gram for spatial characterization in a moment of time is made of the sensor outputs from that very moment. For example, if sensors S1, S2, S3 each give the following output: 1 1 1 0 during four time epochs, we characterize them with the following set of *n*-grams (each *n*-gram contains at the first position the value of S1, the value of S2 at the second and the value of S3 at the third at a certain time epoch): 111 – occurs 3 times, 000 – occurs once, thus the feature value of each *n*-gram is: 111 – 0.75, 000 – 0.25, i.e. the frequencies within the observed period of time.

3.2 Detection of Sybil

The achievement of our design is based on the following two important assumptions:

1. The adversary can capture only a limited number of nodes, which means that most of the output data produced by the sensor nodes is normal. If this is not the case, it means that the adversary has become very powerful, so he is able to subvert any protocol in the network, which would require for the network re-initialization.

2. Output data produced under the influence of an adversary are statistically different from the output produced during the normal operation of the network. For this reason, we establish the detection of anomalies in data as outlier detection (an outlier is an observation that lies an "abnormal" distance from other values in a random sample from a population, i.e. extreme points in the data cloud).

If any of these assumptions is not fulfilled, our model is not able to work properly.

We treat attacks as data outliers and deploy SOM explained in more detail in our previous works [5]. There are two possible approaches for detecting outliers [6] using clustering techniques depending on the following two possibilities: detecting outlying clusters or detecting outlying data that belong to non-outlying clusters. For the first case, we calculate the average distance of each node to the rest of the nodes (or its closest neighborhood) (*MD*). In the latter case, we calculate quantization error (*QE*) of each input as the distance from its group center.

In our case, due to the definition of features and the deployed distance function, the distance can take values from the range [0, 2]. The process of updating cluster centers results in the centers that have all the *n*-grams that appear in the elements that belong to them, and the sum of their values is 1. Thus, in the normal case, *QE* will have values between 0 and 1. However, if an adversary manipulates data, it will result in different *n*-grams, so the corresponding distance will be between 1 and 2. For the same reason, if we have anomalous data in the training, they will form their own clusters. In this case, *MD* will be between 1 and 2, which is taken as anomalous.

3.3 Recovery from Sybil

Every sensor node is being examined by agents that execute SOM algorithm and reside on nodes in its vicinity and listen to its communication. The agents are trained separately. The system of agents is coupled with a reputation system where each node has its reputation value that basically reflects the level of confidence that others have in it based on its previous behavior. In our proposal, the output of an agent affects on the reputation system in the way that it assigns lower reputation to the nodes where it detects abnormal activities and vice versa. We further advocate avoiding any kind of interaction with the low-reputation nodes: to discard any data or request coming from these nodes or to avoid taking them as a routing hop. In this way, compromised nodes remain isolated from the network and have no role in its further performance. After this, additional actions can be performed by the base station, e.g. it can revoke the keys from the compromised nodes, reprogram them, etc.

In this work the reputation is calculated in the following way. We define two reputation values, *repQE* and *repMD* based on the previously defined *QE* and *MD* values and afterwards joint reputation *rep* used for updating overall reputation based on these two values:

```
if (QE<1) repQE = 1;            if (MD<1) repMD = 1;
else repQE=1-QE/2;              else repMD=1-MD/2;
```

For the reasons explained in the previous chapter, the value (*rep*) for updating overall reputation is calculated in the following way:

```
if (QE>1)rep=repQE;   else rep=repMD;
```

There are two functions for updating the overall reputation of the node, depending on whether the current reputation is below or above the established threshold that distinguishes normal and anomalous behavior. If the current reputation is above the threshold and the node starts behaving suspiciously, its reputation will fall quickly. On the other hand, if the reputation is lower than the established threshold, and the node starts behaving properly, it will need to behave properly for some time until it reaches the threshold. In order to achieve this, we use the function $x+log(1.2*x)$ because it provides what we want to accomplish: if x is higher than 0.5, the output rises quickly, so the reputation rises; if x is around 0.5, the output is around 0, so the reputation will not change its value significantly; if x is smaller than 0.4, the output falls below 0. Finally, the reputation is updated in the following way:

```
if (last_reputation[node]>threshold)
new_reputation[node]=last_reputation[node]+rep+log(1.2*rep);
else new_reputation[node]=last_reputation[node]+0.05*(rep+log(1.2*rep));
```

If the final value falls out from the [0, 1] range, it is rounded to 0 if it is lower than 0 or to 1 in the opposite case. In this way, we achieve that once a node start behaving suspiciously, its reputation will fall quickly. Yet, if a malicious node starts behaving properly, it will have to maintain its correct behavior during some time in order to "redeem" itself.

However, if during the testing of temporal coherence, we get normal data different from those that the clustering algorithms saw during the training, it is possible to get high *QE* value as well. On the other hand, the spatial coherence should not detect any anomalies. Thus, the final reputation will fall only if both spatial and temporal algorithms detect anomalies. In the opposite case, its reputation will not change significantly. This is implemented in the following way:

```
  if (value_rep < threshold) {
     if ( space_rep < threshold ) result = value_rep;
      else result = 1 - value_rep; }
   else result = value_rep;
```

where `value_rep` is the reputation assigned by the SOM for temporal characterization and `space_rep` is the reputation assigned by the SOM for spatial characterization.

4 Results

The proposed algorithm has been tested on a simulator of sensor networks developed by our research group. The simulated sensor network contains 200 sensor nodes that can take one of the possible 2000 positions. The network simulates a sensor network for detecting presence in the area of application, i.e. sensors give output 1 if they detect presence of a person or an object, or 0 if they do not detect any presence. The groups for spatial SOM algorithm are formed in a way that close sensors that should give the same output are placed in the same group. The simulation was carried out on a general purpose computer.

In our experiments the Sybil attack impersonates 10 existing sensor IDs. The duration of the experiment is 1000 time ticks. In the following we will present results in different scenarios regarding the presence of Sybil in training data and regarding two different definitions of *MD* value. In the first case, *MD* is defined as the medium distance to the three closest groups, while in the second case *MD* is the maximum of the three closest groups. This can be expressed mathematically in the following way:

Case 1:

$$MD(\mathbf{x}) = \frac{1}{3}\sum_{i=i_1,i_2,i_3}\left\|\mathbf{v}(\mathbf{x})-\mathbf{m}_i\right\|^2$$
$$i_1 = \arg\min_i\left\|\mathbf{v}(\mathbf{x})-\mathbf{m}_i\right\|^2, i_2 = \arg\min_i\left\|\mathbf{v}(\mathbf{x})-\mathbf{m}_i\right\|^2, i_2 \neq i_1, i_3 = \arg\min_i\left\|\mathbf{v}(\mathbf{x})-\mathbf{m}_i\right\|^2, i_3 \neq i_1, i_3 \neq i_2 \quad (1)$$

Case 2:

$$MD(\mathbf{x}) = \max_i\left\|\mathbf{v}(\mathbf{x})-\mathbf{m}_i\right\|^2, i \in \{i_1, i_2, i_3\}$$
$$i_1 = \arg\min_i\left\|\mathbf{v}(\mathbf{x})-\mathbf{m}_i\right\|^2, i_2 = \arg\min_i\left\|\mathbf{v}(\mathbf{x})-\mathbf{m}_i\right\|^2, i_2 \neq i_1, i_3 = \arg\min_i\left\|\mathbf{v}(\mathbf{x})-\mathbf{m}_i\right\|^2, i_3 \neq i_1, i_3 \neq i_2 \quad (2)$$

where $\mathbf{v}(\mathbf{x})$ is the centre to which the current input belongs.

Fig. 1.a and 1.b show the evolution of the reputation of every node after and before introducing the Sybil attack where the training stops at 600th time tick and Sybil starts at 650th. The *MD* value in the first case is calculated according to the formula (1), while in the latter case it is calculated according the formula (2).

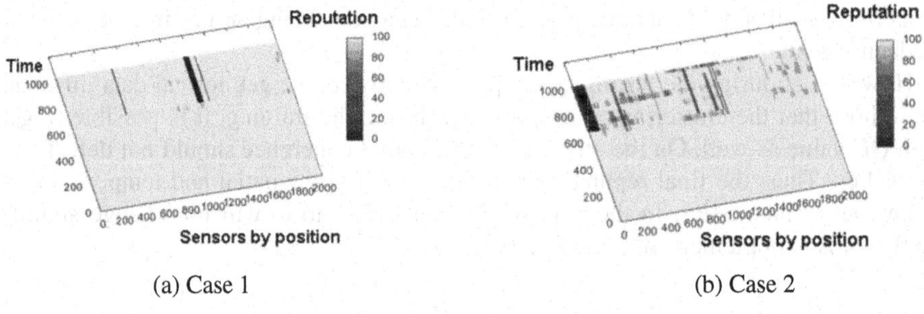

(a) Case 1 (b) Case 2

Fig. 1. Reputation Evolution

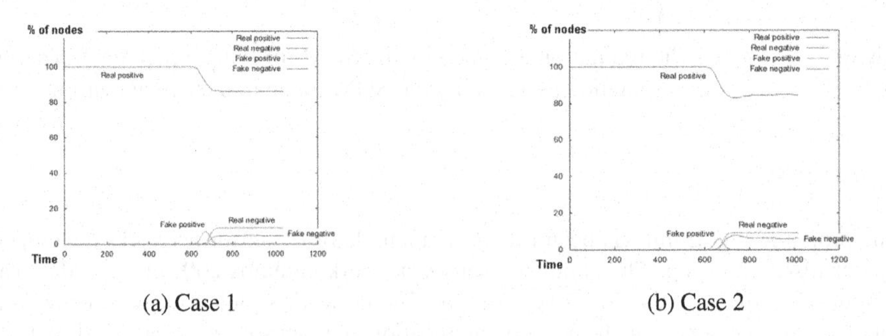

(a) Case 1 (b) Case 2

Fig. 2. Detection Evolution

Fig. 2.a and 2.b show the detection evolution in both of the cases. In these figures real positives are well-behaved nodes, real negatives are the ill-behaved nodes. Fake positives are non-detected ill-behaved nodes, while fake negatives represent the portion of well-behaved nodes falsely detected as ill-behaved.

In both Fig. 2.a and 2.b we can observe a thick dark line, which stands for the group of nodes attacked by Sybil. (Sybil attacks random nodes in each simulation, which is the reason why the dark lines are at different position.) The dark color reflects their low reputation. Fig. 2.a and 2.b confirm that in both cases all the attacked nodes have been detected (Fake Positive line). However, in Case 2 higher number of nodes should be sacrificed in order to confine the attack (Fake negative line). This can also be concluded from Fig. 1.a and 1.b. However, the advantage of Case 2 is its robustness, which will be demonstrated in the following.

In the following experiment Sybil starts at time tick 300. In the Case 2, the detector identifies and confines all the malicious nodes without having to change any of the parameters from the previous case (Fig. 4), while in Case 1 the detector detects the presence of the attack, but it is not able to confine it completely (Fig. 3). Experimenting with various parameters, we concluded that the maximum point to stop the training is 350 in order to completely confine the attack (Fig. 5). It is obvious that Case 1 is more sensitive to the presence of outlying data as minority, while Case 2 is more robust. These experiments also demonstrate that our system functions properly without the limitation of having (or not having) traces of attack in training data. Furthermore, we have demonstrated that detection possibilities of detectors can be enhanced through parameter changing.

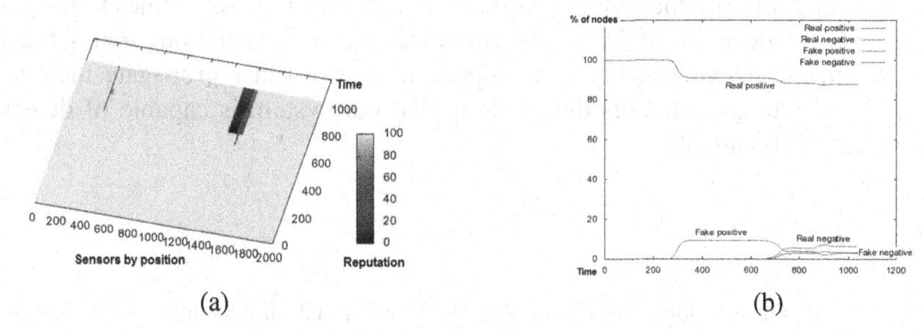

(a) (b)

Fig. 3. Case 1 (a) Reputation Evolution; (b) Detection Evolution

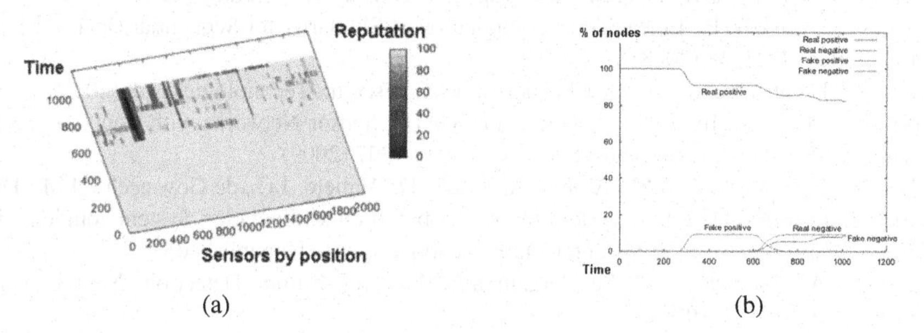

(a) (b)

Fig. 4. Case 2 (a) Reputation Evolution; (b) Detection Evolution

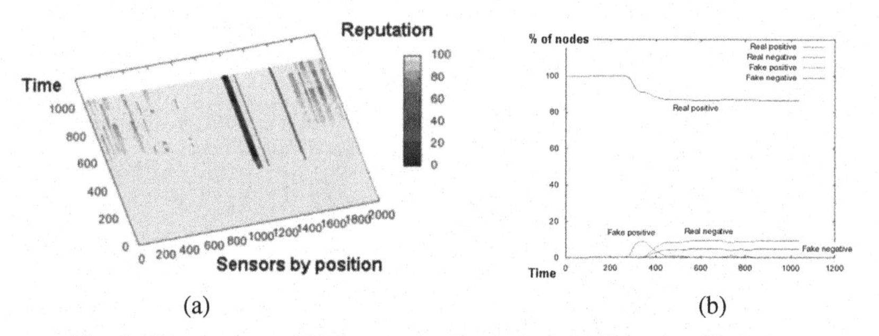

(a) (b)

Fig. 5. Case 1 repeated (a) Reputation Evolution; (b) Detection Evolution

Concerning the time of detection and confinement of the Sybil, our system is capable of detecting and completely confining the attack if up to 30% of the existing IDs have been taken by the Sybil. The presence of the attack is detected at the end of the first testing cycle in all the cases, while the confinement time spans from one to four testing cycles and becomes higher as the Sybil takes more than 15% of the IDs.

5 Conclusions

In this work we have presented a novel approach for coping with the Sybil attack in wireless sensor networks. We have proposed unsupervised machine learning SOM algorithm for detecting outliers in data and deploys a feature set that is more general than those presented by the solutions of the state-of-the-art. Furthermore, it does not depend on the presence (or non-presence) of anomalous data during the training.

The idea of confining the Sybil is based on assigning reputation values to the nodes according to the decision of SOM algorithm. In this way, malicious nodes become isolated from the network which will impede them to further propagate their malicious activity. Our experiments demonstrate that our system is capable of detecting and confining Sybil attack.

References

1. Newsome, J., Shi, E., Song, D., Perrig, A.: The Sybil Attack in Sensor Networks: Analysis & Defenses. In: Proceedings of ACM/IEEE International Conference on Information Processing in Sensor Networks, April 26-27, pp. 259–268. ACM, Berkeley (2004)
2. Rieck, K., Laskov, P.: Linear Time Computation of Similarity for Sequential Data. J. Mach. Learn. Res. 9, 23–48 (2008)
3. Bar El, H.: Introduction to Side Channel Attacks. Discretix Technologies Ltd. (2003)
4. Moya, J.M., et al.: Improving Security for SCADA Sensor Networks with Reputation Systems and Self-Organizing Maps. Sensors 9, 9380–9397 (2009)
5. Banković, Z., Moya, M.J.M., Araujo, A., Fraga, D., Vallejo, J.C., de Goyeneche, J.M.: Distributed Intrusion Detection System for WSNs based on a Reputation System coupled with Kernel Self-Organizing Maps. Int. Comp. Aided Design (to be published)
6. Muñoz, A., Muruzábal, J.: Self-Organizing Maps for Outlier Detection. Neurocomputing 18(1-3), 33–60 (1998)

Statistical Fault Localization with Reduced Program Runs

Lina Hong and Rong Chen*

College of Informational Science and Technology,
Dalian Maritime University, 1 Linghai Road, Dalian 116026, P.R. China
{tsmc.dmu,rchen.cs}@gmail.com

Abstract. A typical approach to software fault location is to pinpoint buggy statements by comparing the failing program runs with some successful runs. Most of the research works in this line require a large amount of failing runs and successful runs. Those required execution data inevitably contain a large number of redundant or noisy execution paths, and thus leads to a lower efficiency and accuracy of pinpointing. In this paper, we present an improved fault localization method by statistical analysis of difference between reduced program runs. To do so, we first use a clustering method to eliminate the redundancy in execution paths, next calculate the statistics of difference between the reduced failing runs and successful runs, and then rank the buggy statements in a generated bug report. The experimental results show that our algorithm works many times faster than Wang's, and performs better than competitors in terms of accuracy.

Keywords: Software Fault Localization, Path Redundancy, Statistical Method, Clustering.

1 Introduction

A typical thinking in fault localization is to compare successful runs and failing runs [2, 3, 4, 5, 6, 7]. There are different ways of comparison, and can be divided into distance measures-based methods [2, 3, 4] and characteristic statistics-based methods [5, 6, 7]. But both methods require a lot of failing runs and successful runs. Our analysis found that many characteristics of the runs are same.

Figure 1 shows a program example with a bug in the assignment statement on line 11, which should be "max = x;" In order to locate this buggy statement by using Wang [3] method, we run manual design of 12 test cases, and obtain 8 successful runs and 4 failing runs, in which there exists exactly same paths, such as max (4, 2, 3) and max (8, 5, 7). To be fair, for sorting methods students write, we execute randomly generated test cases, and statistics showed that the proportion of redundant paths is so much as 20% ~50%.

---

* The corresponding Author. This work was supported by National Natural Science Foundation of China (60775028), Dalian Science &Technology Program (2007A14GX042), IT Industry Development of Jilin Province.

H. Papadopoulos, A.S. Andreou, and M. Bramer (Eds.): AIAI 2010, IFIP AICT 339, pp. 319–327, 2010.

Redundancy leads to many problems: (1) Redundancy contributes nothing to the fault location calculations. If two failing runs are with the same path, it means that they contributions same to fault localization, so there is no need to separately calculate the difference between each of them and successful runs, the same applies two successful runs with two paths the same. (2) Computational efficiency of the system is reduced. Because the system may spend a lot of time doing pointless things. (3) The calculation of too much redundant data may lead to biased results. Thus removing the path redundancy is necessary.

Measures-based method returns executing differences as bug report, however the statements in differences are not ordered by their importance in literature [3]. Techniques based on characteristics statistics rank the statements or predicates in order of suspicious, but it is so many statements and predicates, even including those are not predictive of anything which is usually a majority, statistical work on them are time-consuming and effortless.

To overcome these shortcomings, in this paper, we first propose a clustering method to eliminate the path redundancy, and then through statistical analysis of differences between classes of runs to get and rank suspicious statements. Experimental comparison with Wang's algorithm shows that our algorithm makes a greatly improved efficiency and accuracy of locating. In addition, our algorithm can not only diagnose the reasons for each failing run, it can also get bug report of the whole program through statistical analysis of bug reports of all the failing runs.

```
1  package function;
2  public class ThreeNum{
3  /**
4  * Calculate the maximum of three numbers
5  */
6  public int max(int x, int y, int z)
7  {
8      int max;
9      if (x > y)
10     {
11         max = y; //should be max=x;
12     }
13     else
14     {
15         max = y;
16     }
17     if (z > max)
18     {
19         max = z;
20     }
21     return max;
22  }
23 }
```

Fig. 1. A small program example

Section 2 summarizes the related work; Section 3 introduces our fault localization method based on statistical differences between reduced runs; Section 4 shows comparative analysis of experimental results. Section 5 and gives the next step.

2 Related Work

Program runs statistical method is in essence a statistical intelligent method, how to make statistics and compare successful runs and failing runs attracted interest of many researchers [2, 3, 4, 5, 6, 7]. According to different type of comparison techniques, these methods can be divided into distance measures-based method and feature-based statistical approach.

Distance measures-based method is to find a successful run which is the most similar to the failing run through a certain distance measure technique, and then calculate the difference between it and failing run for fault location. A typical work of these methods is that Renieris [2] et al proposed the "Nearest Neighbor" and Wang [3] et al proposed calculating difference of two runs through alignment based-on control flow information.

In contrast, feature-based statistical approach locates fault-relevant statements (or faulty statements directly) by comparing the statistical information of program elements in these two kinds of run. Such program elements can be statements [5] or predicates [6, 9, 7]. Tarantula [5] statistics the frequency of every statement occurs in failing runs and successful runs, and by analyzing them to get and rank the suspicious statements. Predicate-based statistical techniques, such as CBI [6, 9] and SOBER [7], locate the program predicates related to faults. CBI [6, 9] measures the increase from the probability of a predicate to be evaluated to be true in all failed runs to that in all the runs, This increase is used as the ranking score, which indicates how much the predicate is related to a fault. SOBER [7] defines evaluation bias to estimate the chance that a predicate is evaluated to be true in each run. In brief, CBI and SOBER use similar kinds of statistical mean comparison.

3 Fault Localization with Reduced Run

3.1 Clustering Execution Paths

Next we define execution paths and their clustering.

Definition 1. (Event) An event denotes an execution of a statement of a given program. As [6] we use the line number *i* of a statement to label the event e_i associated with the statement, and the statement is thus denoted as *stmt*(e_i). A test case that executes a given program provides the values of input variables and the expected values of output variables.

Definition 2. (Run) A run π of a program is a sequence of events $<e_0, e_1, e_2, \ldots, e_{n-1}>$ that are sequentially associated with statements executed by a given test case. Moreover, we denote the ith event of a run π as e_i^{π} and number of events of π as $|\pi|$.

Table 1. Test cases and execution runs of program in fig.1

output	Succ0	Succ1	Succ2	Succ3	Fail0	Fail1	Succ4	Succ5	Succ6	Succ7	Fail2	Fail3
TCinputs	2,3,4	2,4,3	3,2,4	3,4,2	4,2,3	4,3,2	5,7,8	5,8,7	7,5,8	7,8,5	8,5,7	8,7,5
Program runs	6_1	6_1	6_1	6_1	6_1	6_1	6_1	6_1	6_1	6_1	6_1	6_1
	8_2	8_2	8_2	8_2	8_2	8_2	8_2	8_2	8_2	8_2	8_2	8_2
	9_3	9_3	9_3	9_3	9_3	9_3	9_3	9_3	9_3	9_3	9_3	9_3
			11_4		11_4	11_4			11_4		11_4	11_4
	15_4	15_4		15_4			15_4	15_4		15_4		
	17_5	17_5	17_5	17_5	17_5	17_5	17_5	17_5	17_5	17_5	17_5	17_5
	19_6		19_6		19_6		19_6		19_6		19_6	
	21_7	21_6	21_7	21_6	21_7	21_6	21_7	21_6	21_7	21_6	21_7	21_6

Consider the program in Fig.1, we can get successful and failing runs showed in Table 1 by executing a set of test cases, where the *TC inputs* row sequentially put down the input values of variables x, y, and z within test cases (e.g., "2,3,4" means a test case having inputs {x=2, y=3, z=4}), while the *output* row describes that fact whether the TC inputs leads to a successful run or a failing run (The indexed Succ for successful runs, and Fail for failing runs), and program runs are put sequentially in columns below the corresponding test cases. For example, the TC inputs "2,3,4" yields a run <6_1, 8_2, 9_3, 15_4, 17_5, 19_6, 21_7>, where the subscripts denotes the event indices (e.g., 8_2 means that the statement on line 8 is executed once and is the second event of this run). Note that some cells in program runs are empty because we align statements for highlighting the difference in program runs.

Definition 3. (Dependence): Given a program P, let a variable v_1 be defined in statement S_1 and is used in statement S_2 defining value of a variable v_2. v_2 is dependent on v_1 if changes to v_1 in the execution will is likely to influence the definition and value of v_2. Equivalently, we also say statement S_2 is dependent on S_1. Moreover, v_2 is *control dependent* on v_1, if v_1 occurs in a condition the truth of which controls the execution from S_1 to S_2; otherwise, we say that variable v_2 is *data dependent* on v_1. Equivalently, we raise control/data dependence to the level of statements [8].

Definition 4. (Dynamic Control Dependence) Given a program run π, an event e_i^π is dynamically control dependent on another event e_j^π if e_j^π is the nearest event coming before e_i^π in π such that $stmt(e_i^\pi)$ is control dependent on $stmt(e_j^\pi)$. Moreover, we use $dep(e_i^\pi, \pi)$ to denote the events on which e_i^π is dynamically control dependent in a run π.

Definition 5. (Alignment [3]) For any pair of events e and e' (e in run π and e' in run π'), we say e and e' are aligned, denoted as $align(e, e')=true$, iff (1) $stmt(e)=stmt(e')$, and (2) either e and e' are the first events appearing in π and π' respectively, or $align(dep(e, \pi), dep(e', \pi'))=true$.

Definition 6. (Alignment Vector) Given two program runs π' and π, a alignment vector, denoted as $\Delta(\pi, \pi')$, is an array of marks for all events in π' and π, obtained as follows: for each event e of π, mark is 1 if there exists e' of π' which can aligned to e, otherwise mark is 0.

By definition 6, the following corollary is obvious:

Corollary 1. Given two program runs π' and π, $\Delta(\pi, \pi')= \Delta(\pi', \pi)$ holds.

We use Algorithm I to compute alignment vectors, where predicate $alignExist(i, \pi_1, \pi_2)$ is used to check whether there exists an event of π_2 that can be aligned to some e_i of π_1, and function $alignIndex(i, \pi_1, \pi_2)$ is used to get the index of such event of π_2. Moreover, function $\Delta.add(i)$ adds a mark i in the alignment vector Δ.

```
Algorithm I: delta(π1, π2)
input: two program runs π1 and π2
output: an alignment vector Δ(π1, π2)
1. t=1; j=0; //temporarily store the index
              //of events of π2
2. Δ={}; //the return result
3. outer:
4. for (i=1; i<=|π1|; i++){
5.   if alignExist(i, π1, π2) {
6.      j=alignIndex(i, π1, π2);
7.      k=j – temp2 – 1;
8.      for (n=0; n<k; n++) Δ.add(0);
9.      Δ.add(1); //add a "1"
10.   } else {//otherwise add a "0"
11.     Δ.add(0); }
12. t=j;
13. continue outer;
13. }
14. return Δ;
```

Definition 7. (Alignment Matrix) An alignment matrix of a program run π is a matrix composed of all alignment vectors $\Delta(\pi, \pi')$ such that π' is any run distinguish from π, and 0s are appended for padding if the length of alignment vectors are not constant.

By Table 1 and definition 6, we have (1) Δ(**fail0**,**fail0**)= <1,1,1,1,1,1,1>, (2) Δ(**fail0**,**fail1**)= <1,1,1,1,1,0,1>, (3) Δ(**fail0**, **fail2**)= <1,1,1,1,1,1,1>, and (4)

Δ(**fail0,fail3**)= <1,1,1,1,1,0,1>. With such an alignment matrix, we cluster the matrix by column and thus four failing runs in Table I can be divided into two classes:

F_I: $fail_0(4,2,3)$, $fail_2(8,5,7)$; F_{II}: $fail_1(4,3,2)$, $fail_3(8,7,5)$

where $fail_i(4,2,3)$ is abbreviated for the ith failing run with a TC inputs (4,2,3)). In the same manner, eight successful runs are divided into three classes:

S_I:$succ_0(2,3,4)$, $succ_4(5,7,8)$; S_{II}:$succ_1(2,4,3)$, $succ_3(3,4,2)$, $succ_5(5,8,7)$ and $succ_7(7,8,5)$;

S_{III}: $succ_2(3,2,4)$, $succ_6(7,5,8)$.

3.2 Difference Metric

Definition 8. (Difference Metric [3]) Given two program runs π and π'. The difference between π and π', denoted as $diff(\pi,\pi')$, is defined as $diff(\pi,\pi')=<e_{i1}^{\pi}, e_{i2}^{\pi},\ldots, e_{ik}^{\pi}>$, such that (1) each event e in $diff(\pi,\pi')$ is a branch event occurrence drawn from run π, (2) the events in $diff(\pi,\pi')$ appear in the same order as in π, that is, for all $1\leq j<k$, $i_j<i_j+1$ (event e_{ij}^{π} appears before event e_{ij+1}^{π} in π), (3) for each e in $diff(\pi,\pi')$, there exists another branch occurrence e' in run π' such that $align(e,e')=true$ (i.e. e and e' can be aligned). Furthermore, the outcome of e in π is different from the outcome of e' in π'. (4) All events in π satisfying (1) and (2) are included in $diff(\pi,\pi')$. (5) As a special case, if execution runs π and π' have the same control flow, then we define $diff(\pi,\pi')=<e_0^{\pi}>$.

```
Algorithm II: diff(π1, π2)
input: two program runs π1 and π2
output: a set Γof event indices representing
        difference between π1 and π2
1. Γ={};
2. Δ=delta(π1, π2);
3. n=0; //to store the total number
        //of 0s coming before 1
4. for (i=0; i <|Δ| – 1; i++) {
5.   if (Δ.get(i)=0) n++;
6.   if (Δ.get(i)=1∧Δ.get(i+1)=0) {
//if the ith event can be aligned, while
//the next event can't be aligned
7.       Γ.add(i+1–n+|Γ|); }
8. }
```

By Definition 8, the statements in the difference are all branching statements from which two runs separately go to different branches. So if there are two adjacent values in the vector such that the former is a "1" and the latter a "0", then the event that "1" represents is added to the difference. In Algorithm II, Note that $\Delta.get(i)$ returns the ith bit of the alignment vector Δ, while $\Gamma.add(i)$ adds a mark i in the alignment matrix Γ. Regarding our running example, differences are summarized in Table2 and Table 3:

Table 2. Difference between each class of failing runs and successful runs

	S_I	S_{II}	S_{III}
F_I	9_3	$9_3,17_5$	null
F_{II}	$9_3,17_5$	9_3	17_5

Table 3. Difference between each class of failing runs

	F_I	F_{II}
F_I	null	17_5
F_{II}	17_5	null

3.3 Fault Localization

For each predicate p in a program P, LIBLIT [6] estimates two conditional probabilities: (1) Pr1=Pr(P fails| p is ever observed); (2) Pr2=Pr(P fails| p is ever observed as true). LIBIT then thinks the difference Pr2–Pr1 as an indicator of how relevant p is to the fault. So we conclude that difference between failing runs and successful runs indicates that faults are more likely to appear in these positions, while difference between failing runs and failing runs indicates that the probability that fault appears in these position is very small, that is, the predicate being true can not increase the possibility of failure. Let $P(e,\pi,S)$ be the probability that event e of run π appears in the difference between π and successful runs, and $P(e,\pi,F)$ between π and failing runs. Then for any event e in a failing run π, we define the following $Score(e,\pi)$ to indicate the suspiciousness of statements $stmt(e)$. the formulas are as follows:

$$P(e,\pi,S) = \frac{\textit{times that e appear in difference between } \pi \textit{ and successful runs}}{\textit{the total number of events in difference between } \pi \textit{ and successful runs}} \tag{2}$$

$$P(e,\pi,F) = \frac{\textit{times that e appear in difference between } \pi \textit{ and failing runs}}{\textit{the total number of events in difference between } \pi \textit{ and failing runs}} \tag{3}$$

$$Score(e,\pi) = \begin{cases} P(e,\pi,S)/P(e,\pi,F), & P(e,\pi,F) \neq 0 \\ \infty, & \textit{otherwise} \end{cases} \tag{4}$$

Given Table 1, Table 2 and (2), (3), and (4), we have:

$P(9_3,F_I,S)=2/3$, $P(9_3,F_I,F)=0$, and $Score(9_3,F_I)=\infty$;
$P(17_5,F_I,S)=1/3$, $P(17_5,F_I,F)=1$, and $Score(17_5,F_I)=1/3$;
$P(9_3,F_{II},S)=1/2$, $P(9_3,F_{II},F)=0$, and $Score(9_3,F_{II})=\infty$;
$P(17_5,F_{II},S)=1/2$, $P(17_5,F_{II},F)=1$, and $Score(17_5,F_{II})=1/2$.

Then we get the rank score of suspiciousness that each event in the differences is the real cause of each failure. We summarize them in Table 4, where $\infty(2/3)$ and the like denote the case of $P(e,\pi,F)=0$, $P(e,\pi,S)=2/3$, in which we compare suspiciousness by $P(e,\pi,S)$. If the rank score of two events are equal, we consider event that appears later has a larger suspiciousness score and will be ranked in the top [3].Assume a certain statement is responsible for a failure; it may be executed several times and appears several times in the differences. So we compute ranking scores of each statement for each failure as (5). So by (5), we get Table 5 of suspicious statements for each class of failures in our running example.

$$Score(stmt,\pi) = \sum Score(e,\pi) \quad (e \in \pi \textit{ and } stmt(e) = stmt) \tag{5}$$

Table 4. Score of suspicious events for failures

Score of suspicious events of F_I		
	9_3	17_5
F_I	$\infty(2/3)$	1/3
Score of suspicious events of F_{II}		
F_I	9_3	17_5
F_{II}	$\infty(1/2)$	1/2

Table 5. Score of suspicious statements for failures

Statements	9	17
F_I	$\infty(2/3)$	1/3
F_{II}	$\infty(1/2)$	1/2

We can now locate faults through the analysis of Table 5. Each row of the table represents the ranking scores of each statement for a certain failing run, if there are some statements the value of which is significantly larger than others, the statements are the real cause of such failures. Similarly each column represents the ranking scores of a certain statement for each failing run,, the relevant statements with significantly larger values are the cause of these failures. As for our example, statement 9 has a large effect on both F_I and F_{II}, indicating that statement 9 leads to the failure with respect to these two classes of failing runs. So we use the following equation to compute the suspiciousness of any statement.

$$Score_p(stmt) = \sum_{\pi \in F} Score(stmt, \pi)(F \text{ is the set of all failing runs}) \quad (6)$$

4 Experiment Results

We chose four middle-sized programs with branches, by manually injecting different errors (some programs injected with an error and some with two errors). In this way, we get 43 buggy programs and 79 failing runs. Table 6 shows the description of buggy programs generated from the original four programs.

Table 6. Description of experiment data

Programs	nested level of branches	Buggy Versions			
		having one bug	*Kinds of failing runs*	*having two bugs*	*Kinds of failing runs*
P1	0	5[a]	14	3	11
P2	1	4	4	4	7
P3	2	8	9	8	16
P4	3	5	5	7	13

To evaluate our algorithm and compared it with Wang's, we consider three cases for each failing run: (1) case "1": for ours the largest suspiciousness statement or for Wang's the whole bug report exactly indicates the actual fault position, (2) case "-1": for two methods the actual faulty location is missed in the bug report, and (3) case "0": for ours the most suspicious statement is not the actual fault position, but the actual fault location is under suspicion. Statistical results obtained from experiments are shown in Fig.2, where the vertical axis is the number of failing runs.

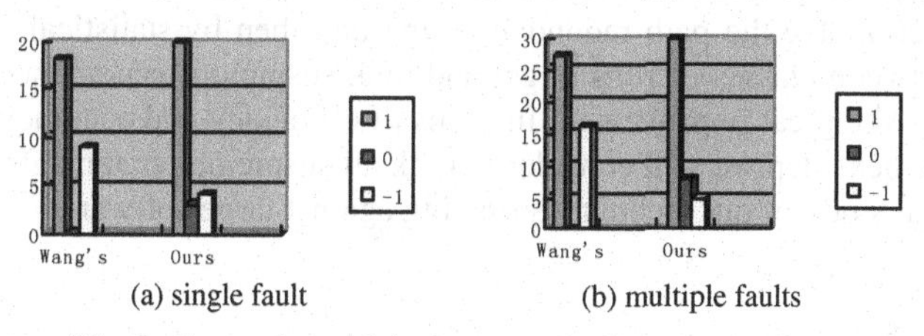

(a) single fault (b) multiple faults

Fig. 2. Comparison of the bug report of two methods

Wang's algorithm returns the smallest difference as the bug report, ignoring that the position of the fault may not be in this minimal difference but in other differences. In contrast, we use statistical methods taking into account all the circumstances, and also rank the suspicious statements according to suspiciousness. The faulty statements are bound to occur in our report unless they do not appear in any of the differences.

For programs with multiple bugs, we use the evaluation criteria in [3] to calculate the *pgm_score(P)*:instead of using each failing run and successful run, we use each class of failing runs and each class of successful runs. The *pgm_score*(*P*) measures the percentage of code that can be ignored for debugging, the algorithm calculates the score of the most suspicious statement and statement with its suspicious in the second place, and finally calculate *pgm_score(P)*, as shown in Table 7.

According to table 7, when there are two errors in the program, the most suspicious statement in our bug report has higher score than that by Wang's algorithm. The second suspicious statement also has a higher score, indicating that it is also likely to be the actual fault position, that is, there are more likely two errors in the program.

It can be seen from Fig.3, the time Wang spending grows exponentially as the redundant data increases since it spends much on calculating differences between each failing run and each successful run. While our algorithm eliminates redundancy by clustering all the runs, and the next difference computation is less time-consuming, so the time grows slightly. Moreover, we not only consider the suspiciousness for each failing run, but also that for the entire program through statistical analysis of all failing runs and rank statements in bug report based on their overall suspiciousness.

Table 7. Comparison of two methods when there are two errors in the program

Score	Wang algorithm	The first place of our algorithm	The second place of our algorithm
0.8-1.0	13.3	33.3	46.7
0.7-0.79	60.0	26.7	6.7
0.6-0.69	6.7	6.7	20.0
0.5-0.59	13.3	26.7	6.7
0-0.49	6.7	6.7	20.0

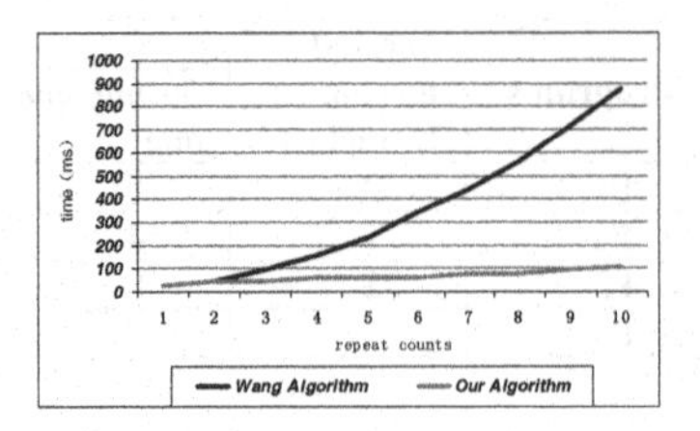

Fig. 3. Comparison of run time of two methods

5 Conclusion

In this paper, we present an improved fault localization method using a clustering method to eliminate the path redundancy first and then by statistical analysis of differences between classes of runs to get and rank suspicious statements. Experimental results show the great improvement in terms of efficiency and accuracy in fault localization. The next step we will consider to rank all suspicious statements and do further experimental study of our techniques running against large software.

References

[1] Jones, J.A., Harrold, M.J., Stasko, J.: Visualization of test information to assist fault localization. In: Proceedings of the 24th International Conference on Software Engineering, pp. 467–477 (2002)

[2] Renieris, M., Reiss, S.P.: Fault localization with nearest neighbor queries (2003)

[3] Guo, L., Roychoudhury, A., Wang, T.: Accurately choosing execution runs for software fault localization. In: Mycroft, A., Zeller, A. (eds.) CC 2006. LNCS, vol. 3923, pp. 80–95. Springer, Heidelberg (2006)

[4] Wang, T., Roychoudhury, A.: Automated path generation for software fault localization. In: ASE 2005: Proceedings of the 20th IEEE/ACM international Conference on Automated software engineering, pp. 347–351. ACM, New York (2005)

[5] Jones, J.A., Harrold, M.J.: Empirical evaluation of the tarantula automatic fault-localization technique. In: ASE 2005: Proceedings of the 20th IEEE/ACM international Conference on Automated software engineering, pp. 273–282. ACM, New York (2005)
[6] Liblit, B., Naik, M., Zheng, A.X., Aiken, A., Jordan, M.I.: Scalable statistical bug isolation. In: PLDI 2005: Proceedings of the 2005 ACM SIGPLAN conference on Programming language design and implementation, pp. 15–26. ACM, New York (2005)
[7] Liu, C., Yan, X., Fei, L., Han, J., Midkiff, S.P.: Sober: statistical model-based bug localization. SIGSOFT Softw. Eng. Notes 30(5), 286–295 (2005)
[8] Ferrante, J., Ottenstein, K.J., Warren, J.D.: The program dependence graph and its use in optimization. ACM Transactions on Programming Languages and Systems 9, 319–349 (1987)
[9] Liblit, B., Aiken, A., Zheng, A.X., Jordan, M.I.: Bug isolation via remote program sampling. In: PLDI 2003: Proceedings of the ACM SIGPLAN 2003 conference on Programming language design and implementation, pp. 141–154. ACM, New York (2003)

Fuzzy Cognitive Map for Software Testing Using Artificial Intelligence Techniques

Deane Larkman[1], Masoud Mohammadian[1],
Bala Balachandran[1], and Ric Jentzsch[2]

[1] Faculty of Information Science and Engineering,
University of Canberra, ACT, Australia
u950777@uni.canberra.edu.au, Masoud.Mohammadian@canberra.edu.au
[2] Business Planning Associates Pty Ltd, ACT, Australia
Bus.Planning.Assoc@gmail.com

Abstract. This paper discusses a framework to assist test managers to evaluate the use of AI techniques as a potential tool in software testing. Fuzzy Cognitive Maps (FCMs) are employed to evaluate the framework and make decision analysis easier. A what-if analysis is presented that explores the general application of the framework. Simulations are performed to show the effectiveness of the proposed method. The framework proposed is innovative and it assists managers in making efficient decisions.

Keywords: Software testing, Fuzzy Cognitive Maps (FCMs), What-if Analysis.

1 Introduction

Software is a key element of systems and devices that support many of the activities that are an accepted part of our modern lifestyle. There is a consequent reliance on the correct behaviour of software, and an expectation that the software will not fail. However, software is an increasingly complex product that requires more and more testing a labour intensive and error prone activity [1]. The consequences of software failure range from the trivial (such as the need to restart a computer program), through the very inconvenient (such as the malfunction of traffic signals), to the catastrophic, where life and property may be affected. To minimise software failure, and its various impacts, high quality software must be created. Testing is a major quality assurance technique to evaluate the quality of software throughout the development cycle [1, 2].

2 Software Testing and Artificial Intelligence

2.1 Research on Software Testing Using Artificial Intelligence

The application of AI techniques to testing software has moved beyond the speculative for many techniques, and now receives widespread attention from the research

H. Papadopoulos, A.S. Andreou, and M. Bramer (Eds.): AIAI 2010, IFIP AICT 339, pp. 328–335, 2010.

community. AI techniques have been used to explore different characteristics of a range of software, such as system software, real time software, embedded software, distributed software, and GUI software. The diverse investigations cover many areas. Various test approaches, at different test levels (such as unit, integration and system testing), have been used; for example, white box or structural testing, black box or functional testing, grey box testing, GUI testing and non-functional testing. Different programming paradigms have been examined, including procedural programs, object-oriented programs and aspect oriented programs. Distinct test aspects have been investigated, such as test data generation and test oracle generation [3, 4, 5, 6]. Software programs have been transformed to facilitate the application of AI techniques. AI techniques have been optimised, modified, hybridized with other AI techniques, and adapted.

2.2 Limitations Using Artificial Intelligence in Software Testing

Although AI techniques are grounded in theory, these theoretical foundations have rarely been developed to address software testing problems. Investigations about the theoretical basis for using AI techniques to test software are limited. Consequently, AI techniques for testing software lack a firm scientific basis. A recent paper analysed the theories underpinning genetic algorithms, developed these theories for structural test data generation, and empirically validated the predictions of the theories against real world programs [3]. The reason for using an AI technique is commonly analytical. When cited, the reason for using an AI technique is usually because of the similarities between the characteristics of the AI technique, and those characteristics of the software testing problems.

Much of the empirical evidence for the use of AI techniques to test software stems from artificial laboratory programs, which are small and simple programs: real world programs are often not considered for validation. Real world programs are not merely scaled up versions of laboratory programs; but present complex test problems, which frequently require an intensive manual test effort to solve. Many researchers have voiced concern about the need to validate AI techniques against real world programs, rather than against simple laboratory programs [3, 5, 6, 7]. The limitations of laboratory programs have been observed, and it has been noted that toy (simple) programs fail to reveal the limitations of some test data generation techniques [6].

3 Framework Software Testing Evaluating AI Techniques

A framework is at a high level generalised perspective of a domain of interest. It is a way to provide an initial understanding of some directed environment and its concepts or constructs. The framework described in this paper is a decision support framework, and was developed for use at an organisational environment level. It is aimed at test managers, or their equivalents – those people who, amongst other things, construct test plans, which encompass or are underpinned by test strategies.

Test managers are decision makers. The purpose of developing this framework is to support one aspect of that essential activity. Decision making is an increasingly complex activity with considerable potential for error. Decision makers need to have at their disposal tools to help reduce the risks inherent in their decisions. The complexity of the real world dictates the requirement to have tools that can be used to help lower the decision risk by selecting and analysing amongst multiple alternatives. Test managers need to have a way to help them decide which test technique will be the most beneficial. The selected technique must be consistent with the test strategy or with the test approach used to evaluate the software artefact, and must help reduce the risk of software defects, especially the critical ones, not being found before the software is released to users, customers, clients, or the public at large.

The components of the AI techniques decision support testing framework are Test Management, Test Information, Test Environment, and Technical Support The framework objective is the possible application of AI Techniques to the particular software to be tested. All the components relate to the objective.

4 Fuzzy Cognitive Maps – Application to the Framework

4.1 Fuzzy Cognitive Maps

Fuzzy Cognitive Maps (FCMs) [9, 10] are graph structures that provide a method to capture and represent complex relationships within an environment (which defines a boundary), to improve understanding of that environment. FCMs have been used to model problems with no data [8, 9, 10]. A FCM can be used for what-if analysis, where several alternative scenarios to a given situation are considered [9, 10]. Concept nodes represent the environment behaviour within a FCM. The concepts are connected using arcs (arrows or edges) showing the relations between concepts. The framework's arcs are the influences between the concepts. The development of the FCM is based on the utilisation of domain experts' knowledge, forming a framework within that environment. Expert knowledge is used to identify concepts and the degree of influence between the concepts.

Use of a FCM to analyse the framework can only demonstrate general application of the framework. In contrast, validation of the framework involves specific, concrete applications of the framework. Concrete applications of the framework require case studies of different organisations in an industry environment. Validation of the framework is outside the scope of this paper.

4.2 FCM Construction

The FCM was constructed as follows. The events and the relationships defined by the proposed framework were used to build the graph structure of the FCM. Thus the FCM concepts are Test Management (C2) Test Information (C3), Test Environment (C4), Technical Support (C5) and the Application of AI Techniques to Software Testing (C1). Relationship weight values were assigned arbitrarily according to the authors' judgement, underpinned by industry experience in software testing of one of the authors. The FCM of the software testing framework is shown in Figure 1.

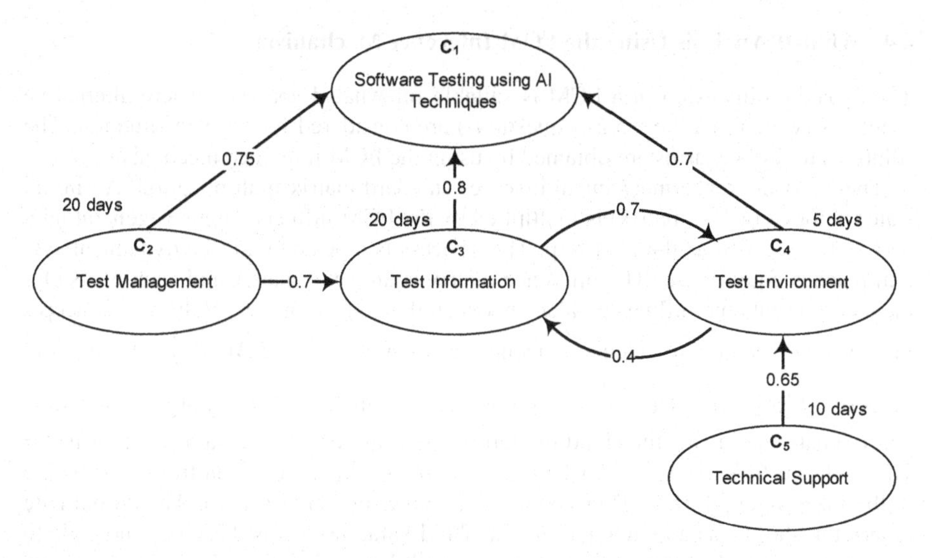

Fig. 1. Fuzzy Cognitive Map with Influences and Temporal

4.3 Static Analysis

If we consider the average effect of the influences from C2 + C3 + C4 (Figure 1) on C1 occurring, there is an estimated up to 75% chance that the software can be tested using an AI technique. Thus there is a 25% chance that some external influence will have an adverse, or even an advantageous effect, on the possibility of using an AI technique to test that particular software. Remember the 75% is based on the current knowledge of the domain expert.

From C2 to C1 shows 0.75. This says that at most (upper boundary) if all the elements that make up Test Management of the software to be tested are positive, then at most there is a 75% chance that AI Techniques will be selected for testing the software.

From C3 to C1 shows 0.80. However, C2 influences C3 by 0.70 and C4 influences C3 by 0.40. If left by itself C3's upper boundary is 80%, but this could be diminished or enhanced by the 70% influence from C2 and the 40% influence from C4.

From C4 to C1 shows 0.70. C4 has two influences on its upper boundary of 0.70: C3 and C5. In this situation C4's upper boundary of 70% is dependent on C3's 70% and C5's 65%. Therefore the influences from the concepts C3 and C5 can potentially provide C4 with constraints that could either adversely or enhance the affect of C4 reaching its upper boundary.

The preceding static analysis shows that a decision maker, in this context, works in a very complex environment. Thus any assistance to reduce the complexity of the environment, or make that complex environment more understandable or easier to interpret, would be a plus for the decision maker. The decision maker needs a way to comprehend the consequences of the interacting concepts of the framework. They also need to be provided guidance for a better understanding of the elements within their decision, and for a better understanding of the consequences of the environment they are working with. The FCM inference mechanism is a technique that can be used to analyse the interacting knowledge captured by the framework.

4.4 What-If Analysis Using the FCM Inference Mechanism

The dynamic dimension of a FCM is suitable for what-if analysis; where alternative scenarios (which involve what-if questions) are considered for a given situation. The solutions to the scenarios are obtained by using the FCM inference mechanism.

The FCM inference mechanism involves standard matrix multiplication. An initial state vector S_0 (a 1xn matrix) is multiplied by the FCM influence matrix, generating a new state vector S_1 as the next step. This process is repeated until the dynamical system reaches equilibrium. The influence matrix is an n x n matrix and each of its elements represents the influence values between the concept nodes. Values of concepts in new state vectors are calculated using the equation $c_i^{t+1} = f\left(W_{ji} c_j^t\right)$. The sigmoid function $f(x) = 1/\left(1+e^{-\lambda x}\right)$ is used as the activation function. Large values of λ approximate the binary threshold or step function [9, 10]. In the step function f(x) = 1if x > T and f(x) = 0 if x ≤ T, where T is the threshold value, taken from somewhere in the fuzzy interval [0,1]. Thus concepts are either on (1) or off (0). An approximate binary threshold was adopted and the threshold value used was 0.5 [10]. The node to be tested is set to 1 in the input vector and in all the result vectors because it is a sustained input.

To develop a what-if analysis, scenarios are defined and the following process will be followed:

1) Create a connection or edge matrix E, which lists the values of the causal links between the nodes;
2) Define the scenario – select a node to test its effect or influence on the potential decision;
3) Create the initial state vector S_0. Set the node to be tested to1 (on) and set all other nodes to 0 (off). This ensures independent analysis of the test node. The node to be tested is modelled as a sustained input, so the test node is set to1 in all the result vectors;
4) Multiply S_0 by E to obtain the result vector S_1;
5) Repeat step 4 with each result vector (S_n * E) until equilibrium is reached – when a vector is repeated, ie the current iteration $S_{n+1} = S_n$;
6) Take the previous result vector S_n for the analysis; and
7) Repeat steps 2 to 6 for each scenario.

The influences among the concepts in Figure 1 can be displayed using the following influence or edge matrix ***E***.

$$E = \begin{array}{c} \\ C_1 \\ C_2 \\ C_3 \\ C_4 \\ C_5 \end{array} \begin{array}{c} \begin{array}{ccccc} C_1 & C_2 & C_3 & C_4 & C_5 \end{array} \\ \begin{pmatrix} 0.00 & 0.00 & 0.00 & 0.00 & 0.00 \\ 0.75 & 0.00 & 0.70 & 0.00 & 0.00 \\ 0.80 & 0.00 & 0.00 & 0.70 & 0.00 \\ 0.70 & 0.00 & 0.40 & 0.00 & 0.00 \\ 0.00 & 0.00 & 0.00 & 0.65 & 0.00 \end{pmatrix} \end{array}$$

The test manager and other domain experts are required to determine the weights of the different links between the concept nodes, and the initial activation level of each concept. Later research will identify a set of questions to assist the test manager to determine weightings. In this test of the framework the authors have carefully considered the system and provided the weights for the FCM shown in Figure 1. Now what-if analysis can proceed.

Let us first analyse the influence of C2; so C2 is set to 1. Thus C2 can be examined independently of the other nodes and their influences. This situation is represented by $\mathbf{S_0}$ = [0, 1, 0, 0, 0].

S0 = [0, 1, 0, 0, 0]
S0 * E = [0.75, 0, 0.70, 0, 0] – this becomes S1 [1, 1, 1, 0, 0]
S1 * E = [1.55, 0, 0.70, 0.70, 0] – this becomes S2 [1, 1, 1, 1, 0]
S2 * E = [2.25, 0, 1.1, 0.70, 0] – this becomes S3 [1, 1, 1, 1, 0]
S3 = S2: equilibrium has been reached.

Let us next set C3 to 1. Thus C3 can be looked at independently of the other nodes and their influences. This situation is represented by $\mathbf{S_0}$ = [0, 0, 1, 0, 0].

S0 = [0, 0, 1, 0, 0]
S0 * E = [0.80, 0, 0, 0.70, 0] – this becomes S1 [1, 0, 1, 1, 0]
S1 * E = [1.5, 0, 0.40, 0.70, 0] – this becomes S2 [1, 0, 1, 1, 0]
S2 = S1: equilibrium has been reached.

Let us next set C4 to 1. Thus C4 can be looked at independently of the other nodes and their influences. This situation is represented by $\mathbf{S_0}$ = [0, 0, 0, 1, 0].

S0 = [0, 0, 0, 1, 0]
S0 * E = [0.70, 0, 0.40, 0, 0] – this becomes S1 [1, 0, 0, 1, 0]
S1 * E = [0.70, 0, 0.40, 0, 0] – this becomes S2 [1, 0, 0, 1, 0]
S2 = S1: equilibrium has been reached.

Let us next set C5 to 1. Thus C5 can be looked at independently of the other nodes and their influences. This situation is represented by $\mathbf{S_0}$ = [0, 0, 0, 0, 1].

S0 = [0, 0, 0, 0, 1].
S0 * E = [0, 0, 0, 0.65, 0] – this becomes S1 [0, 0, 0, 1, 1]
S1 * E = [0.70, 0, 0.40, 0.65, 0] – this becomes S2 [1, 0, 0, 1, 1]
S2 * E =[0.70, 0, 0.40, 0.65, 0] – this becomes S3 [1, 0, 0, 1, 1]
S3 = S2: equilibrium has been reached.

The framework is flexible and the test manager can modify the concepts and relationships of the framework to match their organisational circumstances, and the characteristics of the software being tested. As indicated the influence weights are test manager dependent and may not represent specific organisational settings.

The FCM converges to a fixed point for each scenario – single vectors are the result when equilibrium is reached. All the equilibrium vectors are *not* null, so definite answers can be obtained for each scenario. The decision maker needs to be aware that effort into C4 can provide a more positive use of AI Techniques in testing of software.

C3 and C5 have approximately the same degree of influence on the potential decision to use or not to use an AI technique on the software being tested. The influence of the Test Management node (C2) on the AI techniques decision is more difficult to determine, because its effect on the framework objective is both direct and indirect.

The information provided from what-if analysis of the framework can be used for decision analysis to support improved decision making by test managers. This approach provides a valuable tool for test managers to evaluate different scenarios for individual concepts, or combinations of concepts in the framework, and apply that evaluation in their organisation.

4.5 Temporal Analysis

FCM has introduced quantitative relationships between concepts to describe the strength of influence between elements. Miao in 2000 has shown that fuzzy cognitive maps generally do not provide a temporal mechanism to represent both the strength of influence and the temporal degree of the effect on the overall objective [11]. The FCM within this framework includes a temporal degree as illustrated in Figure 1. The days are computed by the test manager based on their best estimates of the software to be tested and historical data. The temporal aspect conforms to critical path analysis for the FCM. Combined with strength of influence this provides a solid basis for compressing a software testing life cycle.

5 Conclusion and Future Work

A limitation of the AI testing framework is a potential inability to capture all the important concepts from the software testing domain. Therefore the proposed testing framework may be subject to external influences from concepts overlooked in the software testing domain.

Future work involves:

1) Applying increased granular of the AI testing framework, and analysing any differences in the results between representations of the framework;
2) Using an FCM inference method that is able to map or transform concept values of state vectors to any value of the fuzzy interval [0,1], and comparing the results with the less discriminating FCM inference method illustrated in this paper;
3) Extended temporal to include degree of dependency states; and
4) Constructing a set of questions to help test managers more easily determine the weight values for the relationships between the concepts in the AI testing framework.

References

1. Dick, S., Kandel, A.: Series in machine perception and artificial intelligence. In: Computational intelligence in software quality assurance, vol. 63. World Scientific, Hackensack (2005)

2. Hailpern, B., Santhanam, P.: Software debugging, testing, and verification. IBM Systems Journal 41(1), 4–12 (2002)
3. Harman, M., McMinn, P.: A theoretical & empirical analysis of evolutionary testing and hill climbing for structural test data generation. In: Proceedings of the 2007 International Symposium on Software Testing and Analysis, pp. 73–83 (2007)
4. Hermadi, I., Ahmed, M.A.: Genetic algorithm based test data generator. In: The 2003 Congress on Evolutionary Computation, vol. 1, pp. 85–91 (2003)
5. Howe, A.E., Von Mayrhauser, A., Mraz, R.T.: Test case generation as an AI planning problem. Automated Software Engineering 4(1), 77–106 (1997)
6. Michael, C.C., McGraw, G., Schatz, M.A.: Generating software test data by evolution. IEEE Transactions on Software Engineering 27(12), 1085–1110 (2001)
7. Kim, J.-M., Porter, A., Rothermel, G.: An empirical study of regression test application frequency. Software Testing, Verification and Reliability 15(4), 257–279 (2005)
8. Smith, E., Eloff, J.: Cognitive Fuzzy Modeling for Cognitive Fuzzy Modeling for a Health Care Institution. In: IEEE Intelligent Systems, USA, pp. 69–75 (March/April 2000)
9. Kosko, B.: Fuzzy engineering. Prentice Hall, Upper Saddle River (1997)
10. Kosko, B.: Neural networks and fuzzy systems: A dynamical systems approach to machine intelligence. Prentice Hall, Englewood Ciffs (1991)
11. Miao, Y., Zhi-Qiang, L.: On causal inference in fuzzy cognitive map. IEEE Trans. Fuzzy Syst. 8, 107–119 (2000)

Learning User Preferences in Ubiquitous Systems: A User Study and a Reinforcement Learning Approach

Sofia Zaidenberg[1], Patrick Reignier[1], and Nadine Mandran[2]

[1] INRIA
Sofia.Zaidenberg@inrialpes.fr
http://www-prima.imag.fr/prima/people/zaidenberg/index.php
[2] LIG

Abstract. Our study concerns a virtual assistant, proposing services to the user based on its current perceived activity and situation (ambient intelligence). Instead of asking the user to define his preferences, we acquire them automatically using a reinforcement learning approach. Experiments showed that our system succeeded the learning of user preferences. In order to validate the relevance and usability of such a system, we have first conducted a user study. 26 non-expert subjects were interviewed using a model of the final system. This paper presents the methodology of applying reinforcement learning to a real-world problem with experimental results and the conclusions of the user study.

Keywords: Ambient Intelligence, Context-aware Computing, Personnal Assistant, Reinforcement Learning, User Study.

1 Introduction

This work fits into the frame of ambient intelligence (AmI) where a ubiquitous system tries to adapt its behavior to the perceived context. Through the assistant, the system aims at helping the user to perform his everyday tasks and at reducing his cognitive load. The assistant could make sure for instance that its user doesn't miss any important meetings by forwarding him reminders from his calendar. If the reminder pops up while the user is away from his computer, the assistant can find him in the building and find a means of informing him of the upcoming event. Defining preferences for such a system is a fastidious task for the user. It would be difficult for him to describe all the situations of interest and to assign an action to them. The solution proposed by [15] is to learn these user preferences.

We study the automatic acquisition of user preferences in ubiquitous environments by applying known machine learning techniques to AmI. In this paper, we will first introduce a qualitative user study conducted on a model of the system we want to build. We will then present our learning agent approach that takes into account the constraints revealed by the user study.

H. Papadopoulos, A.S. Andreou, and M. Bramer (Eds.): AIAI 2010, IFIP AICT 339, pp. 336–343, 2010.

2 State of the Art

Numerous researchers have worked on *context-aware computing.* For instance, one of the first generic framework for building context aware applications is the *Context Toolkit* [9]. This toolkit is based on *context widgets* and a distributed infrastructure that hosts the widgets. As an example of context-aware application let us cite the *context broker* [6]. The broker maintains a context model of the part of the environment that it is responsible for and shares this information with other agents or services while respecting several confidentiality levels. The context broker is supplied with information from different sources (sensors, other agents, devices, etc.) and merges this information into a consistent model.

This infrastructure enables applications like "EasyMeeting" [7], where the ubiquitous system helps the user Alice to conduct her seminar by automatically projecting her slides in the meeting room. In this work, the user is offered context-aware, ubiquitous services, but these services are hard-wired and not personalized. A similar example is the *PersonisAD* framework [1]. This distributed framework also maintains a hierarchical context model using reasoning on context proofs provided by applications. Components of the system can be notified of context changes in order to adapt themselves to the new context, but user preferences are defined manually.

Some papers try to inject user models into context-aware systems in order to deal with user preferences. For instance, [5] shows how the two closely related concepts of user modelling and context-awareness could be combined. [13] build user profiles of net surfers capturing their preferences and interests for page suggestions. These profiles are learned using classification techniques.

Our work is closely related to the field of *intelligent inhabited environments* which are living spaces equipped with embedded intelligent systems taking care of the users needs. As stated in [12], adaptation to the user in essential in AmI. [10] use fuzzy logic to learn rules representing the user's behavior with the aim of automating his usual actions. But this system does not offer new services to the user. In addition, this system learns rules by observing the user, and not by being trained by the user, which gives him less freedom to customize the system.

3 User Study

The main direction at the beginning of AmI was to build seamless, invisible applications that *weave themselves into the fabric of everyday life until they are indistinguishable from it* [21]. But around 2005, researchers started noticing that *calm computing* is maybe not what people really want [11,17,14]. *UbiComp technologies are designed not to do things for people but to engage them more actively in what they currently do* [17]. To build smart ubicomp applications, [14] suggested that the best approach is to study user habits and their ways of living and create technologies supporting that and not the opposite, as it was done at the time [2]. This approach was, for instance, adopted by [16,20].

For this reason, we believe that conducting user studies to validate our research in AmI is fundamental.

The goal of this user study was to measure the expectations and needs of users with regard to an ambient personal assistant. Subjects were 26 active persons, 12 women and 14 men, distributed in age categories as follows: 9 subjects between 18 and 25, 7 between 26 and 40, 7 between 40 and 60, and 3 over 60. None of the subjects had advanced knowledge in computer science.

The study was based on ~1 hour interviews with every subject. The interviewer followed a predefined script. The script started with a few open questions about information and communication technologies to evaluate the subject's general knowledge, but also his perception and his uses of such technologies. Then, the interviewer presented our ubiquitous system using a model (an interactive power point presentation: some of the slides are shown figure 1). This interacting powerpoint was exposing a simple scenario about the user's laptop. The scenario starts in the morning and the user is at home, browsing for movies and restaurants (figure 1(a)). When he arrives at work, the laptop automatically switches to the user's "work" setting (figure 1(b)). Then, the assistant turns the user's cellphone to vibrate and displays a message about this action. The user can ask for an explanation about this action and choose to undo it or select another action for this situation (figure 1(c)). At the end of the day, the system switched back to the "home" setting. The interviewer explained also orally other examples of services that could be offered by the assistant.

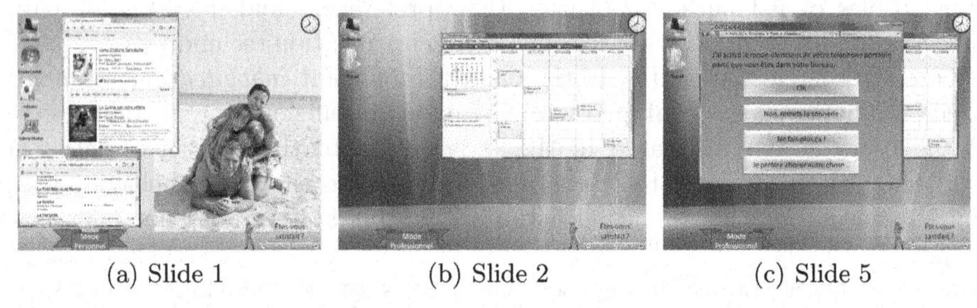

(a) Slide 1 (b) Slide 2 (c) Slide 5

Fig. 1. A few slides from the model used to present our system to the subjects

After the presentation, the subject was asked for his opinion about such a system. He could freely express the advantages and drawbacks of what he saw and the situations in which he thought the assistant was particularly useful or interesting. This gave him the opportunity to talk about ubiquitous assistants in general and about what their usage implies for his everyday life. Another goal of the conversation was to determine the acceptability of the system. The interviewer asked the following questions:

- "If the assistant learns badly, if it offers you wrong services at wrong times, what would be your reaction?"
- "If the assistant makes mistakes, but you know that he is learning to adapt to your behavior, would you give him a chance?"

– "Would you accept to spend some time to answer questions to make the assistant learn more quickly?"
– "What would you gain from getting an explanation about the assistant's decisions?"

We were also interested in finding out if the subjects would feel observed and fear that their privacy was in jeopardy. If they would not bring the subject up themselves, we would ask questions about this.

3.1 Results

After analyzing all the interviews, it appeared that 44% of subjects were interested in our assistant, and 13% were conquered. Interested persons share the same profile: they are very active, very busy in their professional as well as personal lives, they suffer from cognitive overload and would appreciate some help to organize their schedule. Other noticeable observations standing out from the interviews are the following:

– Having a *learning* assistant is considered as a plus by users. In fact, subjects felt a learning system would be more reliable since it would respond to their own training.
– Users prefer a gradual training versus a heavy configuration at the beginning.
– This training must indeed be simple and pleasant ("one click").
– The initial learning phase must be short (one to three weeks).
– It is absolutely necessary for the assistant to be able to explain its decisions. This aspect was particularly discussed by [3].
– The amount of interactions wanted between the user and the assistant varies from one subject to another. Some accept only to give one-click rewards while others would be happy to give more inputs to the system. This confirms that people are interested in *engaging* systems, as stated by [17]. For those users, we could add an optional debriefing phase where the assistant goes through the learned behavior and the user corrects or approves it.
– Mistakes made by the system are accepted to some extent as long as the user knows that the system is learning and as the system is useful enough to the user. But errors must not have critical consequences. Users always want to remain in control, to have the last word over the system and even have a "red button" to stop the whole system at any time.
– Some subjects pointed out that the assistant could even reveal to them their own automatic and subconscious customs.
– A recurrent worry expressed by interviewees is the fear of becoming dependant of a system that cares for them and becoming unable of living without it (what if the system is broken-down?).

This user study justifies our ubiquitous assistant since a sizeable part of interviewed subjects were prone to using it. Points listed above give us constraints to respect in our system. They will be listed in the next section that will present our approach.

4 The Ubiquitous Assistant

4.1 Constraints

We want to build a personal assistant whose behavior is learned from user's inputs. We have to respect several constraints:

(a) The system must not be a black box. As detailed in [3], a context-aware[1] system can not pretend to understand all of the user's context, thus it must be responsible about its limitations. It must be able to explain to the user what it knows, how it knows it, and what it is doing about it. The user will trust the assistant (even if it fails) if he can understand its internal functioning.
(b) The training is going to be performed by the user thus it must be simple, non intrusive and it must not put a burden on the user.
(c) The training period must be short, unless the user changes preferences.
(d) The system must have an initial behavior that is not incoherent.

Several learning techniques have been considered to solve this problem. For instance, [4] uses supervised learning where the user provides offline feedback on the sequence of events that happened during the day, providing for every sequence the correct action. We prefer to gather user feedback online (in context), collecting only positive or negative feedback to simplify the interaction. Reinforcement learning (see [19] for instance) is a possible solution, but it must be adapted to our particular constraints.

4.2 Reinforcement Learning of User Preferences

Our goal is to learn the behavior of a ubiquitous assistant based on the satisfaction or disapproval of the user toward the assistant's actions. Reinforcement learning (RL) approaches are based on state space exploration. It needs many experiences (and many of them will seem inappropriate because of exploration) to converge to the correct solution. As the user is directly implied in those experiences, he may quickly reject the system. To minimize user interaction in the learning process, we have used indirect reinforcement learning.

Indirect RL was proposed by Sutton [18] and consists in learning a model of the environment, a *world model*, in the guise of the transition function ($\mathcal{P}$) and reward function ($\mathcal{R}$). A part of the trials is then done in the world model instead of the real world (this is the DYNA-Q algorithm [18]).

The world model is learned using supervised learning on a history of real interactions between the user, the assistant and the environment. This history contains tree kinds of triplets: (s, a, s'), (s, e, s') and (s, a, r), where s is the previous state of the environment, a is an action taken by the assistant, e is an event beyond the assistant (for instance the user coming inside the office or receiving an email). Both a and e cause a change in the environment's state,

[1] We base our work on Crowley's definition of context [8].

the next state being s'. Finally, r is the reward given by the user when the assistant took action a in state s. The user can give an explicit reward using a slider on a graphical interface. He does not have to worry about the numerical value of r, but only to its relative position between a minimum and a maximum. Implicit reward could also be gathered from clues or from recognizing the user's emotional state after an action was taken. For example, if the assistant opens the email client and the user closes it straight away, the implicit reward is negative. Constraint (b) defined section 4 is thus respected.

Our state is modeled by a set of first order predicates. Each predicate transcribes an observable part of the environment, (the output of one of our sensors). For instance, we have the predicates `entrance(isAlone, friendlyName, btAddress)`[2] and `hasUnreadMail(from, to, subject, body)`. These predicates are readable by humans: the users can understand the internal state of the assistant at all times. This assures the respect of constraint (a) defined section 4. Our states space is huge. For instance, the states "`entrance(friendlyName = Bob, ...)`" and "`entrance(friendlyName = Alice, ...)`" are two different states that must be explored both to learn a behavior. To speed up the learning process, we must generalize the observed examples. We generalize states by replacing values with wildcards: "`<+>`" means any value but "`<null>`" and "`<*>`" means any value. Initially, we merge all these similar states and we deal with "super-states" such as "someone entered the office". Eventually, we will need to split some of these super-states and adapt different behaviors for example when the boss sends an email and when a newletter sends an email. This way, we make best use of every example. Indirect RL and states generalization accelerates the learning phase, respecting constraint (c). More details about this approach are given in [22].

5 Experimental Results

We created a prototype of an AmI system and the assistant. An experimenter is put in the position of the user and creates interactions with the environment

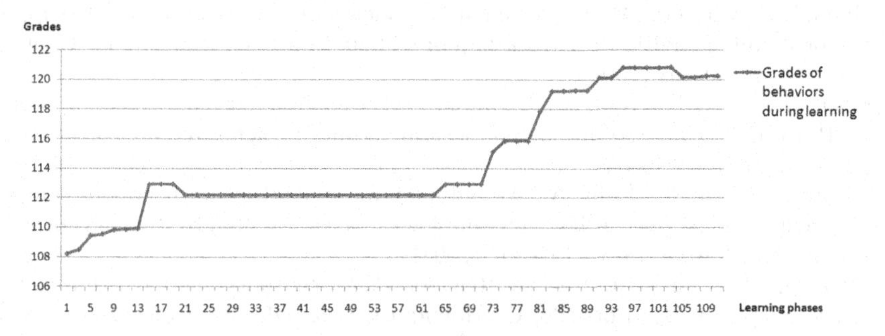

Fig. 2. Grades of behaviors produced by the RL algorithm

[2] The arguments correspond to attributes of a bluetooth device, which is how we detect the presence of users for now.

using a graphical interface for convenience. Based on sensor events sent by the user through the interface, the assistant learns a behavior and puts it into practice. The experimenter, who gives rewards, grades each behavior by indicating if he agrees with every action that has been associated with a situation by the RL algorithm. Those grades are presented figure 2. This curve shows that when the assistant has the opportunity to observe new situations, it quickly learns what to do (phases 1 to 20 and 65 to 110). When nothing new happens, the behavior stays stable (phases 20 to 65).

6 Conclusion

The aim of this research is to investigate AmI systems and their acceptability by users. We exploit a ubiquitous system to provide personalized, context-aware services to users. The personalization of the system is achieved by learning user preferences during interactions. In order to validate the relevance of such an application, we conducted a user study. This study approved our work and revealed constraints for the acceptability of such a system. We developed a method and a prototype respecting those requirements and showed the correct functioning of our ubiquitous assistant.

References

1. Assad, M., Carmichael, D., Kay, J., Kummerfeld, B.: PersonisAD: Distributed, active, scrutable model framework for context-aware services. In: LaMarca, A., Langheinrich, M., Truong, K.N. (eds.) Pervasive 2007. LNCS, vol. 4480, pp. 55–72. Springer, Heidelberg (2007)
2. Barton, J., Pierce, J.: Quantifying magic in ubicomp systems scenarios. In: Position Paper for UbiSys 2006, Orange County, California (2006)
3. Bellotti, V., Edwards, K.: Intelligibility and accountability: human considerations in context-aware systems. HCI 16(2), 193–212 (2001)
4. Brdiczka, O., Crowley, J.L., Reignier, P.: Learning situation models for providing context-aware services. In: Stephanidis, C. (ed.) UAHCI 2007, Part II. LNCS, vol. 4555, pp. 23–32. Springer, Heidelberg (2007)
5. Byun, H.E., Cheverst, K.: Exploiting user models and context-awareness to support personal daily activities. In: Workshop in UM 2001 on User Modeling for Context-Aware Applications (2001)
6. Chen, H., Finin, T., Joshi, A.: A context broker for building smart meeting rooms. In: Proceedings of AAAI Spring Symposium, Stanford, California, pp. 53–60. AAAI Press, Menlo Park (2004)
7. Chen, H., Finin, T., Joshi, A.: An ontology for context-aware pervasive computing environments. Special Issue on Ontologies for Distributed Systems, Knowledge Engineering Review 18(3), 197–207 (2004)
8. Crowley, J.L., Coutaz, J., Rey, G., Reignier, P.: Perceptual components for context aware computing. In: Borriello, G., Holmquist, L.E. (eds.) UbiComp 2002. LNCS, vol. 2498, pp. 117–134. Springer, Heidelberg (2002)
9. Dey, A.K., Abowd, G.D.: The context toolkit: Aiding the development of context-aware applications. In: The Workshop on Software Engineering for Wearable and Pervasive Computing, Limerick, Ireland (June 2000)

10. Doctor, F., Hagras, H., Callaghan, V.: An intelligent fuzzy agent approach for realising ambient intelligence in intelligent inhabited environments. IEEE Tr. on Systems, Man and Cybernetics, Part A 35, 55–65 (2005)
11. Dourish, P.: What we talk about when we talk about context. Personal Ubiquitous Comput. 8(1), 19–30 (2004)
12. Ducatel, K., Bogdanowicz, M., Scapolo, F., Leijten, J., Burgelman, J.-C.: Scenarios for ambient intelligence in 2010. Technical report, ISTAG (2001)
13. Godoy, D., Amandi, A.: User profiling for web page filtering. IEEE Internet Computing 9(4), 56–64 (2005)
14. José, R.: Ubicomp 2.0: From envisioning a future to being part of a new reality. In: Plenary Talk at the UCAmI 2008 conference (2008)
15. Maes, P.: Agents that reduce work and information overload. Commun. ACM 37(7), 30–40 (1994)
16. Pascoe, J., Thomson, K., Rodrigues, H.: Context-Awareness in the Wild: An Investigation into the Existing Uses of Context in Everyday Life. In: Meersman, R., Tari, Z., Herrero, P. (eds.) OTM-WS 2007, Part I. LNCS, vol. 4805, pp. 193–202. Springer, Heidelberg (2007)
17. Rogers, Y.: Moving on from weiser's vision of calm computing: Engaging ubicomp experiences. In: Dourish, P., Friday, A. (eds.) UbiComp 2006. LNCS, vol. 4206, pp. 404–421. Springer, Heidelberg (2006)
18. Sutton, R.S.: Integrated architectures for learning, planning, and reacting based on approximating dynamic programming. In: ICML1990, pp. 216–224 (1990)
19. Sutton, R.S., Barto, A.G.: Reinforcement Learning: An Introduction (Adaptive Computation and Machine Learning). The MIT Press, Cambridge (1998)
20. Taylor, A.S., Harper, R., Swan, L., Izadi, S., Sellen, A., Perry, M.: Homes that make us smart. Personal and Ubiquitous Computing 11(5), 383–393 (2007)
21. Weiser, M.: The computer for the 21st century. Scientific American 265(3), 66–75 (1991)
22. Zaidenberg, S., Reignier, P., Crowley, J.L.: Reinforcement Learning of Context Models for a Ubiquitous Personal Assistant. In: Advances in Soft Computing, vol. 51, pp. 254–264. Springer, Heidelberg (September 2008)

Decision Oriented Programming in HCI: The Multi-Attribute Decision Language MADL

Bjoern Zenker

Chair for Artificial Intelligence
Friedrich-Alexander-University Erlangen-Nuremberg
Haberstraße 2, D-91058 Erlangen
bjoern.zenker@cs.fau.de

Abstract. In Human Computer Interaction (HCI), the computer has to take many decisions to react in a way that human wants. As decisions in HCI are diverse, contradictory, and hard to measure it is hard to study and model them, e.g. finding a mapping from the users preferences to adaptions of the user interface. To ease these tasks, we developed MADL (Multi-Attribute Decision Language). This programming language, based on Multi-Attribute Decision Making (MADM), is designed to model and make hierarchical multi-attributive decisions and is based on the analysis of decisions and goals in HCI. It fosters respecting HCI specific characteristics in development, like uncertainty and risk, can be embedded easily in other applications and allows inclusion of and experimentation with different decision rules. User interface logic can also be modeled by non-programmers and more easily separated from the business logic. The applicability will been shown in three use cases.

Keywords: Multi-Attribute Decision Making, Human Computer Interaction.

1 Introduction

Computers should react the way as humans expect them to. But still in many cases, computers do not react the way that humans expect them to. Thus there is often a discrepancy between the reaction of the computer and the action anticipated by the user. We expect this to be caused in a lacking knowledge of the computer about the users various goals and not taking these goals into account.

As machines and computers are going to be used by more and more users, and become more integrated in our daily lives, methods are needed to further tail the machines behavior to the users preferences. To further study and and undertake a step to close this gap we designed the programming language MADL to easily model tasks in Human Computer Interaction (HCI) and make multi-attribute decisions (for an introduction see e.g. [1]) on them. Our goal was the creation of an environment for decision taking, which hides complexity and different algorithms behind an intuitive programming language, to allow inclusion of decisions

H. Papadopoulos, A.S. Andreou, and M. Bramer (Eds.): AIAI 2010, IFIP AICT 339, pp. 344–351, 2010.

modeled by non-programmers into programs. Furthermore, we wanted to create a platform for experimentation and teaching in multi-attribute decision making.

As far as we know, no general tools for describing and making multi-attribute decisions with risk and uncertainty exist. But two similar systems from the HCI point of view, which also cover modeling of tasks, should be mentioned.

GOMS (Goals, Operators, Methods and Selection) [2], a technique for measuring and analyzing tasks in HCI, is very similar in structure to MADL. GOMS divides the user's interaction into goals, sequences (methods) of actions (operators) and selection rules, which model which sequence would be selected by a user. As we will see, these rules resemble decisions in MADL. But as the focus of GOMS is on analysis of HCI, selection rules are often ignored in GOMS. According to the *Kleindorfer* classification of decision theories [3] GOMS is *descriptive*. In contrast, MADL is *prescriptive*, with the main focus – in GOMS terms – on selection, to shape HCI, rather than to analyze it.

CTT (Concurrent Task Tree) [4] is a formalized task model which models tasks in HCI, with the addition to also allow representation of concurrency besides tasks. CTT is also descriptive, with a focus on modeling tasks.

In contrast to the presented systems, the focus of MADL is on solving decision problems.

After a short overview of decisions in HCI and their relation to human preferences (section 2) this paper introduces some aspects of the MADL language design (section 3) and it's runtime environment. Results of a proof-of-concept study for MADL usage by programming a virtual assistive video-recorder, will be shown (section 4). Finally we summarize our results and give an outlook on future work (section 5).

2 Goals and Decisions In HCI

2.1 Goals

Systems with HCI have despite their variety something in common: the interaction between human and machine serves in all cases the best possible satisfaction of preset goals. According to [5] goals are various, exhibit a hierarchical structure and have complex relations between each other, which means that goals may also be contradictory! Principles and definitions of usability, which is closely related to goals in HCI, are presented by [6], [7] and [8]. Studying the lists of goals from the other authors leads us to the following summarization: The goals are mostly abstract. No methods exist for measuring the goals. There is a deep structure of goals. Relations between goals are unclear and possibly contradictory. This leads us to the conclusion, that there is currently no formal definition of goals and their relations in HCI. To further examine this lack of formality, or rather to bridge the gap between the unformal field of human goals and the formal field of computation, we developed MADL.

2.2 Decisions

With respect to the above mentioned goals, decisions have to be taken in HCI.

Examinations of decisions in HCI lead us to three main kinds of decisions in HCI.

Decisions for interaction control the way, how the machine interacts with the user. [9] gives a list of adaptions for the user interface, like content adaption, selection of information, quantity of information, augmentation of information, layout of information, modality of information and dialog adaption.

Decisions for automation control the technical process of the machine. The main question here is the grade of automation: which processes and tasks should be overtaken be the machine and which not. [7] gives an overview, in which areas are machines are superior to humans and vice versa.

Metadecisions control the act of decision making per se. They include decisions like, what to do in stand-off situations, which risks should be taken and whether more information should be retrieved in order to improve the decision quality. Criteria of these decisions are mostly very special, but due to existing theories like information theory (e.g. [10]), risk theory (e.g. [11]) and Dempster–Shafer theory (e.g. [12]) easy to measure and calculate.

2.3 Erwartungsstrukturen

According to [11] there exist mainly three "Erwartungsstrukturen" (engl. structure of expectance) for decisions: decisions under security, where the future can be predicted; decisions under uncertainty in a closer sense, where no precise probability of the possible future scenarios exist; and decisions under risk, in which probability values can be assigned to the possible future scenarios. For usage in MADL, all three "Erwartungsstrukturen" can be reduced to one: decisions under security, can be seen as decisions under risk with all probability values 100%. Applying the principle of indifference, thus assuming a uniform distribution for the future scenarious, decisions under uncertainty in a closer sense can be expressed as decisions under risk. Thus, MADL will focus on decisions under risk. Furthermore we will only consider decisions with a finite set of alternatives, each alternative consisting of a finite ordered sequence of actions. Closed-loop control systems are not discussed in this paper.

As there are diverse decisions to take in HCI, the language to express them must be general.

3 MADL

3.1 System Overview

A complete system to model and take decisions was built: The MADE (Multi-Attribute Decision Environment)[1] interpreter takes MADL programs as input,

[1] Download, online interpreter and MADL grammar available at
`http://www8.informatik.uni-erlangen.de/en/zenker.html`

evaluates them and returns an ordered list of the best alternatives for a given decision. One of these alternatives can then be passed to a plan execution system. MADE may be used on the command line, as a graphical tool (ShowMADE) or as a Java package.

3.2 MADL Program

A MADL program consists of several MADL-decisions and may be linked to *prognosticating models* (PM). The following code shows a basic MADL-program which will be explained in the subsequent sections.

```
CON examples.HelloWorldModel AS model
CON examples.RiskModel AS risk

ParetoDecision Casino {
  ALT [DO THIS, DO THAT]
  ALT [GO THERE, DO THIS]

  GOAL MAX!(model.fun)
  GOAL model.spendMoney < 1000
  GOAL MAX!(risk.averseToRisky(model.money, 10))
}
```

Prognosticating model. A PM can be a arbitrary system, like Java methods, a Bayesian network or a Hidden Markov models, to simulate the "Erwartungsstruktur", that is to predict values of variables after the application of an alternative. We will call the set of predicted values for one alternative *context*. The two lines starting with `CON` import a model each.

MADL decision. A MADL decision d is a quadruple $d(r, A, G, C)$, where r is the decision rule, A the set of alternatives, G the set of goals and C the set of constraints.

MADL decisions may be inherited from other MADL decisions. The child decision inherits decision rule, alternatives and goals, but more alternatives and goals may be added and the decision rule may be changed. The above example shows a decision named `Casino` with the `ParetoDecision` rule. The decision contains two alternatives, two goals and one constraint.

Actions. A is a ordered sequence of actions $a \in A$. An action can be of several types:

normal action an action from the model with an arbitrary number of parameters, e.g. `DO THIS`

assignement action assignment of a new value to a context variable which will be propagated to the model, e.g. `model.fun = 100%`

best-of decision action import the n best alternatives from another decision, e.g. `!otherDecision n`

mean-of decision action import the mean value of consequences of the n best alternatives, e.g. `?otherDecision n`

expanding action replicates the alternative n times and appends one of the n best alternatives from another decision to each, e.g. `<otherDecision` (see section 3.3 for an example)

If the consequences of an action should only be considered in the decision it is specified in, it can be surrounded by parentheses. That way, the action will not be considered in inherited decisions. Each of the two above alternatives from the example consist of two normal actions with one parameter each.

Goals $g \in G$ is a goal which can be of the type maximize (`MAX!`), minimize (`MIN!`), fix (`FIX!`) or satisfy (`SAT!`). It measures a variable of the alternatives context according to its type. For example, the goal `FIX!(suggestion, user.interest)` states, that the value `suggestion` should be close to `user.interest`. In the above example the variable `model.fun` should be maximized. MADL goals are used by the decision rules to sort alternatives. (Not all types of goals may be used in every decision rule.) Each alternative $a_i \in A$ is assigned a score v_i, which ranks the alternative in respect to the others.

Constraints In addition to goals, MADL constraints eliminate alternatives from the result set, if they do not satisfy the constraint. For example `GOAL model.spendMoney < 1000` will filter out all alternatives which have a context with `costs` ≥ 1000.

Variables Variables change according to the context of the alternatives and are thus calculated by the PM. In the above example you can see the variables `model.spendMoney`, `model.money` and `model.fun`. Note that all variables in MADL are probability distributions. The example `foo = [30%: 1, 70%:2]` states that the value of `foo` is expected to be 1 in 30% and 2 in 70% of all cases.

Functions MADL functions allow altering of variables. As MADL's focus is on decisions, they cannot be implemented in MADL, but may be imported from a MADL model which can be implemented in Java. Note that MADL functions only take MADL variables, which are probability distributions, as arguments. The concept of risk (e.g. see [11]) can be modeled in MADL by integrating a (given or self created) risk function into the goal of the decision. An example for this is the last goal in the above decision.

Decision rules Currently, MADL supports the following decision rules:

Weighted Sum calculates the weighted sum. If no decision rule is denoted, the weighted sum is used.

Pareto Set returns the non dominated set of alternatives.

TakeTheBest identifies the best alternative according to *Gigerenzer et al.*'s TakeTheBest decision rule, which shall resemble human decision behaviour.

MiniMax compares the minimal goal attributes and decides for the alternative with the greatest goal attribute.

MaxiMin compares the maximum goal attributes and decides for the alternative with the smallest goal attribute.

Random takes a random alternative.

Count calculates v_i as the sum of identical alternatives. This is especially usefull when combining multiple strategies as discussed below.
User asks the user to select a set of best alternatives.

Other decision rules may be implemented in Java.

Now we will illustrate the usage of multiple decision rules. The following example shows the decision `routes` with three alternatives and three goals.

```
routes{
  ALT [WAY1, km = 90, costs = 0, jam = 10%]
  ALT [WAY2, km = 80, costs = 3, jam = 40%]
  ALT [WAY3, km = 80, costs = 4, jam = 30%]
  GOAL MIN!(km)
  GOAL MIN!(costs * 5)
  GOAL MIN!(jam * 50)
}
```

To use a decision rule other than `TakeTheBest`, one can inherit from the decision: `TakeTheBestDecision strategy1 EXTENDS routes {}`.

As there exists no best decision rule, *de Almeida Cunha* [13] suggests to minimize the effect of the different decision rules by applying multiple decision rules to the problem. This concept can be implemented in MADL, by selecting the alternative, which has been evaluated best by multiple decision rules and which is included in the Pareto optimum of all alternatives.

3.3 Mode of Operation

The mode of operation will be explained with an short example. The following code shows two decisions `D1` and `D2`.

```
CON test.aModel AS model

SumDecision D1 {
  ALT [a1, <D2]
  ALT [a2]
  GOAL MAX!(model.satisfaction)
  GOAL MIN!(model.costs)
}
SumDecision D2 {
  ALT [b1]
  ALT [b2]
  GOAL MAX!(model.x)
  GOAL MIN!(model.y)
}
```

The MADL program connects to some PM `test.aModel` Each decision contains two alternatives and two goals. Let us assume, one wants to evaluate `D1`. As the first alternative of D1 imports the best alternatives of D2, D2 has to be evaluated first. For each alternative of D2 the result of applying the action to the model is calculated. As the variables are discrete probability distributions, all possible future contexts have to be calculated.

For a context of the first alternative with `x = [50% : 1, 50% 3]` and `y = [50% : 1, 50% 5]` we get $E(x_{b1}) = 2$ and $E(y_{b1}) = 3$. Assume $E(x_{b2}) = 2$ and $E(y_{b2}) = 4$. Based on the resulting contexts, a score v_i is calculated according to it's decision rule, using the expected value of the variables. As the weighted sum of both alternatives is equal, $E(x_{b1}) - E(y_{b1}) = E(x_{b2}) - E(y_{b2})$, both alternatives are imported to D1. This import can be seen as a new MADL decision:

```
SumDecision D1* {
   ALT [a1, b1]
   ALT [a1, b2]
   ALT [a2]
   GOAL MAX!(model.satisfaction)
   GOAL MIN!(model.costs)}
```

4 Evaluation

Three different proof of concept use cases have been implemented in MADL to show its applicability. Decisions, which elements to show in printer dialogs have been implemented in MADL. They take into account whether the document to print contains colour, its length and the estimated costs of the print. Some constraints which only hold for specific alternatives were cumbersome to program. In further development of MADL, alternative specific constraints could be written behind the actions of the alternatives, e.g. `ALT [action] x > 3`, to ease this.

Also an assistant for a virtual digital video recorder has been programmed in MADL. It helps people to delete or compress previously recorded films from the disk, when there is no free space for the recording of an additional film. A user study with nine participants showed that using the MADL assistant, users needed 16.5% less mouse clicks. This experiment also indicates, using MADL you can isolate business logic from "user assistance logic" to foster independent development, testing and maintenance of each.

Route selection for a navigation system by using multiple strategies has also been successfully implemented. Furthermore, MADL was successfully integrated in a program to control a robot from our Chair of Artificial Intelligence and in a system to plan HCI dialogs.

5 Conclusion and Future Work

For closing the gap between abstract goals and concrete adaptions in HCI, we invented the new programming language MADL. Therefore we analysed goals and decisions in HCI: Goals are various, contradictory and hard to measure. Decisions are deeply structured, diverse and include risk and uncertainty.

This led us to design an abstract programming language for multi-attributive decisions called MADL. It allows to model the hierarchy of decisions in HCI and make decisions with multiple goals with this programming language. MADL can handle risk and uncertainty and integrates different decision rules. The language can be extended with own prognosticating models, functions and decision rules and it is possible to integrate MADL in own software. We also showed the applicability of MADL in three different use cases.

In a further step we want to integrate MADL into a route generation algorithm based on the $h_{\epsilon}u$-heuristic [14]. There are still additions like hierarchical composition of goals, expansion of the hiding concept from actions to alternatives and goals, caching of contexts and others to implement. An interesting topic for future research would be to allow recursion by adding alternative specific constraints to gain a Turing-complete system.

References

1. Yoon, P.K., Hwang, C.L., Yoon, K.: Multiple Attribute Decision Making: An Introduction (Quantitative Applications in the Social Sciences). Sage Pubn. Inc., Thousand Oaks (March 1995)
2. Card, S., Moran, T.P., Newell, A.: The Psychology of Human Computer Interaction. Lawrence Erlbaum Associates, Hillsdale (1983)
3. Kleindorfer, P., Kunreuther, H., Schoemaker, P.: Decision Sciences: An Integrative Perspective. Cambridge Univ. Press, New York (1993)
4. Paterno, F., Mancini, C., Meniconi, S.: Concurtasktrees: A diagrammatic notation for specifying task models. In: INTERACT 1997: Proceedings of the IFIP TC13 Interantional Conference on Human-Computer Interaction, London, UK, pp. 362–369. Chapman & Hall, Ltd., Boca Raton (1997)
5. Johannsen, G.: Mensch-Maschine-Systeme. Springer, Heidelberg (1993)
6. Dix, A., Finlay, J., Abowd, G.D., Beale, R.: Human Computer Interaction, 3rd edn. Pearson, Harlow (2003)
7. Shneiderman, B.: Designing the User Interface: Strategies for Effective Human-Computer Interaction, 3rd edn. Addison Wesley Longman, Reading (1998)
8. van Welie, M.: Task-based User Interface Design. PhD thesis, Vrije Universiteit (April 2001)
9. Rothrock, L., Koubek, R., Fuchs, F., Haas, M., Salvendy, G.: Review and reappraisal of adaptive interfaces: toward biologically inspired paradigms. In: Thoretical Issues in Ergonomic Science, pp. 47–84 (2002)
10. Görz, G., Rollinger, C.R., Schneeberger, J. (eds.): Handbuch der Künstlichen Intelligenz, Oldenbourg, München, 4th edn. (2003)
11. Laux, H.: Entscheidungstheorie, 5th improved edn. Springer, Berlin (2003)
12. Shafer, G.: A Mathematical Theory of Evidence. Princeton University Press, Princeton (1976)
13. de Almeida Cunha, C.J.C.: Ein Modell zur Unterstützung der Bewertung und Auswahl von Strategiealternativen. PhD thesis, Reinisch-Westfälische Technische Hochschule Aachen (January 1989)
14. Zenker, B., Ludwig, B.: Rose – an intelligent mobile assistant. In: Proceedings of the 2nd International Conference of Agents and Artificial Intelligence (2010)

Investigating the Role of Mutual Cognitive Environment for End-User Programming

Rémi Barraquand and Patrick Reignier

INRIA Grenoble
Rhônes-Alpes Research Center
655 Ave de l'Europe
38330 Montbonnot, France
{remi.barraquand,patrick.reignier}@inrialpes.fr

Abstract. In this paper we present a situated end user programming approach where user co-constructs, in an iterative process, a mutual cognitive environment with the system. We argue that co-construction of a mutual cognitive environment, between both the human and the system, is a key toward social human-computer interaction. Preliminary results are illustrated with a step by step case study: a user teaches the system new perceptual and abstract concepts using hand gesture and an interactive learning table.

Keywords: End-User Programming, Situation Models, Cognitive Environment.

1 Introduction

Information and communication services pervade nearly all aspect of our daily life, however they have no ability to perceive and understand the social situation or affective reactions of human. Furthermore, they have no sense of social roles played by interacting humans: systems are unaware of human goals and intentions.

Such abilities are beyond systems constructed with current technologies for two main reasons. First, systems lack an ability to acquire the skills for polite interaction using the implicit feedback provided by humans during interaction. Second, systems even lack the most rudimentary understanding of the social order that structures normal human interaction.

Common sense is the collection of shared concepts and ideas that are accepted as correct by a community of people. Social common sense refers to the shared rules for polite, social interaction that implicitly rule behavior within a social group [1]. To a large extent, such common sense is developed using implicit feedback during interaction between individuals. Reeves and Nass demonstrate convincingly in The Media Equation [10] that interactions with computers are identical to real social relationships and they argue that a social interface may be the truly universal interface. Thus our goal in this research is to develop methods to endow an artificial agent with the ability to acquire social common

H. Papadopoulos, A.S. Andreou, and M. Bramer (Eds.): AIAI 2010, IFIP AICT 339, pp. 352–359, 2010.

sense using day to day interaction with people. Our approach focuses on a key aspect of social common sense: the ability to build an appropriate mental model of social situations. We believe that such methods can provide a foundation for socially polite man-machine interaction, and ultimately for other forms of cognitive abilities.

Common knowledge is a phenomenon which underwrites much of social life. In order to communicate or otherwise to collaborate, humans make use of socially shared conventions, common understanding or background knowledge. While theory of mutual knowledge has the characteristic to produce a regression at infinity, Sperber and Wilson [12] developed a weaker but empirically more adequate concept, the mutual manifestness. In this paper we investigate the role that a mutual cognitive environment plays in collaborative social interaction and in the construction of situation models. We believe that building correct social situation models is a key toward designing context aware systems.

2 Related Work

Designing a context aware application is a complex task. There is no widely accepted definition of what is a context. For Dey et al. [4], context is "any information that characterizes a situation related to the interaction between humans, applications and the surrounding environment." Dourish [6], on a more conceptual approach, proposes two categories for context models : representational and interactional. In a representational approach, context and activity are separable. Context is information about the environment where the activity takes place. The main question is how context could be encoded and represented. This is a more "programmer" point of view. In the interactional approach, context is a relational property between objects or activities. It is a more user centered approach. Dourish opposes those two points of view. We believe that those two approaches can be complementary. A software programmer, can use a representational context model to specify a first "general" contextual behavior (not adapted for everyone). The end user then adapt this general contextual model to its own needs. This adaptation phase has to be done while (or after) using and interacting with the system. This is the interactional aspect and corresponds to *End User Programmming.*

End user programming can be seen as offering alternatives to extensive programming for building applications. The question is how the end user can transfer to the system its relevant knowledge so that it can accomplish the required task. In [9], the mobile phone behavior is contextually specialized using feed forward rules. A specific interface allows to easily create new rules based on predefined concepts, brought by the programmer. The end user can specify how those concepts should be used to fulfill his needs. The evaluations showed that the end user appreciates to be able to enter its own knowledge in the system as long as the rules are not too complicated to write. The main problem is that Nokia's system is *bringing its own predefined concepts* which might not correspond to the user's concepts and interpretations.

In [5], the user is doing its task in the environment while the system is recording all the sensors. Using a dedicated human machine interface, the end-user can then annotate the recorded data and a context model is constructed using machine learning approaches. This annotation phase is an interesting way to transfer knowledge and concepts from the user to the system while letting the system building its own internal representation. The main drawback is that it can be rapidly painful and sensor data annotation is done offline (not during the interaction). Ganneau and al. [7] are directly learning (using bayesian networks) the situations' description without a post-annotation phase. The interaction for knowledge injection is more transparent for the end user.

On-line transparent interaction is more natural and corresponds to the interactional aspect of Dourish but can be ambiguous (see section 3). Off-line specification using a dedicated interface can allow the end user to specify more precisely but is painful. We propose a global approach based on both approaches: transparent user interaction and dedicated off-line interactions to disambiguate the injected knowledge.

3 Situated End User Programming

In an idealist scenario people should teach machines through a social and collaborative process. Learning to act appropriately in social situation is a first step toward a better interaction. In a previous work [1] we have sought to train an association between behavior and social situation. However the success of this approach relies on getting both human and system to share a common understanding of social situation.

3.1 Mutual Intelligibility

Salembier et al. [11] review research that point out the importance of the mutual access to contextual information in collaborative work. The better the mutual understanding the greater the collaboration. This mutual understanding is not necessarily achieved by sharing mutual knowledge but rather by the concept of mutual manifestness. Introduced by Sperber and Wilson this concept is weaker but empirically more adequate than the theory of mutual knowledge which has the characteristic to produce regression at infinity.

For Sperber and Wilson "a fact is manifest to an individual at a given time, if and only if, this individual is able at this time to represent this fact mentally and to accept his representation as being true or probably true" [12]. Following this notion, Sperber and Wilson define the one of cognitive environment. A personal cognitive environment (PCA) is defined as whole facts which are manifest for a given individual. A shared cognitive environment (SCE) indicates all the facts which are manifest to several individuals. This simply means that they are able to perceive or deduce the same facts, and not that they share a belief, a knowledge, or a representation concerning those facts. The mutual cognitive environment (MCE) indicates a shared cognitive environment in which the identity of individuals who have access to this environment is manifest. As they share the same environment, they can establish an interaction in relation to their common perception of contextual events. Salembier points out that the notion of cognitive environment does not take into account the activity of individual. He [11] then proposes the definition of shared context, which reduce mutual cognitive environment through activity filtering. Thus shared context is a set of contextual information or events mutually manifest for a set of actors, at a given time in a certain situation, taking into account their perception and cognitive abilities, their task, and current activity.

Considering social learning as a collaborative process and context as a key issue [2] in interaction between human and computer, we argue that mutual intelligibility is the key toward a social human-computer interaction.

3.2 Situation Models

In the previous section, we emphasize the importance of mutual intelligibility in social learning and particularly the needs for a MCE between both human and system. However it is not clear what defined the facts that compose cognitive environment. Johnson-Laird in [8] introduced situation model as cognitive theory for human mental models. A situation model is a mental representation of a described or experienced situation in a real or imaginary world [13]. It is commonly defined as consisting of entities and of relations between those entities. While this model, as well as much of the subsequent literature in this area, has been concerned with spatial reasoning or linguistic understanding, these concepts can be adopted for the construction of software systems and services for understanding social interaction. In [3], we describe the use of situation models for observing and understanding activity in order to provide context aware services. Situations are defined as a set of relations (predicate) between entities (agents, objects or abstract concepts) where entities are sets of properties.

Using this formalism we define a fact as an entity or a relation that can be either observed or deduced by an agent. A cognitive environment is then defined by the set of relations between entities that are manifest by this agent. Thus the shared context between agents (human or machine) is defined by the intersection of their situation models taking into account their perception and cognitive abilities, their task, and current activity. A fact (entity or relation) has the property to exist and is associated to an abstract representation that we call concept. Two facts can be instance of the same concept. Concept is for fact what a class is for object. Observed facts are any instance of concept that can be observed and represented as an entity or relation. Such entities include time of the day, weather, agent, object. Observed relation include spatial relation. Inferred facts are any instance of concept that can be inferred from other facts (observed or inferred). Inferred entities can include group, agent's activity, agent's task. These facts are not directly observable properties but rather interpretation.

3.3 Socially Guided Environment

The concept of mutual manifestness as part of the relevance theory has been concerned with human communication. In the previous section, we argued for the need of such an ability in human-machine interaction. Considering the main differences in term of cognition and perception between human and computer, this theory gives us a clear understanding of why perceiving social context from a human perspective is a difficult problem. Although this let us to propose our approach of end-user programming. In Relevance Theory [12] Sperber and Wilson present communication as the process of making certain facts more manifest than others. Thus in our approach both agents are provided with tools allowing them to alter each other's cognitive environment. Because of the fundamental difference between human an machine, we need to introduce the following condition. A fact is manifest for the system if and only if the related concept is also manifest. Then by definition a concept is manifest if and only if it belongs to the cognitive environment. A concept is then mutually manifest if and only if it has been acknowledged by the user. Therefore, the first step toward making a fact mutually manifest is to teach the system new concepts.

The first tools provides the user with a method to teach the system whatever perceptual concept he thinks is relevant. This concept is tokenized from the environment using bounding volumes such as blob or frustum constructed by user hand gestures. This tool is well adapted to make manifest spatial concept such as a chair, a desk or an area. At the same time, the system is able to display different views of its cognitive environment by displaying the relevant facts it is able to perceive or deduce from the environment. This second tools presents two benefits: first it provides transparency to the user, then it allows the user to respond whenever he did not agree. One of this view is provided by an interactive table. The table provides a window through the system's mind which lets the user browse and interact with the contained concepts and facts. Different actions are then proposed to the user such as teaching a new abstract concept, adding tags to a already learned concept, adding or removing a concept from the mutual cognitive environment. Tags are a simple way for the user to give interpretation to newly extracted concepts (eg. Chair, Desk, Table). The user can also see the current and previous system's situation models and related action, thus enabling the user to tell the system the action it should perform in a given situation. The interface provide also a tools for the correct assignment of reward in the case of social reinforcement learning [1]. Finally we integrated an attentional model to control the manifestness of a fact. The greater the attention of the user toward a fact the more manifest the fact. That way entity closest to the user for example will be more manifest than entity outside the environment.

4 Experimental Evaluation

In this section, we demonstrate our preliminary results with a step by step case study which illustrates a standard scenario in a context aware environment. The goal is to show how a user and a system can build conjointly a mutual cognitive environment.

4.1 Experimental Setup

The experiment takes place in the PRIMA SmartRoom, an environment specifically designed to observe (using cameras and microphones) and interact with users as they go through normal day-life activity such as work, meeting, rest, etc. The system is capable of tracking the position of persons as they evolve through the environment and also provides a rough estimate of the personâĂŹs posture. The system is also equipped with a steerable projector with automatic image rectification able, given a specific region, to transform a planar surfaces in a projection area.

4.2 Case Study

One scenario that comes to mind when working with smart environment is the meeting scenario. In such scenario, smart environments could provide different services such as projecting content of the meeting at the correct place. In the following, we will illustrate the different steps that lead to the construction of a mutual cognitive environment intended to provide our smart meeting system with the needed concepts to characterize/recognize the relevant situations.

This construction process is organized in two stages. The first step is a usual top down reflection where the user builds a mental representation of the situation he wants to explain by selecting relevant assumptions. A possible situation model could be a "**lecturer** *is close to* an **audience**". Then a **lecturer** *is a* "**person** *close to* a **projection area**" and so on. The second step is an iterative process where both human and system interact to make relevant information mutually manifest. Using different modalities the system acknowledges the user of the different facts and related concepts that are manifest in the environment. The user on its side, introduces the system with missing concept. For this experiment the environment already knows some basic concepts, provided by software components, such as the entity-concept **person**, **audience**, **region** and the relational-concept *close* to and *sit on*. The following concepts are "mutually manifest" however **chair** and **lecturer** are not. We could only expect the environment to interpret a correct situation model once **chair** and **lecturer** will be mutually manifest. Then, the steerable projector will project accordingly the content of the meeting at the correct place in the environment.

When the user enters the room, the only facts belonging to the MCE is the user itself. The system is able to perceive the user position and the user knows it, thus this fact is mutually manifest. The first action performed by the user in the environment is to defined the concept of **projection** area which is then integrated to the MCE. This step is done by performing a hand gesture surrounding this area and tagging the extracted region on the interactive table as **projection area**). Then the user pursues by extracting the different sitting areas (see Figure 1) to which he will associate the tag **chair**.

This far, the system is able to observe new facts from its environment. For example any person sitting on the extracted chair or sofa will now be manifest to the system as the related concepts, **person**, **chair** and *sit on* are known. The last

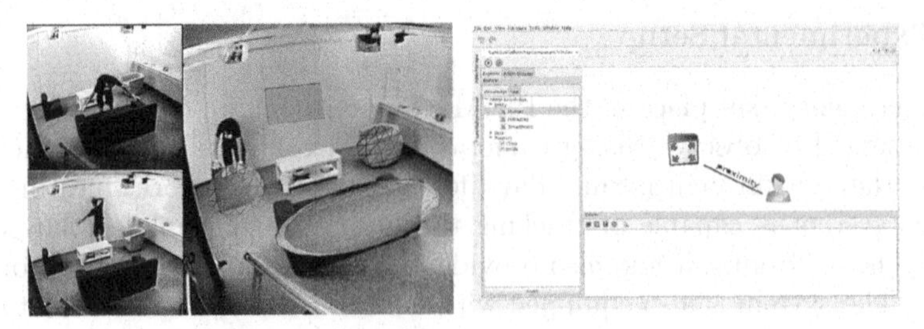

Fig. 1. Teaching concepts (left) and learning abstract concept (right)

missing concept is the one of **lecturer** which is a deduced-like concept. In order to do that, the user interacts with an interface displayed on an interactive table. The interface shows the different concepts mutually manifest as well as observed and deduced facts. Abstract concepts are defined from other set of concepts in relation, thus user has to construct a diagram relating sub-concepts together. This operation is achieved by different drag&drop operation. In our example, the diagram containing a **person** *close to* a **projection area** is labeled as **lecturer** (see right part of Figure 1).

During the real meeting presentation (see Figure 2) the user can, at any time, teach the system news action according to its interpretation of the current situation. The right part of Figure 2 presents the current interpretation of the situation by the system (remark, relation are not present). System can be taught to trigger the steerable projector service, and so the topic of the meeting will be displayed onto the correct projection area.

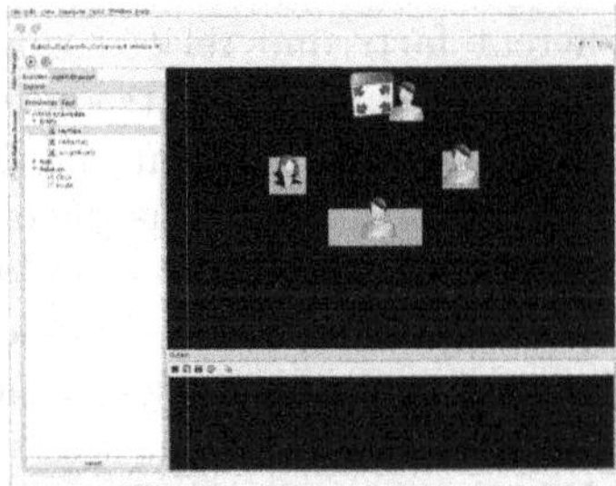

Fig. 2. Presentation meeting (left) and interpretation of the presentation meeting by the system

5 Conclusions

In an attempt to design context aware systems able to acquire social common sense from day to day interaction with non-expert user, a key challenge is to provide systems with methods to perceive social situation from a human perspective. We cannot expect developer to handle all possible social situations a system can encounter in its working life. An alternative inspired by end user programing offers better perspectives.

Our proposed approach provides transparent interaction and learning tools to build a mutual cognitive environment between the user and the system. We argue that mutual cognitive environment plays a significant role in the construction of situation models for social situation. Because system and user do not have the same perceptive and cognitive capabilities we propose a method where user will interactively (when the needs comes) teach to the system the concept that he thinks is relevant. Concepts are learnt by segmenting the environment using simple hand gestures and using an interactive table as a cognitive window through the system mind. The preliminary results are very encouraging and validate our approach.

References

1. Barraquand, R., Crowley, J.L.: Learning polite behavior with situation models. In: HRI 2008: Proceedings of the 3rd ACM/IEEE international conference on Human robot interaction, pp. 209–216. ACM, New York (2008)
2. Coutaz, J., Crowley, J.L., Dobson, S., Garlan, D.: Context is key. Commun. ACM 48(3), 49–53 (2005)
3. Crowley, J.L., Reignier, P., Barraquand, R.: Situation models: A tool for observing and understanding activity. In: Workshop People Detection and Tracking, held in IEEE International Conference on Robotics and Automation, Kobe, Japan (2009)
4. Dey, A.K., Abowd, G.D., Salber, D.: A conceptual framework and a toolkit for supporting the rapid prototyping of context-aware applications. Hum.-Comput. Interact. 16(2), 97–166 (2001)
5. Dey, A.K., Hamid, R., Beckmann, C., Li, I., Hsu, D.: a CAPpella: programming by demonstration of context-aware applications. In: Proceedings of the SIGCHI conference on Human factors in computing systems, Vienna, Austria, pp. 33–40. ACM, New York (2004)
6. Dourish, P.: What we talk about when we talk about context. Personal Ubiquitous Comput. 8(1), 19–30 (2004)
7. Ganneau, V., Calvary, G., Demumieux, R.: Learning key contexts of use in the wild for driving plastic user interfaces engineering. In: Proceedings of the 2nd Conference on Human-Centered Software Engineering and 7th International Workshop on Task Models and Diagrams, Pisa, Italy, pp. 271–278. Springer, Heidelberg (2008)
8. Johnson-Laird, P.N.: Mental models: towards a cognitive science of language, inference, and consciousness. Harvard University Press, Cambridge (1983)
9. Korpip, P.P., Malm, E., Rantakokko, T., Kyllonen, V., Kela, J., Mantyjarvi, J., Hakkil, J., Kansala, I.: Customizing user interaction in smart phones. IEEE Pervasive Computing 5(3), 82–90 (2006)
10. Reeves, B., Nass, C.: The Media Equation: How People Treat Computers, Television, and New Media Like Real People and Places, 1st edn., Cambridge (1996)
11. Salembier, P., Zouinar, M.: Intelligibilité mutuelle et contexte partagé. Inspirations théoriques et réductions technologiques. @CTIVITES 1(2) (2004)
12. Sperber, D., Wilson, D.: Relevance: Communication and Cognition. Blackwell Publishers, Malden (December 1995)
13. Zwaan, R.A., Radvansky, G.A.: Situation models in language comprehension and memory. Psychological Bulletin 123, 162–185 (1998)

On the Quantification of Aging Effects on Biometric Features

Andreas Lanitis and Nicolas Tsapatsoulis

School of Communication and Applied Arts, Cyprus University of Technology,
P.O. Box 50329, 3036, Lemesos, Cyprus
{andreas.lanitis,nicolas.tsapatsoulis}@cut.ac.cy

Abstract. Biometric templates are often used in intelligent human computer interaction systems that include automated access control and personalization of user interaction. The effectiveness of biometric systems is directly linked with aging that causes modifications on biometric features. For example the long term performance of person identification systems decreases as biometric templates derived from aged subjects may display substantial differences when compared to reference templates whereas in age estimation, aging variation allows the age of a subject to be estimated. In this paper we attempt to quantify the effects of aging for different biometric modalities facilitating in that way the design of systems that use biometric features. In this context the homogeneity of statistical distributions of biometric features belonging to certain age classes is quantified enabling in that way the definition of age sensitive and age invariant biometric features. Experimental results demonstrate the applicability of the method in quantifying aging effects.

Keywords: Biometric Templates, Aging Variation, Intelligent Human Computer Interaction.

1 Introduction

Intelligent Human Computer Interaction (HCI) systems with abilities to adapt to individual needs of different users have received increased attention in the research community [9]. In most cases the adaptation of interaction mode is accomplished based on the preferences set by a user. As an alternative, automatic adaptation of the way that a user interacts with a machine can be achieved by analyzing biometrical features of a user. Among other possible scenarios biometric features can be used for identifying the user of a machine allowing in that way automatic access and selection of user preferences and estimating the age of a user enabling the implementation of age-adaptive interaction systems.

Adaptive HCI based on biometric templates is heavily depended on the suitability of features used for the intended application. For example in the case of user authentication discriminatory biometric features invariant to facial expression, face orientation, lighting and aging need to be used. In general intelligent methods are required for selecting the most appropriate biometric features according to the application.

H. Papadopoulos, A.S. Andreou, and M. Bramer (Eds.): AIAI 2010, IFIP AICT 339, pp. 360–367, 2010.

A key issue related to the use of biometric templates is aging variation that causes modifications in biometric features [16]. Despite the fact that aging effects need to be taken into account during the design of biometric systems, so far the topic of assessing the aging invariance of biometric templates did not receive much attention. In this paper we attempt to quantify the effect of aging on different biometric templates. In this context we formulate a metric; called the 'Aging Impact' that reflects the expected intensity of aging effects for different biometric templates. High values of the Aging Impact indicate that the corresponding biometric features are subjected to intense aging-related appearance. According to the application where a biometric feature is intended to be used it may be desirable to select biometric features that exhibit low Aging Impact (i.e. for person identification applications) or high Aging Impact (i.e. for age estimation applications). As part of our study we estimate the Aging Impact for biometric templates of different types and different modalities such as faces, fingerprints, palms, hand movements and face movements. To the best of our knowledge this is a first time that the issue of quantifying the impact of aging for multiple biometric modalities is treated using a unified approach. It is anticipated that the proposed work in the area will influence the design of future intelligent biometric-based HCI systems.

2 Literature Review

A number of surveys and books have been published on the general topic of person authentication using unimodal or multimodal biometric features [12, 13, 21, 22]. However, in those cases only limited discussion related to the effects of aging on biometric templates is presented. Similarly in surveys dedicated to person identification based on specific modalities [2, 23, 25, 26] the topic of aging is not considered in detail. An exception is a survey on iris recognition by Bowyer et al [3] where the effects of aging on iris recognition are explicitly discussed.

Few studies that aim to assess the deterioration of performance of biometric authentication over time appeared in the literature. However, in such studies usually only short term aging effects are considered. Min et al [18] conclude that as the time period between template creation and recognition increases, the performance of a face recognition system decreases dramatically. Poh at al [20] also investigate short-term aging effects on 3D face recognition. Experimental results indicate that aging affects different users in different ways, thus a proper aging invariant biometric authentication system should be user-specific.

All approaches mentioned above aim to highlight the problem of aging rather than to quantify the effects of aging. As an alternative Baker et al [1] attempt quantify the effect of aging on iris templates by comparing the hamming distance between pairs of iris images of the same subject captured with a time difference of four years. According to the results, the differences in the hamming distances recorded indicate that iris patterns may be deformed due to aging.

A limited number of studies try to investigate the use of biometric technology in Human Computer Interaction [6]. The majority of them look into the problem from the biometric perspective; the aim is either to build appropriate interfaces to increase the adoption of biometrics for security reasons [4, 8] or to create authentication systems based on users' interaction with computers in their everyday work [24]. The use of biometrics for creating user-adapted human computer interfaces only recently gained some attention with the creation of special tracks in some conferences [10]. Nevertheless, utilization of biometric based transactions with computers, like the fingerprint based access, for the purpose of age estimation and user profile adaptation has not been examined in other studies so far.

3 Quantification of Aging Effects

Let vector $\mathbf{u}^j_i$ be the jth biometric template of the ith user of a biometric based authentication system. Without loss of generality we assume that there are Ni different biometric templates for the ith user. We can form an extended vector $\mathbf{u}_i$ representing user i by concatenating vectors $\mathbf{u}^j_i$, j= 1,…, N_i. The mean extended template (**M**) among all templates of all users in a pool is calculated and the similarity measure s_i, between user i and the mean template is computed using the correlation coefficient of vectors $\mathbf{u}_i$ and **M**. The homogeneity measure (H) among all templates can be obtained by taking the average value of similarities for all users i. Similarly, a measure of dispersion (D) can be obtained by taking the standard deviation of similarities for all users i. The homogeneity measure gives an indication of the similarity between members of the distribution and dispersion gives an indication of the spread of the correlation coefficients for different users. Although it is possible to calculate the homogeneity and dispersion of a multivariate distribution using other methods, the proposed method was selected because it is standardized, enabling in that way the comparative evaluation of aging effects for different biometric modalities.

Given a set of templates belonging to a number of users, the method outlined above was used for estimating the homogeneity and dispersion among all biometric templates in a dataset. The dataset is then divided into templates corresponding to the younger and older users using a predetermined age threshold and the homogeneity and dispersion measures for each age group are also estimated. The Aging Impact (AI) of a template is defined as the difference between the homogeneity and dispersion metrics of two different groups:

$$AI = \frac{H_i - H_{all}}{\sqrt{\left(\frac{D_i^2}{n_i} + \frac{D_{all}^2}{n_{all}}\right)}} \tag{1}$$

where H_i, D_i are the homogeneity and dispersion measures for the ith age group while H_{all}, D_{all} are the corresponding values computed over all database members. Similarly, n_i and n_{all} are the number of samples in the ith group and all samples in the database respectively. In the case that multiple age groups are considered the overall Aging Impact is the average of the Aging Impact between each group and the distribution of all samples. The largest the value of the AI measure the biggest the difference between the distributions of template values in a certain age group when compared with the distribution of all samples, leading to an increased dependency of the template to aging variation. For example if a biometric feature exhibiting a high Aging Impact factor is used in an identity verification application, it is highly probable that the system will fail to verify the identity of the user as the user grows older. Similarly the use of features with low Aging Impact in an age estimation application will result in a poor performance as the features used will not provide adequate information for discriminating between different ages.

4 Experimental Investigation

The method for assessing the impact of aging presented in the previous section was tested on different modalities and different types of feature vectors, in an attempt to illustrate its applicability in quantifying the impact of aging effects.

4.1 Experiments with Face Templates

Experiments using face templates were carried out using samples from the FG-NET Aging Database [15]. The FG-NET Aging Database is a publicly available image database containing face images showing a number of subjects at different ages. The database contains 1002 images from 82 different subjects with ages ranging between newborns to 69 years old subjects. On average there are 12 images per subject in the database.

In this experiment we consider the use of templates constructed based on an Active Appearance Model (AAM) [5] coding scheme. Since the FG-NET aging dataset contains multiple samples per subject, the calculation of the homogeneity and dispersion measures is done for each subject individually and the overall results are calculated by taking the average values of the two measures among the 82 subjects. For the needs of the experiment face images belonging to the same subject are divided into a set of young and old. Apart from using templates representing the overall facial region, the impact of aging on the upper face, lower face, eyes and nose regions is also investigated. In order to study the effects of aging for different genders, the analysis is performed in a gender-specific style.

According to the results (see Table 1) the impact of aging varies according to the facial region considered, the age group and the gender of the users. For example in old males the impact of aging on the lower face is 22 but the impact of aging on the same region for young males is 13. This is expected as the texture of the lower facial regions of old males undergoes dramatic changes due to the possible appearance of beards and moustaches. The results show that the most dramatically affected facial region is the eye region. Bearing in mind that the FG-NET aging dataset contains unconstrained face images, the results presented contain noisy responses that result from non-aging related variations appearing on faces. Ideally experiments using samples that mainly display aging variation need to be conducted in order to obtain definite conclusions regarding the invariance of different facial regions to aging.

Table 1. Aging Impact for different groups of samples from the FG-NET Aging Database. AIy, AIo, AI are the Aging Impact factors for the group with young samples, old samples and the overall impact respectively.

Facial region	Males			Females		
	AI_y	AI_o	AI	AI_y	AI_o	AI
Overall face	10	25	17	2	21	11
Upper face	8	19	14	16	17	16
Lower face	13	22	18	3	25	14
Nose	12	30	21	7	17	12
Eyes	16	21	18	19	27	23

4.2 Experiments with Palm and Fingerprint Templates

Experimentation for the palm and fingerprint modalities is based on biometric templates constructed with the aid of MPEG-7 visual features [11] computed using the MPEG-7 experimentation software [19]. We have used the POLYBIO database [14] for our experiments. In this database there are four hand images per subject taken using low cost cameras as well as four fingerprint images per finger per subject taken using a light-reflection based fingerprint sensor.

Four different templates were created from every hand image, corresponding to the Color Layout (CL), the Region Shape (RS), the Edge Histogram (EH) and the Homogeneous Texture (HT) descriptors. The above descriptors were chosen to represent the three different types of visual descriptors (color, shape, texture); therefore, the palm images are not used only for computing hand geometry but the color and texture of the upper part of the palm. Each template is examined in isolation and the results are summarized in Table 2. According to the results the highest Aging Impact is recorded in Edge Histogram; that is edge information of the upper part of hand changes significantly with age. Increased aging effects on edge information are caused by reduced skin elasticity, spot appearance or enlargement and veins' pump up mainly encountered in older persons.

For the fingerprint experiments we used only the images of the left index. The templates used are the same as in the hand geometry case with the exception of Region Shape which was excluded due to the absence of any shape information in fingerprint images. According to the results, shown in Table 2, the highest Aging Impact is recorded for features based on Edge Histogram. This is something expected; fingerprint images contain a lot of edge information while aging reduces skin elasticity, and therefore proper finger fitting to the fingerprint sensor reducing in that way the amount of edge information captured by the sensor. In addition, as people get older edges on fingerprints wear out and the probability of finger injuries also increases.

However, we should take into account that both in the case of experiments with palms and fingerprints, the intra class variation (variation across subjects of the same group) is also high and may be higher than the inter class variation (variation across age) smoothing the overall Aging Impact values.

Table 2. Aging Impact for different groups of samples from the POLYBIO multimodal database. AI_y, AI_o, AI are the Aging Impact factors for the group with young samples, old samples and the overall impact respectively

Modality	Template Type	AI_y	AI_o	AI
Palm	Scalable Color	0.19	0.51	0.35
	Region Shape	0.60	0.87	0.73
	Edge Histogram	0.21	1.56	0.89
	Homogeneous Texture	0.22	0.93	0.57
Fingerprints	Scalable Color	0.60	0.03	0.32
	Edge Histogram	1.68	0.35	1.01
	Homogeneous Texture	0.68	0.55	0.61

4.3 Experiments with Templates Based on Hand and Face Movements

In this experiment we consider the use of templates containing information that characterizes hand and head movements of users. In this context, volunteers who participated in this experiment were requested to use the mouse for tracing as accurately as possible the outline of a horizontal line, a rectangle and an ellipse presented on the screen. The ability of each user to trace accurately the shape was calculated by considering the mean Euclidean distance and the difference in the direction of pointer movement when compared with the target shape contour. The three measures for each volunteer and each shape (line, rectangle and ellipse) constitute a user-template. The same experimental set-up was used for assessing templates that contain features extracted from head movements. In this case a tracking algorithm suitable for locating the position of the eyes of a subject while he/she is tracing a shape on the screen was used [7].

Thirty volunteers with ages ranging from 8 years old up to 43 years old participated in the experiment. These volunteers were divided into the group of young (ages 8 to 17) and group of old (ages of 18 to 43). All volunteers who participated in the experiment were regular computer users, familiar and experienced with the operation of the mouse. Because none of the volunteers used face tracking for human computer interaction purposes before, each volunteer was allowed a five-minute training session in order to get familiar with the face tracker operation before the actual data acquisition exercise. The results obtained (see Table 3) demonstrate that the highest Aging Impact is recorded for users tracing circles, either using hand or head movements. These results are consistent with results reported in the literature [17] that demonstrate the superiority of using discrepancy features derived through a circle tracing exercise as the basis for estimating the age of a person. However, the results presented on this experiment are only indicative, due to the small sample of samples considered in the experiment.

Table 3. Aging Impact for hand and face movement based templates. AI_y, AI_o, AI are the Aging Impact factors for the group with young samples, old samples and the overall impact respectively

Modality	Shape Drawn	AI_y	AI_o	AI
Hand Movements	Line	1.26	1.97	1.62
	Rectangle	0.10	0.21	0.15
	Ellipse	4.52	3.63	4.07
Face Movements	Line	2.52	0.17	1.35
	Rectangle	2.6	0.38	1.49
	Ellipse	4.47	3.41	3.94

5 Conclusions

We have presented a method for quantifying aging effects on biometric templates, allowing in that way the selection of appropriate features for use in specific applications. The appropriate selection of features will result in the development of robust systems that can be used for developing biometric-based user adaptive HCI systems. A key aspect of our work is the formulation of a generic Aging Impact measure that can be used for different types of biometric modalities and templates, allowing in that way the derivation of accurate conclusions related to the aging invariance of biometric templates. It is anticipated that the proposed work in the area will influence both the design of future age-invariant biometric authentication systems and it will also provide a standardized framework for assessing the long term permanence of biometric features.

The results obtained prove the potential of the approach in assessing the age invariance of biometric templates using a standardized method. In the future we plan to perform extended experiments that will allow the derivation of concrete results related to the aging invariance of different modalities, different features, different genders and different age groups. In order to accomplice this task we are in the process of collecting suitable multimodal biometric data from subjects with diverse ages, so that the experimentation will cover all age ranges of interests.

References

1. Baker, S.E., Bowyer, K.W., Flynn, P.J.: Empirical Evidence for Correct Iris Match Score Degradation with Increased Time-Lapse between Gallery and Probe Matches. In: Tistarelli, M., Nixon, M.S. (eds.) ICB 2009. LNCS, vol. 5558, pp. 1170–1179. Springer, Heidelberg (2009)
2. Bowyer, K.W., Chang, K., Flynn, P.: A survey of approaches and challenges in 3d and multi-modal 3d+2d face recognition. Computer Vision and Image Understanding 101(1), 1–15 (2006)
3. Bowyer, K.W., Hollingsworth, K., Flynn, P.J.: Image understanding for iris biometrics: A survey. Computer Vision and Image Understanding 110, 281–307 (2007)
4. Briggs, P., Olivier, P.L.: Biometric daemons: authentication via electronic pets. In: CHI 2008 Extended Abstracts on Human Factors in Computing Systems, pp. 2423–2432 (2008)
5. Cootes, T.F., Edwards, G.J., Taylor, C.J.: Active Appearance Models. IEEE Transactions of Pattern Analysis and Machine Intelligence 23, 681–685 (2001)
6. Coventry, L., Johnson, G.I., McEwan, T., Riley, C.: Biometrics in Practice: What Does HCI Have to Say? In: Gross, T., Gulliksen, J., Kotzé, P., Oestreicher, L., Palanque, P., Prates, R.O., Winckler, M. (eds.) INTERACT 2009. LNCS, vol. 5727, pp. 920–921. Springer, Heidelberg (2009)
7. Frangeskides, F., Lanitis, A.: Multi-Modal Contact-less Human Computer Interaction. LNBIP, pp. 405–419. Springer, Heidelberg (2007)
8. Gaboa, H., Fred, A.: A behavioral biometric system based on human-computer interaction. In: Proceedings of SPIE, pp. 381–392 (2004)
9. Germanakos, P., Tsianos, N., Lekkas, Z., Mourlas, C., Samaras, G.: Capturing essential intrinsic user behaviour values for the design of comprehensive web-based personalized environments. Computers in Human Behavior 24(4), 1434–1451 (2008)
10. ICPR 2010: 20th International Conference on Pattern Recognition, Track IV: Biometrics and Human Computer Interaction, Istanbul, Turkey (2010)
11. ISO/IEC 15938-3:2001 Information Technology - Multimedia Content Description Interface - Part 3: Visual, Ver. 1 (2001)
12. Jain, A., Ross, A., Prabhakar, S.: An Introduction to Biometric Recognition. IEEE Transactions on Circuits and Systems for Video Technology 14(1) (2004)

13. Jain, A., Patrick, F., Ross, A. (eds.): Handbook of Biometrics. Springer, Heidelberg (2008)
14. Kounoudes, A., Tsapatsoulis, N., Theodosiou, Z., Milis, M.: POLYBIO: Multimodal Biometric Data Acquisition Platform and Security System. In: Schouten, B., Juul, N.C., Drygajlo, A., Tistarelli, M. (eds.) BIOID 2008. LNCS, vol. 5372, pp. 216–227. Springer, Heidelberg (2008)
15. Lanitis, A.: Comparative Evaluation of Automatic Age Progression Methodologies. EURASIP Journal on Advances in Signal Processing, Article ID 239480 (2008)
16. Lanitis, A.: A Survey of the Effects of Aging on Biometric Identity Verification. International Journal of Biometrics 2(1), 34–52 (2010)
17. Lanitis, A.: Age Estimation Based on Head Movements: A Feasibility Study. In: 4th International Symposium on Communications, Control and Signal Processing (2010)
18. Min, J., Flynn, P.J., Bowyer, K.W.: Assessment of time dependency in face recognition. In: Kittler, J., Nixon, M.S. (eds.) AVBPA 2003. LNCS, vol. 2688, pp. 44–51. Springer, Heidelberg (2003)
19. MPEG-7 Visual Experimentation Model (XM), Version 10.0, ISO/IEC/JTC1/SC29/WG11, Doc. N4063 (2001)
20. Poh, N., Kittler, J., Smith, R., Tena, J.R.: A Method for Estimating Authentication Performance over Time, with Applications to Face Biometrics. In: Rueda, L., Mery, D., Kittler, J. (eds.) CIARP 2007. LNCS, vol. 4756, pp. 360–369. Springer, Heidelberg (2007)
21. Ross, A., Karthik, N., Jain, A.: Handbook of Multibiometrics. Springer, Heidelberg (2006)
22. Woodward, J.D., Orlans, N.M., Higgins, P.T.: Biometrics: Identity Assurance in the Information Age. McGraw-Hill, New York (2002)
23. Yager, N., Anim, A.: Fingerprint Classification: A Review. Pattern Analysis & Applications 7(1), 77–93 (2004)
24. Yampolskiy, R.V., Govindaraju, V.: Direct and Indirect Human Computer Interaction Based Biometrics. Journal of Computers 2(10), 76–88 (2007)
25. Yampolskiy, R.V., Govindaraju, V.: Behavioural biometrics: a survey and classification. International Journal of Biometrics 1(1), 81–113 (2008)
26. Zhao, W., Chellappa, R., Rosenfeld, A., Phillips, P.J.: Face recognition: A literature survey. ACM Computing Surveys, 399–458 (2003)

Fuzzy Inference Systems for Automatic Classification of Earthquake Damages

Petros-Fotios Alvanitopoulos[1], Ioannis Andreadis[1], and Anaxagoras Elenas[2]

[1] Laboratory of Electronics, Department of Electrical and Computer Engineering, School of Engineering, Democritus University of Thrace, Xanthi, Greece
[2] Institute of Structural Mechanics and Earthquake Engineering, Department of Civil Enginnering, School of Engineering, Democritus University of Thrace, Xanthi, Greece
{palvanit,iandread}@ee.duth.gr, elenas@civil.duth.gr

Abstract. This paper presents efficient models in the area of damage potential classification of seismic signals. After an earthquake, one of the most important actions that authorities must take is to inspect structures and estimate the degree of damages. The interest is obvious for several reasons such as public safety, economical recourses management and infrastructure. This approach provides a comparative study between the Mamdani-type and Sugeno-type fuzzy inference systems (FIS). The fuzzy models use a set of artificial accelerograms in order to classify structural damages in a specific structure. Previous studies propose a set of twenty well-known seismic parameters which are essential for description of the seismic excitation. The proposed fuzzy systems use an input vector of twenty seismic parameters instead of the earthquake accelerogram and produce classification rates up to 90%. Experimental results indicate that these systems are able to classify the structural damages in structures accurately. Both of them produce the same level of correct classification rates but the Mamdani-type has a slight superiority.

Keywords: Seismic Parameters, Earthquake Damage Classification, Fuzzy Inference Systems.

1 Introduction

Earthquake engineering can be defined as the field of engineering which deals with earthquakes and their effect on structures such as multistory buildings, bridges, towers, etc. As with any other natural phenomenon, earthquakes and their effects should be considered when designing a structure to ensure its safety. However, it is not economically feasible to design a structure that can withstand any seismic event without damages (elastic response). Sometimes devastating earthquakes hit populated areas. In this case, the ability to estimate the post-seismic damage status of oscillated structures instantly after the earthquake event may optimize the administration measures (optimized management of human and material resources, reduction of loss of lives and injuries).

The thrust of the current research is the automatic approach of the post-seismic status of buildings using intelligent FIS. Two types of fuzzy systems (Mamdani and Sugeno)

H. Papadopoulos, A.S. Andreou, and M. Bramer (Eds.): AIAI 2010, IFIP AICT 339, pp. 368–375, 2010.

are designed and implemented. These models use a set of artificial accelerograms. Numerical results indicate the ability of them to classify the structural damage in the examined structure.

In this research the set of artificial accelerograms is applied as a case study to a sixth floor reinforced concrete frame structure. The proposed classifiers can estimate the degree of structural damage on any other structure under the precondition that the training phase is carried out for the same building.

The key point of this study is to map the input space (artificial accelerograms) to an output space (degree of structural damage). Accelerograms are records of the acceleration versus time measured during an earthquake ground motion. The seismic accelerograms are a useful tool in earthquake engineering because they are able to provide an explicit description of the seismic excitation. However, due to the random sizes and shapes it is very difficult to exploit their similarities. Therefore, a set of twenty seismic parameters has been used to represent the seismic signals. These seismic parameters have been presented in the literature during the last decades and are able to express the damage potential of earthquakes. Table 1 presents those parameters. Previous study [1] reveals the correlation grade between seismic parameters and damage indices of structures.

Table 1. Seismic parameters

Seismic Parameters	
Peak Ground Acceleration	Power ($P_{0.90}$)
Peak Ground Velocity	Central Period
Strong Motion Duration after Trifunac/Brady	Arias Intensity
Root Mean Square Acceleration	Spectral Displacement
Spectrum Intensity after Housner	Spectral Velocity
Spectrum Intensity after Kappos	Spectral Acceleration
Spectrum Intensity after Martinez-Rueda	Seismic Energy Input
Seismic Intensity after Fajfar/Vidic/Fischinger	Effective Peak Acceleration (EPA)
Destructiveness Potential	Maximum EPA
Cumulative Absolute Velocity	PGA to PGV ratio

For the present approach the level of structural damage caused by an earthquake is numerically quantified using damage indices (DIs). In recent years considerable efforts have been devoted to develop different types of damage indices (local-global). In general, a damage index is local when it refers to a restricted area (e.g. a cross section), while it is considered global when it describes the state of the whole structure. In this study the systems were developed based on the maximum inter storey drift ratio (MISDR) [2] and the widely used Park/Ang [3] overall damage indicator.

In order to quantify the consequences of an earthquake, it is necessary to know the physical deterioration (damage) of the building suffered during the seismic event. The examined set of 450 artificial accelerograms generates a wide spectrum of structural damage from negligible to severe. The values of damage indices of each accelerogram are divided into four damage categories (low-medium-large-total) according to Table 2. These classes correspond to non damage or minor damage, repairable damage, irreparable damage and partial or total collapse of the building, respectively. According to this classification of the structural damages it seems at first sight that the classification is

Table 2. Damage Degree

Structural Damage	Damage Degree			
Indicators	Low	Medium	Large	Total
Park/Ang DI_G [-]	≤ 0.3	$0.3 < DI_G \leq 0.6$	$0.6 < DI_G \leq 0.8$	$DI_G > 0.8$
MISDR [%]	≤ 0.5	$0.5 <$ MISDR ≤ 1.5	$1.5 <$ MISDR ≤ 2.5	> 2.5

unequivocal. However, from the engineer's point of view, a damage index value close to the class limits cannot be classified unequivocally to a specific damage category. For example, a structural damage ratio of 0.499 cannot unequivocally classify the accelerogram in the low class, because the option to belong in the medium class should be possible. Thus, the set of damage indices can be considered as a fuzzy set and fuzzy methods are appropriate to be applied to such classification procedures.

Previous studies [4-6] attempt to classify the structural damages in buildings. The first approach is based on the shape similarity of accelerograms. However, due to the random nature of seismic signals this approach led to a poor classification rates. The second approach is based on seismic parameters which can express the damage potential of an earthquake [1]. These seismic parameters were graphically used to represent the seismic load instead of the accelerogram itself. The extracted classification results based on artificial neural networks (ANNs) were better than the results of the first approach. Nevertheless, it was desirable to test the accuracy of the system using an additional damage index and a wide set of seismic signals. This has been achieved by the third technique using a combination of genetic algorithm and ANNs.

This paper studies the potential of two alternate soft computing techniques for earthquake damage classification based on fuzzy logic. Fuzzy classifiers have been successfully and widely applied to many fields. Moreover, recently the evolving fuzzy rule based classifiers have been proved promising tools that offer high performance (classification rates) and computational efficiency [7, 8]. The ability of fuzzy logic to model nonlinear functions of arbitrary complexity makes it even more important to investigate its ability to estimate the earthquake damage in buildings. The classification results show that these systems are able to estimate the seismic vulnerability of buildings avoiding complex mathematical processes.

2 Proposed Method

2.1 The Mamdani-Type Fuzzy Inference System

The proposed fuzzy system has twenty inputs corresponding to the twenty seismic parameters that result from the processing of seismic signals. These values can represent seismic accelerograms and are normalized into a range [0, 1]. The FIS has one output that indicates the level of earthquake damage in structure. A model of the system is shown in Fig. 1.

The fuzzyfication of the inputs is realized using four Gaussian-shaped membership functions (mfs), namely low, medium, large and total. On the other hand the output consists of four triangular-shaped mfs, namely low, medium, large and total. The number and shape of the mfs were selected after an extensive set of preliminary tests. These tests involved simulations with different types and numbers of mfs such as triangular, trapezoidal and Gaussian. The tests confirmed that the below mfs presents good qualitative results. The following equations define the membership functions.

The Gausian function depends on two parameters σ and c as given by:

$$f(x,\sigma,c) = e^{\frac{-(x-c)^2}{2\sigma^2}} \tag{1}$$

where c and σ are the mean value and the variance, respectively.

$$\text{Low: } mf_1(x;0.1416,0) = e^{\frac{-(x-0)^2}{2*(0.1416)^2}} \tag{2}$$

$$\text{Medium: } mf_2(x;0.1416,0.333) = e^{\frac{-(x-0.333)^2}{2*(0.1416)^2}} \tag{3}$$

$$\text{Large: } mf_3(x;0.1416,0.667) = e^{\frac{-(x-0.667)^2}{2*(0.1416)^2}} \tag{4}$$

$$\text{Total: } mf_4(x;0.1416,1) = e^{\frac{-(x-1)^2}{2*(0.1416)^2}} \tag{5}$$

The triangular function depends on two parameters m and σ as given by:

$$f(x;m,\sigma) = \begin{cases} 1-\frac{m-x}{\sigma} & \text{for } m-\sigma \le x \le m \\ 1-\frac{x-m}{\sigma} & \text{for } m \le x \le \sigma+m \\ 0 & \text{otherwise} \end{cases} \tag{6}$$

where m and σ are the center and width of the triangular-shaped function, respectively.

$$\text{Low: } mf_1(x;0,0.3333) = \begin{cases} 1-\frac{0-x}{0.3333} & \text{for } -0.3333 \le x \le 0 \\ 1-\frac{x-0}{0.3333} & \text{for } 0 \le x \le 0.3333 \\ 0 & \text{otherwise} \end{cases} \tag{7}$$

$$\text{Medium: } mf_2(x;0.3333,0.3333) = \begin{cases} 1-\frac{0.3333-x}{0.3333} & \text{for } 0 \le x \le 0.3333 \\ 1-\frac{x-0.3333}{0.3333} & \text{for } 0.3333 \le x \le 0.6667 \\ 0 & \text{otherwise} \end{cases} \tag{8}$$

$$\text{Large: } mf_3(x;0.6667,0.3333) = \begin{cases} 1-\frac{0.6667-x}{0.3333} & \text{for } 0.3333 \le x \le 0.6667 \\ 1-\frac{x-0.6667}{0.3333} & \text{for } 0.6667 \le x \le 1 \\ 0 & \text{otherwise} \end{cases} \tag{9}$$

$$\text{Total: } mf_4(x;1,0.3333) = \begin{cases} 1-\dfrac{1-x}{0.3333} & \text{for } 0.6667 \le x \le 1 \\ 1-\dfrac{x-1}{0.3333} & \text{for } 1 \le x \le 1.3333 \\ 0 & \text{otherwise} \end{cases} \tag{10}$$

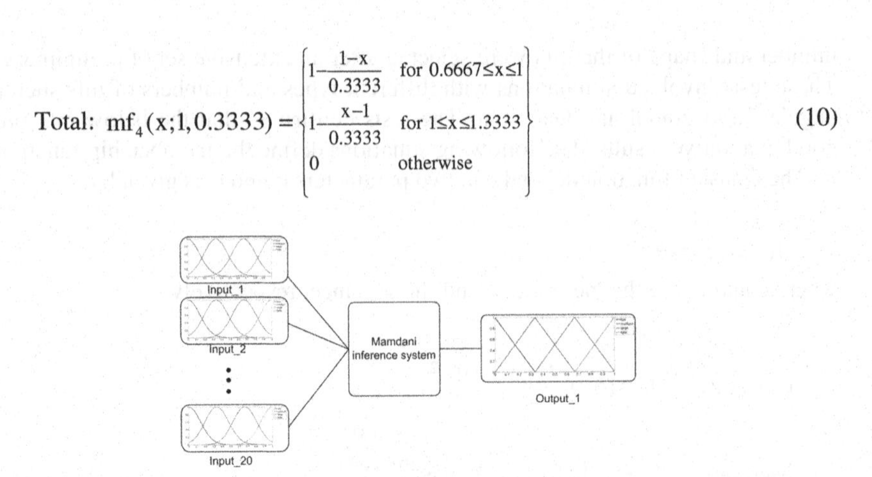

Fig. 1. The Mamdani inference system

The inputs are routed to the output through a Mamdani [9] type of fuzzy inference that uses a set of rules which is not unique and depends on the number of the training samples. In this study two DIs have been used so the structure of the model has a different set of rules for each damage indicator. Knowing the training data points it is possible to extract fuzzy rules [10]. In an effort to generate fuzzy rules the following steps have been considered:

Step 1: Divide the input/output space into fuzzy regions and assign each region a fuzzy membership function. In this approach the number of regions for each variable is equal to four since four are the aforementioned membership functions.

Step 2: Generate fuzzy rules using the training samples. First determine the degree of the given training input X for each membership function, and assign the input to the region with the maximum degree of membership. The extracted rule is as follows:

R_1: If X_1 is Low ^ X_2 is Large . . . X_{20} is Medium, THEN Y is Low

Step 3: Assign a degree to each rule: For each data point a fuzzy rule is generated. The degree of the fuzzy rule is:

$$D(R_1)=\min(\mu Low(X_1),\mu Large(X_2),\ldots\mu Medium(X_{20})) \tag{11}$$

In this model the antecedent of each rule has more than one part. The antecedent consists of twenty terms since twenty are the seismic parameters. The fuzzy operator AND (min) is applied to obtain a number that represents the result of the antecedent of each rule. In Mamdani type of fuzzy inference, the fuzzy sets from the consequent of each rule are combined through the aggregation operator and the resulting fuzzy set is defuzzified to yield the output of the system. The aggregation operator, in this case, is set to maximum and the defuzzification method is that of the center of gravity (CoG) [10].

2.2 The Tagaki–Sugeno–Kang (TSK)-Type Fuzzy Inference System

In the present study a (TSK) fuzzy system was used for the classification of earthquake damage in the examined structure. This methodology was introduced in 1985

[11], and it is similar to the Mamdani method in many respects. The first two parts of the fuzzy inference process, fuzzifying the inputs and applying the fuzzy operator, are exactly the same. The main difference between Mamdani and Sugeno is that the Sugeno output membership functions are either linear or constant.

Let D be the set of input–output data pairs (training data), $X=\{x_1,\ldots,x_n\}\subset\mathbb{R}^p$ be the set of p-dimensional input vectors, and $Y=\{y_1,\ldots,y_n\}\subset\mathbb{R}$ be the associated set of output vectors. The set Z with input–output vectors taken together can be denoted by:

$$Z=\left\{z_k=\begin{pmatrix}x_k\in\Re^p\\ y_k\in\Re\end{pmatrix}\in\Re^{p+1},k=1,\ldots,n\right\} \tag{12}$$

The TSK model comprises of a set of IF-THEN fuzzy rules having the following form:

R_1 : If x_1 is A_{i1} and ... and x_p is A_{ip} Then $y_i=b_{i0}+b_{i1}x_1+\ldots+b_{ip}x_p$

where $i=1,\ldots,M$, $A_{ij}=(j=1,\ldots,p)$ are antecedent fuzzy sets, y_i is the output of the ith rule, and b_{il} $(l=0,1,\ldots,p)$ are the consequent parameters. The overall output $y^{overall}$ of the model is computed as follows:

$$y^{overall}=\frac{\sum_{i=1}^{M}\tau_i y_i}{\sum_{i=1}^{M}\tau_i}=\frac{\sum_{i=1}^{M}\tau_i\left(b_{i0}+b_{i1}x_1+\ldots+b_{ip}x_p\right)}{\sum_{i=1}^{M}\tau_i} \tag{13}$$

where τ_i is the firing strength of R_i.

In our approach the proposed model is a zero-order TSK system. The consequent parameters are set to 0 (except b_{i0}) and the outputs of rules are constant. The rule base is the same with the Mamdani fuzzy model. The mfs of the input vector are set using the previous equations (2-5). All consequent mfs from Mamdani are replaced by singleton spikes. The firing strength of each rule is computed using the AND operator, in this case, is set to minimum. The TSK model is shown in Fig. 2.

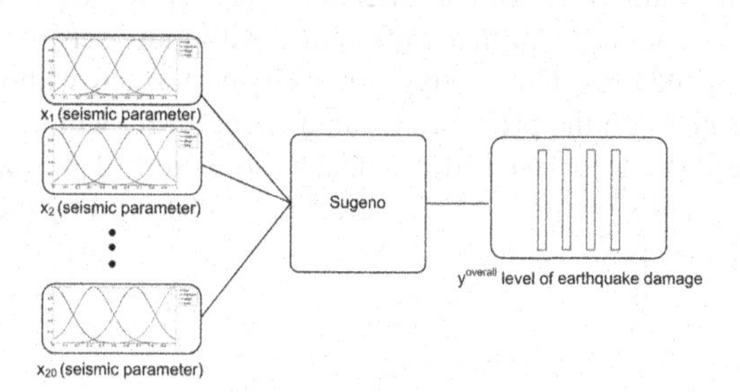

Fig. 2. The TSK fuzzy system

3 Results

After the nonlinear dynamic analysis of the structure, two DIs, namely, the DI of Park/Ang, and the MISDR have been computed. According to these DIs, the damages caused by seismic excitations, were classified into four classes. The next step of our method was to pass the seismic signals through the proposed fuzzy models and examine their effectiveness. Artificial accelerograms were used in order to simulate natural earthquakes. Table 3 presents the classification results for the examined FIS. As it can be seen, from Table 3 the same set of training samples provide different number of rules for each DI. The correct classification rates were better in the case of Park/Ang DI than in the case of MISDR. Moreover, the Mamdani system presents better classification rates than the TSK model. However, both of these systems are able to evaluate the seismic damage potential and produce the same level of correct classification rates.

Table 3. Classification Results Using Mamdani and TSK Fuzzy Systems

Type of FIS	Mamdani		TSK	
	MISDR	DI of Park/Ang	MISDR	DI of Park/Ang
Number of training samples	300	300	300	300
Number of Rules	193	179	193	179
Number of unknown samples	450	450	450	450
Number of well recognized samples	392	407	388	393
Total % of the classification	87.10%	90,44%	86,22%	87,33%

4 Conclusions

The utility of fuzzy logic in pattern classification problems is nowadays quite extensive. This paper attempts to design diagnostic systems with high classification accuracy in the scientific area of earthquake engineering. First a set of 450 artificial accelerograms with known damage effects was required in order to derive the parameters which are able to describe the seismic intensity. Two FIS (Mamdani-type, TSK-type) are presented. The Mamdani-type is slightly better. Experimental results show that the systems developed herein, presents the same level of classification rates (up to 90%). These promising results are the starting point for further exploration with more seismic events applied also to other types of structures. Until today, survey is performed with on-site examination by expert engineers. With the proposed technique engineers will have an additional tool which can guide them to a fast and confident estimation of the post-seismic damage condition of interested structures. Moreover, they can be used by the public administration for the adequate post-seismic distribution of financial and other resources in the case of severe earthquakes.

References

1. Elenas, A., Meskouris, K.: Correlation Study Between Seismic Acceleration Parameters and Damage Indices of Structures. J. Engineering Structures 23, 698–704 (2001)
2. Rodriguez-Gomez, S., Cakmak, A.S.: Evaluation of Seismic Damage Indices for Reinforced Concrete Structures. Technical Report NCEER-90-0022, State University of New York, Buffalo (1990)

3. Park, Y.J., Ang, A.H.S.: Mechanistic Seismic Damage Model for Reinforced Concrete. J. Structural Engineering 111, 722–739 (1985)
4. Andreadis, I., Tsiftzis, Y., Elenas, A.: Intelligent Seismic Acceleration Signal Processing for Damage Classification in Buildings. J. IEEE Transactions on Instrumentation and Measurement 56, 1555–1564 (2007)
5. Alvanitopoulos, P., Andreadis, I., Elenas, A.: A New Algorithm for the Classification of Earthquake Damages in Structures. In: 5th IASTED International Conference on Signal Processing, Pattern Recognition and Applications, Innsbruck, Austria, pp. 151–156 (February 2008)
6. Alvanitopoulos, P., Andreadis, I., Elenas, A.: A Genetic Algorithm for the Classification of Earthquake Damages in Buildings. In: 5th IFIP Conference on Artificial Intelligence Applications & Innovations, Thessaloniki, Greece, pp. 341–346 (April 2009)
7. Angelov, P., Zhou, X.: Evolving Fuzzy-Rule-Based Classifiers from Data Streams. J. IEEE Transactions on Fuzzy Systems 16, 1462–1475 (2008)
8. Angelov, P., Lughofer, E., Zhou, X.: Evolving Fuzzy Classifiers Using Different Model Architectures. J. Fuzzy Sets and Systems 159, 3160–3182 (2008)
9. Mamdani, E.H., Assilian, S.: An Experiment in Linguistic Synthesis with a Fuzzy Logic Controller. J. Man-Machine Studies 7, 1–13 (1975)
10. Kulkarni, A.D.: Computer Vision and Fuzzy-Neural Systems. Prentice Hall PTR, Upper Saddle River (2001)
11. Sugeno, M.: Industrial Applications of Fuzzy Control. Elsevier Science, New York (1985)

A Fuzzy Inference System Using Gaussian Distribution Curves for Forest Fire Risk Estimation

Lazaros Iliadis[1], Stergios Skopianos[1], Stavros Tachos[2], and Stefanos Spartalis[3]

[1] Democritus University of Thrace, Pandazidou 193 str., Orestiada, Greece
liliadis@fmenr.duth.gr
[2] Aristotle University of Thessaloniki, Department of Informatics
[3] Democritus University of Thrace, Xanthi, Greece

Abstract. This paper describes the development of a fuzzy inference system under the MATLAB platform. The system uses three distinct Gaussian distribution fuzzy membership functions in order to estimate the partial and the overall risk indices due to wild fires in the southern part of Greece. The behavior of each curve has been investigated in order to determine which one fits better for the specific problem and for the specific areas. Regardless the characteristics of each function, the risky areas have been spotted from 1984 till 2007. The results have shown a reliable performance over time and they encourage its wider use in the near future.

Keywords. Fuzzy algebra, Gaussian distribution curves, Fuzzy Inference System, wildfire risk.

1 Introduction

Every year, forest fires destroy important forest areas, with enormous consequences. The problem of forest fires occurs mainly in the Mediterranean countries, in Australia, in Canada and the United States [6]. Especially in Greece due to the lack of a proper cadastral survey, the problem of forest fires is related to political elections and also to the cycles of drought [9][10]. More detailed analysis on the problem of forest fires in Greece can be found in [14][16][17].

Fuzzy logic systems are quite flexible and they are used for several purposes such as modeling, forecasting and classification [8]. This research effort aims in the development of a fuzzy inference system, that estimates an integrated fuzzy forest fire risk index offering an overall view of the problem. This is achieved by employing three distinct fuzzy membership functions (of the same family) namely the Gaussian, the Sigmoid and the S-shaped one. The application of fuzzy logic enhances the ability to model proper linguistics that are used in the consequent part of the developed rule-based system. The application area concerns the Greek forest departments of Attica and Peloponnesus which are located in the southern part of the country. The whole research has been performed for the period 1984 to 2007 and the estimation of the integrated risk index has been achieved by considering the annual number of forest fire breakouts and the annual burned forest area for each forest department.

H. Papadopoulos, A.S. Andreou, and M. Bramer (Eds.): AIAI 2010, IFIP AICT 339, pp. 376–386, 2010.

1.1 Literature Review

In order for the appropriate authorities to adopt efficient fire prevention and protection measures, the assistance of modern technology is required towards the evaluation of forest fire risk [6][8][13]. Deeming et al [4], developed the national fire danger rating system (NFDRS), which is still used in the USA after some improvements. It is a system that produces the daily ignition component index and also the man caused risk index. There have been many studies in the United States, in Australia, in New Zealand and in Europe for the long term evaluation of fire risk. A system that evaluates daily as well as long term fire danger is EFFIS (European Forest Fire Information System). It has been developed by the Institute for Environmental and Sustainability Land Management which resides in Italy [1]. In reality EFFIS is an attempt towards the creation of a common European indicator for fire danger. In Greece, the lab of forest informatics of the Democritus University of Thrace has developed decision support systems that are employing heuristic reasoning and fuzzy logic (fuzzy expected intervals) in order to estimate wild fire risk for the following year. The heuristic reasoning is based on the effect of political elections and drought cycles (following a periodicity of 3 to 5 years) in the number of forest fires in Greece [7].

2 Intelligent Information Systems

2.1 Fuzzy Algebra

Fuzzy logic [18],[2],[11], allows the definition of real world concepts, such as "area with high vulnerability to forest fires" in the form of fuzzy sets. Every member of the universe of discourse belongs to a fuzzy set with a different degree of membership, ranging from 0 to 1. The degree of membership can be estimated by the use of proper membership functions (MF). Fuzzy sets correspond to real life linguistics. Fuzzy logic is an human knowledge embodying tool through operational algorithms [6][11]. In fuzzy algebra there are many kinds of membership functions but the most characteristic are the triangular, the trapezoidal ones and the family of the sigmoid functions. Sigmoid MF have been used in several research efforts in order to develop fuzzy inference systems [3]. This research effort has been implemented by applying three distinct fuzzy functions that all belong to the sigmoid family.

2.1.1 The Functions of the Gaussian Distribution Curve Family

The fuzzy toolbox of the Matlab platform has been used in order to obtain the membership degrees for each forest department, given by three MF that belong to the Gaussian distribution curve, namely the Gaussian, the Sigmoid, and the S-shaped one. The Gaussian fuzzy membership function (GFMF) is symmetrical and it depends on two parameters c and σ which represent the center and the width of a fuzzy set [15]. More details can be seen in the following figure 2. Gaussian fuzzy membership functions are quite popular in the fuzzy logic literature, as they are the basis for the connection between fuzzy systems and radial basis function neural networks. A typical form of a GFMF can be seen in the following equation 1 [7].

$$f\left(x,\sigma,c\right)=e^{\frac{-(x-c)^2}{2\sigma^2}} \tag{1}$$

The Sigmoid fuzzy membership function (SFMF) is given by the following equation 2. It is a continuous monotonic mapping of the input into a value between 0.0 and 1.0.

$$f(x;a,c)=\frac{1}{1+e^{-a(x-c)}} \tag{2}$$

In the case of the SFMF the parameters a, c define the schema and the position of the curve [7]. Depending on the sign of the parameter a, it is inherently open to the right or to the left, and thus it is appropriate for representing concepts such as "very large" or "very negative". The S-shaped membership function (SSFMF) is a spline-based curve and it is a special version of the Sigmoid one. The following table 1, presents the commands that were employed in MATLAB in order to perform the membership functions.

Table 1. Fuzzy Membership functions used

Fuzzy membership function	MATLAB's implementation command
GFMF	$y=Gaussmmf\left(x,\left[sig\ c\right]\right)$
SFMF	$y=Sigmf\left(x,\left[a\ c\right]\right)$
SSFMF	$y=Smf\left(x,\left[a\ b\right]\right)$

It should be specified that the parameters a and b locate the extremes of the sloped portion of the curve. The following figure 1 shows the actual shape of the three sigmoid membership function that have been applied.

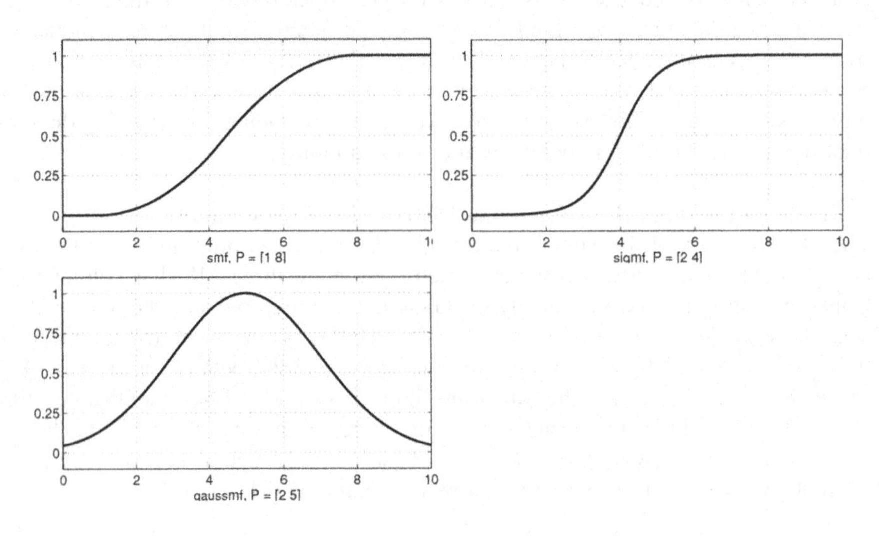

Fig. 1. Graphics of the three employed membership functions

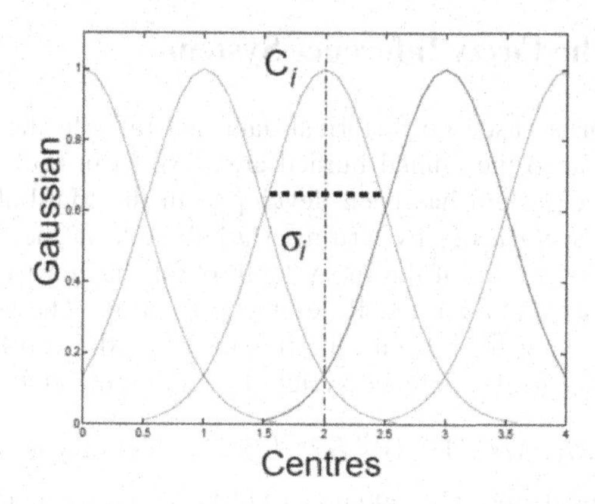

Fig. 2. The center and the width of a Gaussian membership function

2.1.2 Fuzzy Aggregation Operators

According to fuzzy algebra, every element of the universe of discourse can belong to a fuzzy set $\tilde{A}$ and simultaneously to its complement $\tilde{A}_C$ with a degree of membership complementary to 1. T-Norms are the fuzzy conjunction operators between fuzzy sets. Fuzzy algebra contains several T -Norms, each one of them offering a different approach of the fuzzy AND, operation [7],[11]. The following table 2 below contains equations 3 and 4 and it presents a small sample of the most common T-Norms. They have a wide range of applications including the evaluation of Artificial Neural Networks [5].

Table 2. Sample T-Norms

T-Norm name	T-Norm equation
Min (3)	$\tilde{A} \wedge \tilde{B} = \left\{ \left(x, \mu_{\tilde{A} \cap \tilde{B}}(x) \right) \mid \mu_{\tilde{A} \cap \tilde{B}}(x) = \mu_{\tilde{A}}(x) \wedge \mu_{\tilde{B}}(x) = \min\left(\mu_{\tilde{A}}(x), \mu_{\tilde{B}}(x) \right) \right\}$
Drastic Product (4)	$\tilde{A} \wedge \tilde{B} = \min\{\mu_{\tilde{A}}(x), \mu_{\tilde{B}}(x)\}$ if $\max\{\mu_{\tilde{A}}(x), \mu_{\tilde{B}}(x)\}=1$ else $\tilde{A} \wedge \tilde{B} = 0$

The choice of the appropriate norm depends on the perspective under which the problem is examined. For example, if the developer needs an optimistic norm that will assign the least potential overall degree of risk, then the min T-norm should be applied, whereas if the developer wishes to distinguish the cases that are characterized by extreme values for one or more risk parameters, the drastic product is the proper norm to be employed.

3 Developing the Fuzzy Inference System

The risk estimation is based on historical data that refer to the annual forest fire frequency (FFF) and to the annual burned area (ABA) in each area under study. The fuzzy inference system has been developed in the MatlLab integrated environment and more specifically by writing MatLab code in the development environment and also by the use of the fuzzy toolbox for the construction of the fuzzy membership functions and for the Ruleset construction. The primitive data that have been used as input have been initially stored in Microsoft Excel sheets and they are input to the fuzzy inference system by employing commands of the type: $\vec{C} = xlsread\left('data1.xls','1990','B2:B89'\right);$ This sample command imports data concerning burned areas for 1990 into a MATLAB vector named C.

Several commands of this type have been written in order to store each column containing the annual FFF and the ABA in proper one dimensional Matrices (vectors). All of these commands have been stored in a MatLab programming file called *Input_Data.m*. The benefit from the creation of this file is that it runs just by typing the name of the file in the command line. For each one of the two input variables three fuzzy sets have been formed corresponding to proper Linguistics. The following six fuzzy sets have been formed, namely:

$$\tilde{A}_{1}=\left\{(\vec{X}_i\ \mu_i)/i=1..n..\text{“low risky due to FFF”}\right\}$$

$$\tilde{B}_{1}=\left\{(\vec{X}_i\ \mu_i)/i=1..n..\text{“risky due to FFF”}\right\}$$

$$\tilde{C}_{1}=\left\{(\vec{X}_i\ \mu_i)/i=1..n..\text{“extremely risky due to FFF”}\right\}$$

$$\tilde{A}_{2}=\left\{(\vec{X}_i\ \mu_i)/i=1..n..\text{“low risky due to ABA”}\right\}$$

$$\tilde{B}_{2}=\left\{(\vec{X}_i\ \mu_i)/i=1..n..\text{“risky due to ABA”}\right\}$$

$$\tilde{C}_{2}=\left\{(\vec{X}_i\ \mu_i)/i=1..n..\text{“extremely risky due to ABA”}\right\}$$

where n is the number of case under study. Each forest department belongs to all six fuzzy sets with a distinct degree of membership. Of course the characteristic linguistic that describes a case is the one with the highest degree of membership. This approach not only allows the proper definition of linguistics that show clearly the status of an area due to one factor, but it also enables the distinction between two cases of the same linguistic. For example if two forest departments are both high risky, their degree of membership determines which one is more risky than the other.

During the design of the fuzzy inference system, the developer has to input the maximum and the minimum annual values ever recorded in Greece for the forest fire frequency and for the total burned area. Then the fuzzy toolbox determines the values of the σ and c parameters of the Gaussian function and the parameter values α, c, used by the Sigmoid and the S-shaped membership functions. For the output of the system also three fuzzy sets corresponding to three linguistics have been defined namely:

$$\tilde{A}_{3}=\left\{(\vec{X}_i, \mu_i)/i=1..n..\text{“low risky due to integrated risk”}\right\}$$

$$\tilde{B}_{3}=\left\{(\vec{X}_i, \mu_i)/i=1..n..\text{“risky due to integrated risk”}\right\}$$

$$\tilde{C}_{3}=\left\{(\vec{X}_i, \mu_i)/i=1..n..\text{“extremely risky due to integrated risk”}\right\}$$

After the storage of the data in vectors and the definitions of the fuzzy membership functions the Ruleset has been constructed. Figure 3, presents the design of the Ruleset. Several approaches have been proposed lately in order to design decision tables and decision trees [12]. In this case, due to the small number of parameters a rather heuristic approach has been followed in order to minimize the number of required rules as much as possible and to produce a quite rational system.

forest fire frequency	LR	R	LR	R	ER	R	ER	ER	LR
annual burned area	LR	LR	R	R	R	ER	ER	LR	ER
OUTPUT									
	LR	LR	R	R	R	ER	ER	R	R
The Rule Set									

Fig. 3. A sample of the Ruleset of the fuzzy inference system

The unification of the partial risk degrees due to the number of forest fire frequencies and to the total burned area to an integrated risk index has been performed by the system by the use of the minimum T-Norm.

Three distinct systems have been developed, each one using a specific sigmoid membership function and they have been saved as a *ffriskGaus.fis*, ffriskSigm.fis and ffriskSm.fis using the Gaussian, the Sigmoid and the S-shaped membership functions respectively. Each one of the developed fuzzy inference systems can be executed from the Matlab command line by using the commands *readfis* and *evalfis*.

For example the commands: *fffrisk = readfis('ffriskGaus.fis'); and intergrated_risk = evalfis(ffriskGaus, fffrisk);* have been used to execute the Gaussian system and they store the output (unified degrees of membership taking values from 0 to 1) into a proper matrix named integrated_risk. The results have been exported to sheets of Microsoft Excel again by using the MatLab command *xlswrite*.

4 Results – Discussion

The system has been tested and evaluated with actual data related to forest fire incidents in Greece for the period 1984-2007 which is a quite long time interval.

Table 3. PRI based on the FFF using the Gaussian and S shaped functions

Year	forest depart-ment	GFMF Ex-tremely risky index FFF	SFMF Extremely risky in-dex FFF	Year	forest depart-ment	GFMF Ex-tremely risk index FFF	SFMF Ex-tremely risky index FFF
1985	Pyrgos	0.9737	1	1999	Hleia	1	1
	Olympia	0.9538	0.966		Messinia	1	1
	Patra	0.8275			Argolida	0.9967	1
1986	Olympia	0.9802	1	2000	Messinia	1	1
	Pyrgos	0.995	1		Korinthos	1	1
	Patra	0.956	0.9938		Hleia	0.9994	1
	Korinthos	0.9231	0.9753		Lakonia	0.9508	1
1987	Pyrgos	1	1	2001	Messinia	1	1
	Olympia	0.9692	1		Korinthos	1	1
1988	Olympia	0.9523	1		Arkadia	1	1
	Argolida	0.9692	0.962		Lakonia	0.9569	1
1989	Pyrgos	0.9802	1	2003	Messinia	1	1
	Olympia	0.9802	1		Hleia	1	1
	Tripolh	0.9802	0.9988		Arkadia	0.9999	1
	Patra	0.956	1		Korinthos	0.9928	1
1990	Pyrgos	0.9756	1	2004	Messinia	1	1
	Olympia	0.9756	0.9882		Hleia	1	1
1991	Argolida	0.9692	0.9644		Korinthos	0.9889	0.9828
	Korinthos	0.9692	0.9644		Arkadia	0.9595	0.946
	Olympia	0.9321	1		Lakonia	0.9406	0.92
1992	Korinthos	0.9978	1		Argolida	0.9231	1
	Olympia	0.9802	1	2005	Messinia	1	1
1993	Kalamata	0.9396	1		Hleia	1	1
1994	Argolida	0.9287	1		Korinthos	0.9996	0.9929
	Kalamata	0.9737	0.9444	2006	Hleia	1	1
1995	Kalamata	1	1		Messinia	1	1
	Amaliada	0.9577	1		Arkadia	0.9928	1
1996	Kalamata	1	1		Argolida	0.9851	0.9812
	Pyrgos	0.96	0.933	2007	Hleia	1	1
	Pendeli		1		Korinthos	1	1
	Parnitha		1		Patra	0.96	1

Table 3. (*continued*)

1997	Kalamata	1	1	Messinia	0.9675	
	Olympia	1	1	Arkadia	0.8	1
	Pyrgos	0.9596				

The above Table 3 presents clearly the obtained partial risk indices based on the number of forest fire incidents' frequencies, with the use of both Gaussian an S-shaped membership functions only for the cases of the Extremely risky forest departments.

Table 4. PRI for the ABA and overall risk indices for the combination of GFMF and Min (1985-2007)

Year	Forest department	GFMF Ex-tremely risky Partial index ABA	SSFMF Ex-tremely risky Partial index ABA	OVERALL risk inde GFMF and Min Norm	Year	Forest depart-ment	GFMF Ex-tremely risky Partial index ABA	SSFMF Ex-tremely risky Partial index ABA	OVERALL risk index GFMF and Min Norm
1984	Kapandriti	1	1		1994	Patra	1	1	XR 0.181
	Aigaleo	0.9817	1			Pendelh	1	1	XR 0.606
	Korinthos	0.9247	1			Gytheio	1	1	R 0.323
1985	Aigaleo	1	1	R 0.150	1995	Patra	1	1	R 0.969
	Megara	1	1	XR 0.110		Gytheio	1	1	XR 0.369
	Patra	0.9999	1	XR 0.827		Molaoi	0.9584	1	XR 0.310
	Lavrio	0.9903	1	XR 0.582		Kalamata	0.9525	0.9191	XR 0.952
	Amaliada	0.9522	1	XR 0.457	1996	Molaoi	1	1	XR 0.230
1986	Korinthos	1	1	XR 0.923		Patra	0.9718	1	XR 0.234
	Molaoi	0.9867	1	R 0.946		Korin-thos	0.9969	0.9904	XR 0.774
	Olymbia	0.9356	0.9249	XR 0.935	1997	Olympia	1	1	XR 1
	Pyrgos	0.9097	0.8943	XR 0.909		Gytheio	0.9886	1	R 0.946
1987	Patra	1	1	XR 0.681	1998	Korin-thos	1	1	XR 0.700
	Kalamata	0.9868	1	XR 0.754		Pendeli	1	1	XR 0.290
	Moaloi	1	1	XR 0.457		Kapan-driti	0.95	1	R 0.950

Table 4. (*continued*)

	Argolida	0.9712	0.9675	XR 0.339	1999	Lakonia	1	1	XR 0.910
1988	Kalamata	1	1	R 0.216		Messhnia	0.9679	1	XR 0.967
	Tripolh	0.9024	1	XR 0.153		Argolida	0.9998	1	XR 0.996
1989	Pyrgos	1	1	XR 0.980	2000	Korinthos	1	1	XR 1
	Xylokastro	0.9201	1	XR 0.920		Xylokatsro	1	1	XR 0.433
	Patra	1	1	XR 0.956		Arkadia	0.8479	1	XR 0.847
	Sparti	0.9978	0.9835	XR 0.882	2001	Kapandriti	1	1	XR 0.244
1990	Kapandriti	1	1	XR 0.657		Lakonia	1	1	XR 0.965
	Korinthos	1	1	R 0.324		Arkadia	0.9708	1	XR 0.970
	Molaoi	0.9737	1	XR 0.673	2003	Arkadia	1	1	XR 0.999
1991	Kapandriti	1	1	XR 0.135		Hleia	1	1	XR 1
	Korinthos	0.8667	1	XR 0.866		Argolida	0.9166	1	XR 0.928
	Argolida	0.9704	1	XR 0.969	2004	Lakonia	1	1	XR 0.940
	Olympia	0.9584	0.9349	XR 0.932		Lavrio	1	1	XR 0.740
1992	Pendeli	1	1	XR 0.766		Hleia	0.9136	1	XR 0.913
	Kapandriti	1	1	R 0.1543	2005	Pendeli	1	1	XR 0.432
	Sparti	0.9065	1	R 0.2226		Lakonia	1	1	XR 0.8795
	Lavrio	0.9794	1	XR 0.657		Messhnia	0.9393	1	XR 0.939
	Kalamata	0.9850	0.9771	XR 0.882		Arkadia	0.9761	1	XR 0.841
	Argolida	0.9845	0.9764	XR 0.748	2006	Lakonia	1	1	XR 0.718
1993	Kalamata	1	1	XR 0.939		Xylokatsro	0.8276	1	XR 0.102
	Tripolh	1	1	R 0.923		Arkadia	0.9874	0.9892	XR 0.992
	Molaoi	0.8604	1	XR 0.439	2007	Hleia	1	1	XR 1
	Xylokastro	0.9258	1	XR 0.354		Arkadia	1	1	XR 0.798
						Korinthos	0.9337	1	XR 0.933

The determination of the most risky areas based on partial risk indices (due to forest fire frequencies or due to the annual burned area) is similar regardless the use of the Gaussian, the Sigmoid or the S-shaped membership functions. However the S-shaped function assigns the highest risk value more often whereas the Gaussian one differentiates the forest departments more clearly by assigning them with a wide range of risk indices.

Table 4 presents the partial risk indices for the extremely risky areas according to the annual burned area, in the cases of the Gaussian and of the S-shaped membership functions. The estimated overall risk indices' vector (obtained as the output of the fuzzy inference system) for the Gaussian membership function and for the Min T-Norm is also presented clearly in the above Table 4. It contains the most risky forest departments based on partial risk indices and also on the overall risk index (considering risk due to both FFF and ABA). It estimates their corresponding linguistics (risky R, or extremely risky XR) only for the Gaussian function due to the fact that it offers a more rational production of risk indices, whereas the other two functions assign the highest index with the value 1 for too many cases and thus there is no proper distinction between the cases. The determination of the unified overall risk indices has been performed by the fuzzy inference system based on the Ruleset which is presented in figure 3.

5 Conclusions

The annual 'extremely risky' PRIs and the overall integrated 'extremely risky' indices obtained by the system for each year were compared to the actually most risky areas based on the number of FFF and on the ABA for the following year in order to determine the validity of the system. As far as the Gaussmf curve is concerned, for the parameter FFF the percentages of successful risk characterization reaches as high as 70%. The same function for the parameter of the annual burned area offers a percentage of success equal to 48%. The Sigmoid function (Sigmf), for the parameter FFF has also a percentage of successful risk characterizations as high as 70%, but for the burned area the percentage is 51%. Finally Smf has a percentage of success as high as 74% for the number of forest fires and 46% for the burned area. During the period 1985-2007 the overall integrated risk index obtained by employing the min T-Norm offers an average accuracy of 65%, 54% and 52% for the cases of the Gaussian, Sigmoid and S-shaped membership functions respectively. The Gaussian function achieves the highest percentages of success towards overall risk estimation, whereas the other two functions give also satisfying results. The system will be extended by applying weights in the unification of the risk indices according to the importance of each factor.

References

1. Ayanz, J.S.M., Barbosa, P.M., Schmuck, G., Liberta, G.: The European Forest Fire Information System (EFFIS), European Commission. JIR Institute for Environment and sustainability (2003), `http://natural-hazards.jrc.it/documents/fires/2003-publications/EFFIS-earsel-paper.pdf`

2. Cox, E.: Fuzzy Modeling and Genetic Algorithms for Data Mining and Exploration. Elsevier Inc., USA (2005)
3. Duprey, B., Taheri, S.: A Fuzzy Based Stability Index Using a Right Sigmoid Membership Function SAE. International Journal of Commercial Vehicles 2(2) (March 2010)
4. Deeming, J., Burgan, R., Cohen, J.: The National Fire Danger Rating System USDA (1978)
5. Iliadis, L., Spartalis, S., Tachos, S.: Application of fuzzy T-norms towards Artificial Neural Networks evaluation: A case from wood industry. Journal Information Sciences, Informatics and Computer Science Intelligent Systems Applications 178(20), 3828–3839 (2008)
6. Iliadis, L.: A decision support system applying an integrated Fuzzy model for long - term forest fire risk estimation. Environmental Modeling and Software 20(5), 613–621 (2005)
7. Iliadis, L.: Intelligent Information Systems in Risk Estimation Stamoulis Editors (2007) (Greek)
8. Iliadis, L., Papastavrou, A., Lefakis, P.: A computer-system that classifies the prefectures of Greece in forest fire risk zones using fuzzy sets. Journal Forest policy and Economics 4(1), 43–54 (2002)
9. Kailidis, D.: Forest fires, 3rd edn., Giahoudi-Giapouli editions. Thessaloniki, Greece, 510 p. (1990) (in Greek)
10. Kailidis, D., Xanthopoulos, G.: The forest fire problem in Greece. Aristotelian University of Thessaloniki, Greece, Department of Forestry and Natural Environment, Forest Protection Laboratory. No. 310 (1991)
11. Kecman, V.: Learning and Soft Computing. MIT Press, Cambridge (2001)
12. Kuncheva, L.I., Rodriguez, J.J.: An experimental study on Rotation Forest ensembles. In: Haindl, M., Kittler, J., Roli, F. (eds.) MCS 2007. LNCS, vol. 4472, pp. 459–468. Springer, Heidelberg (2007)
13. Lin, C.: The Development, Systems, and Evaluation of Forest Fire Danger Rating: A Review. Taiwan Forestry Research Institute (2002), `http://fire.tfri.gov.tw/DocRef14.pdf`
14. Markalas, S., Pantelis, D.: Forest fires in Greece in 1993 Aristotelian Univ. of Thessaloniki, Department of Forestry and the Natural Environment, Forest Protection Laboratory. No. 3, 40 p. (1996) (in Greek)
15. Pieczynski, A., Obuchowicz, A.: Application of the General Gaussian Membership Function for the Fuzzy Model Parameters Tunning. In: Rutkowski, L., Siekmann, J.H., Tadeusiewicz, R., Zadeh, L.A. (eds.) ICAISC 2004. LNCS (LNAI), vol. 3070, pp. 350–355. Springer, Heidelberg (2004)
16. Xanthopoulos, G.: Greek forest fires and property damage: A brief history. In: Proceedings - Symposium and Workshop on Protecting People and Homes from Wildfire in the Interior West, October 6-8, Missoula, Montana, USA. USDA For. Serv. Gen. Tech. Rep. INT-251, pp. 199-200, 213 (1988)
17. Xanthopoulos, G.: Forest fires in Greece: Past, present and future. Epikentra (Foundation for Political Research and Advanced Education) 6, 62–71 (1998) (in Greek); Special issue titled "Greek forests: myths and reality - Proposals for the present and for the future
18. Zadeh, L.A.: Inform Control, vol. 12, pp. 94–102 (1968)

Evolutionary Prediction of Total Electron Content over Cyprus

Alexandros Agapitos[1], Andreas Konstantinidis[2], Haris Haralambous[2], and Harris Papadopoulos[2]

[1] School of Computer Science and Informatics, University College Dublin, Dublin, Ireland
alexandros.agapitos@ucd.ie
[2] Computer Science and Engineering department, Frederick University, Nicosia, Cyprus
{com.ca,eng.hh,h.papadopoulos}@frederick.ac.cy

Abstract. Total Electron Content (TEC) is an ionospheric characteristic used to derive the signal delay imposed by the ionosphere on trans-ionospheric links and subsequently overwhelm its negative impact in accurate position determination. In this paper, an Evolutionary Algorithm (EA), and particularly a Genetic Programming (GP) based model is designed. The proposed model is based on the main factors that influence the variability of the predicted parameter on a diurnal, seasonal and long-term time-scale. Experimental results show that the GP-model, which is based on TEC measurements obtained over a period of 11 years, has produced a good approximation of the modeled parameter and can be implemented as a local model to account for the ionospheric imposed error in positioning. The GP-based approach performs better than the existing Neural Network-based approach in several cases.

Keywords: Evolutionary Algorithms, Genetic Programming, Global Positioning System, Total Electron Content.

1 Introduction

The ionosphere is defined as a region of the earth's upper atmosphere where sufficient ionisation can exist to affect the propagation of radio waves. It ranges in height above the surface of the earth from approximately 50 km to 1000 km. The influence of this region on radio waves is accredited to the presence of free electrons. The impact of the ionosphere on communication, navigation, positioning and surveillance systems is determined by variations in its electron density profile and total electron content along the signal propagation path [1,2]. As a result satellite systems for communication and navigation, surveillance and control that are based on transonospheric propagation may be affected by complex variations in the ionospheric structure in space and time leading to degradation of the accuracy, reliability and availability of their service. Total Electron Content (TEC) is an important parameter in trans-ionospheric links since when multiplied by a factor which is a function of the signal frequency, it yields an estimate of the delay imposed on the signal by the ionosphere due to its dispersive nature.

H. Papadopoulos, A.S. Andreou, and M. Bramer (Eds.): AIAI 2010, IFIP AICT 339, pp. 387–394, 2010.

This paper describes an attempt to develop a model to predict TEC above Cyprus to encapsulate its variability on a diurnal, seasonal and long-term scale. The model development is based on around 60000 hourly TEC measurements recorded above Cyprus from 1998 to 2009. The practical application of this model lies in its possible use as an alternative candidate local model to the existing Klobuchar global model [3] that is currently being used in single frequency GPS navigation system receivers to improve positioning accuracy.

Metaheuristics and more specifically Evolutionary Algorithms were proven efficient and effective in dealing with difficult-to-solve real-life problems [4]. Particularly, Genetic Programming (GP) based approaches performed well in evolving computer programs, controllers and models [5] in the past. In this paper, we have adopted a Pareto-based Genetic Programming approach for dealing with the TEC problem. The proposed GP is a panmictic, generational, elitist genetic algorithm with an expression-tree representation [5]. To the best of our knowledge this is the first time that a GP-based approach is applied to the proposed problem. The main contribution of our paper is: a GP-based prediction model designed for the TEC over Cyprus that outperforms the previously proposed Neural Network-based model [6]. The rest of this paper is structured as follows. Section 2 describes the measurements and characteristics of TEC values used for the model and section 3 discusses the model parameters involved in the model development. Section 4 gives a description of the proposed evolutionary algorithm and section 5 outlines the experimental results. Finally, section 6 gives the concluding remarks of the paper.

2 Measurements and Characteristics of TEC

Dual-frequency GPS data recorded by GPS receivers enable an estimation of the TEC measured in total electron content units, (1 TECU = 10^{16} elecrons m^{-2}). This is the total amount of electrons along a line of sight between the receiver and a GPS satellite in a column of 1 m^2 cross-sectional area and represents a parameter of interest to GPS users (see Figure 1a). TEC is the integral of the electron density (Ne see Figure 1b) from the ground to an infinite height (the height of the satellite).

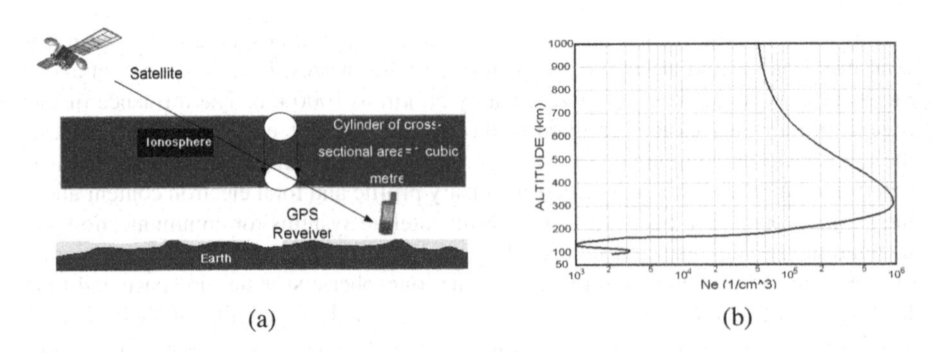

(a) (b)

Fig. 1. TEC representation and typical electron density profile of the ionosphere over Cyprus

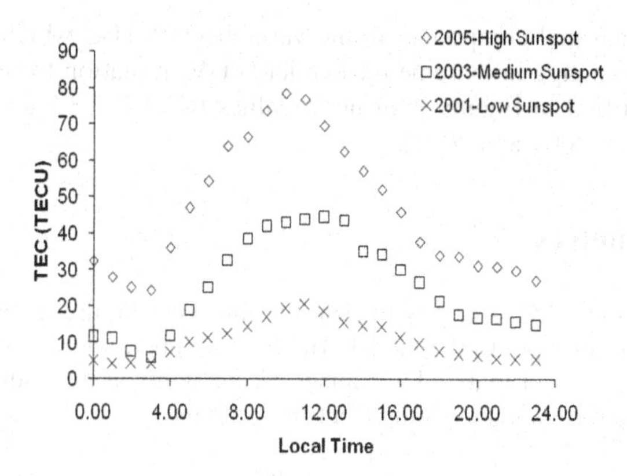

Fig. 2. Diurnal variability of TEC for low, medium and high solar activity

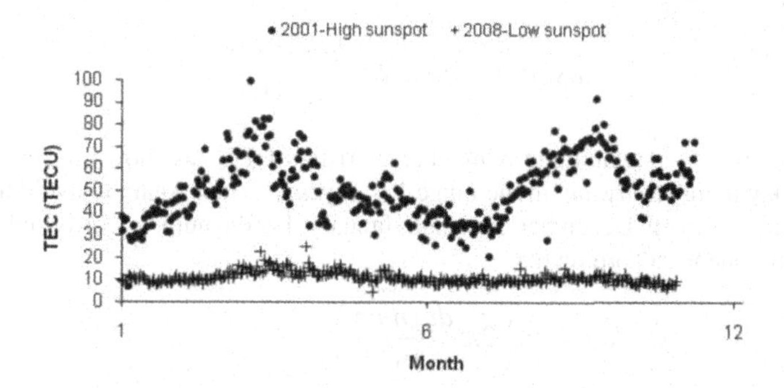

Fig. 3. Seasonal variation of TEC at 12:00

The electron density of free electrons within the ionosphere and therefore TEC depend upon the strength of the solar ionizing radiation which is a function of time of day, season, geographical location and solar activity [1], [2]. Since solar activity has an impact on ionospheric dynamics which in turn influence the electron density of the ionosphere, TEC also exhibits variability on daily, seasonal and long-term time scales in response to the effect of solar radiation. It is also subject to abrupt variations due to enhancements of geomagnetic activity following extreme manifestations of solar activity disturbing the ionosphere from minutes to days on a local or global scale. The most profound solar effect on TEC is reflected on its daily variation as shown in Figure 2. As it is clearly depicted, there is a strong dependency of TEC on local time which follows a sharp increase of TEC around sunrise and gradual decrease around sunset. This is attributed to the rapid increase in the production of electrons due to the photo-ionization process during the day and a more gradual decrease due to the recombination of ions and electrons during the night. The long–term effect of solar activity on TEC is also clearly shown as we can observe a correlation of the mean level of TEC and sunspot number which is an established index of solar activity.

There is also a seasonal component in the variability of TEC which can be attributed to the seasonal change in extreme ultraviolet (EUV) radiation from the Sun. This can be clearly identified in Figure 3 for noon values of TEC for high and low solar activity periods (years 2001 and 2008).

3 Model Parameters

The diurnal variation of TEC is clearly evident by observing Figure 2. We therefore include hour number as an input to the model. The hour number, *hour*, is an integer in the range $0 \le hour \le 23$. In order to avoid unrealistic discontinuity at the midnight boundary, hour is converted into its quadrature components according to:

$$sinhour = \sin\left(2\pi \frac{hour}{24}\right) \tag{1}$$

and

$$coshour = \cos\left(2\pi \frac{hour}{24}\right) \tag{2}$$

A seasonal variation is also an underlying characteristic of TEC as shown in fig. 3 and is described by day number *daynum* in the range $1 \le daynum \le 365$. Again to avoid unrealistic discontinuity between December 31st and January 1st daynum is converted into its quadrature components according to:

$$sinday = \sin\left(2\pi \frac{daynum}{365}\right) \tag{3}$$

and

$$cosday = \cos\left(2\pi \frac{daynum}{365}\right) \tag{4}$$

Long-term solar activity has a prominent effect on TEC. To include this effect in the model specification we need to incorporate an index, which represents a good indicator of solar activity. In ionospheric work the 12-month smoothed sunspot number is usually used, yet this has the disadvantage that the most recent value available corresponds to TEC measurements made six months ago. To enable TEC data to be modelled as soon as they are measured, and for future predictions of TEC to be made, the monthly mean sunspot number values were modeled using a smooth curve defined by a summation of sinusoids.

4 The Proposed Evolutionary Algorithm

Genetic Programming (GP) is an Evolutionary Computation (EC) technique that evolves populations of computer programs as solutions to problems. It is a general purpose evolutionary search technique that can be applied in both regression and classification problems. In contrast to linear and non-linear regression, the technique does not presuppose a functional form as it generally proceeds for a definition of lower-level building blocks (the function set). Therefore Koza [5] speaks of symbolic regression or system identification as the object of search is both the functional form and the optimal coefficients.

The term evolutionary algorithm describes a class of stochastic search procedure inspired by principles of natural genetics and survival of the fittest. They operate through a simulated evolution process on a population of solution structures that represent candidate solutions in the search space. Evolution occurs through (1) a selection mechanism that implements a survival of the fittest strategy, and (2) diversification of the selected solutions to produce offspring for the next generation.
In GP, programs are usually expressed using hierarchical representations taking the form of syntax-trees. It is common to evolve programs into a constrained, and often problem-specific user-defined language. The variables and constants in the program are leaves in the tree (collectively named as terminal set), whilst arithmetic operators are internal nodes (collectively named as function set). It is common in the GP literature to represent expressions in the prefix notation similar to that used in LISP or Scheme. For example, `x+3*x` becomes `(+ x (* 3 y)))`. This representation eases the expression-tree data structure formation, and its manipulation during the application of variation operators.

GP finds out how well a program works by running it, and then comparing its behaviour to some ideal. We might be interested, for example, in how well a program predicts a time series or controls an industrial process. This comparison is quantified to give a numeric value called *fitness*. Those programs that do well are chosen to breed, and produce new programs for the new generation. The primary variation operators to perform transitions within the space of computer programs are crossover and mutation. Once a stopping criterion has been met the algorithm terminates and the best program is designated as the output of the run.

The most commonly used form of crossover is subtree crossover [5]. Given two parents, subtree crossover randomly (and independently) selects a crossover point (a node) in each parent tree. Then, it creates the offspring by replacing the subtree rooted at the crossover point in a copy of the first parent with a copy of the subtree rooted at the crossover point in the second parent. Copies are used to avoid disrupting the original individuals. This way, if selected multiple times, they can take part in the creation of multiple offspring programs. Note that it is also possible to define a version of crossover that returns two offspring, but this is not commonly used. Often crossover points are not selected with uniform probability. Typical GP primitive sets lead to trees with an average branching factor of at least two, so the majority of the nodes will be leaves. Consequently the uniform selection of crossover points leads to crossover operations frequently exchanging only very small amounts of genetic material (i.e., small subtrees); many crossovers may in fact reduce to simply swapping two leaves. To counter this, Koza [5] suggested the widely used approach of choosing functions 90% of the time and leaves 10% of the time.

The most commonly used form of mutation in GP (which we will call subtree mutation) randomly selects a mutation point in a tree and substitutes the subtree rooted there with a randomly generated subtree. Another common form of mutation is point mutation, which is GP's rough equivalent of the bit-flip mutation used in genetic algorithms. In point mutation, a random node is selected and the primitive stored there is replaced with a different random primitive of the same arity taken from the primitive set. When subtree mutation is applied, this involves the modification of exactly one subtree. Point mutation, on the other hand, is typically applied on a per-node basis. That is, each node is considered in turn and, with a certain probability, it is altered as explained above. This allows multiple nodes to be mutated independently in one application of point mutation.

Like in other evolutionary algorithms, in GP the individuals in the initial population are typically randomly generated. Two dominant methods are the *full* and *grow* and the widely used combination of the two known as *Ramped half-and-half* [5]. In both the *full* and *grow* methods, the initial individuals are generated so that they do not exceed a user-specified maximum depth. The depth of a node is the number of edges that need to be traversed to reach the node starting from the tree's root node (the depth of the tree is the depth of its deepest leaf). The *full* method generates full tree-structures where all the leaves are at the same depth, whereas the *grow* method allows for the creation of trees of more varied sizes and shapes.

The evolutionary algorithm employed is a standard elitist (i.e. the best is always preserved), generational (i.e. populations are arranged in generations, not steady-state), panmictic (i.e. no program mating restrictions) genetic algorithm [8]. The algorithm uses tournament selection with a tournament size of 7. Evolution proceeds for 50 generations, and the population size is set to 1000 individuals. Ramped-half-and-half tree creation with a maximum depth of 6 is used to perform a random sampling of rules during run initialisation. Throughout evolution, expression-trees are allowed to grow up to depth of 12. The evolutionary search employs a mixture of mutation-based variation operators, where subtree mutation is combined with point-mutation; a probability governing the application of each, set to 0.6 in favour of sub-tree mutation. Neither recombination, nor reproduction was used. The primitive language consisted of the basic arithmetic operators (+, -, *, /) serving as the function set, whereas the terminal set consisted of the five independent variables *sinhour*, *coshour*, *sinday*, *cosday* and modeled monthly mean sunspot number.

5 Experimental Results and Discussion

The primary goal of our experimental studies is to investigate the performance of our GP-based approach in designing a prediction model for the TEC over Cyprus with a good approximation to the measured values, compared to the previously developed Neural Network based model. The NN used in [6] had a fully connected two-layer structure, with 5 input, 10 hidden and 1 output neurons. Both its hidden and output neurons consisted of hyperbolic tangent sigmoid activation functions. The number of hidden neurons was determined by trial and error. The training algorithm used was the Levenberg-Marquardt back propagation algorithm.

The data-set was segmented in 10 continuous folds similarly to the case of NNs [6]. In each cross-validation cycle, 9 folds were used as the training set, whereas the evolved model was tested on the remaining 10th fold. The training set was further randomly divided into two data-sets (with no overlapping): the fitness evaluation data-set, with 67% of the training data, and the validation data-set with the remaining

33%. The fitness measure consists of minimising the RMSE on the fitness evaluation data-set. At each generation, a two-objective sort is conducted in order to extract a set of non-dominated individuals [7] (Pareto front) with regards to the lowest fitness evaluation data-set RMSE, and the smallest model complexity in terms of expression-tree size, as measured by the number of tree-nodes [9]. The rationale behind this is to create a selection pressure towards simpler prediction models that have the potential to generalise better. These non-dominated individuals are then evaluated on the validation data-set, with the best-of-generation prediction model selected as the one of these with the smallest RMSE. During tournament selection based on the fitness evaluation data-set performance, we used the model complexity as a second point of comparison in cases of identical error rates. The two approaches were coded in Java and run on an Intel® Pentium 4 3.2 GHz Windows XP server with 1.5 GB RAM. We performed 50 independent evolutionary runs for each test fold, in order to account for the stochastic nature of the adaptive search algorithm, and obtain statistically meaningful results.

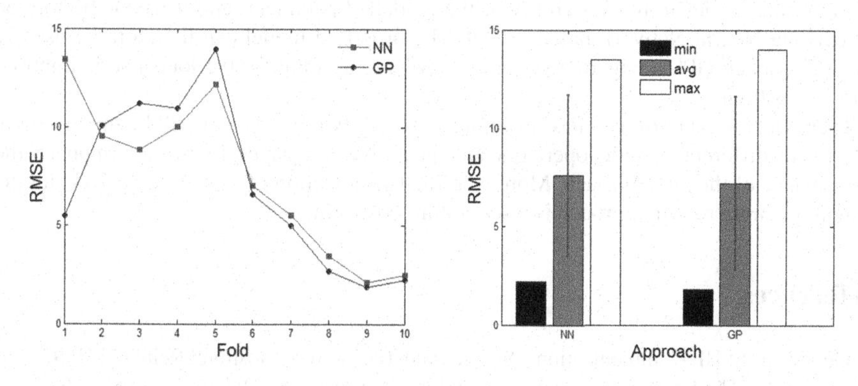

Fig. 4. GP versus NN in terms of Root Mean Square Error (RMSE) per fold and average Root Mean Square Error (RMSE)

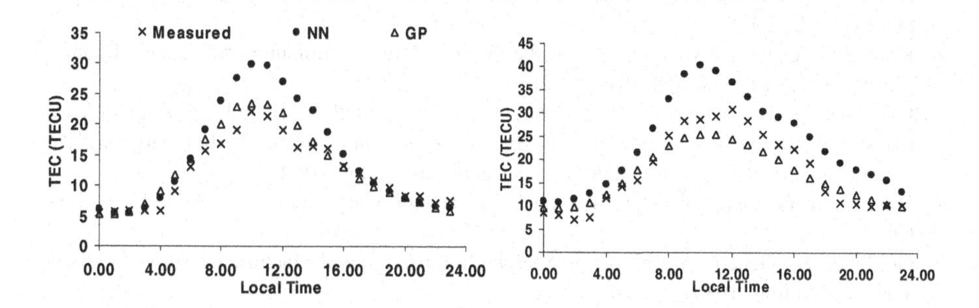

Fig. 5. Examples of measured and predicted TEC values

Figure 4 shows a comparison between the GP and the NN approaches in terms of min RMSE per fold and average RMSE, respectively. The results in Figure 4 show that the GP model provides a lower RMSE compared to the NN-based model in 6 out of 10 folds, giving a better prediction and consequently a better approximation in around 60% of the TEC measured values. However, the NN-based model outperforms the GP-model in folds 2, 3, 4 and 5. The average RMSE obtained by the GP approach is around 4% less than the average RMSE obtained by the NN approach. The two approaches have a similar standard deviation of around 4 RMSE. Some examples of measured and predicted TEC values are given in Figure 5. These demonstrate both the good performance of the developed GP and its superiority over the NN model.

6 Conclusions and Future Work

In this paper, a Genetic Programming based approach is used to design a prediction model for the Total Electron Content over Cyprus. Particularly, a panmictic, generational, elitist genetic algorithm with an expression-tree representation is used. A prediction model is developed based on a data set obtained during a period of eleven years covering a full sunspot cycle. The GP-model has shown a good approximation of the different time-scales in the variability of the modelled parameter and it has outperformed the existing Neural Network based model. The proposed model can therefore be used in single frequency GPS navigation system receivers to account for the ionospheric imposed error in positioning.

There are a number of avenues for future research. For example, it will be interesting to investigate different genetic operators and primitive languages to further improve the performance of the GP approach. Moreover, the hybridization of the GP with NNs and the design of a more robust approach is also a future possibility.

References

1. Goodman, J.: HF Communications, Science and Technology, Nostrand Reinhold (1992)
2. Maslin, N.: The HF Communications, a Systems Approach, San Francisco (1987)
3. Klobuchar, J.A.: Ionospheric Time-Delay Algorithm for Single-Frequency GPS Users. IEEE Trans. on AES 23(3), 325–331 (1987)
4. Reeves, C.: Genetic algorithms. In: Handbook of Metaheuristics, pp. 65–82. Kluwer, Dordrecht (2003)
5. Koza, J.R.: Genetic Programming: on the programming of computers by means of natural selection. MIT Press, Cambridge (1992)
6. Haralambous, H., Vrionides, P., Economou, L., Papadopoulos, H.: A local Total Electron Content Neural Network model over Cyprus. In: Proceedings of the 4th International Symposium on Communications, Control, and Signal Processing (2010)
7. Deb, K.: Multi-Objective Optimization Using Evolutionary Algorithms. Wiley and Sons, Chichester (2002)
8. Goldberg, D.: Genetic Algorithms in Search Optimisation and Machine Learning. Addison-Wesley, Reading (1989)
9. Gagne, C., Schoenauer, M., Parizeau, M., Tomassini, M.: Genetic Programming, Validation Sets, and Parsimony Pressure. In: Proceedings of the 9th European Conference on Genetic Programming, April 10-12. Springer, Heidelberg (2006)

A Multi-layer Perceptron Neural Network to Predict Air Quality through Indicators of Life Quality and Welfare

Kyriaki Kitikidou and Lazaros Iliadis

Democritus University of Thrace, Department of Forestry and Management of the Environment and Natural Resources, Pandazidou 193, 68200, Orestiada, Greece
kkitikid@fmenr.duth.gr

Abstract. This paper considers the similarity between two measures of air pollution/quality control, on the one hand, and widely used indicators of life quality and welfare, on the other. We have developed a multi-layer perceptron neural network system which is trained to predict the measurements of air quality (emissions of sulphur and nitrogen oxides), using Eurostat data for 34 countries. We used life expectancy, healthy life years, infant mortality, Gross Domestic Product (GDP) and GDP growth rate as a set of inputs. Results were dominated by GDP growth rate and GDP. Obtaining accurate estimates of air quality measures can help in deciding on distinct dimensions to be considered in multidimensional studies of welfare and quality of life.

Keywords: Air quality, Economic welfare, Artificial Neural Networks, Multi-Layer Perceptron, Quality Of Life.

1 Introduction

Quantitative analysis of quality of life (QOL) across countries, and the construction of summary indices for such analyses have been of interest for some time [15]. Most early work focused on largely single dimensional analysis based on such indicators as per capita GDP, the literacy rate, and mortality rates. Maasoumi (1998) [11] and others called for a multidimensional quantitative study of welfare and quality of life. The argument is that welfare is made up of several distinct dimensions, which cannot all be monetized, and heterogeneity complications are best accommodated in multidimensional analysis. Hirschberg et al. (1991) [8] and Hirschberg et al. (1998) [9] identified similar indicators, and collected them into distinct clusters which could represent the dimensions worthy of distinct treatment in multidimensional frameworks.

In this research effort we have considered the role of air quality indicators in the context of economic and welfare life quality indicators, using artificial neural networks (ANN). Therefore in this presentation we have obtained the key variables (life expectancy, healthy life years, infant mortality, Gross Domestic Product (GDP) and GDP growth rate) and developed a neural network model to predict the air quality

H. Papadopoulos, A.S. Andreou, and M. Bramer (Eds.): AIAI 2010, IFIP AICT 339, pp. 395–402, 2010.

outcomes (emissions of sulphur and nitrogen oxides). Sustainability and quality of life indicators have been proposed recently by Flynn et al. (2002) [6] and life quality indices have been used to estimate willingness to pay [12]. The innovative part of this research effort lies in the use of a soft computing machine learning approach like the ANN to predict air quality.

2 Materials and Methods

It is well known that the quality of the air in a locale influences the health of the population and ultimately affects other dimensions of that population's welfare and its economy. As a simple example, in cities where pollution levels rise significantly in the summer, worker absenteeism rates rise commensurately and productivity is adversely impacted. Other dimensions of the economy are influenced on "high pollution days" as well. For example, when outdoor leisure activity is restricted this may have serious consequences for the service sector of the economy [2]. In this paper, we have introduced two measures of environmental quality or air quality as quality of life factors. A feature of these indices is the fact that these types of pollution are created by some of the very activities that define economic development. The two factors under investigation here are sulfur oxides (SOx) and nitrogen oxides (NOx) (million tones of SO_2 and NO_2 equivalent, respectively). They are both produced as byproducts of fuel consumption as in case of the generation of electricity. Vehicle engines also produce a large proportion of NOx. SOx is primarily produced when high sulfur coal is burned which is usually in large-scale industrial processes and power generation. Thus, the ratio of these emissions to the population is an indication of pollution control.

The attributes of quality of life used in this paper are the following:

- Life expectancy at birth: The mean number of years that a newborn child can expect to live if subjected throughout his life to the current mortality conditions (age specific probabilities of dying).
- Healthy life years: The indicator Healthy Life Years (HLY) at birth measures the number of years that a person at birth is still expected to live in a healthy condition. HLY is a health expectancy indicator which combines information on mortality and morbidity. The data required are the age-specific prevalence (proportions) of the population in healthy and unhealthy conditions and age-specific mortality information. A healthy condition is defined by the absence of limitations in functioning/disability. The indicator is also called disability-free life expectancy (DFLE). Life expectancy at birth is defined as the mean number of years still to be lived by a person at birth, if subjected throughout the rest of his or her life to the current mortality conditions.
- Infant mortality: The ratio of the number of deaths of children under one year of age during the year to the number of live births in that year. The value is expressed per 1 000 live births.

- Gross Domestic Product (GDP) per capita: GDP is a measure of the economic activity, defined as the value of all goods and services produced less the value of any goods or services used in their creation. These amounts are expressed in PPS, i.e. a common currency that eliminates the differences in price levels between countries allowing meaningful volume comparisons of GDP between countries.
- GDP growth rate: The calculation of the annual growth rate of GDP volume is intended to allow comparisons of the dynamics of economic development both over time and between economies of different sizes. For measuring the growth rate of GDP in terms of volumes, the GDP at current prices are valued in the prices of the previous year and the thus computed volume changes are imposed on the level of a reference year; this is called a chain-linked series. Accordingly, price movements will not inflate the growth rate.

Data were extracted for 34 European countries, for the year 2005, from the Eurostat database [4]. Descriptive statistics for all variables are given in Table 1.

Table 1. Descriptive statistics for all variables used in the analysis

	Emissions of sulphur oxides (million tones of SO_2 equivalent)	Emissions of nitrogen oxides (million tones of NO_2 equivalent)	Infant Mortality	GDP (PPS)	GDP Growth Rate	Life Expectancy At Birth (years)	Healthy Life Years
Valid N	34	34	34	33	33	33	27
Missing	0	0	0	1	1	1	7
Mean	0.503	0.372	5.721	95.921	4.206	77.535	60.448
Std. Deviation	0.648	0.482	4.227	46.620	2.521	3.244	5.443
Min	0.00	0.00	2.30	28.50	0.70	70.94	50.10
Max	2.37	1.63	23.60	254.50	10.60	81.54	69.30

For the performance of the analysis, multi-layer perceptron (MLP) network models were used, employing the back propagation (BP) optimization algorithm. As it is well known in BP the weighted sum of inputs and bias term are passed to the activation level through the transfer function to produce the output ([1], [5], [7], [14]). The sigmoid transfer function was employed ([3], [10]), due to the fact that the algorithm requires a response function with a continuous, single valued with first derivative existence [13]. These networks were trained in an iterative process. The number of hidden layers is chosen to be only one to reduce the network complexity, and increase the computational efficiency [7]. The schematic representation of the neural network is given in Fig. 1.

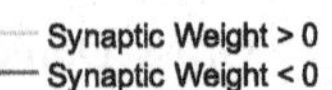

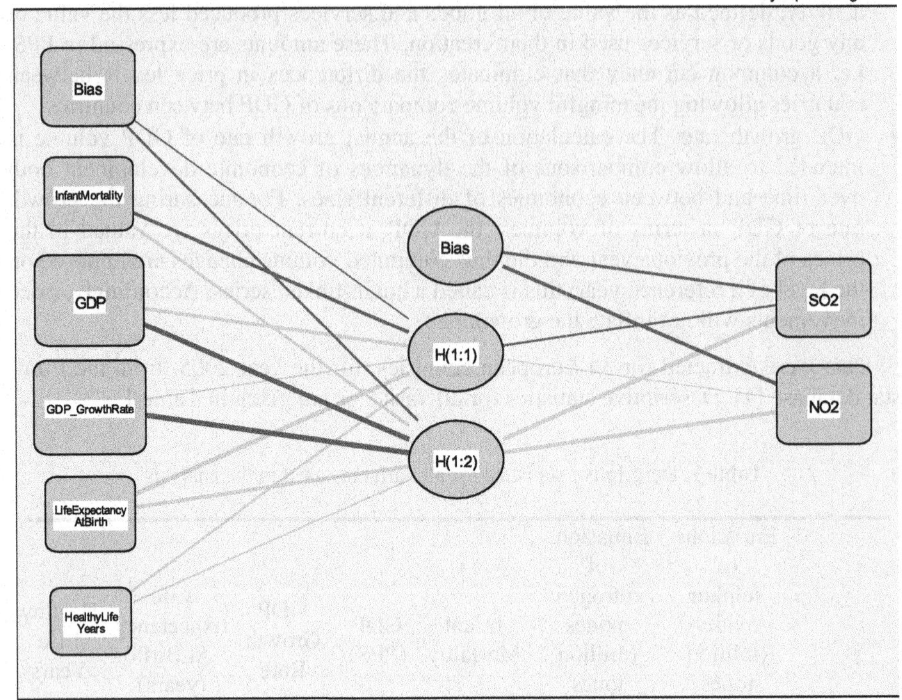

Hidden layer activation function: Sigmoid

Output layer activation function: Identity

Fig. 1. Multi-layer perceptron network structure

3 Results-Discussion

From the analysis, 19 cases (70.4%) were assigned to the training sample, 2 (7.4%) to the testing sample, and 6 (22.2%) to the holdout sample. The choice of the records was done in a random manner. The whole effort targeted in the development of an ANN that would have the ability to generalize as much as possible. The seven data records which were excluded from the analysis were countries that did not had available data on Healthy Life Years. Two units were chosen in the hidden layer.

Table 2 displays information about the results of training and applying the final network to the holdout sample. Sum-of-squares error is displayed because the output layer has scale-dependent variables. This is the error function that the network tries to minimize during training. One consecutive step with no decrease in error was used as stopping rule. The relative error for each scale-dependent variable is the ratio of the sum-of-squares error for the dependent variable to the sum-of-squares error for the "null" model, in which the mean value of the dependent variable is used as the predicted value for each case. There appears to be more error in the predictions of emissions of sulphur oxides than in emissions of nitrogen oxides, in the training and holdout samples.

The average overall relative errors are fairly constant across the training (0.779), testing (0.615), and holdout (0.584) samples, which give us some confidence that the model is not overtrained and that the error in future cases, scored by the network will be close to the error reported in this table.

Table 2. Model Summary

Training	Sum of Squares Error		14.029
	Average Overall Relative Error		0.779
	Relative Error for Scale Dependents	Emissions of sulphur oxides (million tones of SO_2 equivalent)	0.821
		Emissions of nitrogen oxides (million tones of NO_2 equivalent)	0.738
Testing	Sum of Squares Error		0.009
	Average Overall Relative Error		0.615
	Relative Error for Scale Dependents	Emissions of sulphur oxides (million tones of SO_2 equivalent)	0.390
		Emissions of nitrogen oxides (million tones of NO_2 equivalent)	0.902
Holdout	Average Overall Relative Error		0.584
	Relative Error for Scale Dependents	Emissions of sulphur oxides (million tones of SO_2 equivalent)	0.603
		Emissions of nitrogen oxides (million tones of NO_2 equivalent)	0.568

In the following Table 3 parameter estimates for input and output layer, with their corresponding biases, are given.

Table 3. Parameter Estimates

Predictor		Predicted			
		Hidden Layer 1		Output Layer	
		H(1:1)	H(1:2)	SO_2	NO_2
Input Layer	(Bias)	-0.119	-0.537		
	Infant Mortality	-0.805	0.752		
	GDP	1.033	-3.377		
	GDP Growth Rate	0.318	-3.767		
	Life Expectancy At Birth	1.646	1.226		
	Healthy Life Years	0.567	0.358		
Hidden Layer 1	(Bias)			-0.635	-0.877
	H(1:1)			-0.518	0.116
	H(1:2)			1.396	1.395

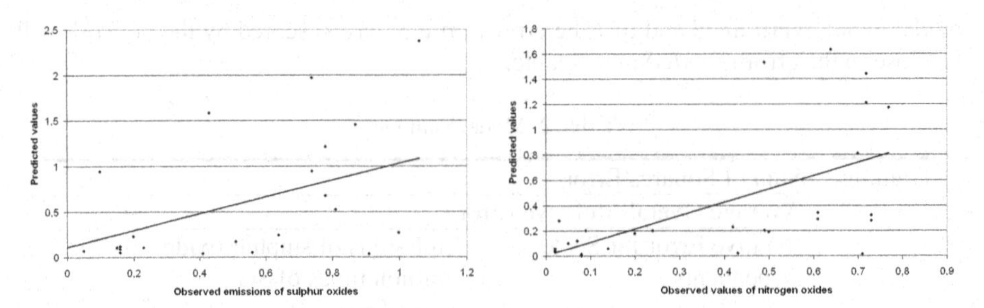

Fig. 2. Linear regression of observed values for emissions of sulphur oxides by predicted values

Fig. 3. Linear regression of observed values for emissions of nitrogen oxides by predicted values

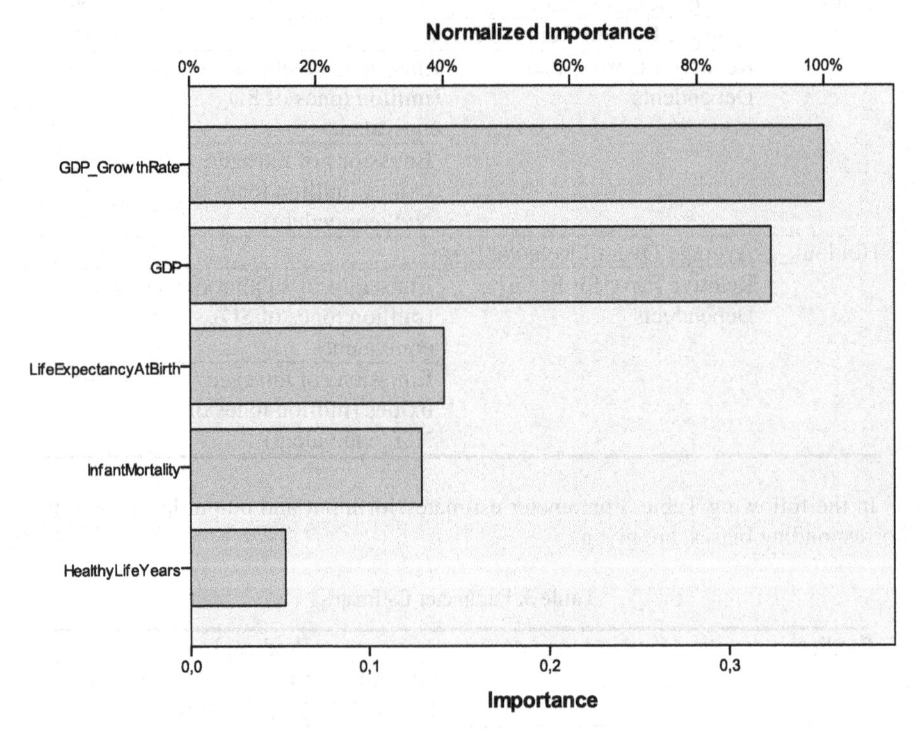

Fig. 4. Independent variable importance chart

Linear regression between observed and predicted values ($SO_2 = a + b\hat{SO_2} + error$, $NO_2 = a + b\hat{NO_2} + error$) showed that the network does a reasonably good job of predicting emissions of sulphur and nitrogen oxides. Ideally, linear regression parameters a and b should have values 0 and 1, respectively, while values of the observed-by-predicted chart should lie roughly along a straight line. Linear regression gave results for the two output variables $SO_2 = 0.114 + 0.918\hat{SO_2} + error$ (Fig. 2) and $NO_2 = 0.005 + 1.049 \hat{NO_2} + error$ (Fig. 3), respectively. There appears to be more error in the predictions of emissions of sulphur oxides than in emissions of nitrogen oxides, something that we also pointed out in Table 2. Figs 2 and 3 actually seem to suggest that the largest errors of the ANN are overestimations of the target values.

The importance of an independent variable is a measure of how much the network's model-predicted value changes for different values of the independent variable. A sensitivity analysis to compute the importance of each predictor is applied. The importance chart (Fig. 4) shows that the results are dominated by GDP growth rate and GDP (strictly economical QOL indicators), followed distantly by other predictors.

4 Conclusions

The multi-layer perceptron neural network model, that was trained to predict air quality indicators, using life quality and welfare indicators, appears to perform reasonably well. Results showed that GDP growth rate and GDP influenced mainly air quality predictions, while life expectancy, infant mortality and healthy life years followed distantly.

One possible way to ameliorate performance of the network would be to create multiple networks. One network would predict the country result, perhaps simply whether the country increased emissions or not, and then separate networks would predict emissions conditional on whether the country increased emissions. We could then combine the network results to likely obtain better predictions. Note also that neural network is open ended; as more data is given to the model, the prediction would become more reliable.

References

1. Bishop, C.: Neural Networks for Pattern Recognition, 3rd edn. Oxford University Press, Oxford (1995)
2. Bresnahan, B., Mark, D., Shelby, G.: Averting behavior and urban air pollution. Land Economics 73, 340–357 (1997)
3. Callan, R.: The Essence of Neural Networks. Prentice-Hall, UK (1999)
4. Eurostat, `http://epp.eurostat.ec.europa.eu`
5. Fine, T.: Feedforward Neural Network Methodology, 3rd edn. Springer, New York (1999)
6. Flynn, P., Berry, D., Heintz, T.: Sustainability & Quality of life indicators: Towards the Integration of Economic, Social and Environmental Measures. The Journal of Social Health 1(4), 19–39 (2002)
7. Haykin, S.: Neural Networks: A Comprehensive Foundation, 2nd edn. Prentice-Hall, UK (1998)
8. Hirschberg, J., Esfandiar, M., Slottje, D.: Cluster analysis for measuring welfare and quality of life across countries. Journal of Econometrics 50, 131–150 (1991)
9. Hirschberg, J., Maasoumi, E., Slottje, D.: A cluster analysis the quality of life in the United States over time. Department of Economics research paper #596, University of Melbourne, Parkville, Australia (1998)

10. Kecman, V.: Learning and Soft Computing. MIT Press, London (2001)
11. Maasoumi, E.: Multidimensional approaches to welfare. In: Silber, L. (ed.) Income Inequality Measurement: From Theory to Practice. Kluwer, New York (1998)
12. Pandey, M., Nathwani, J.: Life quality index for the estimation of social willingness to pay for safety. Structural Safety 26(2), 181–199
13. Picton, P.: Neural Networks, 2nd edn. Palgrave, New York (2000)
14. Ripley, B.: Pattern Recognition and Neural Networks. Cambridge University Press, Cambridge (1996)
15. Slottje, D., Scully, G., Hirschberg, J., Hayes, K.: Measuring the Quality of Life Across Countries: A Multidimensional Analysis. Westview Press, Boulder (1991)

Author Index